THE
MATH WORKSHOP
ELEMENTARY
FUNCTIONS

THE MATH WORKSHOP
ELEMENTARY FUNCTIONS

Deborah Hughes-Hallett
HARVARD UNIVERSITY

W • W • NORTON & COMPANY • INC • New York • London

Library of Congress Cataloging in Publication Data

Hughes-Hallett, Deborah.
   The math workshop: elementary functions.
   Includes index.
   1. Functions.   I. Title.
QA331.H795   1980   515   79-28482
ISBN 0-393-09033-7   3-2-82

W. W. Norton & Company Inc., 500 Fifth Avenue, New York N.Y. 10110
W. W. Norton & Company Ltd., 25 New Street Square, London EC4A 3NT

234567890

085116

To everyone who saw the sunrise over Route 2:
Barbara, Bruce, Ken, Rob, Ken

CONTENTS

PREFACE

Origin

The Math Workshop has grown out of eight years' experience in teaching algebra and precalculus at Harvard. Here, as at a great many colleges and universities, there are significant numbers of freshmen who need to study elementary functions before going on to calculus, as well as students who are stricken with "math anxiety" when required to undertake anything more than basic arithmetic. *The Math Workshop* evolved in response to the needs of such students, and consists of two independent texts: *Algebra* and *Elementary Functions*. The material on which these books are based has been used very successfully in the classroom for several years, including use by teaching fellows who are initially inexperienced and therefore rely very heavily on it.

Purpose

The goal of this book is to show students that they *can* understand math—that whether they enjoy it or not, they can learn it. I write for ordinary people, not for mathematicians, and I attempt to talk to my readers, rather than lecture them. I hope that by the end the student will feel that math is an ordinary human activity—some parts are easy, some parts are not, but all of it is possible, and perhaps even enjoyable.

Audience

The Math Workshop: Elementary Functions was designed for a one-semester or two-quarter course for students expecting to go on to a standard calculus sequence. It presupposes a good command of basic algebra (such as that provided by *The Math Workshop: Algebra*). By omitting certain topics, *The Math Workshop: Elementary Functions* can also be used for a terminal course.

Approach

Many students enrolling in introductory college math courses approach the subject with anxiety and little confidence. I am convinced that the best way to restore such students' confidence is to start them on material they feel comfortable with and to explain new ideas in terms of ones they know well. For example, I first treat graphing and then explain functions in terms of graphs.

I am also convinced that in order to hold the students' attention (or suspend their disbelief), a book must not only explain what is correct but also dispel misconceptions and explain why what is correct has to be that way. I have tried to anticipate potential questions and discuss common mistakes, particularly in the "Things You Can and Can't Do" sections.

Exercises

The problems have been carefully designed and tested. They move gradually from routine ones designed to build the student's confidence to those that are in less standard form. Working through a reasonable selection of the problems in this book should build a thorough understanding of the subject.

The first chapter is a summary review, with exercises designed to test preparation for the course. Answers to all these problems appear at the end of the book; thereafter only answers to the odd-numbered problems are given.

Throughout the text, there are a great many exercises calling for the student to sketch or plot a graph, since I believe that facility with graphing (and the understanding that goes with it) is important to success in calculus. Though it adds somewhat to the length of the book, illustrations of the answers to virtually all of these (odd-numbered) graphing problems are given so that the student can verify his or her work.

Treatment of Specific Topics

Graphing Students going into calculus need to be able to plot graphs quickly by recognizing key features of an equation. The first few chapters show how to look for and use these features systematically. It is assumed that students will begin the text knowing how to plot simple graphs by using a table of values.

Functions In this text, students come to functions after several chapters of graphing. By then they are very comfortable calculating the value of a function from the value of x, so the idea that a function shows the dependence of one quantity on another is quite natural. We have found that this approach does much to quell vague feelings of uncertainty about what functions really are, and what they are for.

However, there is a good deal of flexibility here: all the graphing doesn't have to be done first. For example, the instructor may wish to cover lines (Chapter 2), then introduce functional notation (Chapter 7), and then return to Chapter 3 for more graphing.

Logarithms To enable students to develop a firm understanding of logs, many numerical examples are provided. These make log manipulations seem much less frightening and unintuitive than is usually the case: the log laws appear merely as rules grounded in solid numerical fact. I find that by working with the actual numerical values of logarithms, students come to see what does and doesn't make sense much more quickly than when numbers are not used. In other words, log calculations are here because they have proved an effective teaching tool, not as an end in themselves.

Trigonometry   Again, to make a new subject as concrete and easy to visualize as possible, the trigonometric functions are introduced and defined in terms of right triangles.

Acknowledgments

To thank properly everyone who helped with *The Math Workshop* would take another book. But for their particular contributions of ideas, problems, coffee, and cheerful good humor in the middle of the night, I would like to thank Barbara Peskin, Ken Manning, Ken Argentieri, Rob Olian, Bruce Molay, Reed Eichner, Adele Peskin, Charlie Klippel, Brian Leverich, Steve Ballmer, Mike Graceffo, John Maggio, Mark Robbins, Caren Jahre, Paul Segel, Lynn Smolik, Jim Rhodes, Karen Fifer, Dorris Woolery, Margie Oettinger, Betsy Griscom, Keith Salkowski, Bob Gibbons, John Holcombe, Mark Karlan, Janny Leung, Ellen Simons, Tom Atwater, Leonora Tortorella, Frank Tortorella, Mac Jones, Dan Davis, Nancy Gittelson, Bill Bishai, John van Wye, Don Young, Steve Tilles, Rick Pearce, Chris O'Hare, George Hodakowski, Ricardo Alvarez, Joe Halpern, Bob Sutor, and everyone else who hasn't helped yet, but who will.

The publisher and author would also like to thank the following reviewers for their helpful critiques of *The Math Workshop*: Louise Raphael, Clark College, Atlanta; Douglas Burke, Malcolm X College; Andrew J. Berner, Allegheny College; A. W. Goodman, Univ. of South Florida; Stanley M. Lukawecki, Clemson Univ.; Deborah T. Haimo, Univ. of Missouri, St. Louis; Douglas Brown, Nassau Community College; Maurice Monahan, South Dakota State Univ.; Ignacio Bello, Hillsborough Community College; Calvin Lathan, Monroe Community College; Cleon R. Yohe, Washington Univ.; Ward Bouwsma, Southern Illinois Univ.; Robert Donaghey, Baruch College, CUNY; David Cohen, UCLA; Daniel Marks, Stanford Univ.; Jerry Karl, Golden West College.

1 ALGEBRA REVISITED

1.1 A LOOK IN THE REAR-VIEW MIRROR

The purpose of this book is to show you everything you need to know about functions to understand and enjoy calculus. But as you undoubtedly know, math has a way of depending on other math. So before starting on functions let's take a minute to see where we are, where we're going, what we know, and what good it's going to be in the future. This chapter will make clear exactly how much algebra it is assumed that you remember, and what it has to do with the topics in this book.

First, let's get an aerial view of the math on which calculus is based. If you look back at algebra, you will see that it is really just a generalization of arithmetic. In arithmetic you add, subtract, multiply, and divide different kinds of numbers (decimals, fractions, negatives, etc.), in algebra you do exactly the same operations, but with letters instead of numbers. Since these letters can stand for any number, each algebraic statement stands for a whole crowd of arithmetical statements. For example, $2x + 3x = 5x$ tells you that $2 + 3 = 5$ (when $x = 1$); that $4 + 6 = 10$ (when $x = 2$); that $(-0.6) + (-0.9) = -1.5$ (when $x = -0.3$), and so on.

The fact that one can manipulate an algebraic expression without knowing what numbers the letters in it represent makes algebra extremely powerful. It is this that allows us to solve equations for an unknown, and to summarize a mass of arithmetical facts into one compact algebraic statement, such as the distributive law.

Learning algebra consists, first, of learning to manipulate expressions containing all manner of letters and symbols; and second, of using these skills to solve equations. A great many practical problems (and a great many not so practical ones, too many of which find their way into math books as word problems) can be solved by using equations.

After looking at equations that can be used to solve for one unknown, it is interesting to look at equations such as $y = 2x$ that give the relationship between two unknowns. It is not possible to solve such an equation for the unknowns, because there are infinitely many values that will fit the equation. Indeed, any value is possible for the first variable, but once the value of the first variable has been chosen, the value of the second is fixed. In the equation $y = 2x$, for example, x can be any number, but the value of y must always be twice the value chosen for x. Similarly, y can be any number, but the value of x must always be half the value of y. Therefore, the equation does not determine the value of either variable, but it *does* specify the relationship between them.

There are a vast number of situations in which it is not the value of either variable alone that is important, but the relationship between them. One of the best ways of displaying such a relationship is on a graph, and the first part of this book will concentrate on graphing a multitude of different equations. Then we come to functions, whose definition grows naturally out of this idea of a relationship between two variables. Calculus is a study of functions and their properties, so functions clearly occupy a rather important place in mathematics.

Last we come to logarithms and trigonometry. Both of these arose originally as aides to calculation, and this is how they will be introduced in this book. Then you will see them as functions, which is how they appear in calculus.

1.2   WHAT YOU SHOULD REMEMBER ABOUT ALGEBRA

This section contains a brief summary of the algebra you need to remember. If any of it is unfamiliar, you can look it up wherever you originally learned it or in *The Math Workshop: Algebra*. I strongly suggest doing this before starting on this book, not while you're going along. The exercises at the end of this chapter will give you a clear idea of exactly what is expected; the answers to all these problems are at the back of the book.

Now for the review:

Scientific Notation

Powers of 10 $10^5 = 100{,}000$; $10^1 = 10$; $10^0 = 1$; $10^{-5} = 1/10^5 = 0.00001$.

Scientific Notation $a \cdot 10^c$ (also written $a \times 10^c$), where $1 \le a < 10$, c an integer.

Multiplying:

EXAMPLE: $(3 \cdot 10^{-5}) \cdot (5.1 \cdot 10^9)$

$= 15.3 \cdot 10^{9-5} = 1.53 \cdot 10^5$

Dividing:

EXAMPLE: $\dfrac{4.2 \cdot 10^{-17}}{8.4 \cdot 10^{-3}}$

$= 0.5 \cdot 10^{-17+3} = 5 \cdot 10^{-15}$

Adding and subtracting:

EXAMPLE: $5.1 \cdot 10^{12} + 1.2 \cdot 10^{13}$

$= 0.51 \cdot 10^{13} + 1.2 \cdot 10^{13} = 1.71 \cdot 10^{13}$

Polynomials

Factoring out a Common Factor

EXAMPLE: $3x^2 - 9x^2y^2 - 3x^6$

$= 3x^2(1 - 3y^2 - x^4)$

Factoring a Quadratic Polynomial   Use the systematic method or guess-work:

EXAMPLE: $6x^2 + x - 15$

$= (2x - 3)\,(3x + 5)$

Factoring a Higher-Degree Polynomial   Use the methods for quadratics:

EXAMPLE: $x^4 - 5x^2 + 4$

$= (x^2 - 1)\,(x^2 - 4) = (x + 1)\,(x - 1)\,(x + 2)\,(x - 2)$

Find factors of the form $(lx - k)$. Possible factors are those in which k divides the constant term and l divides the coefficient of the highest power of x. Find out which actually are factors by seeing which values of k/l give zero when substituted into the polynomial:

EXAMPLE: $2x^3 - 7x^2 + 8x - 3$

$= (2x - 3)(x - 1)^2$

Completing the Square

EXAMPLE: *Add 9 to $x^2 - 6x$ to give*

$x^2 + 6x + 9 = (x - 3)^2$

EXAMPLE: *Add $\frac{9}{8}$ to $2x^2 - 3x$ to give*

$$2x^2 - 3x + \tfrac{9}{8} = 2(x^2 - \tfrac{3}{2}x + \tfrac{9}{16}) = 2(x - \tfrac{3}{4})^2$$

Note: Completing the square is needed only for the section on conics.

Algebraic Fractions

Multiplying and Simplifying

EXAMPLE:

$$\frac{x^2 + 3x}{x^2 - 3x + 2} \cdot \frac{x^2 - 5x + 6}{x^2 - 9}$$

$$= \frac{x(x + 3)}{(x - 1)(x - 2)} \cdot \frac{(x - 3)(x - 2)}{(x + 3)(x - 3)}$$

$$= \frac{x}{x - 1}$$

Dividing

EXAMPLE:

$$\frac{\dfrac{4b^2 - a^2}{a}}{\dfrac{a^2 - 2ab}{a^2 b}} = \frac{(2b + a)(2b - a)}{a} \cdot \frac{a^2 b}{a(a - 2b)} = -b\,(2b + a)$$

Adding and Subtracting

EXAMPLE:

$$\frac{1}{T^2 - T} - \frac{1}{T^2 + 2T} = \frac{1}{T(T - 1)} - \frac{1}{T(T + 2)}$$

$$= \frac{(T + 2)}{T(T - 1)\,(T + 2)} - \frac{(T - 1)}{T(T + 2)\,(T - 1)}$$

$$= \frac{(T + 2) - (T - 1)}{T(T - 1)\,(T + 2)}$$

$$= \frac{3}{T(T - 1)\,(T + 2)}$$

Simplifying Complex Fractions
Combine top and bottom separately into a single fraction and divide:

EXAMPLE:

$$\frac{\dfrac{1}{p+5} - \dfrac{p}{p^2-25}}{\dfrac{1}{p-5} - \dfrac{1}{p}} = \frac{\dfrac{(p-5)-p}{p^2-25}}{\dfrac{p-(p-5)}{p(p-5)}}$$

$$= \frac{\dfrac{-5}{p^2-25}}{\dfrac{5}{p(p-5)}}$$

$$= \frac{-5}{(p-5)(p+5)} \cdot \frac{p(p-5)}{5}$$

$$= \frac{-p}{p+5}$$

Linear Equations

EXAMPLE:

$$4(2x+3) - x = 3(1-x)$$
$$8x + 12 - x = 3 - 3x$$
$$8x - x + 3x = 3 - 12$$
$$10x = -9$$
$$x = -\tfrac{9}{10}$$

Quadratic Equations

By factoring

EXAMPLE:

$$(10x+1)(x-1) = 6$$

Multiply out, put everything on the left, and refactor with zero on the right:

$$10x^2 - 9x - 1 = 6$$
$$10x^2 - 9x - 7 = 0$$
$$(2x-1)(5x+7) = 0$$
$$x = \tfrac{1}{2} \quad \text{or} \quad x = -\tfrac{7}{5}$$

[*Note:* You may not cancel factors containing the variable, or you may loose a root. For example, cancelling $(x-1)$ in $x(x-1) = 2(x-1)$, giving $x = 2$, means that you have lost the second root, namely $x = 1$, to the original equation.]

By the quadratic formula   The solution to $ax^2 + bx + c = 0$ is

$$x = \frac{-b \pm \sqrt{b^2 - 4ac}}{2a}$$

EXAMPLE: $4x = 1 - 2x^2$

Rearranging gives $2x^2 + 4x - 1 = 0$, so

$$x = \frac{-4 \pm \sqrt{16 + 8}}{4} = \frac{-4 \pm \sqrt{24}}{4} = \frac{-4 \pm 2\sqrt{6}}{4}$$

$$= \frac{-2 \pm \sqrt{6}}{2}$$

Polynomial Equations

By factoring

EXAMPLE: $x^4 - 16 = 0$

$$(x^2 + 4)\,(x^2 - 4) = 0$$

$$(x^2 + 4)\,(x + 2)\,(x - 2) = 0$$

$$x = 2 \quad \text{or} \quad x = -2$$

[*Note:* The factor $(x^2 + 4)$ does not give any real roots; and we will not be using complex roots.]

Fractional Equations

Multiply through by the least common denominator (L.C.D.) of the fractions occuring in the equation; solve the equation you get by the appropriate method. Remember to check for extraneous roots.

EXAMPLE: $$1 = \frac{1}{x - 2} - \frac{2}{x^2 - 2x}$$

$$\text{L.C.D.} = x(x - 2)$$

$$x(x - 2) \cdot 1 = x(x - 2)\left(\frac{1}{x - 2}\right) - x(x - 2)\frac{2}{x(x - 2)}$$

$$x(x - 2) = x - 2$$

$$x^2 - 2x = x - 2$$

$$x^2 - 3x + 2 = 0$$

This is a quadratic that can be solved by factoring:

$$(x - 1)\,(x - 2) = 0$$

$$x = 1 \quad \text{or} \quad x = 2$$

Check: For $x = 1$,

Left side $= 1$

Right side $= \dfrac{1}{x-2} - \dfrac{2}{x^2 - 2x}$

$\qquad = \dfrac{1}{-1} - \dfrac{2}{-1}$

$\qquad = -1 + 2 = 1$

So $x = 1$ is a solution.

For $x = 2$,

Left $= 1$

Right $= \dfrac{1}{x-2} - \dfrac{2}{x^2 - 2x}$

$\qquad = \dfrac{1}{0} - \dfrac{2}{0}$

$\qquad =$ undefined

so $x = 2$ is an extraneous root.

Therefore $x = 1$ is the only solution.

Literal Equations

Literal equations are equations involving several letters in which we are told to solve for one letter in terms of the others.

EXAMPLE: *Solve $pq - r^2 t = 4t - (q - r)(t - p)$ for t.*

This is linear in t, so solve by the usual method for linear equations.

$$pq - r^2 t = 4t - qt + qp + rt - rp$$

$$qt - rt - 4t - r^2 t = qp - rp - pq$$

$$(q - r - 4 - r^2)t = -rp$$

$$t = \frac{-rp}{q - r - 4 - r^2} = \frac{rp}{r^2 + 4 + r - q}$$

Simultaneous Equations

By elimination

EXAMPLE: $\begin{cases} 2x - 3y = 8 \\ x - 5y = 11 \end{cases}$

Subtract twice the second equation from the first:

$$-3y + 2(5y) = 8 - 2(11)$$

$$7y = -14$$

$$y = -2$$

Substitute into the second equation and get $x = 1$.

So the solution is $x = 1$, $y = -2$.

By substitution

EXAMPLE: $\begin{cases} x + y = 7 \\ \quad xy = 10 \end{cases}$

Solve for y in the first equation and substitute into the second:

$$y = 7 - x$$
$$x(7 - x) = 10$$
$$x^2 - 7x + 10 = 0$$
$$(x - 2)(x - 5) = 0$$
$$x = 2 \quad \text{or} \quad x = 5$$

Substitute back to find that when $x = 2$, $y = 5$; and when $x = 5$, $y = 2$.

Inequalities

Linear Inequalities   Linear inequalities are solved like linear equations except that the inequality sign must be reversed when you multiply or divide by a negative number.

EXAMPLE: $3 - 2x < 4x + 15$

$$-6x < 12$$
$$x > 2$$

Nonlinear Inequalities

EXAMPLE: $x^2 < 9$

Solution: $-3 < x < 3$.

EXAMPLE: $x^2 - 4x + 3 > 0$

$$(x - 1)(x - 3) > 0$$

The product of two factors is positive if both are positive or both are negative, so *either* $(x - 1) > 0$ and $(x - 3) > 0$ *or* $(x - 1) < 0$ and $(x - 3) < 0$, which reduces to

$$x < 1 \quad \text{or} \quad x > 3$$

Absolute Values

$$|a| = \text{magnitude of } a$$
$$|a - b| = \text{distance between } a \text{ and } b$$

Note that the absolute value is never negative, and

$$\sqrt{a^2} = |a|$$

Equations involving absolute value usually have two solutions.

EXAMPLE: $|x - 3| = 7$

means that the distance between x and 3 is 7, so

$$x = 10 \text{ or } x = -4$$

Inequalities Involving Absolute Values

EXAMPLE: $|2x - 1| > 5$

means that the distance between $2x$ and 1 is more than 5, so

$$2x < -4 \quad \text{or} \quad 2x > 6$$

giving

$$x < -2 \quad \text{or} \quad x > 3$$

Graphs

The basic shapes are shown in Figure 1.1.

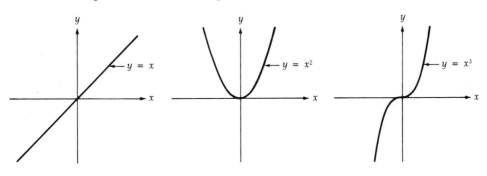

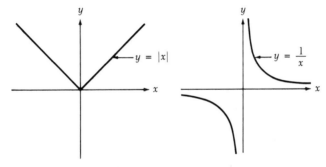

FIG. *1.1*

Distance Formula

$$d = \sqrt{(x_1 - x_2)^2 + (y_1 - y_2)^2}$$

where (x_1, y_1) and (x_2, y_2) are points.

Midpoint Formula

$$M = \left(\frac{x_1 + x_2}{2}, \frac{y_1 + y_2}{2} \right)$$

Geometric Formulas

See appendix.

EXERCISES

Evaluate:

1. $2 + \frac{1}{2}(\frac{1}{10} - \frac{1}{5})$
2. $(0.4 - 0.5) - [0.5 - (2.3 - 3.2)]$
3. $\sqrt{0.04}$
4. $\sqrt{\frac{1.21}{16}} \cdot \frac{16}{11}$
5. $\frac{\frac{2}{540} - \frac{1}{324}}{\frac{1}{162}}$

6. $\dfrac{1}{2 + \dfrac{1}{3 + \frac{1}{4}}}$

7. $(2 \cdot 10^{-1})^2 - 0.02$
8. $15 \cdot 10^{-24} - 7 \cdot 10^{-22}$

9. 0.2% of 15 is three-fifths of what number?
10. By what percent must 55 be decreased to give 50?
11. Find the average of $\frac{3}{7}$, 0.025, and 10^{-3} correct to three decimal places.
12. Arrange in ascending order: $-\frac{1}{20}$, -0.052, $-5.1 \cdot 10^{-3}$.

Find the perimeter of:

13. A square of side $\frac{1}{32}$ inch
14. The shape in Figure 1.2

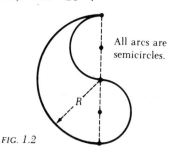

All arcs are semicircles.

FIG. 1.2

Find the area of:

15. A rectangle of sides $5a$ and $0.6a$
16. A circle of radius $2 \cdot 10^{-5}$
17. The surface of a can of radius 2 inches and height 3 inches (ends included)

Find the volume of:
   18. A rectangular box of sides x, $2x$, and $3x$
   19. A cone of radius 10 inches and height 30 inches
   20. A sphere of radius 2π

Factor fully:
   21. $16 - 9a^4$ 25. $x^3 - 6x^2 + 11x - 6$
   22. $6x^2 - x^4 - x^3$ 26. $10A^2T^2 - 9AT\pi - \pi^2$
   23. $35L^2 - 74L + 35$ 27. $3(c + d)^2 - 10(c + d) + 3$
   24. $A^4 - B^4$ 28. $p^2 - 2pq + q^2 - r^2$

Complete the square:
   29. $x^2 + x$
   30. $3x - 2x^2$

Simplify:

31. $\dfrac{4a^2 - 4a + 1}{2 - 3a - 2a^2}$ 32. $\dfrac{(x - y)^2 + 4xy}{x^2 - y^2}$

33. $(z^2 + 4zw + 4w^2)\left(\dfrac{10z + w}{2w^2 + 21wz + 10z^2}\right)$

34. $\dfrac{\dfrac{p^2 - 1}{p^4 - 1}}{\dfrac{(p - 1)^2}{p^2 + 1}}$ 35. $\dfrac{1 - \dfrac{1}{d}}{1 - \dfrac{1}{d^2}}$

36. $\left(K - \dfrac{6}{K - 1}\right)\left(\dfrac{4}{K - 3} + \dfrac{4}{K + 2}\right)$

37. $\dfrac{x - y}{c - d} + \dfrac{y}{2d - 2c} + \dfrac{x - y}{d - c}$

38. $\dfrac{2z}{9z^2 - 6z + 1} - \dfrac{z}{1 - 9z^2} + \dfrac{1}{1 - 3z}$

39. $\dfrac{\dfrac{x - a}{x + a} - \dfrac{x + a}{x - a}}{\dfrac{1}{x - a} - \dfrac{1}{x + a}}$ 40. $\dfrac{\dfrac{1}{t^2 - 4} + \dfrac{1}{4t^2 - 1}}{\dfrac{1}{2t^2 - 5t + 2} - \dfrac{1}{4t^2 - 4t + 1}}$

Multiply out:

41. $(\sqrt{a} + \sqrt{b})^2$

42. $\sqrt{x^2 + a^2}\left(\sqrt{x^2 + a^2} - \dfrac{a^2}{\sqrt{x^2 + a^2}}\right)$

Simplify:

43. $b^2\sqrt{A} - \sqrt{b^4 A}$

44. $v^7\left(2 - \dfrac{1}{v}\right)^7 + (2v - 1)^7$

45. $\sqrt{p^2 - 1} \cdot \sqrt{\dfrac{p + 1}{p - 1}}$

Solve for the variables:

46. $4x - 6 = 3 - 6x$

47. $2 - [12 - (2 - x)] = 1 - (2x - 1)$

48. $\dfrac{5 - t}{5 + t} - \dfrac{1}{5} = 0$

49. $\dfrac{p - \frac{1}{10}}{1.1} = \dfrac{p + 1.1}{\frac{1}{10}}$

50. $6p^2 + 13p + 6 = 0$

51. $x(x + 3) = 2(x + 3)$

52. $10 = (a - 2)(2a + 1)$

53. $\dfrac{1 - 2t}{4t - 2} + \dfrac{1}{t} = 1$

54. $\dfrac{p}{p^2 - 5p + 6} + \dfrac{1}{p^2 - 3p} = 0$

55. $x^3 - 3x^2 - x + 3 = 0$

56. $\begin{cases} 5x + 4y = 28 \\ 3y - x = 21 \end{cases}$

57. $\begin{cases} \dfrac{1}{x} + \dfrac{2}{y} - 2 = 0 \\ \dfrac{3}{x} + \dfrac{4}{y} - 5 = 0 \end{cases}$

58. $\begin{cases} AB = 0.09 \\ A + B = 1 \end{cases}$

59. $\begin{cases} 2p^2 - q^2 = 7 \\ p + \dfrac{q^2}{p} - \dfrac{5}{p} = 0 \end{cases}$

60. $\begin{cases} x + y = 5 \\ 4x - 5y = \dfrac{10}{x + y} \end{cases}$

Solve for the indicated variable:

61. $\dfrac{p - At}{p + At} + At = 0$ for p

62. $\dfrac{y - z}{2y + z} = \left(\dfrac{y + z}{y + 2z}\right)w - \dfrac{2z}{y + 2z}$ for w

63. $\dfrac{p}{n} + (q + re)n = \dfrac{p + r}{n}$ for n

64. $b^2 + 1 - (ax - b)^2 = 2b$ for x

65. $\left(\dfrac{p}{d} - 1\right)\left(\dfrac{2p}{d} + 3\right) = 7\left(\dfrac{p}{d} - 1\right)$ for p

Solve and graph on the number line:

66. $2x - 1 > 5 - x$

67. $6 - x^2 \geqslant -26 + x^2$

68. $(x + 1)(x - 1) > 0$

69. $|1 + x| < 7$

70. $|2x - 1| > 11$

Solve:

71. $\left|\dfrac{1}{z} - \dfrac{1}{2}\right| = \dfrac{7}{2}$

72. $3\left(\dfrac{x + 1}{2x - 1}\right)^2 + 2\left(\dfrac{x + 1}{2x - 1}\right) - 1 = 0$

73. $x^2 + 12 < 7x$

74. $ax + c < d$ for x

75. $2(x - 1)^4 - 5(x - 1)^2 + 2 = 0$

76. As an official of the Justice Department's antitrust section, it is your job to investigate mergers of companies that might cause a monopoly. If a company formed by merger forms 45% or more of the market, you can sue and prevent the merger. Currently Amanita's Ice Cream wants to merge with E. Coli's Frozen Yogurt. You have been able to determine that Amanita's sales are $12 million a year, which is 40% of the ice cream market, and that E. Coli's sales are $6 million a year or 60% of the frozen yogurt market.

Calculate:
   (a) The total sales of the other ice cream manufacturers.
   (b) The total sales of the other frozen yogurt makers.
   (c) If the Justice Department considers ice cream and frozen yogurt as one market, can it bring suit to stop the merger?

77. Ben Killdare was a quack Hollywood plastic surgeon who limited his practice to facelifts. For each successful operation he charged a fee of $8 \cdot 10^4$, but every time he did one wrong he lost $1.2 \cdot 10^6$ in a malpractice suit. One morning near the end of his career, Ben realized that he had done exactly 30 times as many successful lifts as unsuccessful ones. After that day he did 10 more successful operations, and then retired, netting $9.2 \cdot 10^6$ from his surgeries. How many times had Ben been sued for malpractice?

78. The trip up the Amazon from Belén to Manáus takes 8 days by riverboat. The return trip takes only 5 days. The towns are 2000 miles apart along the river. How fast is the current? How fast is the boat in still water?

79. About all we know of Diophantus' personal life is contained in the following summary of his epitaph: "Diophantus passed one-sixth of his life in childhood, one-twelfth in youth, and one-seventh more as a bachelor. Five years after his marriage was born a son who died four years before his father, at half his father's [final] age." How old was Diophantus when he died?

80. The mixed price of nine citrons and seven fragrant wood apples is 107; again, the mixed price of seven citrons and nine fragrant wood apples is 101. O you arithmetician, tell me quickly the price of a citron and of a wood apple here, having distinctly separated those prices well. (From Mahāvīra, Hindu, ca. 850.)

81. A bamboo 18 cubits high was broken by the wind. Its top touched the ground 6 cubits from the root. Tell the lengths of the segments of the bamboo. (From Brahmagupta, Hindu ca. 630)

Find the distance between these points:
   82. $(0, 0.2)$ and $(-0.3, 1)$
   83. $\left(\pi, -\dfrac{\pi}{2}\right)$ and $\left(-\dfrac{3\pi}{2}, -\dfrac{3\pi}{2}\right)$

Find the midpoint of:
   84. $(1.1, -0.2)$ and $(0.1, -7.8)$
   85. $\left(a + 1, \dfrac{1}{a}\right)$ and $(1 - a, a)$

Graph:
   86. $y = x$ 92. $x = 3$
   87. $y = -x$ 93. $y > 2$
   88. $y = x + 2$ 94. $y = x^3$
   89. $y = -x + 3$ 95. $y = \dfrac{1}{x}$
   90. $y = x^2$
   91. $y = 2x^2$

2 LINES

2.1 "STEEPNESS" AND "POSITION" OF A LINE

Imagine that you're trying to tell me—without pointing—how to draw a particular line on the Cartesian plane. What do you need to tell me in order that I may draw exactly the line that you are thinking of? One way is this. First tell me what angle the line is at. Is it a "steep" line, or does it climb slowly? Does it climb from left to right, or from right to left? But knowing what angle the line is at isn't enough, because I can draw lots of different lines at the same angle. For example, see the lines in Figure 2.1. But if you also tell me something about where the line is—a point it goes through or one of the intercepts—then I can draw your line.

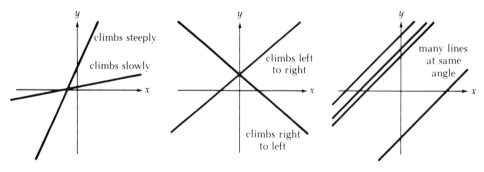

FIG. 2.1

So it seems that a line is characterized by two things: its "steepness" and its "position." Once we know those, we can draw the line, and given a line, we can say how "steep" it is, and what its "position" is.

The point of this section is to look at the equations whose graphs are lines, and to see what in the algebraic equation makes a line "steep" or "not steep," and what determines its "position." Before we can do this, however, we must know what kinds of equations to look at. As it turns out, *any linear equation in x and y has a graph that is a line*—which is why such equations are called linear.

What Determines the "Position" of a Line?

Look at the graphs of $y = x$ and $y = x + 1$, which are shown in Figure 2.2. Comparing these graphs, you can see that $y = x + 1$ is everywhere one unit higher up than $y = x$. The reason is that $y = x + 1$ has all its y's one greater than the corresponding y's on $y = x$.

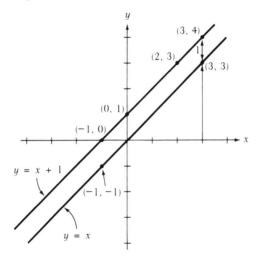

FIG. 2.2

Changing the 1 in $y = x + 1$ will move the graph up or down. For example, look on p. 16 at the graphs of $y = x + 2$, $y = x + 3$, or $y = x - 1$. The first of these, $y = x + 2$, has all its y's two greater than the corresponding y's for $y = x$, and so the graph of $y = x + 2$ is the graph of $y = x$ shifted up two. (If you don't believe this, please do the graph of $y = x + 2$ with a table of values.) By similar arguments you can figure out the graphs of all of the lines shown in Figure 2.3. All these lines have the same "steepness," and the only thing that varies is their position. Also, all of these lines have equations of the form $y = x + b$. Looking at the graphs will show you that b tells you where the line cuts the y axis—so *b is the y-intercept of $y = x + b$.* To check this algebraically, realize that finding the y intercept means finding y when $x = 0$. Substituting $x = 0$ into $y = x + b$ gives $y = 0 + b$ or $y = b$, which is the y intercept.

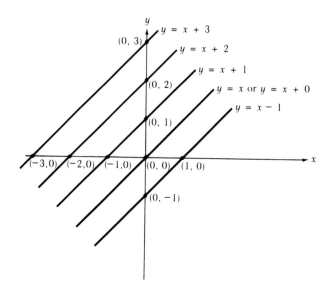

FIG. 2.3

So it is the b in $y = x + b$ that is responsible for the "position" of the line.

What Determines the "Steepness" of a Line?

Now let's look at some other equations: $y = 2x + 1$, $y = 3x + 1$, $y = -x + 1$, $y = -2x + 1$, and $y = -\frac{1}{2}x + 1$. Since at the moment we have no idea what changing the coefficient of the x will do to the graph, we will have to use a table of values. For $y = 2x + 1$, the table of values is given in Table 2.1. By doing similar tables of values for the others, we can draw the graphs of the other equations. They are shown in Figure 2.4. Interestingly enough, they all turn out to be lines through (0,1) but with different "steepness."

Table 2.1

x	$y = 2x + 1$
-2	-3
-1	-1
0	1
1	3
2	5
3	7

The coefficient of x seems to determine the "steepness" of the line. In the equation $y = mx + 1$, if m is positive, the line climbs from left to right; if nega-

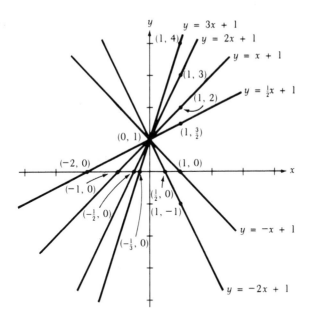

FIG. 2.4

tive, it climbs from right to left. The larger the magnitude of m, the "steeper" the line is.

Therefore, *the m in $y = mx + 1$ is responsible for the "steepness" of the line.*

If you plot $y = mx + 2$ for the same values of m, you get exactly the same picture but moved up by one. Each line now crosses the y axis at $(0, 2)$ but has the same "steepness" as before. So the m in $y = mx + 2$ still represents steepness.

What about changing the coefficient of y? What does putting a 2 in front of the y do? Luckily, nothing new: The equation $2y = x + 1$ is the same as $y = \frac{1}{2}x + \frac{1}{2}$, whose graph is a line parallel to $y = \frac{1}{2}x + 1$ but moved down $\frac{1}{2}$, as shown in Figure 2.5.

No matter what nonzero coefficient y is given, you can always divide through by it and get an equation of the form $y = mx + b$ in which m represents steepness.

In general:

> Any equation of the form $y = mx + b$ gives a line where
>
> b is the y intercept
>
> m is the "steepness"

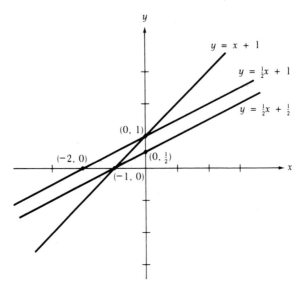

FIG. 2.5

PROBLEM SET 2.1

Graph and label the following lines on the same coordinate system.

1. $y = x$
2. $y = -x$
3. $y = 3x$
4. $y = -3x$
5. $y = \frac{1}{3}x$
6. $y = -\frac{1}{3}x$

Graph and label the following lines on the same coordinate system.

7. $y = x$
8. $y = x + 1$
9. $y = x + \frac{1}{2}$
10. $y = x + 3$
11. $y = x - 2$
12. $y = -x - 2$

Graph the line $y = 2x + b$ when:

13. $b = 2$
14. $b = -1$
15. $b = 0$
16. $b = 1$
17. $b = \frac{2}{23}$
18. $b = 7$
19. $b = \frac{1}{30}$

Graph the line $y = mx + 3$ when:

20. $m = 1$
21. $m = 0$
22. $m = -1$
23. $m = 2$
24. $m = 3$
25. $m = 5$
26. $m = \frac{1}{100}$

Graph the line $ay = 3x + 1$ when:

27. $a = 1$ 31. $a = 400$
28. $a = 3$ 32. $a = \frac{1}{10}$
29. $a = 0$ 33. $a = -\frac{1}{10}$
30. $a = -1$

2.2 DEFINITION OF SLOPE

From the previous discussion it must be obvious that the idea of the "steepness" of a line is of great importance. However, it is hard to compare the "steepness" of two lines or see why m should represent the "steepness" without knowing a good deal more exactly what we mean by "steepness." We need a precise definition—although it must certainly be one that fits in with our intuitive ideas.

The "steepness" of a line is meant to measure how fast it is climbing (or descending). If, for example, you wanted to convey to someone the steepness of a road, you might tell them that it rises 1 foot for every 10 feet of horizontal distance. See Figure 2.6. (This is sometimes called a "1 in 10" hill, and is, by the way, an extremely steep one.) The ratio $\frac{1}{10}$, which is called the *slope*, is a measure of the steepness of the hill. A road of slope $\frac{3}{20}$ would therefore be climbing 3 feet for every 20 horizontal feet or 3 meters for every 20 horizontal meters, and so on. See Figure 2.7. As another example, a mountain face of slope $\frac{7}{3}$ would look like Figure 2.8.

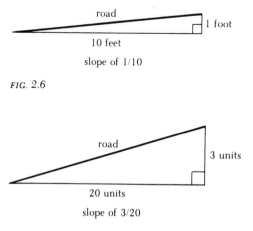

road
1 foot
10 feet
slope of 1/10

FIG. 2.6

road
3 units
20 units
slope of 3/20

FIG. 2.7

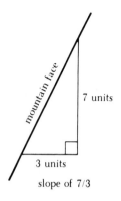

mountain face
7 units
3 units
slope of 7/3

FIG. 2.8

So we will define the slope of a line as:

$$\text{slope} = \frac{\text{vertical change between two points on line}}{\text{corresponding horizontal change}}$$

Suppose the points $P = (x_1, y_1)$ and $Q = (x_2, y_2)$ are on the line whose slope we want to find. See Figure 2.9. T is the point (x_2, y_1) because T has the same y coordinate as P and the same x coordinate as Q.

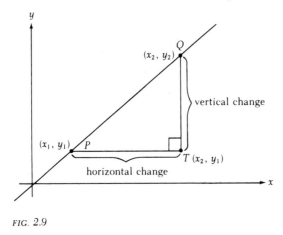

FIG. 2.9

The vertical change between P and $Q = y_2 - y_1$.
The corresponding horizontal change $= x_2 - x_1$.
So:

$$\text{slope} = \frac{y_2 - y_1}{x_2 - x_1}$$

There are three things that you should notice about this formula:

1. We are interested in the *change* in y from P to Q, represented by $y_2 - y_1$, rather than by $|y_2 - y_1|$, which represents a distance. Only by looking at changes rather than distances can we see whether y is increasing or decreasing, and knowing this is essential to seeing whether a line climbs or descends as we move from left to right.

2. *It doesn't matter which point you call (x_1, y_1) and which (x_2, y_2).*

$$\text{slope} = \frac{y_2 - y_1}{x_2 - x_1} = \frac{-(y_1 - y_2)}{-(x_1 - x_2)} = \frac{y_1 - y_2}{x_1 - x_2}$$

Therefore, if you interchange x_1 and x_2, and y_1 and y_2, you still get the same value for the slope. The only thing that is important is to be consistent: x_1 and y_1 must be the coordinates of the same point.

3. *It doesn't matter which two points on the line you pick—the slope will always come out the same.*

Suppose $P = (x_1, y_1)$, $Q = (x_2, y_2)$, $R = (x_3, y_3)$, $S = (x_4, y_4)$ are all points on the same line. We will find the slope by using P and Q and by using R and S. See Figure 2.10

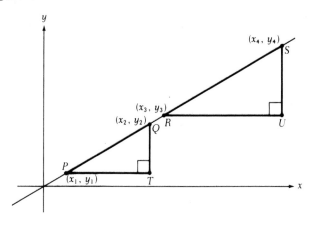

FIG. 2.10

The two triangles PTQ and RUS are similar because both contain a right angle, and angles QPT and SRU are equal—both being the angle the line makes with the x axis. Also, angles PQT and RSU are equal, so the triangles have the same angles and are similar. Therefore,

$$\frac{SU}{QT} = \frac{RU}{PT} = \text{``magnification factor''}$$

Hence,

$$\frac{y_4 - y_3}{y_2 - y_1} = \frac{x_4 - x_3}{x_2 - x_1}$$

Multiplying through by $\dfrac{y_2 - y_1}{x_4 - x_3}$ we get

$$\frac{(\cancel{y_2 - y_1})}{(x_4 - x_3)} \cdot \frac{(y_4 - y_3)}{(\cancel{y_2 - y_1})} = \frac{(y_2 - y_1)}{(\cancel{x_4 - x_3})} \cdot \frac{(\cancel{x_4 - x_3})}{(x_2 - x_1)}$$

So

$$\frac{y_4 - y_3}{x_4 - x_3} = \frac{y_2 - y_1}{x_2 - x_1}$$

but the slope from P to Q is $\dfrac{y_2 - y_1}{x_2 - x_1}$

and the slope from R to S is $\dfrac{y_4 - y_3}{x_4 - x_3}$

So the slope using R and S = the slope using P and Q.

EXAMPLE   *Find the slope of the line through the points $(-2, 1)$ and $(3, -1)$.*

If we let $P = (-2, 1)$ and $Q = (3, -1)$, the formula tells us that

$$\text{slope} = \frac{(-1) - (1)}{(3) - (-2)} = -\frac{2}{5}$$

Looking at Figure 2.11 gives us an idea what such a slope means. You can see that in going from $(-2, 1)$ to $(3, -1)$, y drops by 2 and x increases by 5. The fact that y drops while x increases gives us the minus sign, the fact that y changes only two-fifths as fast as x gives us the $\frac{2}{5}$.

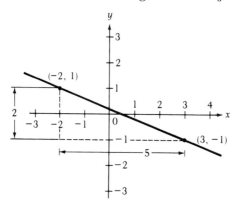

FIG. 2.11

EXAMPLE   *Find the slope of the line $y = \frac{1}{3}x + 2$.*

In order to use the slope formula we must find two points on the line, which can be done in the same way that we create a table of values—by picking x values and plugging them into the formula to get the corresponding y values. If $x = 0$, $y = 2$, and if $x = 1$, $y = \frac{7}{3}$, so $(0, 2)$ and $(1, \frac{7}{3})$ are on the line.

$$\text{slope} = \frac{\frac{7}{3} - 2}{1 - 0} = \frac{\frac{1}{3}}{1} = \frac{1}{3}$$

If we had picked two different points, say $(2, \frac{8}{3})$ and $(3, 3)$, on the line, the slope would have been

$$\frac{3 - \frac{8}{3}}{3 - 2} = \frac{\frac{1}{3}}{1} = \frac{1}{3}$$

which is the same as before.

As you might expect, a graph shows that the line climbs from left to right and gains 1 unit of height for every 3 in a horizontal direction. See Figure 2.12.

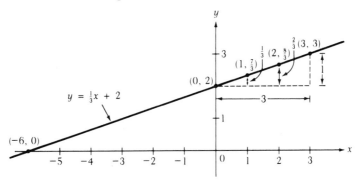

FIG. 2.12

However, the most noticeable thing about the slope of the line $y = \frac{1}{3}x + 2$ is that it comes out to be the coefficient of x—namely, $\frac{1}{3}$. This is great, because we know that m (here $\frac{1}{3}$) measures the "steepness" of the line and we intended slope to measure the same thing. A proof that the slope always turns out to be m is given in the next section.

EXAMPLE   *Find the slope of a horizontal line.*

Along a horizontal line the x coordinate changes, but the y coordinate does not. Suppose the line cuts the y axis at $(0, b)$; then the line must also go through $(1, b)$, as well as $(2, b)$, $(3, b)$, $(-1, b)$, and so on. See Figure 2.13.

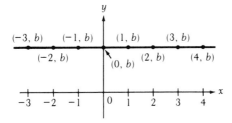

FIG. 2.13

Using the points $(0, b)$ and $(1, b)$, we get

$$\text{slope} = \frac{b - b}{1 - 0} = \frac{0}{1} = 0$$

So the slope of a horizontal line is 0.

EXAMPLE   *Find the slope of a vertical line.*

On a vertical line the y coordinate changes whereas the x coordinate is constant. Suppose the x intercept is k; then the points $(k, 0)$, $(k, 1)$, $(k, 2)$, and so on, are on the line. See Figure 2.14.

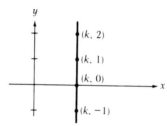

FIG. 2.14

Using the points $(k, 0)$ and $(k, 1)$, we get

$$\text{slope} = \frac{1 - 0}{k - k} = \frac{1}{0}$$

which is undefined.

So the slope of a vertical line is undefined.

PROBLEM SET 2.2

What is the slope of the line that passes through the points:

1. $(3, 2)$ and $(7, -6)$
2. $(-1, -2)$ and $(6, -2)$
3. $(12, 5)$ and $(-1, -1)$
4. $(4, 2)$ and $(4, 6)$
5. $(-1, 5)$ and $(-3, 0)$
6. $(1, 2)$ and $(19, 2)$
7. $(0, 0)$ and $(4, -5)$
8. $(1, 0)$ and $(\frac{4}{3}, 7\frac{1}{8})$
9. $(-1, 3\frac{1}{2})$ and $(6, -4\frac{3}{5})$
10. $(4, 3\frac{9}{23})$ and $(4, \frac{81}{23})$
11. $(0.03, 7.20)$ and $(0.61, 6.31)$
12. $(1.22, 8.40)$ and $(-2.19, -1.57)$

Graph and label the following lines and find the slope in each case.

13. $y = x + 1$
14. $y = x - 1$
15. $y = -x + 1$
16. $y = -x - 1$

17. $y = 2x - 3$ 19. $x + 2y = 4$
18. $y = -3x + 2$ 20. $3x - 2y = 1$

21. Take the line $y = 2x + 2$ and pick any three pairs of points on the line. For each pair of points compute the slope between them. Show that the slope in each case is the same.

22. Can you draw a straight line through the points:

$$A = (100, 106) \qquad B = (107, 112) \qquad C = (114, 118)$$

2.3 SLOPE-INTERCEPT FORM OF A LINE

An equation of the form

$y = mx + b$ where m and b are constants is the equation of a line in *slope-intercept form.* For any such line, m *is the slope* and b *is the y intercept.*

Since m can be varied to give any slope, and b can be varied to give any y intercept, any line (except a vertical one) has an equation of this form. For completeness, we will now demonstrate officially that m and b really are what we claim they are.

To show the slope of the line $y = mx + b$ is m. Suppose (x_1, y_1) and (x_2, y_2) are two points on the line. This means that the coordinates of these points satisfy the equation; that is, $y_1 = mx_1 + b$ and $y_2 = mx_2 + b$. Now

$$\text{slope} = \frac{y_2 - y_1}{x_2 - x_1} = \frac{(mx_2 + b) - (mx_1 + b)}{x_2 - x_1}$$

$$= \frac{mx_2 + b - mx_1 - b}{x_2 - x_1}$$

$$= \frac{mx_2 - mx_1}{x_2 - x_1}$$

$$= \frac{m(x_2 - x_1)}{x_2 - x_1}$$

$$= m$$

To show the y intercept of the line $y = mx + b$ is b. On the y axis, $x = 0$, so the y intercept is given by

$$y = m \cdot 0 + b = b$$

To show the x intercept of the line $y = mx + b$ is $\left(-\dfrac{b}{m}\right)$ (provided $m \neq 0$). On the x axis, $y = 0$, so the x intercept is given by

$$0 = mx + b$$

or

$$x = -\frac{b}{m}$$

Note: This assumes $m \neq 0$, that is, that the line is not horizontal.

Examples of Use of the Slope-Intercept Form of a Line

EXAMPLE   *Find the slope of the line*

$$\frac{x}{2} + \frac{y}{3} = 1$$

and draw its graph.

Rewrite

$$\frac{x}{2} + \frac{y}{3} = 1 \quad \text{as} \quad \frac{y}{3} = -\frac{x}{2} + 1$$

so that

$$y = \left(-\frac{3x}{2}\right) + \underset{m}{\overset{b}{\textcircled{3}}}$$

Now you can see that the slope is $-\frac{3}{2}$. The y intercept is 3, and the x intercept is

$$\frac{-b}{m} \quad \text{or} \quad \frac{-3}{\left(-\frac{3}{2}\right)} = 2.$$

The intercepts allow you to plot the graph very easily, since you know it goes through the points (0, 3) and (2, 0). The line is graphed in Figure 2.15.

The graph also shows that in moving from P to Q, y decreases by 3, and x increases by 2—as you would expect of a line that has slope $-\frac{3}{2}$.

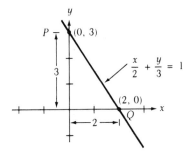

FIG. 2.15

EXAMPLE *Graph the family of lines $y = mx$.*

Any line of the form $y = mx$ goes through the origin, since $b = 0$. As m varies, the equation $y = mx$ gives a whole collection of lines; this collection is called a family of lines and is shown in Figure 2.16.

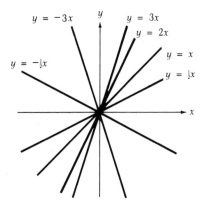

FIG. 2.16

You should notice how m affects the angle of the line $y = mx$. *If $|m|$ is large, the line is steep* (e.g., $y = 3x$, $y = -3x$); if $|m|$ is small, it is not steep (e.g., $y = \frac{1}{2}x$). If m is positive, the line climbs to the right (e.g., $y = 3x$); if m is negative, it descends to the right (e.g., $y = -3x$).

On any of these lines y and x are connected by an equation of the form:

$$y = (\text{constant})x \qquad \text{where constant} = m$$

and so y is *proportional* to x, and the constant of proportionality is the slope m. Therefore, if you know that one quantity is proportional to another and you draw their graph, it will come out as a straight line through the origin, with the constant of proportionality as the slope.

Finding the Equation of a Line Given the Slope and a Point

EXAMPLE   *Find the equation of the line of slope −2 through the point (−1, 3).*

Since the slope is −2, the equation must be of the form

$$y = -2x + b$$

Now the coordinates of the point (−1, 3) must satisfy the equation since the point lies on the line. So

$$3 = -2(-1) + b$$

This equation can be solved for *b*:

$$b = 3 - 2 = 1$$

The equation of the line is $y = -2x + 1$. Its graph is shown in Figure 2.17. In this example the slope fixes the direction of the line, and the point (−1, 3) determines its position.

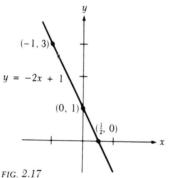

FIG. 2.17

Finding the Equation of a Line Given Two Points

EXAMPLE   *Find the equation of the line through the points (2, 6) and (−2, 8).*

These two points can be used to find the slope of the line:

$$\text{slope} = m = \frac{8 - 6}{-2 - 2} = \frac{2}{-4} = -\frac{1}{2}$$

Therefore, the equation is of the form

$$y = -\frac{1}{2}x + b$$

To find b—which determines the position of the line—we use the fact that the point $(2, 6)$ lies on the line. (The point $(-2, 8)$ could equally well be used.)   Substituting $x = 2$, $y = 6$ into $y = -\frac{1}{2}x + b$:

$$6 = \left(-\frac{1}{2}\right)(2) + b$$

So

$$b = 7$$

Therefore, the line has the equation

$$y = -\frac{1}{2}x + 7$$

Alternative Method of Finding the Equation of a Line Given Two points

Instead of using the two points to find the slope, we could have used the fact that *both* points must satisfy the equation of the line. This would give us simultaneous equations for m and b.

Specifically, the fact that the points $(2, 6)$ and $(-2, 8)$ lie on the line $y = mx + b$ means that

$$6 = 2m + b$$
and
$$8 = -2m + b$$

We can now treat these as simultaneous equations for m and b: Add the equations to solve for b:

$$14 = 2b$$

or

$$b = 7$$

Subtract the equations to solve for m:

$$-2 = 4m$$

or

$$m = -\frac{1}{2}$$

Therefore, the equation of the line is

$$y = -\frac{1}{2}x + 7$$

as before.

PROBLEM SET   2.3

Find the equation of the lines satisfying the following conditions:
1. Passes through the points (3, 5) and (1, 1)
2. Passes through (−1, 5) and has slope 2
3. Has y intercept −31 and slope 2
4. Passes through (−1, −5) and has a slope −$\frac{1}{2}$
5. Passes through (2, −5) and (−3, 3)
6. Passes through (3, 0) and (0, 3)
7. Has x intercept 4 and slope 3
8. Passes through (3, 1) and (5, 4)
9. Has x intercept 5 and y intercept 5
10. Passes through (0.01, −0.03) and (4.3, 2.02)
11. Passes through (1, −16) and has slope m
12. Passes through (a, 0) and (0, a)
13. Passes through (a, b) and has slope 2
14. Passes through (a, 0) and (b, c)
15. Passes through (a, b) and has slope c
16. Graph and label the lines in Problems 1–10.

Find the slope, x intercept, and y intercept of:

17. $2x + y = 1$

18. $\frac{x}{2} + \frac{y}{2} = 1$

19. $-x - 4 = 5$

20. $0.2y - 0.04x = 1.3$

21. $\frac{2y + 1}{x + 1} = 3$

22. $2x = \frac{3(y - x + 1)}{2}$

23. $3y + x = 2$

24. $4 = y - x$

25. $\frac{7}{x + y} = \frac{2}{x}$

26. $0 = 10y - 7x - 4$

27. $\frac{2}{2x + y} = \frac{3}{x + y}$

28. $\frac{y}{4} - x = \frac{3}{2}$

29. $\sqrt{2}x - y = \sqrt{18}$

30. $ax + by + c = 0$

31. Graph and label the lines in Problems 17–19.

Find the equations of the following lines (32–35).

32.

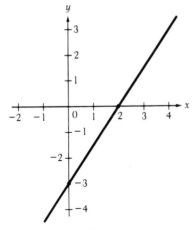

33.

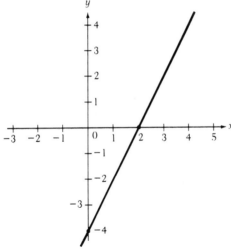

34.

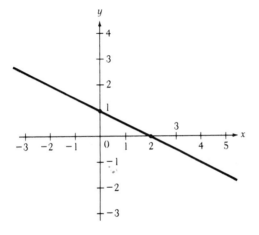

35.

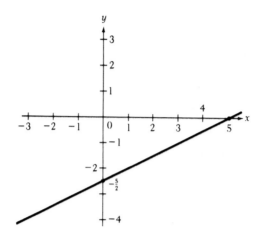

36. Match the following graphs with the equations below.

    (a) $y = x - 4$ (d) $y = -3x - 4$

    (b) $-2x + 3 = y$ (e) $y = x + 5$

    (c) $4 = y$ (f) $y = \frac{1}{2}x$

(A)

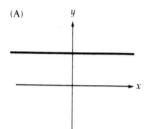

(B)

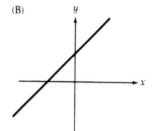

(C)

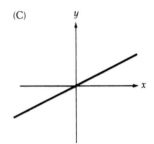

(D)

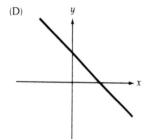

(E)

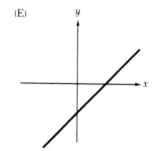

(F)

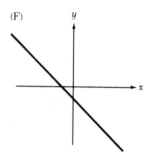

37. At which points do the following lines intersect the x and y axes?

    (a) $x - 2y = 3$ (c) $0 = x + 2y - 7$

    (b) $1 = \frac{x}{4} + \frac{y}{5}$

38. What is the length of that portion of the line between the x and y axes in each of the lines in Problem 37?

39. In Figure 2.18, if P is the graph of $y = mx + c$, and Q is the graph of $y = nx + d$:

    (a) Which is larger, m or n? Why?

    (b) Which is larger, c or d? Why?

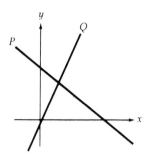

FIG. 2.18

40. Here are three equations:

$$2x + 3y = 12$$

$$-4y = 24x + 3$$

$$x = 2y - 3$$

    (a) Which of the above equations has the graph with the greatest slope?

    (b) Which has the greatest y intercept?

41. Consider a graph of Fahrenheit temperature, y, against Centigrade temperature, x, and assume that the graph is a line. You know that when $y = 212$, $x = 100$ (because 212°F and 100°C both represent the temperature at which water boils). Similarly when $y = 32$, $x = 0$ (water's freezing point). Hence $(100, 212)$ and $(0, 32)$ are two points on the line.

    (a) What is the slope of the graph?

    (b) What is the equation of the line?

    (c) Use the equation to find what temperature in Fahrenheit corresponds to 20°C.

    (d) At what temperature are the centigrade and Fahrenheit temperatures equal?

42. Mark out on graph paper the axes and the lines $y = 12$ and $x = 6$. This is a pool table. The *pool rule* states that if a ball travels along a line with slope m and strikes the side of the table, it will bounce back along a line with slope $-m$.

    (a) A ball starts at $(3, 8)$ with slope 2 toward the y axis. Where does it strike the *y* axis? What slope does it have after bouncing off the *y* axis? What is the equation describing the new path?

(b) Follow the ball in (a) for two more bounces.

(c) Assume the pockets are located at the corners [e.g., (6, 12)]. Where are the others? If you hit a ball from (3, 8) to $(0, \frac{16}{3})$, will it go into a pocket?

(d) Start two balls at (3, 8), one toward the *y* axis at slope of $\frac{3}{2}$ and the other toward (12, $5\frac{1}{3}$). On the second bounce, which is closer to a pocket?

2.4   HORIZONTAL AND VERTICAL LINES

Let's look at horizontal and vertical lines in a bit more detail, given what we know about slopes and intercepts.

Horizontal Lines

At the end of the section on slope (Section 2.2), you saw that the slope of a horizontal line was 0—as you might expect, since it does not climb at all.

If you want to find the equation of a horizontal line at a height of 4 above the *x* axis, you know its slope is 0 and its *y* intercept is 4, so the equation is

$$y = 0 \cdot x + 4$$

or

$$y = 4$$

This equation looks funny because it does not contain an *x*. However, you are meant to think of it as meaning that *x* can be anything, but *y* must be restricted to 4, which gives the points on the horizontal line in Figure 2.19. Therefore the equation *y* = 4 does specify exactly those points that lie on the line.

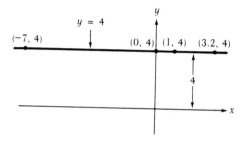

FIG. 2.19

In general, a *horizontal line at a distance b from the x axis has equation* *y* = *b*. If *b* is positive, the line is above the *x* axis, and if *b* is negative, it is below the x axis. In the special case that $b = 0$, we have the equation $y = 0$, which must be the x axis. If you think about it, this is reasonable because the x axis does consist of all the points whose y coordinates are 0. So $y = 0$ *is the equation of the x axis.*

Vertical Lines

We showed earlier that the slope of a vertical line is not defined. For example, the slope of a line between (1, 2) and (1, 4) involves calculating

$$\frac{4-2}{1-1} = \frac{2}{0}$$

which is not defined, and so the slope is not defined either. A vertical line is therefore the one and only case that cannot be described by the slope-intercept form of the equation.

Looking at the equation of a horizontal line, however, gives you an idea of what the equation of a vertical line might look like. Since a horizontal line consists of all the points with a certain y coordinate, say 3, and has equation $y = 3$, it would be reasonable for a vertical line that consists of all the points with a certain x coordinate, say 5, to have an equation $x = 5$. And indeed, if you understand the equation $x = 5$ to mean that the y coordinate can be anything but the x coordinate is restricted to 5, then the equation specifies exactly those points on a vertical line at a distance of 5 to the right of the y axis. See Figure 2.20.

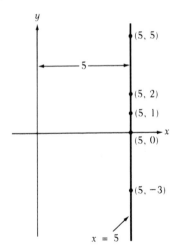

FIG. 2.20

In general, the equation of a *vertical line at a distance k from the y axis is* $x = k$. If k is positive, the line is to the right of the y axis, and if negative, to the left. If $k = 0$ the equation reads $x = 0$ and represents the y axis—which does, after all, consist of all those points whose x coordinate is 0. So *x = 0 is the equation of the y axis.*

2.5   **PARALLEL AND PERPENDICULAR LINES**

Parallel Lines

A set of parallel lines all climb at the same rate and in the same direction (see Figure 2.21, and therefore you will not be surprised to learn that:

Parallel lines have equal slopes.

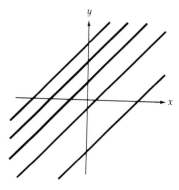

FIG. 2.21

Proof that if $y = m_1 x + b_1$ **and** $y = m_2 x + b_2$ **are parallel, then** $m_1 = m_2$ Suppose the two lines in Figure 2.22 are parallel. In order to go on from here, we have to get some notation straight.

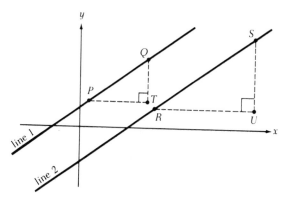

FIG. 2.22

$|TQ|$ will mean the distance between T and Q. It is always a positive number, and

$$|TQ| = |QT|$$

T and Q are on the same vertical line, and without the absolute value signs TQ will mean the change in y in going from T to Q.

QT therefore means the change in y in going from Q to T, and so

$$TQ = -QT.$$

For two points on the same horizontal line, such as P and T, PT means the change in x in going from P to T.

TP means the change in x in going from T to P and so, again,

$$PT = -TP.$$

Now let's go back to Figure 2.22. Since two parallel lines cut the horizontal at the same angle, the angles QPT and SRU must be equal. In the same way, the angles PQT and RSU must be equal. Therefore the triangles PTQ and RUS are similar, and so the ratios of corresponding sides are equal, giving us:

$$\frac{|TQ|}{|US|} = \frac{|PT|}{|RU|}$$

But, looking at the graph, TQ and US must have the same signs. We have no way of knowing if both are positive or both negative, but we do know that either way the ratio $\dfrac{TQ}{US}$ will come out positive. Therefore,

$$\frac{|TQ|}{|US|} = \frac{TQ}{US}$$

By exactly the same reasoning,

$$\frac{|PT|}{|RU|} = \frac{PT}{RU}$$

Therefore, the equation

$$\frac{|TQ|}{|US|} = \frac{|PT|}{|RU|}$$

can be replaced by:

$$\frac{TQ}{US} = \frac{PT}{RU}$$

Multiplying this through by $\dfrac{US}{PT}$ we get:

$$\frac{TQ}{\cancel{US}} \cdot \frac{\cancel{US}}{PT} = \frac{\cancel{PT}}{RU} \cdot \frac{US}{\cancel{PT}}$$

$$\frac{TQ}{PT} = \frac{US}{RU}$$

But

$$\frac{TQ}{PT} = \frac{\text{vertical change along line 1}}{\text{corresponding horizontal change}} = \text{slope of line 1.}$$

Similarly,

$$\frac{US}{RU} = \text{slope of line 2}$$

So

$$\frac{TQ}{PT} = \frac{US}{RU}$$

becomes

Slope of line 1 = slope of line 2

EXAMPLE   *Find the equation of a line parallel to $2y = 3x + 2$ but through the origin.*

Rewriting $2y = 3x + 2$ as $y = \frac{3}{2}x + 1$, you can see that its slope is $\frac{3}{2}$. Therefore the slope of the line that we are looking for is $\frac{3}{2}$, and so its equation must be of the form

$$y = \frac{3}{2}x + b$$

Our line is to go through the origin and so has a y intercept of 0, so $b = 0$.

Therefore the equation of our line is $y = \frac{3}{2}x$.

Perpendicular Lines

It is easy to tell whether or not two lines are parallel to one another—if they are, they have the same slope. Since lines that are perpendicular to one another are important, it would also be useful to know how to tell if two lines were perpendicular, or how to find a line that is perpendicular to some other line. As it turns out, *perpendicular lines have slopes that are negative reciprocals of one another*, or, in other words:

If $y = m_1x + b_1$ and $y = m_2x + b_2$ are perpendicular, then

$$m_1 = -\frac{1}{m_2}$$

(or, equivalently,

$$m_1m_2 = -1)$$

Note: These formulas don't hold for horizontal and vertical lines because, although they are perpendicular, the slope of a vertical line is not defined.

Justification of $m_1 = -\dfrac{1}{m_2}$ If two lines are perpendicular, in general one will have to be climbing as you go from left to right while the other is descending, and therefore they will have to have slopes of opposite signs. The fact that they are perpendicular also means that if one line is climbing (or falling) fast—that is, is nearly vertical—then the other line must not be climbing or falling fast, and so must be nearly horizontal. In other words, if one line has a slope of large magnitude, the other has a slope of small magnitude. These facts certainly fit in with a formula like $m_1 = -\dfrac{1}{m_2}$. Of course, this certainly doesn't prove that such a formula is valid, but merely suggests it.

Proof that $m_1 = -\dfrac{1}{m_2}$ for Perpendicular Lines Suppose that L_1 and L_2 are any two perpendicular lines and that p is their point of intersection. Draw any vertical line cutting them at the points Q and R as shown in Figure 2.23, and draw in the horizontal line PS also.

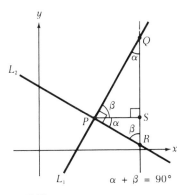

FIG. 2.23

The angle QPR is a right angle because L_1 and L_2 are perpendicular. This means that angles PQS and RPS are equal because both make 90° when added to angle QPS. Similarly, angles QPS and PRS are equal, since both make 90° when added to angle SPR.

Therefore the triangles PQS and RPS are similar.

This means that the ratios of lengths of corresponding sides are equal:

$$\frac{|SQ|}{|PS|} = \frac{|PS|}{|SR|}$$

or

$$\frac{|SQ|}{|PS|} \cdot \frac{|SR|}{|PS|} = 1 \quad \left(\text{multiplying both sides by } \frac{|SR|}{|PS|}\right)$$

Looking at Figure 2.23, we can see that SQ and SR must have opposite signs. Which is positive and which negative is impossible to say, but we can say definitely that

$$SQ \cdot SR = -|SQ|\,|SR|$$

since one of SQ and SR must be negative, and $|SQ| \cdot |SR|$ is positive because it is the product of two lengths. Therefore,

$$|SQ| \cdot |SR| = -SQ \cdot SR$$

PS may be positive or negative, but either way it is true to say that:

$$|PS| \cdot |PS| = PS \cdot PS$$

since both sides are squares and therefore positive. Therefore,

$$\frac{|SQ|}{|PS|} \cdot \frac{|SR|}{|PS|} = 1$$

becomes

$$-\frac{SQ}{PS} \cdot \frac{SR}{PS} = 1$$

or

$$\frac{SQ}{PS} \cdot \frac{SR}{PS} = -1$$

But

$$\frac{SQ}{PS} = \frac{\text{vertical change along } L_1}{\text{corresponding horizontal change}} = \text{slope of } L_1 = m_1$$

and, similarly,

$$\frac{SR}{PS} = \text{slope of } L_2 = m_2$$

Therefore, we have

$$m_1 \cdot m_2 = -1 \quad \text{or} \quad m_1 = -\frac{1}{m_2}$$

EXAMPLE   *Find the equation of the line perpendicular to $3y + x = 7$ and through (a, a).*

The equation $3y + x = 7$ can be rewritten as $y = -\frac{1}{3}x + \frac{7}{3}$, and so its slope is $-\frac{1}{3}$. Therefore the slope of a perpendicular line is

$$-\frac{1}{-\frac{1}{3}} = 3$$

and so its equation must be of the form

$y = 3x + b$

If the point (a, a) is to lie on the line, its coordinates must satisfy the equation—in other words, b must satisfy

$a = 3a + b$

so

$b = -2a$

Therefore the equation is

$y = 3x - 2a$

PROBLEM SET   2.5

Find the equations of the lines satisfying the following conditions:
1. A vertical line through $(0, 4)$
2. A horizontal line 2 units below the x axis
3. Parallel to the x axis and through $(0, -4)$
4. Parallel to the y axis and through $(\sqrt{2}, 0)$
5. Parallel to $3x + 5y = 8$ and through the origin
6. Perpendicular to the y axis and through $(3, 7)$
7. Perpendicular to $y + 3x = 7$ and through the origin
8. Perpendicular to $2y - x - 1 = 0$ and through $(1, -1)$
9. Parallel to $3x + y = 15$ and through $(1, 4)$
10. Perpendicular to $x = 5y - 5$ and through $(-2, -3)$
11. Perpendicular to $y - 1 = -3(x - 4)$ and through $(1, 1)$
12. Perpendicular to $2x + 3y = 4$ and through $(2, 1)$
13. Parallel to $2x - 3(y - 2x) - 3 = 0$ and through $(7, 5)$
14. Perpendicular to $x = zy$ and through $(3, 0)$
15. Perpendicular to $\dfrac{x + y}{3} - 5 = 0$ and through $(-0.2, 3.3)$
16. Perpendicular to $y + ax = 0$ and through the origin
17. A horizontal line through $(4, k)$

18. Parallel to $\dfrac{ay + bx}{c} = n$ and through the origin

19. A vertical line through (a, b)

20. A horizontal line through the point of intersection of $3x + 5y = 8$ and $x = 16$.

21. Sketch the lines in Problems 1–20.

22. Show that the line segments joining the points $(-1, 1)$, $(5, 3)$, and $(1, 5)$ form a right triangle.

23. The equations of a side of a triangle are $2x - y + 5 = 0$; $y + 5 = 3x$; $x + 2y = 6$. Show that this is a right triangle.

24. Show algebraically that the line through $(2, 3)$ and $(4, 9)$ is perpendicular to the line passing through $(1, 2)$ and $(4, 1)$.

25. Let A be the line through $(1, -2)$ and $(4, 2)$. Let B be the line perpendicular to A and through $(1, -2)$. If a square were made with line A as one side, line B as another side, and the two points $(1, -2)$ and $(4, 2)$ as vertices, what would be the coordinates of the other two vertices?

26. The vertices of a quadrilateral are $(-4, 1)$, $(0, -2)$, $(6, 6)$, and $(2, 9)$.
    (a) Show that this figure is a parallelogram.
    (b) Show that this figure is actually a rectangle.

27. An ant is crawling along the line $y = 2x - 2$ in the direction of increasing y. When he comes to the point $(3, 4)$ he takes a left turn ($90°$) and keeps going in a straight-line path. What is the equation of his new path?

28. Show that the diagonals of a square are perpendicular. *Hint:* Draw a general square on a coordinate system as in the figure.

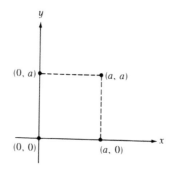

In Problems 29–32, give the equations of line L and line M.

29.

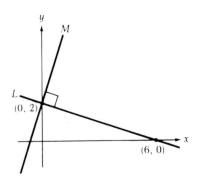

30.

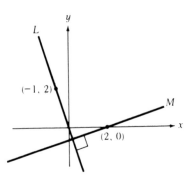

31.

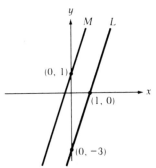

(*Note:* M and L are parallel)

32.

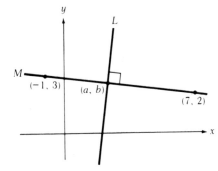

33. Find the equation of the line tangent at the point $\left(\frac{\sqrt{2}}{2}, \frac{\sqrt{2}}{2}\right)$ to a circle of unit radius centered at the origin. *Hint:* A tangent and the radius drawn to the same point are perpendicular lines.

34. The line segment from $(1, -2)$ to $(-2, 3)$ in the figure is the diameter of a circle.

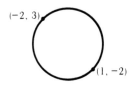

(a) Find the coordinates of the center of the circle.
(b) Find the area of the circle. (You may leave π in your answer.)
(c) Find the equation of the line through the point $(-4, -5)$ and perpendicular to the line connecting the points $(-2, 3)$ and $(1, -2)$.

For Problems 35–43 you will need to find the point of intersection of two lines. If a point lies on two lines, its coordinates satisfy the equations of both lines. Therefore, to find a point of intersection, solve the equations of the lines simultaneously.

35. Write the equation of the line that passes through the origin (0, 0) and through the point of intersection of the following two lines:

$$2x - y + 5 = 0$$

$$y + 5 = -3x$$

36. An electric current is running through a wire along the line $y = 2x + 1$. An electron at (4, 4) is attracted toward the wire and moves toward the closest point on the wire (i.e., it moves along a line perpendicular to the wire).

(a) What is the equation of the line along which the electron travels?
(b) Where does it hit the wire?
(c) How far does it travel before hitting the wire?

37. You know you can find the distance between two points by using the formula

$$D = \sqrt{(x_1 - x_2)^2 + (y_1 - y_2)^2}$$

It is also possible to find the distance between a point and a line, and between two parallel lines. For example, take the line $y = 2x - 1$ and the point $(-2, 5)$. The shortest distance between the point and the line is the length of the line segment through $(-2, 5)$ and perpendicular to $y = 2x - 1$. The procedure to follow is this:

(a) Find the equation of the line perpendicular to $y = 2x - 1$ and through $(-2, 5)$.
(b) Find the point at which the two lines intersect, say (a, b).
(c) Find the distance between $(-2, 5)$ and (a, b). This is the distance we want.

38. What is the distance between (1, 6) and $y = 2x + 3$?

39. What is the distance between (6, 9) and $y = -\frac{3}{4}x + 1$?

40. What is the distance between (11, 0) and $y = 2x + 3$?

41. Given the two parallel lines $y = -3x - 5$ and $y = -3x + 2$, find the distance between them. *Hint:* Pick a point on one line and proceed as in Problem 37.

42. Find the distance between $y = 2x + 5$ and $y = 2x - 5$.

43. Can you find a general formula to express the distance between any point (x_0, y_0) and any line $y = mx + b$?

CHAPTER 2 REVIEW

Find equations for the following lines:
1. The line passing through $(-2, -3)$ and $(7, 9)$
2. The line passing through $(-3, -2)$ with slope $-\frac{1}{3}$
3. The line passing through $(0.02, -0.01)$ and $(-0.02, -0.04)$
4. The line with x intercept $\frac{1}{5}$ and y intercept -3
5. The line through $(4, 5)$ with slope 0
6. The line passing through $(-\frac{2}{5}, 3)$ and $(1, -6)$
7. The line through $(0, \frac{5}{8})$ with slope $\frac{98}{89}$
8. The line with x intercept -4 and y intercept $-\frac{1}{4}$
9. The line with slope 2 passing through (a, b)
10. The line with slope $\dfrac{1}{m}$ passing through $(x_0 + t, x_0 - t)$

Graph the following:
11. $\dfrac{x}{4} + \dfrac{y}{-6} = 1$
12. $2x - \frac{1}{3}y = 4$
13. $0.2(x - 3) = (y + 1)$
14. $y = -\frac{1}{2}x - 1$
15. $\dfrac{x - y + 3}{4} = \dfrac{2x + y + 1}{6}$
16. $\frac{3}{8}x = \frac{2}{3}y$
17. $2x = -42$

18. Find the equation of the line perpendicular to the line in Problem 2 but passing through $(3, 2)$. Graph it.

19. Find the equation of the line perpendicular to the line in Problem 3 but passing through $(0.02, -0.01)$. Graph it.

20. Find the equation of the line parallel to $2y - 7x = y + 4$ but passing through $(-1, 2)$. Graph it.

21. If a line passes above the point $(-2, 3)$ and below the point $(4, -2)$, what is the greatest slope a line perpendicular to it could have?

22. Start at $(-1, 2)$ and head directly for $(2, -2)$.
    (a) Where do you hit the y axis?
    (b) What is the equation of the line travelled on?
    (c) Two lines might be described as skewed if the ratio of the products of their slopes and the sum of these slopes $= -1$. At $(2, -2)$, what is the line skewed to the one that you are on?

23. In an attempt to move into the sports field, the Gallup poll compiled Table 2.2 about the players for a well-known football team. It shows the

number of times each player got to carry the ball (this is the number of attempts), the total distance he carried it (this is the yards), and the average distance he carried it, per attempt.

Table 2.2
Rushing Statistics—Five Game Totals

Player	Attempts	Yards	Average
Ron	60	334	5.6
Jon	39	201	5.2
Al	62	192	3.1
Tom	36	120	3.3
Burke	18	49	2.7

(a) Putting "attempts" along the x axis and "yards" up the y axis, plot a point for each player, and draw a line from that to the origin.

(b) What is the equation of the line joining Al to the origin?

(c) What is the slope of the line perpendicular to that joining Burke to the origin?

24. I. A. Rich owns an estate on the corner of Ivy Street (which runs east–west) and Vine Avenue (which runs north–south). The driveway on the estate runs in a straight line from a point on Vine Avenue 300 yards north of the corner directly to the house. If this driveway runs northeast–southwest and is 1500 yards long, how far is the house from Ivy Street?

25. A wire is laid on a piece of graph paper in such a way that it forms the graph of $y = 3x + 2$. If this wire is moved vertically up by 2 units, where does it then cross the x axis?

26. Able is a crook, fleeing from a crime by motorboat going at a constant speed in a straight line across Axis Lake. He moves from $(-5, 10)$ to $(-3, 10)$ in 10 minutes. Just at the time he reaches $(-3, 10)$ Baker, a policeman, leaves the Axis Lake Police Station on an island at $(3, 2)$ to catch him. Baker wants to catch Able as soon as possible, but the maximum speed of a police boat is 10 units per hour. Give the equation of the straight-line path Baker should follow in order to catch Able.

27. Suppose that
   (i) Two lines are "reciprocal" to each other if their slopes are reciprocals (for example, $y = 3x + 2$, $y = \frac{1}{3}x - 1$)
   and
   (ii) Two lines are "negative" to each other if their slopes are of the same magnitude but opposite in sign (for example, $y = 2x + 1$, $y = -2x + 3$)
   then:

(a) Can a line ever have the same slope as a line "reciprocal" to it? Give an example, or explain why not.

(b) Is it possible that a line could be both perpendicular *and* "negative" to another line? Give an example, or explain why not.

(c) Is it possible that a line could be both "negative" and "reciprocal" to another line? Give an example, or explain why not.

28. Two lines, A and B, intersect at the point (2, 3). Both lines have positive slopes, and the slope of line A is greater than the slope of line B.

(a) Which line will have the greater *y* intercept?

(b) Which line will have the greater *x* intercept?

(c) Which line will have its *x* intercept closer to the origin (0, 0)?

(d) If the problem had stated that both lines had *negative* slopes, would your answers have been different? (Be careful!)

Problems 29–30 require you to find the point of intersection of two lines by solving the equations simultaneously. See note before Problem 35, in Problem Set 2.5.

29. Given a triangle with vertices (0, 0), (0, 4), and (4, 0), show that the three lines connecting each vertex to the midpoint of the opposite side meet in a single point.

30. A radio beacon is located at $(1, -1)$ and another at $(-\frac{2}{3}, \frac{3}{4})$. Your navigation equipment tells you that the line joining you to the first beacon has slope $-\frac{1}{2}$ and the line joining you to the second has slope $= \frac{1}{2}$. Where are you?

3 NEW GRAPHS FROM OLD

3.1 WHY BOTHER?

The Cartesian plane is an extremely simple idea, but, as we will see in the next chapters, a staggeringly powerful one. Its power stems from the fact that it enables us to combine two fields, algebra and geometry, which together can reach far further than either can reach alone.

The bridge between the two fields is graphing. When we draw a graph of an equation, we are turning an algebraic object into a geometric one; when we find the equation of a line or curve, we are doing the opposite.

We are going to need to draw the graphs of a great many equations, and we are going to need to be able to do it fast. You may have noticed that in the section on lines we stopped doing a table of values and just plotted the two intercepts and looked at the slope. With some practice, it is possible to get a great deal of information about any graph just by looking at the form of the equation and doing a few well-chosen calculations. The point of this chapter is to show you how to graph an equation by doing as little work as possible, and in particular without using a table of values.

3.2 SYMMETRY

Noticing that a graph is symmetric about a point or a line is helpful because it means that you only have to draw half the graph, and can get the other half by reflection. There are two main kinds of symmetry: symmetry about a line and symmetry about a point.

Symmetry About the y Axis: Even Symmetry

Even symmetry occurs when the part of the graph to the left of the y axis is the reflection of that on the right in an imaginary mirror along the y axis.

To find an algebraic test for this kind of symmetry, look at the graph of $y = x^2$. Draw up a table of values:

x	y
-2	4
-1	1
0	0
1	1
2	4

and think about what happens when you plot the graph. Notice that $x = 2$ and $x = -2$ lead to the same y value (namely 4) and so to two points the same height above the x axis. Similarly $x = 1$ and $x = -1$ lead to the points $(1, 1)$ and $(-1, 1)$ which are the same height above the axis. See Figure 3.1.

Indeed, for every point to the right of the y axis there is a companion point to the left of the y axis at the same height.

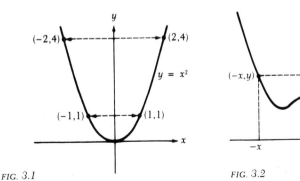

FIG. 3.1   FIG. 3.2

In general, *you get even symmetry if x and* $-x$ *always give the same y value.* This means that whenever some point (x, y) is on the graph, so is $(-x, y)$. See Figure 3.2 *This kind of symmetry occurs when the equation is unaltered by replacing x by* $-x$. It occurs in equations in which y is equal to a polynomial containing only constants and *even* powers of x (an *even* polynomial, hence the name *even symmetry*). An algebraic fraction that is the quotient of two odd or two even polynomials is always even. (An odd polynomial contains only odd powers of x and no constant term.) For example,

$$y = 3x^4 - 4x^2 + 3 \qquad y = \frac{x^2 + 1}{x^2 - 1} \qquad y = \frac{x^3}{x + x^5}$$

are all even.

EXAMPLE: $y = x^2 - 2$

This has even symmetry because

the y you get from $-x$ is $(-x)^2 - 2 = x^2 - 2$

the y you get from x is $x^2 - 2$ } equal

See Figure 3.3.

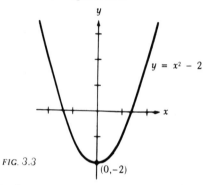

$y = x^2 - 2$

FIG. 3.3
$(0,-2)$

Note: In the same way, the graph of an equation in which y and $-y$ both lead to the same value of x has symmetry about the x axis.

Symmetry About the Origin: Odd Symmetry

Symmetry about the origin occurs when the part of the graph in the third (or fourth) quadrant is obtained from that in the first (or second) by a double reflection, as shown in Figure 3.4.

Examples with this kind of symmetry are $y = x^3$ and $y = \dfrac{1}{x}$. The graph of $y = x^3$ has this kind of symmetry because substituting $x = 1$ gives $y = 1$, while $x = -1$ gives $y = -1$; substituting $x = 2$ gives $y = 8$, while $x = -2$ gives $y = -8$. Therefore, for every point in the first quadrant there is a companion point in the third, as you can see in Figure 3.5.

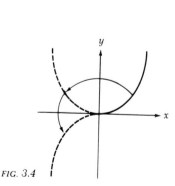

FIG. 3.4

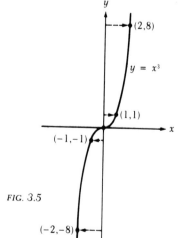

$(2,8)$

$y = x^3$

$(1,1)$

$(-1,-1)$

FIG. 3.5

$(-2,-8)$

You get odd symmetry if x and −x give opposite y values. This means that whenever some point (x, y) is on the graph, then so is $(-x, -y)$, as in Figure 3.6. *This kind of symmetry occurs when the sign of y is reversed by replacing*

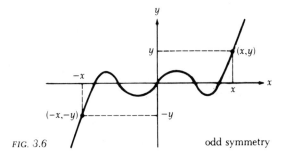

FIG. 3.6   odd symmetry

x by $-x$. It occurs in equations in which y is equal to a polynomial containing only *odd* powers of x and no constant term (hence the name *odd symmetry*). An algebraic fraction that is the quotient of one even and one odd polynomial is always odd. For example,

$$y = 4x^7 - 6x^5 + x^3 \qquad y = \frac{x^4 + 1}{x^3} \qquad y = \frac{x^5 + x}{x^4 + 4}$$

are all odd.

EXAMPLE: $y = \dfrac{1}{x}$

This has odd symmetry because

the y you get from $-x$ is $\dfrac{1}{(-x)} = -\left(\dfrac{1}{x}\right)$

the y you get from x is $\dfrac{1}{x}$ $\Big\}$ opposite

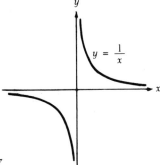

FIG. 3.7

See Figure 3.7.

3.3   SHIFTS: VERTICAL AND HORIZONTAL

Vertical Shifts

If you compare the graphs of $y = x^2$ and $y = x^2 + 1$, you can see that the graph of $y = x^2 + 1$ is exactly that of $y = x^2$ shifted up by one unit. The reason for this is that if you imagine substituting the same x into $y = x^2$ and $y = x^2 + 1$, the y you get in the second case is exactly one more than the y you get in the first (for example, if $x = 2$, $y = 4$ in the first equation and $y = 5$ in the second). So it is really not necessary to draw a table of values to graph $y = x^2 + 1$: Given that you know the graph of $y = x^2$, you can just move it up by one unit, as shown in Figure 3.8. Similarly, the graph of $y = x^2 - 4$ is the graph of $y = x^2$ moved down by four; see Figure 3.8.

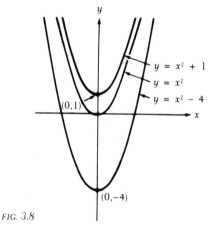

FIG. 3.8

This kind of reasoning can make the drawing of many graphs much easier.

EXAMPLE: *Graph* $y = 1 + \dfrac{1}{x}$.

We have already drawn the graph of $y = \dfrac{1}{x}$; it looks like Figure 3.9.

The graph of $y = 1 + \dfrac{1}{x}$ is the same thing moved up by one, as shown in Figure 3.10.

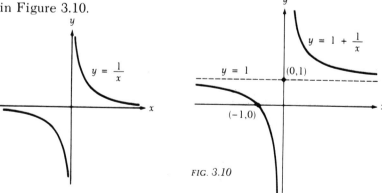

FIG. 3.9   FIG. 3.10

EXAMPLE: *Graph* $y = x^3 - 2$.

The graph of $y = x^3$ looks like Figure 3.11. Therefore the graph of $y = x^3 - 2$ is as shown in Figure 3.12.

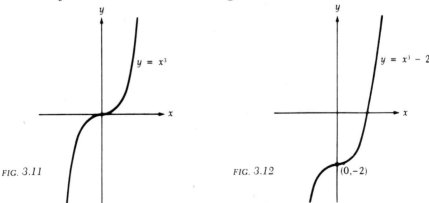

FIG. 3.11   FIG. 3.12

There is another way of looking at all this, which turns out to be useful in considering horizontal shifts. Let us look at $y = x^2 + 1$ again. Suppose you rewrite this as $y - 1 = x^2$, and compare it with $y = x^2$. In order to get the same value for x^2 (and therefore for x) from both equations, you must make y one larger in the first equation. So for a fixed x, the points that lie on the graph of the first equation must have y one larger than those that lie on the graph of the second. Therefore, *replacing y by $(y - 1)$ has the effect of shifting the graph up by one.* See Figure 3.13.

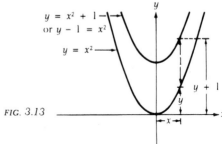

FIG. 3.13

Horizontal Shifts

Suppose we now consider the graph of $y = (x - 1)^2$. Since we are not yet sure what this will look like, we should do a table of values:

x	y
-3	16
-2	9
-1	4
0	1
1	0
2	1
3	4

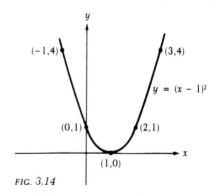

The graph, then, looks like Figure 3.14.

FIG. 3.14

Now you notice that this is just the graph of $y = x^2$ shifted over to the right by one. To see why this is so, compare the graphs of $y = (x - 1)^2$ and $y = x^2$. In order to get the same value of y from both equations, you must have x one larger in $y = (x - 1)^2$ than in $y = x^2$. Therefore, for a fixed value of y, a point lying on the graph of $y = (x - 1)^2$ has x one larger than that lying on the graph of $y = x^2$. This means that *replacing x by $(x - 1)$ has the effect of shifting the graph one to the right;* see Figure 3.15.

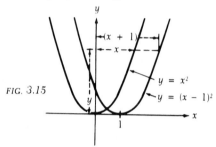

FIG. 3.15

EXAMPLE: *Graph $y = (x - 2)^3$.*

This is the graph of $y = x^3$ shifted to the right by 2. The graph is shown in Figure 3.16.

EXAMPLE: *Graph $y = (x + 3)^2$.*

The $+$ sign may throw you until you realize that you can write this equation in the form:

$$y = (x - (-3))^2$$

So the graph we are looking for is that of $y = x^2$ moved to the *left* by 3. The graph is shown in Figure 3.17.

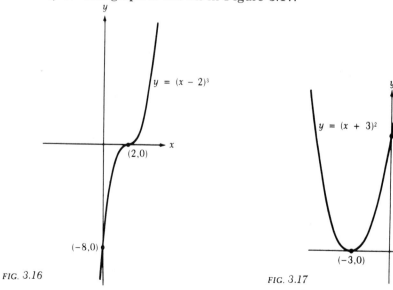

FIG. 3.16

FIG. 3.17

If you aren't entirely sure why this moves to the left, reason it out by comparing $y = (x + 3)^2$ and $y = x^2$. To get the same value of y from both equations, the x value that you put into the first must be 3 less than—that is, 3 to the left of—the x value you put into the second. Therefore at a fixed y the point on $y = (x + 3)^2$ is 3 to the left of the point on $y = x^2$. Hence the whole graph is moved 3 to the left.

EXAMPLE: *Graph* $(y + 2) = (x - 3)^2$.

This is the graph of $y = x^2$ moved to the right by 3 and down by 2 [since $y + 2 = y - (-2)$]. The graph is shown in Figure 3.18.

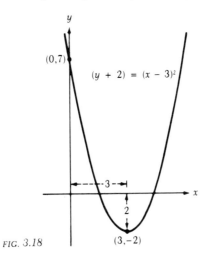

FIG. 3.18

PROBLEM SET   3.3

Graph, and label, on the same Cartesian plane:
1. $y = x^2$
2. $y = x^2 + 1$
3. $y = x^2 + 3$
4. $y = x^2 + 5$
5. $y = x^2 - 2$
6. $y = x^2 - 4$

Graph, and label, on the same Cartesian plane:
7. $y = x^3$
8. $y = x^3 + 2$
9. $y = x^3 + 4$
10. $y = x^3 - 1$
11. $y = x^3 - 3$

Graph, and label, on the same Cartesian plane:
12. $y = \dfrac{1}{x}$
13. $y = \dfrac{1}{x} + 1$
14. $y = \dfrac{1}{x} + 3$
15. $y = \dfrac{1}{x} - 2$
16. $y = \dfrac{1}{x} - 5$

Graph, and label, on the same Cartesian plane:
17. $y = x^2$
18. $y = (x - 1)^2$
19. $y = (x - 2)^2$
20. $y = (x + 2)^2$
21. $y = (x + 4)^2$

Graph, and label, on the same Cartesian plane:

22. $y = \dfrac{1}{x}$ 24. $y = \dfrac{1}{(x+3)}$ 26. $\dfrac{1}{(x-2)}$

23. $y = \dfrac{1}{(x+1)}$ 25. $y = \dfrac{1}{(x-1)}$

Graph, and label, on the same Cartesian plane:

27. $y = x^3$ 29. $y = x^3 - 1$ 31. $y = (x+1)^3$
28. $y = x^3 + 1$ 30. $y = (x-1)^3$

Graph, and label, on the same Cartesian plane:

32. $y = |x|$ 34. $y = |x+1|$ 36. $y = |x| - 1$
33. $y = |x| + 1$ 35. $y = |x-1|$

Graph, and label, on the same Cartesian plane:

37. $y = x^2$ 39. $(y+2) = (x+3)^2$
38. $y = (x+3)^2$ 40. $(y-3) = (x-1)^2$

Graph, and label, on the same Cartesian plane:

41. $y = |x^2|$
42. $y = |x^2 - 1|$ 45. Graph $y = \dfrac{1}{(x+2)^2}$
43. $(y-2) = |x^2 - 1|$
44. $(y+3) = |x^2 + 2|$ 46. Graph $y = \dfrac{1}{x-2} + 3$

3.4   STRETCHING AND FLIPPING: POSITIVE AND NEGATIVE COEFFICIENTS

If you use a table of values to draw graphs of $y = x^2$, $y = 2x^2$, $y = \frac{1}{2}x^2$, $y = -x^2$, and $y = -2x^2$ on the same axes, you get the curves shown in Figure 3.19. This shows that the effect of *a factor such as 2 in front of the x^2 is to make the graph climb faster*, while *a factor such as $\frac{1}{2}$ makes it climb slower*. *A negative sign flips the whole graph over*, so a factor of -2 makes the graph descend more quickly. In general, factors greater than 1 make the graph climb faster, while factors between 0 and 1 make it climb slower.

x	$y = 2x^2$
-2	8
-1	2
0	0
1	2
2	8

x	$y = x^2$
-2	4
-1	1
0	0
1	1
2	4

x	$y = \frac{1}{2}x^2$
-2	2
-1	$\frac{1}{2}$
0	0
1	$\frac{1}{2}$
2	2

x	$y = -x^2$
-2	-4
-1	-1
0	0
1	-1
2	-4

x	$y = -2x^2$
-2	-8
-1	-2
0	0
1	-2
2	-8

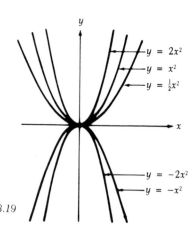

FIG. 3.19

EXAMPLE: *Graph $y + \frac{1}{2} = -2(x - 1)^2$.*

You want the graph of $y = x^2$ moved to the right by one [for the $(x - 1)$], down by a half (for the $y + \frac{1}{2}$), and then flipped over and stretched downward (the factor of -2). The result is shown in Figure 3.20.

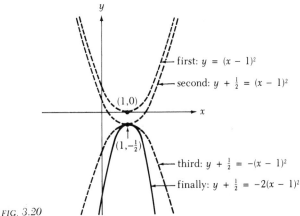

first: $y = (x - 1)^2$

second: $y + \frac{1}{2} = (x - 1)^2$

$(1,0)$

$(1,-\frac{1}{2})$

third: $y + \frac{1}{2} = -(x - 1)^2$

finally: $y + \frac{1}{2} = -2(x - 1)^2$

FIG. 3.20

PROBLEM SET 3.4

Graph, and label, on the same Cartesian plane:

1. $y = x^4$
2. $y = 2x^4$
3. $y = 4x^4$
4. $y = \frac{1}{2}x^4$
5. $y = -x^4$
6. $y = -2x^4$

Graph, and label, on the same Cartesian plane:

7. $y = x^3$
8. $y = \frac{1}{2}x^3$
9. $y = 3x^3$
10. $y = -x^3$
11. $y = -2x^3$

Graph, and label, on the same axes.

12. $y = |x|$
13. $y = 2|x|$
14. $y = 3|x|$
15. $y = \frac{1}{2}|x|$
16. $y = -2|x|$

Graph, and label, on the same Cartesian plane:

17. $y = |x|$
18. $(y - 1) = |x + 1|$
19. $(y + 2) = \frac{1}{2}|x + 3|$
20. $(y - 3) = -2|x - 2|$
21. $2y = |x - 1| + 4$

Graph, and label, on the same Cartesian plane:

22. $y = \dfrac{1}{x}$
23. $2y = \dfrac{1}{x} + 1$

Graph each of the following equations:

24. $y - 2 = \dfrac{1}{(x - 1)}$
25. $y + 1 = 2\left(\dfrac{1}{x - 2}\right)$
26. $-y + 3 = \dfrac{1}{2}\left(\dfrac{1}{x + 1}\right)$
27. $(y - 1) = \dfrac{2}{(x - 2)^2}$

28. The curve shown in Figure 3.21 has been shifted down and to the right. The equation of the original graph is $y = 2x^2 + 12x + 18$.

    (a) Note how the coordinates of points A, B, and C have changed and write the equation of the *shifted* graph.

    (b) By plotting some points, sketch the graph of

$$y = 2\left(\frac{x}{3}\right)^2 + 12\left(\frac{x}{3}\right) + 18$$

Be sure to indicate the new locations of the points A, B, and C.

    (c) How is the graph of

$$y = 2\left(\frac{x}{3}\right)^2 + 12\left(\frac{x}{3}\right) + 18$$

related to the original curve,

$$y = 2x^2 + 12x + 18?$$

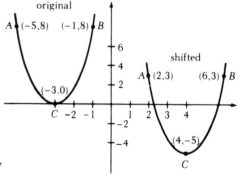

FIG. 3.21

3.5   INTERCHANGING X AND Y

Suppose that we interchange x and y in $y = x^2$, giving the equation $x = y^2$, and then graph that. We must use a table of values to be sure what the graph looks like, but, not surprisingly, we find that the roles of the x and y axes have simply been interchanged. In other words, the graph of $x = y^2$ is the same shape as that of $y = x^2$, but is centered around the x axis instead of the y axis.

    When you are making up the table of values, you will notice that every x value gives rise to two y values. In this case it is more convenient to start with the y values and calculate the x values from them, rather than the other way round. What you get is shown in Figure 3.22. You should notice that interchanging x and y in the equation amounts to reflecting the graph in the diagonal line $y = x$.

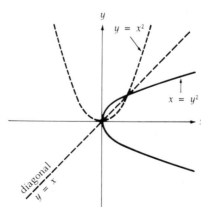

x	y
4	-2
1	-1
0	0
1	1
4	2
9	3

FIG. 3.22

EXAMPLE: *Graph $x = -2y^2$.*

Starting with the graph of $y = x^2$, we have to interchange x and y, which means centering the graph around the positive x axis, then flip it over, making it centered around the negative x axis, and then stretch it. The result is shown in Figure 3.23.

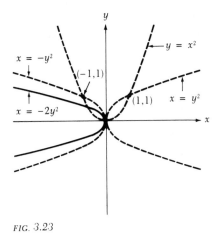

FIG. 3.23

Note: When you have to interchange x and y and do several other things to a graph, you must do the interchange first.

EXAMPLE: *Graph $x = y^3$.*

To convert $y = x^3$ into $x = y^3$, x and y must be interchanged, which is graphically equivalent to reflection in the diagonal $y = x$. Therefore the graph of $x = y^3$ is as shown in Figure 3.24.

In case you are not convinced by the reflection argument, you can plot the graph using a table of values, which is again easier if you start with a y value and calculate x from it:

x	y
-8	-2
-1	-1
0	0
1	1
8	2

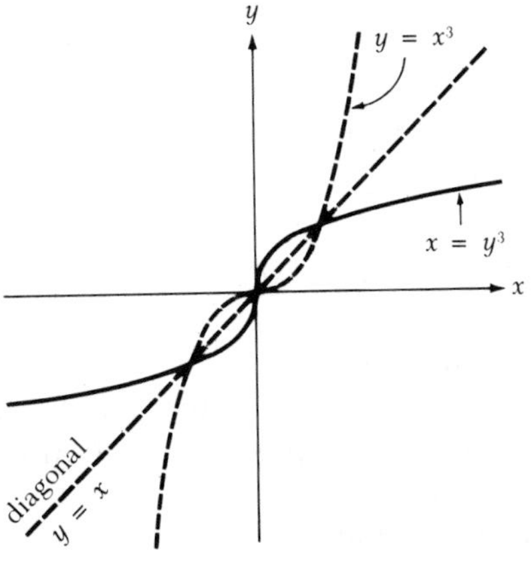

FIG. 3.24

EXAMPLE: *Graph $2x = (y + 1)^3$.*

Since x and y have reversed roles, the "stretch" factor must be in front of the y term here. We therefore rewrite the equation as

$$x = \tfrac{1}{2}(y + 1)^3.$$

The graph of this equation is derived from that of $x = y^3$ by moving it down by one unit, and then pushing it back toward the x axis for the $\frac{1}{2}$. The graph is shown in Figure 3.25.

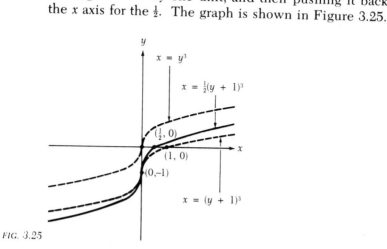

FIG. 3.25

PROBLEM SET 3.5

Graph, and label, on the same Cartesian plane:
1. $y = x^2$
2. $x = y^2$
3. $y = x^2 + 1$
4. $x = y^2 + 1$
5. $y = \tfrac{1}{2}x^2$
6. $x = -\tfrac{1}{2}y^2$

Graph, and label, on the same Cartesian plane:
7. $y = |x|$
8. $|y| = x$
9. $|y - 1| = x$
10. $|x - 1| = y$
11. $2y = |x| + 1$
12. $|y| = 2x - 1$

13. Graph on the same axes:

   (a) $y - 2 = \dfrac{2}{x - 3}$

   (b) $x - 2 = \dfrac{2}{y - 3}$

14. Suppose v and w are connected by the equation $3v + w + 3 = 0$
   (a) Draw a graph of v against w (i.e., putting the v axis vertical and the w axis horizontal).
   (b) Draw a graph of w against v (i.e., putting the w axis vertical).
   (c) Compare the graphs you drew in (a) and (b).
15. If $v = w^3$, draw graphs of v against w and of w against v.
16. If s is proportional to t^2, sketch a graph of s against t. Then sketch a graph of t against s.

3.6   ADDING GRAPHS

Suppose you have the graphs for $y = x^2$ and for $y = x$ and that you want the graph for $y = x^2 + x$. Then the y coordinate for a point on the graph of $y = x^2 + x$ is the sum of the y coordinates of the corresponding points on $y = x^2$ and $y = x$, as shown in Figure 3.26.

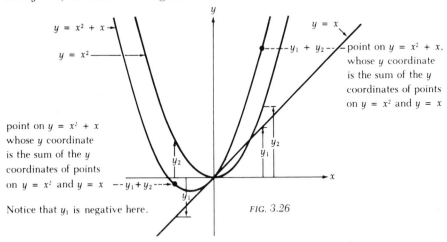

FIG. 3.26

   You should notice that since both the original graphs go through the origin, so does their sum. Also, when the original graphs are both positive or both negative, their sum has the same sign; when one of the original graphs is positive and one is negative, their sum is somewhere in between them.

PROBLEM SET 3.6

   1. Show how you can get the graph of $y = x^2 + 1$ by adding the graphs of $y = x^2$ and $y = 1$.

   2. Show how adding the graphs of $y = x$ and $y = 2x$ gives the graph of $y = 3x$

   3. Graph $y = |x| - x$ by adding the graphs of $y = |x|$ and $y = -x$.

   4. Graph $y = x^3 + x$ by adding graphs.

   5. Graph $y = \dfrac{1}{x} + x$ by adding the graphs of $y = \dfrac{1}{x}$ and $y = x$.

   6. Graph $y = 2|x| - x$ by adding the graphs of $y = 2|x|$ and $y = -x$.

   7. Which is larger, x or x^2, when x is between 1 and 2? When x is between $\tfrac{1}{2}$ and 1? Draw the graphs of $y = x$ and $y = x^2$ on the same Cartesian plane and show how the graphs illustrate your answers.

   8. For what values of x is $x^3 + 2x^2 - 3$ positive? Negative? Draw a graph of $y = x^3 + 2x^2 - 3$ and show how you could get the answers from that. Then draw the graphs of $y = x^3$ and $y = 2x^2 - 3$ on the same Cartesian plane and show how you could get the answer from that.

3.7   GRAPHING RECIPROCALS

Suppose you have the graph of $y = x$, and want to derive the graph of $y = \dfrac{1}{x}$ from it. The y value of any point on the second graph is the reciprocal of the y value of the corresponding point on the first graph.

Remember that:

1. A number and its reciprocal have the same sign; therefore both graphs are positive together and both are negative together.

2. The larger a number, the smaller its reciprocal, and vice versa.

3. If a number is zero, its reciprocal is undefined. Therefore when one of the graphs cuts the x axis, the other one is undefined.

Compare the two graphs in Figure 3.27.

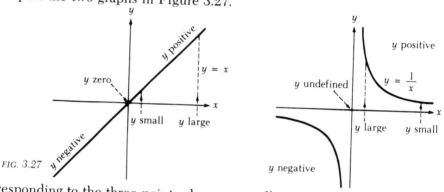

FIG. 3.27

Corresponding to the three points above you will notice that:

1. Both graphs are in the first and third quadrants.

2. When $y = x$ is near the x axis (i.e., when x is near zero), $y = \dfrac{1}{x}$ is far away.

When $y = x$ is far from the x axis (as $x \to \pm\infty$), $y = \dfrac{1}{x}$ is near the x axis.

3. When $y = x$ crosses the x axis (at the origin), $y = \dfrac{1}{x}$ is undefined.

EXAMPLE: *Use the graph of $y = x - 1$ to get the graph of $y = \dfrac{1}{x - 1}$.*

First graph $y = x - 1$, as shown in Figure 3.28.

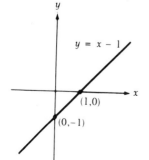

FIG. 3.28

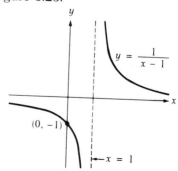

FIG. 3.29

1. The graph of $y = \dfrac{1}{x-1}$ must be positive for $x > 1$, negative for $x < 1$.

2. When x gets very large, the graph of $y = x - 1$ gets far from the x axis, so the graph of $y = \dfrac{1}{x-1}$ gets close. As x gets close to 1, the graph of $y = x - 1$ gets close to the x axis, and therefore the graph of $y = \dfrac{1}{x-1}$ is far away from the x axis.

Putting (1) and (2) together:

As x gets close to 1 from above, the graph of $y = \dfrac{1}{x-1}$ goes upward away from the x axis.

As x gets close to 1 from below, the graph of $y = \dfrac{1}{x-1}$ goes downward away from the x axis.

3. At $x = 1$, the graph of $y = \dfrac{1}{x-1}$ does not exist at all.

So we get the graph shown in Figure 3.29 above.

Check: Since $y = \dfrac{1}{x-1}$ is $y = \dfrac{1}{x}$ with x replaced by $x - 1$, you can also get the graph of $y = \dfrac{1}{x-1}$ by shifting $y = \dfrac{1}{x}$ one to the right. It does come out the same!

PROBLEM SET   3.7

1. Use the graph of $y = x^2$ to get the graph of $y = \dfrac{1}{x^2}$.

2. Get the graph of $y = \dfrac{1}{(x-3)^2}$ both by shifting $y = \dfrac{1}{x^2}$ and from the graph of $y = (x-3)^2$.

3. Use the graph of $y = (x-1)(x-2)$ to get the graph of $y = \dfrac{1}{(x-1)(x-2)}$.

4. Graph $y = \dfrac{1}{x^2-4}$.

Graph, and label, each of the following pairs of equations on different Cartesian planes:

5. $y = |x|$

$y = \dfrac{1}{|x|}$

6. $y = x^2 - 2$

$y = \dfrac{1}{x^2-2}$

7. $y = \dfrac{1}{x} - 3$

$y = \dfrac{1}{\dfrac{1}{x} - 3}$

8. (a) Draw graphs of $y = x^2 + 1$ and $y = x^2 - 1$; then draw graphs of $y = \dfrac{1}{x^2 + 1}$ and $y = \dfrac{1}{x^2 - 1}$

(b) What are the major differences between the graphs of $y = \dfrac{1}{x^2 + 1}$ and $\dfrac{1}{x^2 - 1}$ and how are they related to the differences between the graphs of $y = x^2 + 1$ and $y = x^2 - 1$?

This is the graph of an equation of the form

$$y = \text{some polynomial in } x$$

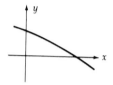

If we call this polynomial p, then:

9. Which of the graphs below is the graph of $y = \dfrac{1}{p}$?

10. Which is the graph of $y = -p$?

(a)

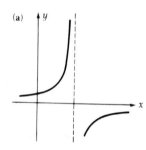

(b)

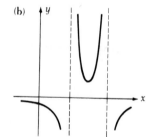

(c)

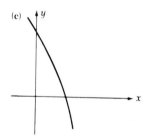

(d)

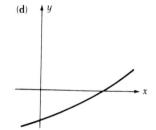

**3.8   INTERCEPTS OF GRAPHS AND
           SOLUTIONS OF EQUATIONS**

Rather than showing how to graph new equations, this section will show how
to get alebraic information from a graph that you already have.

   Let us go back and look at the graphs of $y = x^2 - 1$ and of $y = x^2 - 4$ again.
See Figure 3.30. The y intercept of $y = x^2 - 1$ is -1 because the graph cuts
the y axis where $x = 0$ and at that point $y = 0^2 - 1 = -1$. Similarly, the y intercept of $y = x^2 - 4$ is -4 because it cuts the y axis where $y = 0^2 - 4 = -4$.

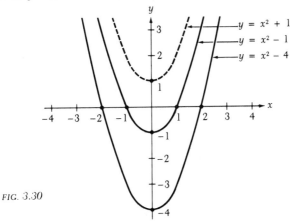

FIG. 3.30

In general:

To find the y intercepts of a graph, substitute $x = 0$ in the equation and
solve for y.

   Now for the x intercepts. From the graph you can see that the x intercepts of
$y = x^2 - 1$ are 1 and -1. You can also find the x intercepts by realizing that the
graph crosses the x axis when $y = 0$, and therefore $0 = x^2 - 1$. The solutions
to this equation are $x = \pm 1$, the x intercepts. Similarly, from the graph the x
intercepts of $y = x^2 - 4$ are 2 and -2, the solutions to $0 = x^2 - 4$.
   In general:

To find the x intercepts of a graph, substitute $y = 0$ into the equation and
solve for x.

   Now look at the graph of $y = x^2 + 1$. Knowing that the equation $0 = x^2 + 1$
has no real solutions tells you that the graph of $y = x^2 + 1$ can't have any x intercepts. Conversely, seeing that the graph of $y = x^2 + 1$ has no x intercepts
tells you that the equation $0 = x^2 + 1$ has no real roots.

This relationship between intercepts and solutions of equations can have remarkably powerful consequences when the equation is more complicated.

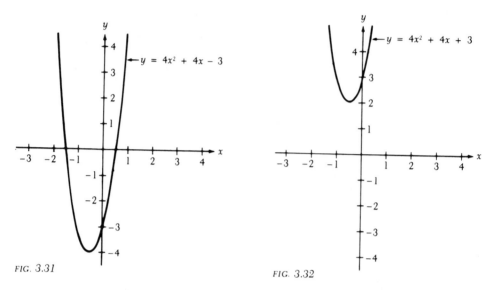

FIG. 3.31

FIG. 3.32

Suppose I told you that the graph of $y = 4x^2 + 4x - 3$ looks like Figure 3.31, and that the graph of $y = 4x^2 + 4x + 3$ looks like Figure 3.32. Then, without a single calculation, and without messing around with the quadratic formula, you could tell me that the equation $0 = 4x^2 + 4x - 3$ has two roots, and even that one is between -2 and -1 and the other is between 0 and 1, because the graph of $y = 4x^2 + 4x - 3$ cuts the x axis at such points. At the same time you could tell me that $0 = 4x^2 + 4x + 3$ has no roots, because the graph of $y = 4x^2 + 4x + 3$ does not cut the x axis at all. It's amazing that you can get so much information about an algebraic problem just by looking at the geometry of a graph!

Our results can even be generalized. The graph of $y = x^2 - 1$ can give us information about the solution of other equations besides $0 = x^2 - 1$. For example, *consider the equation* $3 = x^2 - 1$. Since 3 has replaced y in $y = x^2 - 1$, we are looking for points on the graph of $y = x^2 - 1$ that have $y = 3$. Since all points have $y = 3$ lie on a horizontal line through 3 on the y axis, we are looking for the points where $y = x^2 - 1$ cuts that line.

From Figure 3.33 you can see that the line $y = 3$ cuts the graph at two points, where $x = 2$ and -2. Therefore the solutions to $3 = x^2 - 1$ are $x = 2$ or $x = -2$. (You can check this algebraically: $3 = x^2 - 1$, so $4 = x^2$, giving $x = 2$ or $x = -2$.)

What about the solutions to $-3 = x^2 - 1$? This time we want the values of x making $y = -3$, which means finding the points at which the graph cuts the line $y = -3$. But the graph just doesn't go down that far, and so it doesn't cut the line at all. (See Figure 3.33 again.) This means there are no x's that make $x^2 - 1 = -3$; in other words, $-3 = x^2 - 1$ has no solution. (Solving al-

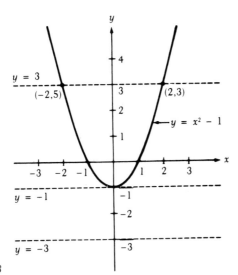

FIG. 3.33

gebraically, in case you don't believe all this, we see that if $-3 = x^2 - 1$, then $-2 = x^2$, which indeed does not have any real solutions.)

Now, *what about* $-1 = x^2 - 1$? A horizontal line through -1 on the y axis cuts the graph just once, at its lowest point which is where $x = 0$. Therefore the only x value that makes $x^2 - 1 = -1$ is $x = 0$, and so the equation $-1 = x^2 - 1$ has only one solution, namely, 0.

Therefore, in general:

> *If k is a constant, the solutions to the equation $k = x^2 - 1$ are the x coordinates of the points at which the horizontal line $y = k$ cuts the graph $y = x^2 - 1$.*
>
> If the graph cuts the line twice, the equation has two roots.
> If the graph cuts the line once, the equation has one root.
> If the graph does not cut the line, the equation has no roots.

PROBLEM SET   3.8

1. For what values of x is $x^2 - 9$ positive? Negative? Draw a graph of $y = x^2 - 9$ and show how you could get the answer from that.

Determine the values of x that make each of the following expressions (a) positive, (b) negative, and (c) zero. Use a graph to illustrate your answers.

2. $x^2 - 3x - 4$

3. $x^2 - 3x - 1$

4. $2x - x^3$

5. $\dfrac{1}{x-1} + 2$

6. $4 - |x|$

Find the x and y intercepts of the graphs of the following equations:

7. $y = (x - 1)(x - 2)(x - 3)$

8. $y = \dfrac{x + 5}{x + 6}$

9. $y - x^2 + 6x - 8 = 0$

10. $x^2 + 2x + y^2 - 2y + 1 = 0$

11. By writing $y = x^2 - 6x + 8$ as $y + 1 = (x - 3)^2$ draw a graph of $y = x^2 - 6x + 8$. Use the graph to show why the equation

(a) $0 = x^2 - 6x + 8$ has two real roots

(b) $-1 = x^2 - 6x + 8$ has one real root

(c) $-2 = x^2 - 6x + 8$ has no real roots

12. Using a graph, show why the equation $k = x^3$ (k a constant) has exactly one solution for every value of k.

13. For what value(s) of k (if any) does the equation

$$k = \frac{1}{x}$$

have **(a)** one solution   (b) two solutions   (c) no solutions?

14. Use the graph of $y = |x - 1| - 3$ to solve the inequality

$$|x - 1| - 3 < 0$$

CHAPTER 3 REVIEW

Graph and label on the same Cartesian plane:

1. (a) $y = x^2$ (b) $y = x^2 - 2$ (c) $y = x^2 + 1\frac{1}{2}$

Which of the following have even symmetry? Odd?

2. $y = 4x^2 - 3x^6$

3. $y = 7x + 2x^3$

4. $y = 7x^5 + 5x^4 + 3x^3 + x^2 - x$

Which of the following graphs display even symmetry? Odd?

5.

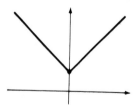

6.

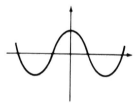

7. Graph and label on the same Cartesian plane:

(a) $y = \dfrac{1}{x}$ (b) $y = \dfrac{1}{x + 1}$ (c) $y = \dfrac{1}{x - 3}$

8. Graph and label on the same Cartesian plane:

(a) $y = x^4 - 6$ (b) $7 - y = 3 - x^4$ (c) $2 + y = x^4$

9. Graph and label on the same Cartesian plane:
    (a) $y + 7 = x^3$ (b) $y - 1 = x^3 + 2$ (c) $3 + y = 3 - x^3$

10. For what values of x is $x^3 + 2$ positive?

11. Graph and label the following on the same Cartesian plane:
    (a) $\dfrac{1}{2x} = y$ (b) $-y = \dfrac{5}{x}$ (c) $y = \dfrac{9}{2x}$

12. Graph and label on the same Cartesian plane:
    (a) $3x = -y^2$ (b) $-2x = \tfrac{1}{4}y^2$ (c) $y^2 - 2 = -x$

13. Add the graphs of

    $$y = x + 2 \quad \text{and} \quad 3 - y = 2x + 7$$

    to get a graph of their sum.

14. Use the graph of $y = 2 + x$ to get the graph of $y = \dfrac{1}{2 + x}$.

15. Given the equation $3x - 6 + 9y = 0$, graph:
    (a) y against x (i.e., with the y axis vertical and the x axis horizontal)
    (b) x against y (i.e., with the y axis horizontal and the x axis vertical)

16. Graph on the same Cartesian plane:
    (a) $y = \sqrt{3 - x}$ (b) $-x = y^2 - 3$

17. Graph and label on the same Cartesian plane:
    (a) $y = \dfrac{1}{x^2}$ (b) $y = \dfrac{1}{x^2} - 4$

Which of the following have even symmetry? Odd?

18. $y = |3x|$ 19. $y = |x| - 6$

Which of the following graphs display even symmetry? Odd?

20. 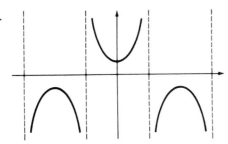   21.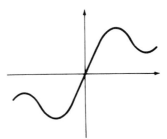

22. Label and graph on the same Cartesian plane:
    (a) $y = |x^3 - 1|$ (b) $y = |2 + x^3|$ (c) $y = |5 - x^3|$

23. Graph and label on the same Cartesian plane:
    (a) $(y - 1) = 3 + (x + 4)^2$ (c) $y + 3 = (x - 1)^2 + 1$
    (b) $2 - y = 5 - (x + 4)^2$

24. For what values of x is $x^3 - 7x^2 + 14x - 8$ positive?

Graph and label the following equations given that $a > 1$, $b < -1$, and $-1 < c < 0$:

25. $y = -ax^2$ 27. $abx^2 = y$ 29. $(a - x) = y^2$

26. $b + y = cx^2$ 28. $bx = -cy^2$ 30. $x = -(y - a)^2$

31. Use the graph of $0 = 2 - x - y$ to get the graph of $y = \dfrac{1}{2 - x}$.

32. Given the equation $x - 2 = \dfrac{1}{3 + y}$, graph:

    (a) y against x (i.e., with the y axis vertical and the x axis horizontal)

    (b) x against y (i.e., with the y axis horizontal and the x axis vertical)

33. Graph on a pair of Cartesian planes:

    (a) $y - 1 = \dfrac{1}{x^2 - 2}$

    (b) $y^2 - 1 = \dfrac{1}{x - 2}$

34. Given the equation $\dfrac{1}{r^2} = 3Q$, graph Q versus r (i.e., Q axis vertical)

35. If $\dfrac{1}{r^2} = 3Q$, graph r versus Q.

36. By drawing a graph of $y = 2t + 3$, show for what t values $2t + 3 > 7$.

37. By graphing $y = |x - 3|$, solve $|x - 3| < 6$.

38. Draw a graph of $y = x^2 - 4x + 7$ and use it to show why $x^2 - 4x + 7 = 0$ has no real solution, whereas $x^2 - 4x + 7 = 9$ has two real solutions. For what value of k does $x^2 - 4x + 7 = k$ have exactly one real solution?

Determine the values of x that make each of the following expressions (a) positive, (b) negative, and (c) zero. (Hint: Use #8 of problem set 3.6.)

39. $|x| - \dfrac{3}{|x|}$ 40. $x^3 + \dfrac{1}{x - 1}$

4 SYSTEMATIC GRAPHING

4.1 GRAPHING POLYNOMIALS

We already know how to graph lines by using just their x and y intercepts, and in the last chapter we saw how a complicated graph can sometimes be obtained very quickly from simpler ones. The point of this section is to develop a systematic method of graphing equations *without* using a table of values. Of course, if you get stuck, or you are not sure how something goes, you should use a table of values to check your graph, but otherwise you should manage without.

It turns out that it is possible to get a rough idea of the graph of any polynomial—or at least, any one that factors—without using a table of values at all. The idea is to make a few well-chosen calculations to give you a few crucial points, and to use your common sense to see what happens between them. This can give you a rough graph *far* more quickly than a table of values, and a rough graph is usually all you need.

To graph an equation of the form $y = $ some polynomial in x, here are the things to look for:

1. *Symmetry: Odd*—if all powers of x are odd (and there is no constant term.
   Even—if all powers of x are even (and there may be a constant term)

   Use symmetry to check your graph (or if you're very confident, to get the left half from the right half). See Section 3.2.

2. *Factors* of the polynomial

3. *y intercept:* the point at which the graph cuts the y axis, found by substituting $x = 0$. This should be easy to find.

4. *x intercept(s):* the point(s) at which the graph cuts the x axis, found by substituting $y = 0$ and solving for x. This can be a nuisance because you have to solve a polynomial equation for x.

5. *Positiveness-Negativeness:* When y is positive, the graph is above the x axis; when y is negative, the graph is below. As y changes from positive to negative, or vice versa, the graph either has to make a sudden jump or has to cross the x axis. The graph of a polynomial doesn't make sudden jumps, and so it must cross the x axis. Therefore, y can only change sign at an x intercept. Between x intercepts y is always positive or always negative. You can find out which by looking at the signs of each factor. *Note:* The graph can't cut the axis between two intercepts or we would have another root—which we don't.

6. *Tendencies* as $x \to \pm\infty$: Look at what happens to y as x gets very large positive (written as $x \to \infty$) or as x gets very large negative ($x \to -\infty$).

EXAMPLE: *Graph* $y = x^3 - x$.

1. This is symmetric about the origin, since it has only odd powers of x.

2. Factor: $y = x(x^2 - 1) = x(x - 1)(x + 1)$

3. y-intercept: When $x = 0$, $y = 0$, so the graph cuts the y axis at the origin and nowhere else.

4. x-intercept: When $y = 0$,

$$0 = x(x - 1)(x + 1)$$

So $x = 0, x = 1, \text{ or } x = -1$

The graph cuts the x axis at $(0, 0), (1, 0), (-1, 0)$ and nowhere else.

5. Positiveness-negativeness: The x intercepts -1, 0, and 1 divide the axis into four sections, shown in Figure 4.1, and the

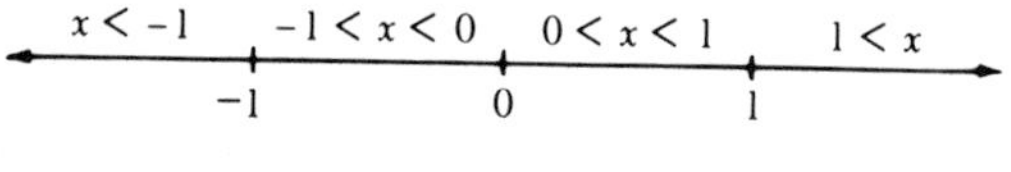

FIG. 4.1

graph cannot cut the axis between these points (or else the equation would have another root, which it doesn't). We can find out whether y is positive or negative as follows:

If $x < -1$, suppose for example that $x = -2$. Then

$$\left. \begin{array}{l} x = -2 \\ x - 1 = -3 \\ x + 1 = -1 \end{array} \right\} \text{ so } y = (-2)(-3)(-1) = -6, \text{ which is negative.}$$

Now the actual values of these factors and of y are irrelevant—all that matters is the signs. So we could have written:

If $x < -1$ (imagine $x = -2$ or any other value), then:

x is negative

$x - 1$ is negative $\Big\}$ and $(-)(-)(-) = (-)$, so y is negative.

$x + 1$ is negative

Or, abbreviating right away for the next region:

If $-1 < x < 0$ (imagine $x = -\frac{1}{2}$, or any other value between -1 and 0):

x is $(-)$

$(x - 1)$ is $(-)$ $\Big\}$ and $(-)(-)(+) = (+)$, so y is $(+)$

$(x + 1)$ is $(+)$

If $0 < x < 1$ (imagine $x = \frac{1}{2}$, or any other value between 0 and 1):

x is $(+)$

$(x - 1)$ is $(-)$ $\Big\}$ and $(+)(-)(+) = (-)$, so y is $(-)$

$(x + 1)$ is $(+)$

if $1 < x$ (imagine $x = 2$, or any other value greater than 1):

x is $(+)$

$(x - 1)$ is $(+)$ $\Big\}$ and $(+)(+)(+) = (+)$, so y is $(+)$

$(x + 1)$ is $(+)$

So altogether y has the signs shown in Figure 4.2.

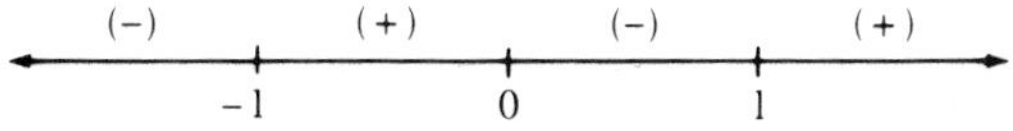

FIG. 4.2

6. Tendencies as $x \to \pm\infty$: As $x \to \infty$, meaning as x gets very large positively, x becomes insignificant in comparison with x^3. For example, if $x = 10$, $x^3 = 1000$ or, more clearly, if $x = 100$, $x^3 = 1,000,000$. So for very large x, y is pretty nearly equal to x^3. As x grows very large and positive, so does x^3, and therefore so does y. So

as $x \to \infty$, $y \to \infty$ also

As $x \to -\infty$, x can again be ignored in comparison with x^3, since if $x = -100$, $x^3 = -1,000,000$, and so on. As x gets very large negatively, so does x^3, and therefore so does y. So

as $x \to -\infty$, $y \to -\infty$

This gives us Figure 4.3, and putting all this together we get the graph in Figure 4.4.

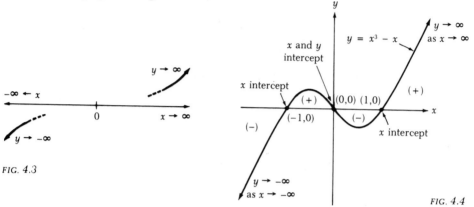

FIG. 4.3

FIG. 4.4

Check: This is symmetric about the origin, as we expected.

EXAMPLE: *Graph $y = x^4 - 5x^2 + 4$.*

1. This is symmetric about the y axis since the equation contains only even powers of x and constants.

2. Factor: Fortunately you can think of this as $y = (x^2)^2 - 5(x^2) + 4$, which gives:

$$y = (x^2 - 1)(x^2 - 4)$$

Since we now have two differences of squares, we can factor further:

$$y = (x - 1)(x + 1)(x - 2)(x + 2)$$

3. y intercept: When $x = 0$, $y = 4$.

4. x intercept: When $y = 0$,

$$0 = (x - 1)(x + 1)(x - 2)(x + 2)$$

$$x = 1, -1, 2, \text{ or } -2$$

5. Positiveness-negativeness: To make this business more concise, notice that a factor of the form $(x - k)$ is positive to the right of the root it causes, and negative to the left. Therefore we have the situation in Figure 4.5.

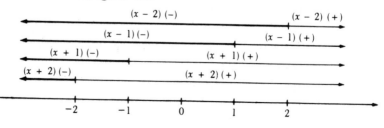

FIG. 4.5

This means that when

$x < -2$:   all four factors are ($-$), so y is ($+$)

$-2 < x < -1$:   three factors are ($-$), one ($+$), so y is ($-$)

$-1 < x < 1$:   two factors are ($-$), two ($+$), so y is ($+$)

$1 < x < 2$:   one factor ($-$), three ($+$), so y is ($-$)

$2 < x$:   all four factors ($+$), so y is ($+$)

So we get the situation in Figure 4.6.

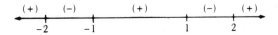

FIG. 4.6

6. Tendencies: As $x \to \infty$, the x^4 term drowns the other two and so y is approximately x^4. Now x^4 grows large and positive as x does, so

as $x \to \infty$, $y \to \infty$

Similarly, as $x \to -\infty$, y behaves like x^4, which becomes large and positive as x becomes large and negative. So

as $x \to -\infty$, $y \to \infty$

Putting all this together, we get the graph in Figure 4.7.

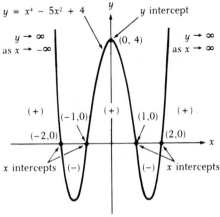

FIG. 4.7

Check: This is symmetric about the y axis.

EXAMPLE: *Graph $y = x^2 - 2x^3$.*

1. No symmetry.

2. Factor by taking out x^2:

$$y = x^2 - 2x^3$$

$$= x^2(1 - 2x)$$

It is very much easier to tell if a factor is positive or negative if it is written in the form $(x - k)$ or $(lx - k)$, so we will factor a (-1) out of the $(1 - 2x)$ giving:

$$y = -x^2(2x - 1)$$

3. y intercept: When $x = 0$, $y = 0$.

4. x-intercept: When $y = 0$,

$$0 = -x^2(2x - 1) \quad \text{so}$$

$$x = 0 \text{ or } x = \tfrac{1}{2}$$

Notice that there are only two roots because there was a repeated factor of x.

5. Positiveness-negativeness: If $l > 0$, factors of the form $(lx - k)$ are positive to the right of the root they cause, and negative to the left, so we have the situation in Figure 4.8.

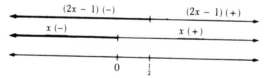

FIG. 4.8

Remember that $y = -x^2(2x - 1)$, so we get the sign of y from the sign of the factors x, x, $(2x - 1)$, *and* an extra $(-)$ sign. So

if $x < 0$: all three factors $(-)$, so, with the extra $(-)$, y is $(+)$

if $0 < x < \tfrac{1}{2}$: two factors $(+)$, one $(-)$, so, with the extra $(-)$, y is $(+)$

if $\tfrac{1}{2} < x$: all three factors $(+)$, so, with the extra $(-)$, y is $(-)$

Notice that y does not change sign as you go past zero—the root from the repeated factor of x.

6. Tendencies:

As $x \to \pm\infty$, the $-2x^3$ becomes the dominant term.

As x gets large positively, $-2x^3$ gets large negatively, so

as $x \to \infty$, $y \to -\infty$

As x gets large negatively, $-2x^3$ gets large positively, so

as $x \to -\infty$, $y \to +\infty$

Putting all this together we get the graph in Figure 4.9.

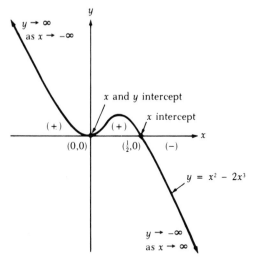

FIG. *4.9*

Notice that because y does not change sign at $x = 0$, the graph does not cross the axis there but merely touches it. This is characteristic of a root coming from a squared factor.

PROBLEM SET   4.1

Graph the following:

1. $y = x^2 + x$
2. $y = x^2 - 4$
3. $y = x^2 - x$
4. $y = 3x^3 - 8$
5. $y = (x + 1)(x + 2)(x - 3)$
6. $y = -x^4 + 9x^2$
7. $y = x^2 - 6x + 5$
8. $y = 2x^2 - 7x + 3$
9. $y = -3x^2 + 5x + 2$
10. $y = 4x^2 - 10x + 6$
11. $y = (x - 3)(x + \frac{1}{2})(x - 4)$
12. $y = 2x^3 + 16$
13. $y = 2x^2 - 8x - 42$
14. $y = -x^3 - 4x^2 + 4x + 16$
15. $y = -x^4 + 41x^2 - 400$
16. $y = x^4 - 2x^3 + x - 2$

Match the following graphs to the corresponding equation:
17. $y + 3 = (x - 4)^2$
18. $y - 2 = -2(x - 3)^2$
19. $y - 3 = -x^2$
20. $y - 2 = \frac{1}{2}(x + 1)^2$
21. $y + 3 = -3x^2$
22. $y - 2 = 2(x - 3)^2$

(a)
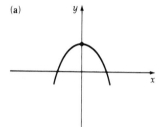

(b)
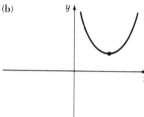

(c)
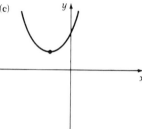

(d)
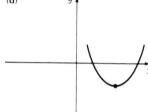

(e)
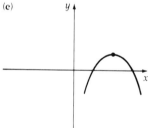

(f)
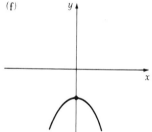

For each of the following graphs, write an equation, for y in terms of x, whose graph would resemble that shown. Indicate your reasoning clearly.

23.
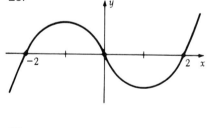

24.
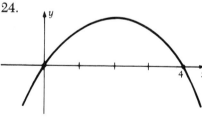

25.
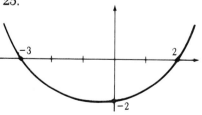

26.
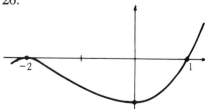

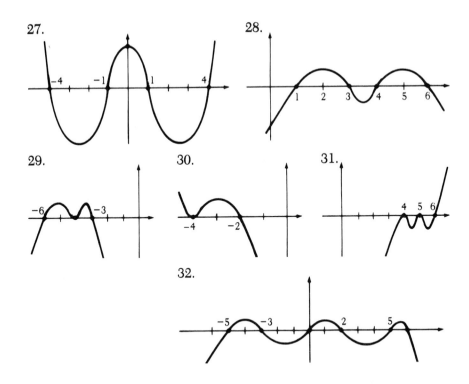

4.2   GRAPHING ALGEBRAIC FRACTIONS

The purpose of this section is to provide a systematic method for graphing equations in which y is an algebraic fraction. You should be able to do this—at least roughly—without a table of values, and by the end of the section I hope that you will be able to tell how the various parts of a fraction produce various aspects of the graph.

To graph an equation in which $y =$ an algebraic fraction, you should look for:

1. *Symmetry: Odd*—if quotient of one odd and one even polynomial
             Even—if quotient of two odd polynomials or two even polynomials
   Use symmetry to check graph, or to get left half from right half. See Section 3.2.

2. *Factors* of the top and the bottom of the fraction

3. *y intercept:* by substituting $x = 0$

4. *x intercept(s):* by substituting $y = 0$ and solving for x. This amounts to setting the numerator equal to zero, since $y =$ a fraction and a fraction is zero only when its numerator is zero.

5. *Vertical asymptotes:* where the denominator equals zero. These are vertical lines that the graph approaches, but never touches. They divide the graph up into several pieces and are the only points at which a sudden jump is made.

Think, for example, of the graph of $y = \dfrac{1}{x}$, which has the y axis as a vertical asymptote. Notice that the vertical asymptotes occur at x values that make the fraction "blow up." At such points the denominator is zero and the fraction is not defined. Therefore there is *no* point on the graph at such an x value, and so the graph cannot cross the asymptote. At nearby values the denominator is very small, and so the whole fraction is very large. Therefore as you get near an asymptote, the graph shoots up or down to $+\infty$ or $-\infty$, and you have to find out which way by looking at the signs of the factors.

6. *Horizontal or diagonal asymptotes:* These are lines or curves which limit the graph as x gets very large. To find out what happens there, notice that all but the highest-powered terms of x in the top and bottom of the fraction become insignificant when x gets very large (positively or negatively). Therefore, use just these terms of highest power to see what happens for large x.

    If the denominator has a higher degree than the numerator, then y will go toward zero. For example, if $y = \dfrac{1}{x}$, $y \to 0$ as $x \to \pm\infty$.

    If the degrees of the numerator and the denominator are equal, then y goes toward the fraction that is the ratio of the highest-powered terms. For example, if $y = \dfrac{2x - 3}{x + 1}$, then $y \to \dfrac{2x}{x} = 2$ as $x \to \pm\infty$.

    If the denominator has a lower degree than the numerator, the situation is more complicated. As x gets very large our curve grows parallel to the curve given by the ratio of the highest-powered terms. However it need not get arbitrarily close to the latter. For example, if $y = \dfrac{3x^3 - 2}{2x + 3}$, then y grows like $y = \dfrac{3x^3}{2x}$ or $y = \dfrac{3}{2}x^2$ as $x \to \pm\infty$. To find out which side of the asymptote the graph is on, you have to check some particular points.

    You should note that, unlike vertical asymptotes, horizontal or diagonal asymptotes *can* be crossed *for small x*. This is because the asymptote represents the trend for large x and is not necessarily binding for small values.

7. *Positiveness-negativeness* (not usually necessary): By the time you have found out what the graph is doing near each of the asymptotes, you probably know everything you need to about where it is positive and where it is negative.

    If not, realize that y can only change from positive to negative at an x intercept, where the graph crosses the axis, or at a vertical asymptote, where it makes a sudden jump. Therefore between x intercepts and vertical asymptotes, the graph must be all above or all below the x axis. You can find out which by looking at the signs of the factors.

EXAMPLE: *Graph* $y = \dfrac{x-1}{x-2}$.

1. Not symmetric.

2. Can't be factored.

3. y intercept: Substitute $x = 0$. $\qquad y = \dfrac{0-1}{0-2} = \dfrac{-1}{-2} = \dfrac{1}{2}$

4. x intercept: $\qquad 0 = \dfrac{x-1}{x-2}$

$\qquad\qquad\qquad 0 = x - 1 \qquad$ [if fraction $= 0$,
$\qquad\qquad\qquad\qquad\qquad\qquad$ its numerator $= 0$]

$\qquad\qquad$ So $x = 1$

5. Vertical asymptotes: Set denominator $= 0$. $\qquad 0 = x - 2$

So line $x = 2$ is the vertical asymptote.

Therefore, so far we have the information in Figure 4.10.

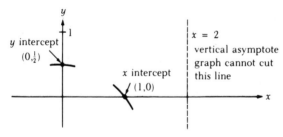

FIG. 4.10

What happens near a vertical asymptote?

When $x > 2$,

$(x - 1)$ and $(x - 2)$ are both $(+)$

and

$y = \dfrac{(+)}{(+)} \quad$ so $\quad y$ is $(+)$

Suppose x is greater than 2, but decreasing toward 2 (which can be written $x \to 2^{+}$).

Let $x = 2.1$; then $y = \dfrac{2.1 - 1}{2.1 - 2} = \dfrac{1.1}{0.1} = 11$

Let $x = 2.01$; then $y = \dfrac{2.01 - 1}{2.01 - 2} = \dfrac{1.01}{0.01} = 101$

Let $x = 2.001$; then $y = \dfrac{2.001 - 1}{2.001 - 2} = \dfrac{1.001}{0.001} = 1001$

Therefore as x decreases toward 2, y increases wildly.

Symbolically:

As $x \to 2^+$, $y \to \infty$

When $x < 2$, if x is also greater than 1,

$(x - 1)$ is $(+)$ and $(x - 2)$ is $(-)$

$y = \dfrac{(+)}{(-)}$ so y is $(-)$

Suppose x is less than 2, but increasing toward 2 (written $x \to 2^-$).

Let $x = 1.9$; then $y = \dfrac{1.9 - 1}{1.9 - 2} = \dfrac{0.9}{-0.1} = -9$

Let $x = 1.99$; then $y = \dfrac{1.99 - 1}{1.99 - 2} = \dfrac{0.99}{-0.01} = -99$

Let $x = 1.999$; then $y = \dfrac{1.999 - 1}{1.999 - 2} = \dfrac{0.999}{-0.001} = -999$

Therefore as x increases toward 2, y becomes very large negatively.

Symbolically:

As $x \to 2^-$, $y \to -\infty$

This means that the behavior near the vertical asymptote looks like Figure 4.11.

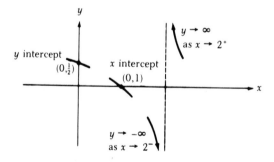

FIG. 4.11

Now we have to find what happens at the ends of this graph, which means finding the horizontal or diagonal asymptotes.

6. Horizontal or diagonal asymptotes: Suppose x gets very large positively (i.e., $x \to \infty$).

Let $x = 10$; then $y = \dfrac{10 - 1}{10 - 2} = \dfrac{9}{8}$ ... close to 1

Let $x = 100$; then $y = \dfrac{100 - 1}{100 - 2} = \dfrac{99}{98}$ ...   closer to 1

Let $x = 1000$; then $y = \dfrac{1000 - 1}{1000 - 2} = \dfrac{999}{998}$ ...   *very* close to 1

Therefore the larger x gets, the more insignificant the "-1" and the "-2" become, and y gets closer and closer to 1, though always remaining larger than 1. In other words,

As $x \to \infty$, $y \to 1^+$

Suppose x gets very large negatively (i.e., $x \to -\infty$).

Let $x = -10$; then $y = \dfrac{-10 - 1}{-10 - 2} = \dfrac{11}{12}$ ...   close to 1

Let $x = -100$; then $y = \dfrac{-100 - 1}{-100 - 2} = \dfrac{101}{102}$ ...   closer to 1

Let $x = -1000$; then $y = \dfrac{-1000 - 1}{-1000 - 2} = \dfrac{1001}{1002}$ ...   *very* close to 1

Again the -1 and the -2 become insignificant and y gets closer to 1, though this time y remains always smaller than 1. In other words,

As $x \to -\infty$, $y \to 1^-$

Therefore, near the horizontal asymptote we have the situation shown in Figure 4.12.

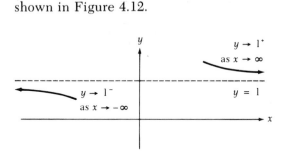

FIG. 4.12

In general, whenever x is very large, positively or negatively, all but the terms involving the highest powers of x become insignificant. Therefore, in this case,

As $x \to \pm\infty$, $y \to \dfrac{x}{x} = 1$

The line $y = 1$ is the horizontal asymptote. Had we found the asymptote this way, we would have had to substitute values for x in order to see if the graph approached the asymptote from above or below.

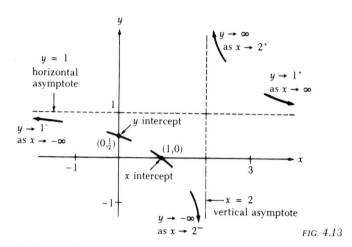

FIG. 4.13

So far we have the information shown in Figure 4.13. Joining all this up, we see that the graph must look like Figure 4.14.

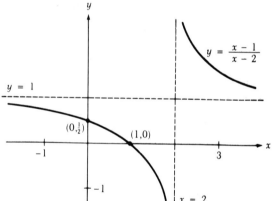

FIG. 4.14

CHECK: See where y is positive and where it is negative. Looking at Figure 4.15, we see that:

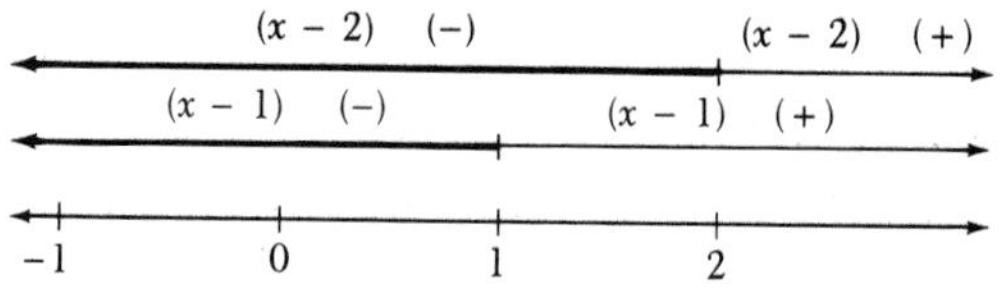

$$
\begin{array}{cccc}
(x-2) & (-) & (x-2) & (+) \\
(x-1) & (-) & (x-1) & (+) \\
\end{array}
$$

```
 +----+----+----+----+----+----
 -1   0    1    2
```

FIG. 4.15

If $x < 1$: $y = \dfrac{(-)}{(-)}$ so y is $(+)$

If $1 < x < 2$: $y = \dfrac{(+)}{(-)}$ so y is $(-)$

If $2 < x$: $y = \dfrac{(+)}{(+)}$ so y is $(+)$

So the sign of y is as shown in Figure 4.16, which agrees with the graph.

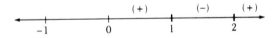

FIG. 4.16

EXAMPLE: $Graph\ y = \dfrac{1}{x^2 - 1}.$

1. Symmetric about the y axis since it is a quotient of two even polynomials.

2. Factor: $y = \dfrac{1}{x^2 - 1}$

$$y = \dfrac{1}{(x - 1)\,(x + 1)}$$

3. y intercept: $y = \dfrac{1}{0^2 - 1} = -1$

4. x intercept: $0 = \dfrac{1}{x^2 - 1}$

There is no solution to this equation because the numerator is never zero; therefore there is no x intercept.

5. Vertical asymptotes: where denominator $= 0$; that is,

$0 = (x - 1)\,(x + 1)$

So, the vertical asymptotes are $x = 1$ and $x = -1$.

To find behavior near the asymptotes, try out some values and see that:

If x slightly > 1, $y = \dfrac{1}{(+)\,(+)} = (+)$

If x slightly < 1, $y = \dfrac{1}{(-)\,(+)} = (-)$

If x slightly > -1, $y = \dfrac{1}{(-)\,(+)} = (-)$

If x slightly < -1, $y = \dfrac{1}{(-)\,(-)} = (+)$

Therefore near the asymptotes we have the situation shown in Figure 4.17 on the next page. (If you aren't sure why it looks this way, try out a whole lot of x values, including 1.1, 1.01, 0.9, 0.99, $-1.1, -1.01, -0.9, -0.99$, etc.)

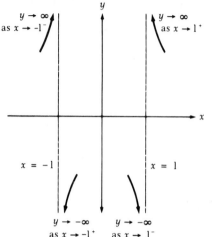

FIG. 4.17

6. Horizontal or diagonal asymptotes: As $x \to \pm\infty$, ignore all but the highest powers of x. So

$$y \to \frac{1}{x^2}$$

As x gets large positively or negatively, $\frac{1}{x^2}$ gets very close to zero. So,

As $x \to \pm\infty$, $y \to 0$.

This tells us that the x axis is an asymptote.

Now for large $|x|$, $x^2 - 1$ is always positive, so y is positive too. Therefore the graph approaches the asymptote from above. Hence we have the situation shown in Figure 4.18.

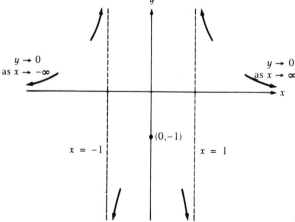

FIG. 4.18

Joining all this up, we get the graph in Figure 4.19.

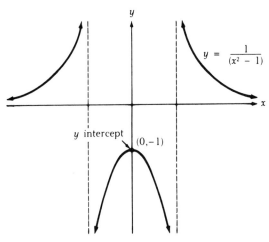

FIG. 4.19

Check: This graph does have even symmetry as expected.

EXAMPLE: *Graph* $y = \dfrac{1}{x^2 + 1}$.

1. Symmetric about the y axis because it is the quotient of even polynomials.

2. Doesn't factor.

3. y intercept: $y = \dfrac{1}{0^2 + 1} = 1$

4. x intercept: $0 = \dfrac{1}{x^2 + 1}$

No solution, so no x intercept.

5. Vertical asymptotes: where denominator $= 0$.

$0 = x^2 + 1$

No solution, so no vertical asymptotes.

6. Horizontal asymptotes:

As $x \rightarrow \pm\infty$, ignoring all but the highest powers of x gives

$y \rightarrow \dfrac{1}{x^2} \rightarrow 0$

so $y = 0$ is the horizontal asymptote.

Since y is always positive (because $x^2 + 1$ is always positive), the graph approaches the asymptote from above. Therefore we have the situation shown in Figure 4.20 on the next page.

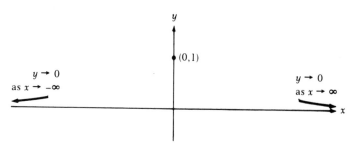

FIG. 4.20

x^2 grows steadily as x moves away from zero, so $y = \dfrac{1}{x^2 + 1}$ decreases steadily as x moves away from zero. Therefore the graph should slope downward on either side of $(0, 1)$. Graphs of fractions, however, like graphs of polynomials, are smooth, so there won't be a spike at $(0, 1)$. Hence there is really only one reasonable way to join this up—namely to make it look like a hump-backed bridge as in Figure 4.21.

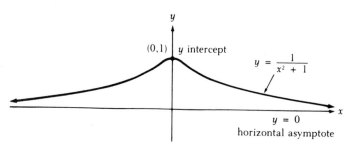

FIG. 4.21

Note: It is interesting to compare the graphs of $y = \dfrac{1}{x^2 - 1}$ and $y = \dfrac{1}{x^2 + 1}$. The equations look similar, but the graphs are extremely different. The reason is that the different sign means that one denominator factors and has roots, while the other does not. Therefore one fraction "blows up" and the other does not.

EXAMPLE: *Graph* $y = \dfrac{4x^2 - 1}{x + 1}$.

1. Not symmetric.

2. Factor: $y = \dfrac{4x^2 - 1}{x + 1}$

$$y = \frac{(2x - 1)(2x + 1)}{(x + 1)}$$

3. y intercept: $y = \dfrac{4 \cdot 0^2 - 1}{0 + 1} = -1$

4. x intercept: $0 = \dfrac{(2x - 1)\,(2x + 1)}{(x + 1)}$

$$0 = (2x - 1)\,(2x + 1)$$

So $x = \frac{1}{2}$ or $-\frac{1}{2}$.

5. Vertical asymptote: when denominator $= 0$.

$0 = x + 1$

So $x = -1$ is the vertical asymptote.

Behavior near $x = -1$:

When x slightly > -1: $y = \dfrac{(-)\,(-)}{(+)} = (+)$

When x slightly < -1: $y = \dfrac{(-)\,(-)}{(-)} = (-)$

(If you're unsure, try specific values for x).

So we have the situation in Figure 4.22.

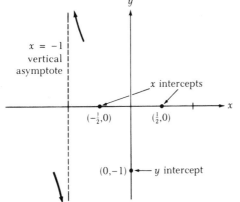

FIG. 4.22

6. Horizontal/diagonal asymptote:

When $x \to \pm\infty$, y grows like $y = \dfrac{4x^2}{x} = 4x$

So $y = 4x$ is the diagonal asymptote.

Behavior near $y = 4x$:

When $x \to \pm\infty$, $y = \dfrac{4x^2 - 1}{x + 1}$ is less than $\dfrac{4x^2}{x}$

[because $(4x^2 - 1)$ is smaller than $4x^2$ and $(x + 1)$ is bigger than x, so $\dfrac{4x^2 - 1}{x + 1}$ has a smaller top and a bigger bottom than $\dfrac{4x^2}{x}$, so $\dfrac{4x^2 - 1}{x + 1}$ is a smaller fraction than $\dfrac{4x^2}{x}$].

Therefore as $x \to \infty$, our curve approaches $y = 4x$ from underneath.

So we have the information in Figure 4.23, and joining up; we get the graph in Figure 4.24.

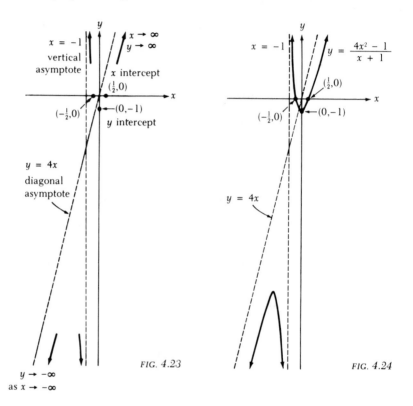

FIG. 4.23   FIG. 4.24

EXAMPLE: *Graph* $y = \dfrac{2}{x^2 - 2x + 1}$.

1. Not symmetric.

2. Factor: $y = \dfrac{2}{x^2 - 2x + 1}$

$$y = \frac{2}{(x - 1)^2}$$

3. y intercept: $y = \dfrac{2}{(0 - 1)^2} = 2$

4. x intercept: $0 = \dfrac{2}{(x - 1)^2}$

No solution, so no x intercept.

5. Vertical asymptote: When $(x - 1)^2 = 0$.

Therefore, $x = 1$ is vertical asymptote.

y is always positive, so behavior near $x = 1$ is as shown in Figure 4.25.

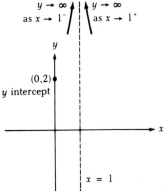

FIG. 4.25

6. Horizontal asymptote:

As $x \to \pm\infty$, $\quad y \to \dfrac{2}{x^2} \to 0$

Therefore, $y = 0$ is the horizontal asymptote.

Since y is always positive, the graph approaches $y = 0$ from above; as shown in Figure 4.26. The final graph is Figure 4.27.

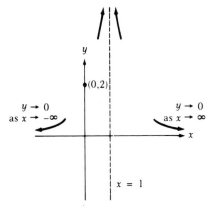

FIG. 4.26

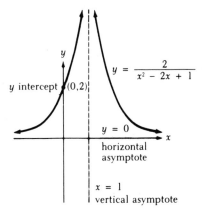

FIG. 4.27

PROBLEM SET   4.2

For problems 1 and 2 you may use a calculator if you wish.

1. Let $y = \dfrac{1}{(x+1)(x-3)}$

   (a) For what values of x is y undefined?

   (b) Find y when $x = -2, 0, 1, 2, 4$.

   (c) Find approximate decimal values for y when $x = 100$, when $x = 1000$, when $x = -100$, and when $x = -1000$.

   (d) Find approximate values for y when $x = 3.1$, when $x = 3.01$, when $x = 3.001$, and again when $x = 2.9$, $x = 2.99$, and $x = 2.999$.

   (e) Do the same thing around $x = -1$: Find approximate values for y when $x = -0.99$ and $x = -0.999$, and when $x = -1.01$ and $x = -1.001$.

   (f) Draw a graph of $y = \dfrac{1}{(x+1)(x-3)}$ using all of the information deduced above. Draw the graph clearly, and label all intercepts and asymptotes.

2. Let $y = \dfrac{2x-1}{x-2}$

   (a) For what values of x is y undefined?

   (b) For what values of x is y zero?

   (c) What is the approximate value of y when $x = 1000$? When $x = -1000$?

   (d) What is the approximate value of y when $x = 2.001$? When $x = 1.999$?

   (e) Draw a graph of $y = \dfrac{2x-1}{x-2}$.

For each of the following, draw a large, clear graph, labeling y intercepts, x intercepts, vertical asymptotes, and horizontal asymptotes.

3. $y = \dfrac{1}{x(x-1)}$

4. $y = \dfrac{2}{(x-2)(x+2)}$

5. $y = \dfrac{-1}{(x-1)(x-3)}$

6. $y = \dfrac{-1}{(x-4)}$

7. $y = \dfrac{1}{(x-4)^2}$

8. $y = \dfrac{1}{x^2+1}$

9. $y = \dfrac{x}{x-2}$

10. $y(x+3) = x+1$

11. $y = |x+1|$

12. $y = \dfrac{1}{x^2-2x-3}$

13. $y = \dfrac{(x+2)}{(x-4)}$

14. $y = \dfrac{(x-3)}{(x+1)(x-1)}$

15. $y = \dfrac{(x+1)}{x(x-3)(x+2)}$

16. $y = \dfrac{x^2-5x+6}{x^2-16}$

17. $y = \dfrac{(2x+3)(x-1)}{(x+\frac{1}{2})(x-\frac{7}{2})}$

18. $y = \dfrac{x^2-7x+6}{(x+1)(x+4)(x-2)}$

19. $y = \dfrac{2(x+2)}{(x-2)}$

20. $y = \dfrac{(x-1)\,(x+1)}{3(x^2-4)}$

21. $y = \dfrac{-2x^2+7x-5}{x^2-x-12}$

22. $y = \dfrac{(3-x)\,(2-x)\,(1-x)}{(x+3)\,(x+2)\,(x+1)}$

23. Write an equation for y in terms of x whose graph would resemble that shown in Figure 4.28. Indicate your reasoning clearly.

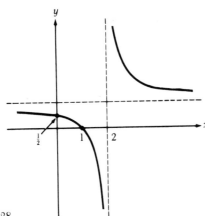

FIG. 4.28

Graph each of the following equations:

24. $2y + 4 = (x-3)^2$

25. $\dfrac{1}{(x+2)} = -3y + 1$

26. $\dfrac{1}{(x+2)} = \dfrac{-y}{4}$

27. $(y-2)^3 = -x + 3$

28. $|y + \tfrac{3}{2}| = 2x + 4$

29. $y = \dfrac{-1}{|x|+3} - 3$

30. $2y = \dfrac{4}{x+1} - 6$

31. $4(y-3) + \dfrac{2}{x} = 1$

32. $3 - 2(x+4) - \dfrac{2}{(y-2)} = 0$

33. $3|y| = \dfrac{2}{(x+4)}$

CHAPTER 4 REVIEW

Graph the following:

1. $y = x^2 - 3x$
2. $y = (x+1)\,(x-3)\,(x+4)$
3. $y = 4x^2 - 13x + 3$
4. $-y + 3 = -(x-1)\,(x-2)$

Give the equation for each of the following. Indicate your reasoning.

5.

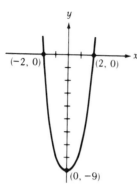

6.

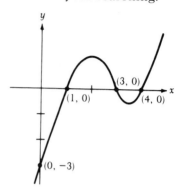

7. Let $y = \dfrac{3x - 1}{x - 1}$.

(a) For what values of x is y undefined?
(b) For what values of x is y zero?
(c) What is the approximate value of y when $x = 1000$?   When $x = -1000$?
(d) What is the approximate value of y when $x = 1.001$?   When $x = 0.999$?
(e) Draw a graph of $y = \dfrac{3x - 1}{1 - x}$.

Draw a large, clear graph, labeling y intercepts, x intercepts, vertical asymptotes, and horizontal asymptotes:

8. $y = \dfrac{1}{x^2 - 7x + 12}$

9. $2x - 1 = \dfrac{4 + x}{y}$

10. $y = \dfrac{x^2 - 9}{5 - 6x + x^2}$

11. $y = \left| \dfrac{1}{x} \right|$

12. Write an equation for y in terms of x whose graph would resemble that shown in Figure 4.29. Indicate your reasoning.

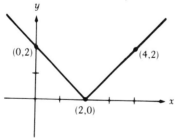

FIG. 4.29

Draw a large, clear graph, labeling x and y intercepts, vertical asymptotes, and horizontal asymptotes:

13. $\dfrac{3x - 1}{y} = \dfrac{1}{5 + x}$

14. $\dfrac{-y}{x - 1} = x^2 - x - 2$

15. $y = \dfrac{(1 + x)(2 - x)(2 + x)}{(3 - x)(4 + x)(x - 1)}$

Graph the following:

16. $y = x^3 + 6x^2 + 12x + 8$
17. $y = (x + 1)(x - 1)(x + 5)(x - 2)$
18. $y = 2x^3 - 5x^2 - 7x$
19. $3 = y - (2 - x)(x - 5)$

Give an equation for each of the following (there may be more than one). Indicate your reasoning.

20.

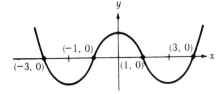

22.

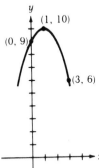

21.

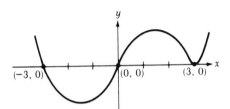

Draw a large, clear graph, labeling y intercepts, x intercepts, vertical asymptotes, and horizontal asymptotes:

23. $y = \dfrac{(x - 1)(x + 2)}{3 - x}$

24. $|x|y = x$

25. $y = \dfrac{(x^2 - 6x + 8)(x - 1)}{9 + 6x + x^2}$

26. $y = \dfrac{3}{|x + 1|}$

27. $y = \dfrac{x}{1 + |x|}$

28. $y = \dfrac{x^3 - x}{2 + x}$

29. $\dfrac{1}{|x^3 + 1|} = \dfrac{1}{y}$

30. $y = \dfrac{2 - 3x + x^2}{(x + 1)(4 + x)(x - 2)}$

5   THE CONIC SECTIONS

5.1   THE PARABOLA

There are four kinds of graphs that have been studied in great detail, and are of particular importance in astronomy. They are the parabola, the circle, the ellipse and the hyperbola, together called the *conic sections*.

We have seen parabolas already. The graph of $y = x^2$ is an example of a parabola, as are the graphs of $y = 2x^2$, $y - 1 = 2(x - 3)^2$, and $y - 1 = -2(x - 3)^2$ in Figure 5.1.

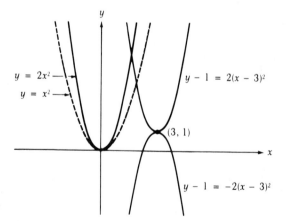

FIG. 5.1

Since the graph of $y - 1 = 2(x - 3)^2$ is the graph of $y = 2x^2$ moved over by 3 and up by 1, the lowest point or *vertex* of the parabola $y - 1 = 2(x - 3)^2$ is at the point (3, 1). The factor of 2 in this equation tells you how fast the parabola is climbing (as you can see by comparing $y = x^2$ and $y = 2x^2$). When a negative number is put in this position, as in $y - 1 = -2(x - 3)^2$, the parabola is flipped over. The vertex is then still (3, 1), but this is now the highest point.

The standard form of a parabola is

$$y - k = p(x - h)^2$$

where (h, k) is the vertex. If p is positive, the parabola opens upward; if p is negative, it opens downward. If $(x - h)$ and $(y - k)$ are interchanged to give

$$x - h = p(y - k)^2$$

the parabola opens sideways instead of up or down.

Putting a Parabola in Standard Form by Completing the Square

Any equation of the form

$$y = ax^2 + bx + c \qquad (a, b, c \text{ constants}, \quad a \neq 0)$$

has a graph that is a parabola. The reason for this is that such an equation can always be put in the standard form above by completing the square.

EXAMPLE: *Graph and find the vertex of the parabola* $y = x^2 - 6x + 7$.

To complete the square on the $x^2 - 6x$, we must add 9 (see Section 1.2).

When working on the equation:

$$y = x^2 - 6x + 7$$

we must add 9 to both sides to preserve the equality:

$$y + 9 = x^2 - 6x + 9 + 7$$

$$y + 9 = (x - 3)^2 + 7$$

Moving the 7 to the other side of the equation puts it in standard form:

$$y + 2 = (x - 3)^2$$

or

$$y - (-2) = (x - 3)^2$$

So the vertex is $(3, -2)$. Since $p = 1$, the graph of this parabola is the same shape as $y = x^2$ but moved over by 3 and down by 2. The result is shown in **Figure 5.2** on the next page.

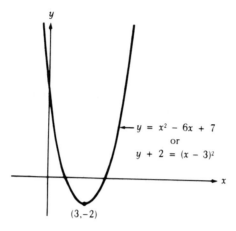

$y = x^2 - 6x + 7$

or

$y + 2 = (x - 3)^2$

$(3, -2)$

FIG. 5.2

EXAMPLE: *Graph* $2y + 4x^2 + 4x + 3 = 0$.

First put the x's on the other side of the equation and divide through by the coefficient of y, because in the standard form y always has a coefficient of 1:

$$2y + 3 = -4x^2 - 4x$$

so

$$y + \frac{3}{2} = -2x^2 - 2x$$

Now we have to complete the square on $-2x^2 - 2x$. We first factor out the -2:

$$-2x^2 - 2x = -2(x^2 + x)$$

and then add $\frac{1}{4}$ inside the parentheses to complete the square on $x^2 + x$ [because $\frac{1}{4} = $ (half the coefficient of $x)^2$]. This amounts to subtracting $\frac{1}{2}$ from $-2x^2 - 2x$, since:

$$-2x^2 - 2x - \frac{1}{2} = -2\left(x^2 + x + \frac{1}{4}\right) = -2\left(x + \frac{1}{2}\right)^2$$

Therefore we will subtract $\frac{1}{2}$ from both sides of the original equation. Hence:

$$y + \frac{3}{2} = -2x^2 - 2x$$

becomes

$$y + \frac{3}{2} - \frac{1}{2} = -2x^2 - 2x - \frac{1}{2}$$

$$y + 1 = -2\left(x + \frac{1}{2}\right)^2$$

The graph of this parabola is that of $y = -2x^2$ with the vertex moved to $(-\frac{1}{2}, -1)$, as shown in Figure 5.3. The coefficient of -2 means that it opens downward.

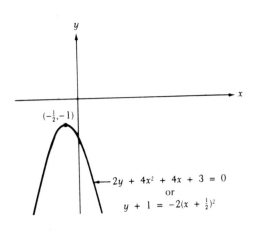

FIG. 5.3

$(-\frac{1}{2}, -1)$

$-2y + 4x^2 + 4x + 3 = 0$
or
$y + 1 = -2(x + \frac{1}{2})^2$

PROBLEM SET   5.1

Graph the following equations, labeling the vertex and all intercepts. All are parabolas except 17.

1. $y = x^2 - 6x$
2. $y = x^2 - 4$
3. $y = \dfrac{x^2}{9}$
4. $15x - 6y^2 = 0$
5. $y = x^2 - 2x - 8$
6. $y - x^2 + 2x + 1 = 0$
7. $3x^2 + 5x - y - 2 = 0$
8. $2x^2 + 4x + y - 10 = 0$
9. $y = (x - 4)^2 + 1$
10. $y = \frac{1}{2}x^2 - 5x + 3$

11. $y = 2x^2 + 7$
12. $y^2 - 4y + 3 - x = 0$
13. $x + y + 5 - 4y^2 = 0$
14. $x = (2y - 3)(y + 2)$
15. $2y^2 - 4y - x = -5$
16. $2y - x^2 + 2x = 5$
17. $y = |x^2 - 4x|$
18. $x = |(y + 3)^2|$
19. $3y^2 - 12y + 3 + x = 0$
20. $y^2 = x + 3y - 3$

5.2   THE CIRCLE

In the last section we started with the equation of a parabola and derived its graph from that. We are going to look at circles from the other direction, by starting with a graph and deriving the equation from that.

The graph of a circle is all the points at a given distance from the center. Suppose, to be specific, that the center of the circle is the origin and its radius is 2, as shown in Figure 5.4. Suppose (x, y) is a typical point on the circle.

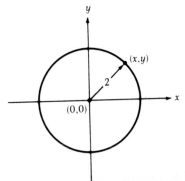

(x, y)

2

$(0, 0)$

FIG. 5.4

Using the distance formula to express the fact that the distance from (x, y) to the origin is 2,

$$\sqrt{(x - 0)^2 + (y - 0)^2} = 2$$

we see that

$$\sqrt{x^2 + y^2} = 2$$

or

$$x^2 + y^2 = 4$$

This is the equation of a circle. Any point whose coordinates satisfy this equation will be at a distance of two from the origin and so will be on the circle. More generally, a circle of radius r and center at the origin has equation

$$x^2 + y^2 = r^2$$

The equation

$$(x - h)^2 + (y - k)^2 = 4$$

represents the circle $x^2 + y^2 = 4$ moved to the right by h and up by k, putting its center at (h, k).

A *circle in standard form* has equation

$$(x - h)^2 + (y - k)^2 = r^2$$

where radius $= r$ and center $= (h, k)$.

EXAMPLE: *Graph $(x + 1)^2 + (y - 1)^2 = 1$.*

Since $(x + 1) = (x - (-1))$, this is a circle of radius 1 centered at $(-1, 1)$. It touches both axes, as shown in Figure 5.5.

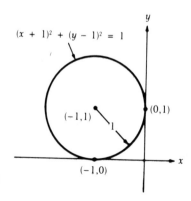

FIG. 5.5

Putting Circles in Standard Form

Any equation of the form

$$x^2 + ax + y^2 + by + c = 0$$

can be converted into the standard form above by completing the square on both x and y. Provided r^2 comes out positive (or r won't be real), the graph of such an equation is a circle.

EXAMPLE: *Find the center and radius of $x^2 + 4x + y^2 - 2y - 2 = 0$.*

To complete the square on $x^2 + 4x$ we must add 4, and on $y^2 - 2y$ we must add 1:

$$x^2 + 4x + y^2 - 2y - 2 = 0$$

$$x^2 + 4x + 4 + y^2 - 2y + 1 - 2 = 4 + 1 \qquad \text{(adding } 4 + 1 \text{ to both sides)}$$

$$(x + 2)^2 + (y - 1)^2 - 2 = 5$$

Moving the 2 to the other side puts the equation in standard form:

$$(x + 2)^2 + (y - 1)^2 = 7$$

so the center is $(-2, 1)$ and the radius is $\sqrt{7}$.

EXAMPLE: *Graph $4x^2 - 4x + 4y^2 + 4y - 7 = 0$.*

This is not in the form $x^2 + ax + y^2 + by + c = 0$ until we divide through by 4, giving

$$x^2 - x + y^2 + y - \frac{7}{4} = 0$$

Completing the square:

$$x^2 - x + \frac{1}{4} + y^2 + y + \frac{1}{4} - \frac{7}{4} = \frac{1}{4} + \frac{1}{4}$$

$$\left(x - \frac{1}{2}\right)^2 + \left(y + \frac{1}{2}\right)^2 - \frac{7}{4} = \frac{1}{2}$$

so

$$\left(x - \frac{1}{2}\right)^2 + \left(y + \frac{1}{2}\right)^2 = \frac{9}{4}$$

This is a circle of radius $\frac{3}{2}$ and center $(\frac{1}{2}, -\frac{1}{2})$, as shown in Figure 5.6 on the next page.

Notice that to get a circle, the coefficients of x^2 and y^2 must be equal, so that you can divide through by it to make the coefficients of x^2 and y^2 both 1. What happens when the coefficients of x^2 and y^2 are not equal is the point of the next two sections.

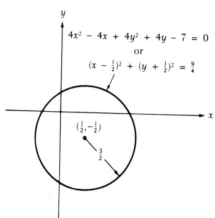

$$4x^2 - 4x + 4y^2 + 4y - 7 = 0$$

or

$$(x - \tfrac{1}{2})^2 + (y + \tfrac{1}{2})^2 = \tfrac{9}{4}$$

$(\tfrac{1}{2}, -\tfrac{1}{2})$

$\tfrac{3}{2}$

FIG. 5.6

EXAMPLE: *Graph $4x^2 - 4x + 4y^2 + 4y + 3 = 0$.*

Apart from the constant term, this is the same equation as in the previous example. It can be rewritten in the same way:

$$x^2 - x + y^2 + y + \frac{3}{4} = 0 \qquad \text{(dividing by 4)}$$

$$\left(x - \frac{1}{2}\right)^2 + \left(y + \frac{1}{2}\right)^2 + \frac{3}{4} = \frac{1}{2} \qquad \text{(completing the square)}$$

so

$$\left(x - \frac{1}{2}\right)^2 + \left(y + \frac{1}{2}\right)^2 = -\frac{1}{4}$$

Now no amount of fiddling will make that $-\tfrac{1}{4}$ on the right into a positive number while leaving both terms on the left positive. Therefore $r^2 = -\tfrac{1}{4}$ and there is no real value for r. What, then, does this equation represent?

The answer is nothing—and this is why: Both terms on the left are perfect squares and so the left side is positive or zero for every value of x and y. The right side, on the other hand, is always negative. Therefore there are no values for x and y that could possibly satisfy the equation, and therefore there are no points on the graph.

EXAMPLE: *Graph $4x^2 - 4x + 4y^2 + 4y + 2 = 0$.*

This equation again differs from the two above only in its constant term, and so can be written:

$$x^2 - x + y^2 + y + \frac{1}{2} = 0$$

$$\left(x - \frac{1}{2}\right)^2 + \left(y + \frac{1}{2}\right)^2 + \frac{1}{2} = \frac{1}{2}$$

so

$$\left(x - \frac{1}{2}\right)^2 + \left(y + \frac{1}{2}\right)^2 = 0$$

Here $r^2 = 0$, so $r = 0$. A circle of radius zero is a point—the center. So the graph of this equation is a single point, namely the point $(\frac{1}{2}, -\frac{1}{2})$.

PROBLEM SET   5.2

Write in standard form the equation of the circle with:
 1. Center $(3, 4)$ and radius 5
 2. Center $(0, 1)$ and radius 2
 3. Center $(-1, -3)$ and radius 6
 4. Center $(-\frac{1}{2}, \frac{1}{2})$ and radius $\frac{3}{2}$
 5. Center $(9, 6)$ and radius 8

Graph the following circles, labeling the center and stating the radius:
 6. $(x - 0)^2 + (y - 0)^2 = 4$
 7. $(x - 0)^2 + (y - 3)^2 = 9$
 8. $(x + 2)^2 + (y - \frac{3}{2})^2 = \frac{25}{4}$
 9. $(x - 0)^2 + (y + 2)^2 = 4$
 10. $x^2 + y^2 - 10x + 9 = 0$
 11. $x^2 + y^2 - 4x + 6y - 12 = 0$
 12. $x^2 + y^2 + 6x + 5y = 0$
 13. $x^2 + y^2 - 2x - 4y + 5 = 0$
 14. $x^2 + y^2 - 6x - 8y + 16 = 0$
 15. $x^2 + y^2 - 2x - 2y = -1$
 16. $4x^2 + 4y^2 + 20x + 10y = 1$
 17. $4x^2 - 12x + 1 = -4y^2 - 4y$
 18. $2x^2 + 2y^2 - 6x + 10y = 1$
 19. $2x^2 + 2y^2 + 8x + 40y = 0$
 20. $4x^2 + 4y^2 + 80x + 12y = -265$

5.3   THE ELLIPSE

Now we'll go back to the approach we used for the parabola, and go from an equation to its graph. Suppose we had

$$x^2 + 4y^2 = 4$$

This is not a circle because no amount of dividing through by 4 or anything else will make the coefficients of x^2 and y^2 both 1 at once. Therefore the equation cannot be put in the form $x^2 + y^2 = r^2$.

Let's plot $x^2 + 4y^2 = 4$ and see what shaped graph it has. First the intercepts:

y intercept: setting $x = 0$ gives

$$4y^2 = 4 \quad \text{so} \quad y = \pm 1$$

x intercept: setting $y = 0$ gives

$$x^2 = 4 \quad \text{so} \quad x = \pm 2$$

Replacing x by $-x$ doesn't change the equation and neither does replacing y by $-y$, so the graph is going to be symmetric both about the y and about the x axes. Before plotting, let's calculate some more points to give us some idea of what is happening in the first quadrant. When $x = 1$, $y = \pm \sqrt{3}/2 \approx \pm 1.7/2 = \pm 0.85$; $x = \frac{1}{2}$ gives $y = \pm 0.97$; and $x = \frac{3}{2}$ gives $y = \pm 0.66$. We end up with a thing that looks like a squashed circle which is called an *ellipse*. See Figure 5.7.

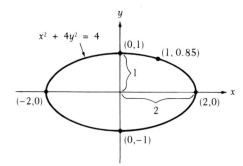

FIG. 5.7

Notice that the equation for the ellipse can be written in the form

$$\frac{x^2}{4} + y^2 = 1$$

Now we'll change this a bit and look at

$$\frac{x^2}{4} + \frac{y^2}{9} = 1$$

This has the same sort of shape as the ellipse above, except that it is "fatter." It has different intercepts,

y intercepts: $\dfrac{x^2}{9} = 1$ so $y = \pm 3$

x intercepts: $\dfrac{x^2}{4} = 1$ so $x = \pm 2$

and it is oriented vertically instead of horizontally. See Figure 5.8.

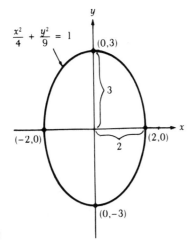

FIG. 5.8

You can see that the 4 and the 9 in the equation tell you where the graph cuts the axes. Not surprisingly, therefore, the equation

$$\frac{x^2}{a^2} + \frac{y^2}{b^2} = 1$$

is an ellipse that crosses the axes at $x = \pm a$ and at $y = \pm b$. See Figure 5.9.

The longer axis of symmetry—in this figure the line from $(-a, 0)$ to $(a, 0)$—is called the *major axis*. In this case its length is $2a$. The shorter axis of symmetry—here the line from $(0, -b)$ to $(0, b)$—is called the *minor axis*, which here has length $2b$.

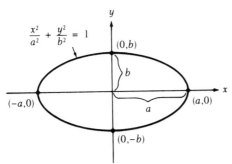

FIG. 5.9

The center of the ellipse in Figure 5.9 is the origin. If the graph were shifted h to the right and k up, its center would be (h, k) and its equation becomes

The standard form of the ellipse:

$$\frac{(x - h)^2}{a^2} + \frac{(y - k)^2}{b^2} = 1$$

You should notice that the standard form of the circle,

$$(x - h)^2 + (y - k)^2 = r^2$$

can be rewritten

$$\frac{(x - h)^2}{r^2} + \frac{(y - k)^2}{r^2} = 1$$

Hence a circle is just the particular case of an ellipse in which $a = b = r$. This is confirmed by looking at the graph of an ellipse and noticing that if $a = b$, the ellipse is no longer a "squashed" circle, but is a real proper circle.

EXAMPLE: *Graph $x^2 + 12y^2 - 3 = 0$.*

First put the 3 on the right-hand side and divide through by it to get the 1:

$$x^2 + 12y^2 = 3$$

$$\frac{x^2}{3} + 4y^2 = 1$$

This ellipse is centered at the origin because h and k are zero here. To see what a and b are, we have to rewrite

$$\frac{x^2}{3} + 4y^2 = 1$$

as

$$\frac{x^2}{(\sqrt{3})^2} + \frac{y^2}{\frac{1}{4}} = 1$$

or

$$\frac{x^2}{(\sqrt{3})^2} + \frac{y^2}{(\frac{1}{2})^2} = 1$$

so $a = \sqrt{3}$ and $b = \frac{1}{2}$ and the graph looks like Figure 5.10.

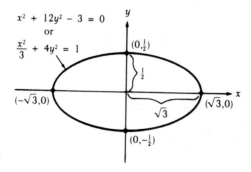

$x^2 + 12y^2 - 3 = 0$
or
$\frac{x^2}{3} + 4y^2 = 1$

$(0, \frac{1}{2})$

$(-\sqrt{3}, 0)$

$(\sqrt{3}, 0)$

$\sqrt{3}$

$(0, -\frac{1}{2})$

FIG. 5.10

Putting Ellipses in Standard Form

Equations with an x^2 and a y^2 but no higher terms can be put into the standard form of the ellipse if the coefficients of these terms have the same sign when put on the same side of the equation, and if a positive number appears on the right after completing the square.

EXAMPLE: *Graph $2x^2 + 9y^2 + 4x - 18y = 7$.*

Rearranging the terms gives:

$2x^2 + 4x + 9y^2 - 18y = 7$

Completing the square on $2x^2 + 4x$ means adding 2 because

$2x^2 + 4x + 2 = 2(x^2 + 2x + 1) = 2(x + 1)^2$

9 is needed to complete the square on $9y^2 - 18y$ because

$9y^2 - 18y + 9 = 9(y^2 - 2y + 1) = 9(y - 1)^2$

So we will add $2 + 9$ to both sides of the equation:

$2x^2 + 4x + 2 + 9y^2 - 18y + 9 = 7 + 2 + 9$

$$2(x + 1)^2 + 9(y - 1)^2 = 18$$

Dividing through by 18 to get the 1 on the right:

$$\frac{(x + 1)^2}{9} + \frac{(y - 1)^2}{2} = 1$$

or

$$\frac{(x + 1)^2}{3^2} + \frac{(y - 1)^2}{(\sqrt{2})^2} = 1$$

So $a = 3$, $b = \sqrt{2}$, and the center is $(-1, 1)$. Before we plot the graph, it would be helpful to have the intercepts as well.

y intercept: $\dfrac{1}{9} + \dfrac{(y - 1)^2}{2} = 1$
(setting $x = 0$)

So $y = \dfrac{7}{3}$ or $-\dfrac{1}{3}$.

x intercept: $\dfrac{(x + 1)^2}{9} + \dfrac{1}{2} = 1$
(setting $y = 0$)

Solving gives $x = -1 + \dfrac{3}{\sqrt{2}}$ or $-1 - \dfrac{3}{\sqrt{2}}$

So $x \simeq 1$ or -3.

Plotting gives the graph in Figure 5.11 on the next page.

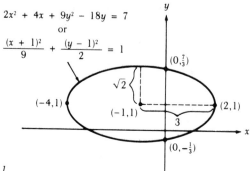

$2x^2 + 4x + 9y^2 - 18y = 7$

or

$$\frac{(x+1)^2}{9} + \frac{(y-1)^2}{2} = 1$$

FIG. 5.11

PROBLEM SET 5.3

Graph the following ellipses, labeling the center and vertices:

1. $\dfrac{x^2}{9} + \dfrac{y^2}{16} = 1$

2. $\dfrac{x^2}{4} + \dfrac{y^2}{9} = 1$

3. $\dfrac{x^2}{121} + \dfrac{y^2}{144} = 1$

4. $\dfrac{x^2}{169} + \dfrac{y^2}{64} = 1$

5. $\dfrac{(x-2)^2}{36} + \dfrac{(y-4)^2}{81} = 1$

6. $\dfrac{(x-\frac{1}{2})^2}{\frac{9}{4}} + \dfrac{(y+\frac{3}{2})^2}{\frac{25}{4}} = 1$

7. $\dfrac{(x-2)^2}{13} + \dfrac{(y+4)^2}{8} = 1$

8. $9(x-1)^2 + 36(y-3)^2 = 36$

9. $16(x+\frac{1}{2})^2 + 100(y-\frac{5}{2})^2 = 400$

10. $4(x-3)^2 + 9(y+2)^2 = 25$

11. $x^2 + 9y^2 + 4x - 18y - 23 = 0$

12. $x^2 + 4y^2 - 8x - 16y - 68 = 0$

13. $2x^2 + y^2 - 12x - 4y = -21$

14. $x^2 + 2y^2 - 10x + 8y + 29 = 0$

15. $4x^2 + y^2 - 8x - 2y + 1 = 0$

16. $9x^2 + 25y^2 - 36x - 150y + 36 = 0$

17. $3x^2 + 2y^2 - 30x - 4y + 23 = 0$

18. $x^2 + 2y^2 + 4x - 12y + 20 = 0$

19. $36x^2 + 11y^2 - 216x + 110y + 203 = 0$

20. $3x^2 + 12x + 4y^2 - 8y = 176$

5.4   THE HYPERBOLA

An equation such as

$$\frac{x^2}{4} - y^2 = 1$$

is extremely like some of those in the last section. Yet it cannot be made into the form

$$\frac{x^2}{a^2} + \frac{y^2}{b^2} = 1$$

because of the minus sign in the middle, and so it can't be an ellipse. There's even less hope of its being a circle, because that would require the coefficients of x^2 and y^2 to be equal. We will have to graph it to see what we've got.

First of all, the intercepts:

y intercepts: setting $-y^2 = 1$ gives

$$y^2 = -1 \quad \text{No solution, so no } y \text{ intercepts}$$

x intercepts: setting $\dfrac{x^2}{4} = 1$ gives

$$x = \pm 2$$

The graph of this equation will have symmetry about both axes, since replacing x by $-x$ and y by $-y$ doesn't change the equation.

We're much more used to plotting an equation in the form:

$$y = \text{some expression containing } x\text{'s}$$

so let's solve our equation for y in terms of x and see if that helps:

$$\frac{x^2}{4} - y^2 = 1$$

$$y^2 = \frac{x^2}{4} - 1 = \frac{x^2 - 4}{4}$$

$$y = \pm\sqrt{\frac{x^2 - 4}{4}}$$

$$= \pm\frac{1}{2}\sqrt{x^2 - 4}$$

If y is to be real, $x^2 - 4$ must be positive or zero, so

$$x^2 - 4 \geqslant 0$$

that is,

$$x \geqslant 2 \quad \text{or} \quad x \leqslant -2$$

This means that the graph exists only to the right of $x = 2$ and to the left of $x = -2$.

What happens as $x \to \infty$? Or as $x \to -\infty$? For very large x (either negative or positive), x^2 is so much bigger than 4 that

$$y = \pm\frac{1}{2}\sqrt{x^2 - 4}$$

is nearly

$$y \simeq \pm\frac{1}{2}\sqrt{x^2} \qquad \text{[ignoring the 4]}$$

or

$$y \simeq \pm\frac{x}{2}$$

Now $y = \pm\frac{x}{2}$ are a pair of diagonal lines through the origin, and these are the *asymptotes* to the curve.

For the moment let's just consider positive values of x and y. Since $x^2 - 4$ is less than x^2, the real value of y is less than $y = \frac{x}{2}$, and so the graph approaches the asymptote from below. Therefore, so far we have the situation shown in Figure 5.12.

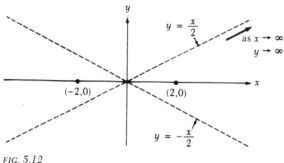

FIG. 5.12

Now as x increases, so do $x^2 - 4$ and $y = \pm\frac{1}{2}\sqrt{x^2 - 4}$. Therefore y increases steadily from $y = 0$ when $x = 2$ to just below the asymptote $y = \frac{x}{2}$ for large x. The graph in the other four quadrants is obtained by reflection and looks like Figure 5.13.

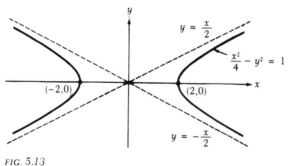

FIG. 5.13

Now let's look at

$$\frac{x^2}{4} - \frac{y^2}{9} = 1$$

This equation turns out to have x intercepts of ± 2 and no y intercepts. Since solving for y gives

$$y = \pm\frac{3}{2}\sqrt{x^2 - 4}$$

the graph again exists only to the right of $x = 2$ and to the left of $x = -2$, and the asymptotes are $y = \pm\frac{3}{2}x$, as shown in Figure 5.14.

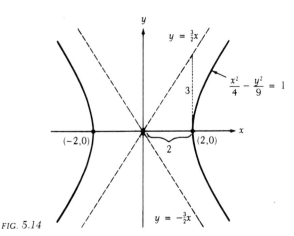

FIG. 5.14

Now, what happens if we interchange the x and y terms, getting:

$$\frac{y^2}{9} - \frac{x^2}{4} = 1$$

This equation has y intercepts of ± 3 and no x intercepts. Also,

$$y = \pm\frac{3}{2}\sqrt{x^2 + 4}$$

which for large x is nearly

$$y = \pm\frac{3}{2}x$$

so this equation has the same asymptotes as the previous one. The difference is that since $x^2 + 4$ is larger than x^2, this equation approaches the asymptote from *above*. The graph ends up looking like Figure 5.15.

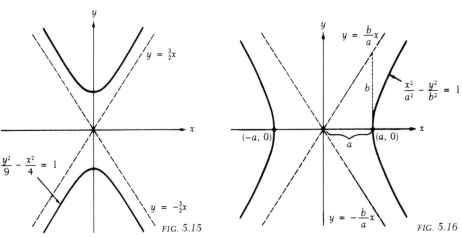

FIG. 5.15

FIG. 5.16

The equation

$$\frac{x^2}{a^2} - \frac{y^2}{b^2} = 1$$

has a graph exactly like that of $\dfrac{x^2}{4} - \dfrac{y^2}{9} = 1$; see Figure 5.16.

The graph of

$$\frac{y^2}{b^2} - \frac{x^2}{a^2} = 1$$

opens vertically, just like $\frac{y^2}{9} - \frac{x^2}{4} = 1.$ It has the same asymptotes as the graph above, but its intercepts are $(0, b)$ and $(0, -b)$.

If the center of the hyperbola is moved from the origin to (h, k), we get

---

the standard form of the hyperbola:

$$\frac{(x - h)^2}{a^2} - \frac{(y - k)^2}{b^2} = 1$$

This is a hyperbola opening horizontally. A hyperbola opening vertically has the form:

$$\frac{(y - k)^2}{b^2} - \frac{(x - h)^2}{a^2} = 1$$

The asymptotes of both these hyperbolas are lines of slope $\pm\frac{b}{a}$ through the center (h, k). The equations of the asymptotes are the lines $y = \pm\frac{b}{a}x$ shifted h horizontally and k vertically, namely

$$y - k = \pm\frac{b}{a}(x - h)$$

Putting Hyperbolas in Standard Form

Any equation containing an x^2 and a y^2 term but no higher terms can be put into the standard form of the hyperbola if the coefficients of these terms have opposite signs when put on the same side of the equation.

EXAMPLE: *Graph $16x^2 - 64x + 109 = 3y^2 + 6y$.*

Since the x^2 and y^2 terms would have opposite signs if moved to the same side of the equation, this is going to be a hyperbola.

Now for completing the square. We must add 64 to $16x^2 - 64x$ [because $16x^2 - 64x + 64 = 16(x^2 - 4x + 4) = 16(x - 2)^2$] and 3 to $3y^2 + 6y$ [because $3y^2 + 6y + 3 = 3(y^2 + 2y + 1) = 3(y + 1)^2$]. So we add 64 and 3 to each side:

$$16x^2 - 64x + 109 + 64 + 3 = 3y^2 + 6y + 64 + 3$$

giving

$$16(x-2)^2 + 109 + 3 = 3(y+1)^2 + 64$$

so

$$16(x-2)^2 - 3(y+1)^2 = -48$$

Dividing through by -48, we get the 1 on the right:

$$-\frac{(x-2)^2}{3} + \frac{(y+1)^2}{16} = 1$$

which can be rewritten as

$$\frac{(y+1)^2}{4^2} - \frac{(x-2)^2}{(\sqrt{3})^2} = 1$$

This hyperbola opens vertically, has $a = \sqrt{3}$, $b = 4$, and center $(2, -1)$. The asymptotes have slope $\pm\dfrac{b}{a} = \pm\dfrac{4}{\sqrt{3}}$ and go through $(2, -1)$. The graph is shown in Figure 5.17.

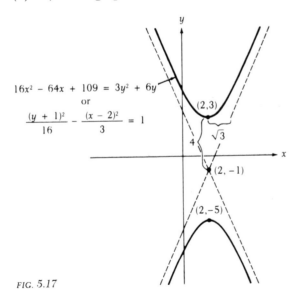

FIG. 5.17

The Rectangular Hyperbola

We have in fact seen a hyperbola before, though in a different context and with a completely different equation—namely $y = \dfrac{1}{x}$. See Figure 5.18. This graph has all the characteristics of a hyperbola: It has two symmetric branches that tend toward the asymptotes, which cross at the center. Since here the asymptotes cross at right angles, this is called a *rectangular hyperbola*.

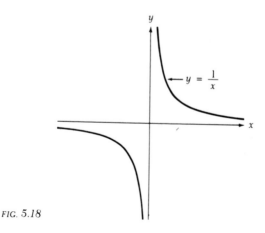

$y = \frac{1}{x}$

FIG. 5.18

The difference between this rectangular hyperbola and the ones we were looking at before is that this one opens neither horizontally nor vertically but diagonally. However, if it were rotated through 45° to the right or left, this rectangular hyperbola would be in the standard form.

PROBLEM SET 5.4

Graph the following hyperbolas, labeling the center and asymptotes:

1. $\dfrac{x^2}{16} - \dfrac{y^2}{4} = 1$

2. $\dfrac{y^2}{9} - \dfrac{x^2}{25} = 1$

3. $\dfrac{(x+3)^2}{4} - \dfrac{(y-1)^2}{25} = 1$

4. $\dfrac{(x-\frac{1}{2})^2}{16} - \dfrac{(y+\frac{3}{2})^2}{36} = 1$

5. $\dfrac{(y-4)^2}{9} - \dfrac{(x-2)^2}{4} = 1$

6. $\dfrac{y^2}{4} - \dfrac{x^2}{12} = 1$

7. $\dfrac{(x-3)^2}{16} - \dfrac{(y-2)^2}{25} = 1$

8. $16y^2 - 9x^2 = 144$

9. $25(x+3)^2 - 9(y-1)^2 = 225$

10. $4(y-\frac{1}{2})^2 - 16(x+\frac{3}{2})^2 = 64$

11. $9x^2 - 4y^2 - 36x + 32y + 8 = 0$

12. $4x^2 - y^2 + 24x + 4y + 28 = 0$

13. $x^2 - y^2 - 2x - 4y - 4 = 0$

14. $4x^2 - y^2 + 8x - 4y - 4 = 0$

15. $36y^2 - x^2 - 24y + 6x - 41 = 0$

16. $9y^2 - 4x^2 + 54y + 8x + 45 = 0$

17. $x^2 - 9y^2 - 4x - 90y = 222$

18. $4y^2 + 24y - x^2 - 2x + 19 = 0$

19. $16y^2 - 25x^2 + 96y - 50x = 281$

20. $9x^2 - 36x - y^2 - 10y = 35$

5.5   A SUMMARY OF STANDARD FORMS

Given a quadratic equation in x and y, but without an xy term, you can try to get it into one of the standard forms.

No x^2 or y^2 term:

$$\left.\begin{array}{l} y - k = p(x - h)^2 \\ x - h = p(y - k)^2 \end{array}\right\} \text{parabola}$$

Coefficients of x^2 and y^2 equal:

$$(x - h)^2 + (y - k)^2 = r^2: \quad \text{circle}$$

Coefficients of x^2 and y^2 same sign:

$$\frac{(x - h)^2}{a^2} + \frac{(y - k)^2}{b^2} = 1: \quad \text{ellipse}$$

Coefficients of x^2 and y^2 opposite sign:

$$\left.\begin{array}{l} \dfrac{(x - h)^2}{a^2} - \dfrac{(y - k)^2}{b^2} = 1 \\[2ex] \dfrac{(y - k)^2}{b^2} - \dfrac{(x - h)^2}{a^2} = 1 \end{array}\right\} \text{hyperbola}$$

Note on Why These Are Called Conic Sections

By now you may be wondering how the conic sections got their name. It turns out that if you cut a double cone by a plane at different angles, you will get all the conic sections. See Figure 5.19.

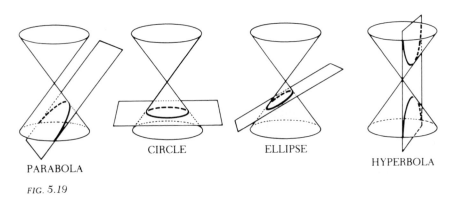

PARABOLA   CIRCLE   ELLIPSE   HYPERBOLA

FIG. 5.19

CHAPTER 5 REVIEW

Graph the following conics, labeling relevant points (e.g., centers, asymptotes, etc.):

1. $4x^2 + y^2 = 16$
2. $4x^2 - y^2 = 16$
3. $4y^2 - x^2 = 16$
4. $x^2 + 4y^2 = 16$
5. $3x^2 + 2y^2 - 45 = 0$
6. $2y^2 + x^2 - 4 + x = -x^2$
7. $2x^2 = y + x + 3$
8. $2x^2 = -y + x + 3$
9. $(y - 2)(y + 2) = x^2 + 2$

10. $y^2 + 6y - 8x - 31 = 0$
11. $4x^2 - 4y - 4x + 24 = 0$
12. $9x^2 - 16y^2 - 144 = 0$
13. $16(x + 3)^2 + 25(y - 7)^2 = 400$
14. $y^2 - 6x - 4y + 7 = 0$
15. $x^2 - 20y + 4x + 64 = 0$
16. $9x^2 - 4y^2 + 54x + 16y - 79 = 0$
17. $9x^2 - 25y^2 - 36x - 150y - 414 = 0$
18. $5x^2 + 9y^2 - 20x - 25 = 0$

What equations have the following graphs?

19.

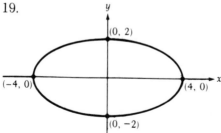

20.

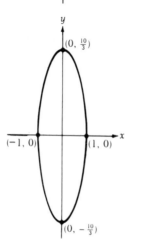

21.

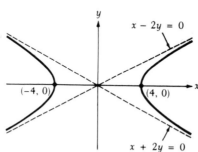

22.

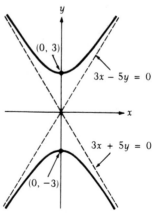

23.

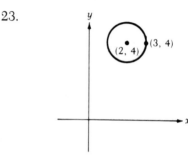

24.

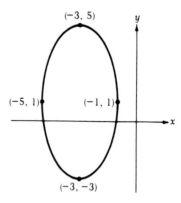

Graph the following equations, labeling all relevant points:

25. $x^2 - 4y^2 - 4x + 32y + 4 = 0$

26. $144y^2 - 25x^2 - 576y + 200x + 3776 = 0$

27. $16x^2 + 9y^2 + 32x - 36y - 92 = 0$

28. $2y^2 + 4x - y = 0$ 33. $x^2 - 3y - 6x - 7 = 0$

29. $x^2 + 4y^2 - 6x + 32y + 69 = 0$ 34. $2x^2 - 4y^2 + 8x + 6y = 0$

30. $x^2 + y^2 + 25x + 10y + 12 = 0$ 35. $x + y + x^2 + y^2 = 0$

31. $(x^2 + 8x - 15) + (y^2 - 6y) = 0$ 36. $x(x - 1) + 2y(y - 3) = 1$

32. $x^2 + 25x + y^2 - 2y + 26 = 75$ 37. $x^2 + y^2 + 8x - 2y + 15 = 0$

38. A bridge looks like the graph of $y = -3x^2 + 12x + 8$. You have a 10-foot-high truck and a load 8 feet wide. Will you make it under the bridge?

The four equations below correspond to the four parabolas on the graph in Figure 5.20.

39. $(y - a) = bx^2$
40. $(y + a) = cx^2$

41. $(y - a)^2 = dx$
42. $(y - a) = ex^2$

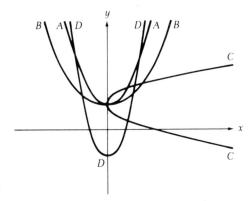

FIG. 5.20

You are also given that $a > 0$ and $b > e$. Which equation belongs to which parabola? Please give reasons.

The four equations below correspond to the four circles on the graph in Figure 5.21.

43. $(x - a)^2 + (y - b)^2 = c^2$
44. $(x - a)^2 + (y - d)^2 = e^2$

45. $(x + d)^2 + (y - b)^2 = f^2$
46. $(x - a)^2 + (y - b)^2 = f^2$

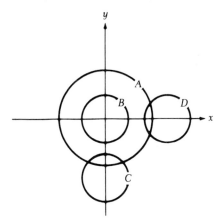

FIG. 5.21

You are also given that $b > d$. Which equation belongs to which circle? Why?

47. On a set of coordinate axes, draw a circle; then draw an ellipse which encloses that entire circle and touches it at two points only. Now write the equations of the two conic sections you have just drawn, indicating which equation belongs to which figure. (There are many possible answers.)

48. The orbit of a planet around the sun can be depicted as shown in Figure 5.22. If C is at the origin of a Cartesian plane, write the equation for this ellipse. (You may assume PR and QC are horizontal and vertical, respectively.)

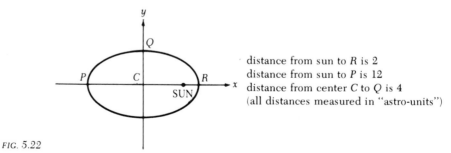

distance from sun to R is 2
distance from sun to P is 12
distance from center C to Q is 4
(all distances measured in "astro-units")

FIG. 5.22

49. If the trajectory of a basketball is described by the equation $y = x + 6 - x^2$, and the ball is thrown from the point $(-2, 0)$, will the ball make the basket if

(a) $(1, 6)$ is the center of the basket? Why?

(b) $(2, 5)$ is the center of the basket? Why?

50. A McDonalds "golden arch" is shaped like a parabola as shown in Figure 5.23. Choose your own axes, and write an equation to describe the arch. (There are many possible answers.)

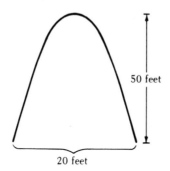

50 feet

FIG. 5.23   20 feet

6 A GRAPHICAL LOOK AT EQUATIONS AND INEQUALITIES

6.1 SIMULTANEOUS LINEAR EQUATIONS AND THEIR GRAPHS

Algebra can be terrifically useful in giving you information about graphs. For example, solving the equation $x^2 - 1 = 0$ tells you at what points the graph of $y = x^2 - 1$ cuts the x axis; solving $x^2 - 1 = 3$ gives the x coordinates of the points at which $y = x^2 - 1$ cuts the line $y = 3$ (see Section 3.8). This chapter is to show you what else algebra can tell us about graphs, and, equally important, what graphs can tell us about algebra. The answer is a lot!

For example, suppose that we want to find the solution to the system of equations

$$\begin{cases} x + y = 3 \\ x - y = 1 \end{cases}$$

which means finding values for x and y that satisfy both equations simultaneously. Now you can also look at $x + y = 3$ and $x - y = 1$ as the equations of lines (rewriting them as $y = -x + 3$ and $y = x - 1$ makes them look more normal). Now remember that a point (x, y) lies on a line if its coordinates satisfy the equation of that line. Therefore finding values of x and y that satisfy $x + y = 3$ and $x - y = 1$ simultaneously means finding the coordinates of a point that lies on both lines at once. In other words, *the x and y values that satisfy a pair of simultaneous linear equations are the coordinates of the point of intersection of the lines.*

Let us see the example through:

$$\begin{cases} x + y = 3 \\ x - y = 1 \end{cases}$$

Algebra shows you that the solution to these equations is $x = 2$, $y = 1$. Now let us plot graphs of these lines. Rewrite the equations as

$$y = -x + 3: \quad \text{slope } -1, \text{ } y \text{ intercept } 3, \text{ } x \text{ intercept } 3$$

and

$$y = x - 1: \quad \text{slope } 1, \text{ } y \text{ intercept } -1, \text{ } x \text{ intercept } 1$$

and then graph, as in Figure 6.1.

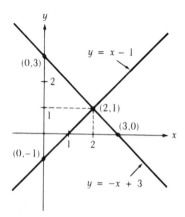

FIG. 6.1

You can see that the lines seem to intersect at the point with coordinates $(2, 1)$, as you would expect since the solution to the equations is $x = 2, y = 1$. Notice that if we had not known that the solution was $x = 2$, $y = 1$, the graph could not have told us this exactly. From the graph you can learn that the solution is roughly $x = 2$, $y = 1$, but it might also be $x = 2.1$, $y = 1.1$, or any other nearby values. The only way to be sure is to check the values you read off the graph by substituting into the original equations and making sure they satisfy them.

EXAMPLE: *Solve graphically*

$$\begin{cases} 2x - y - 1 = 0 \\ x + y + 4 = 0 \end{cases}$$

Solving simultaneous equations graphically means drawing the graphs of each line accurately enough to be able to read off the coordinates of the points of intersection. These are the x and y values we're looking for, provided they are checked by substitution.

$2x - y - 1 = 0$ can be written $y = 2x - 1$, which has slope 2, y intercept -1, x intercept $\frac{1}{2}$.

$x + y + 4 = 0$ can be written $y = -x - 4$, which has slope -1, y intercept -4, x intercept -4.

This gives us the graphs in Figure 6.2.

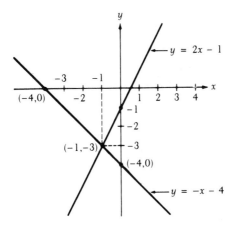

FIG. 6.2

The point of intersection seems to be $(-1, -3)$, telling us the solution to the simultaneous equations is $x = -1$, $y = -3$.

CHECK: Substitute $x = -1$ and $y = -3$ into the equations.

First equation:

LHS $= 2x - y - 1 = 2(-1) - (-3) - 1 = -2 + 3 - 1 = 0$

RHS $= 0$

Second equation:

LHS $= x + y + 4 = (-1) + (-3) + 4 = 0$

RHS $= 0$

Therefore $x = -1$, $y = 3$ does satisfy both equations.

Alternatively, you can check this result by solving the equations algebraically.

The Number of Solutions to Simultaneous Equations A pair of simultaneous equations in x and y usually has exactly one solution. This is because we are looking for the point of intersection of a pair of lines, and two lines usually intersect at exactly one point. However, you may know that it is possible for a pair of equations to have either no solution or infinitely many solutions. Looking at the lines that are the graphs of these equations offers an interesting way of seeing why this is so.

Let us consider

$$\begin{cases} x + y = 1 \\ x + y = 3 \end{cases}$$

This pair of equations has no solution, because it is impossible for $x + y$ to equal 1 and 3 at the same time; the system is *inconsistent*. Suppose we graph the lines $x + y = 1$ and $x + y = 3$ to see what is going on. Rewriting the equations as $y = -x + 1$ (slope -1, y intercept 1, x intercept 1) and $y = -x + 3$ (slope -1, y intercept 3, x intercept 3) gives the graphs in Figure 6.3.

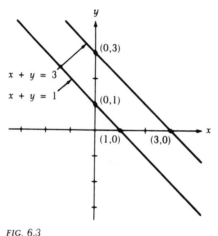

FIG. 6.3

You can see immediately that the lines are parallel and therefore have no point of intersection. No wonder there is no solution to the equations!

What about equations like

$$\begin{cases} x + y = 1 \\ 2x + 2y = 2 \end{cases}$$

This pair of equations is *degenerate*, because the second equation tells us no more than the first (since it is just the first equation with everything doubled). Any values of $x + y$ that satisfy the first equation satisfy the second. Therefore there are infinitely many solutions to the system. If we were to graph these two equations, we would rewrite $x + y = 1$ as $y = -x + 1$ and $2x + 2y = 2$ as $y = -x + 1$. This immediately makes it clear what the trouble is: the equations give rise to the *same* line. Therefore all a point has to do for its coordinates to satisfy *both* equations is to lie on *one* line. There are infinitely many such points, and therefore infinitely many solutions.

Now, having seen how a graph can illuminate the algebraic problem of solving simultaneous equations, let's look at how a geometric problem can be solved using algebra.

EXAMPLE: *In a fog your navigation equipment tells you that the line joining you to a lighthouse at $(1, -3)$ has slope -5, and the line joining you to a buoy at $(\frac{3}{4}, 0)$ has slope 2. Where are you?*

There are two lines mentioned in this problem, one joining you to the lighthouse and one joining you to the buoy, and clearly you are at the intersection of these lines. If you can find the equations of these lines, you can solve them simultaneously to find their point of intersection, and hence your position. Therefore the problem has become finding the equations of the lines.

For each of the two lines you know two pieces of information: the slope and one point the line passes through—the position of either the lighthouse or the buoy. Using this information and the slope-intercept form of a line ($y = mx + b$, where $m = $ slope and $b = y$ intercept), you can find their equations.

Line from you to the lighthouse: The slope $= -5$ so the line is of the form

$$y = -5x + b.$$

Since the lighthouse lies on the line, the point $(1, -3)$ must satisfy the equation of the line, and this will give an equation for b. Substituting $x = 1$ and $y = -3$ into $y = -5x + b$ gives

$$-3 = -5(1) + b$$

so that $b = -3 + 5 = 2$ and therefore the line is

$$y = -5x + 2$$

Line from you to the buoy: The slope $= 2$, so the line has the form

$$y = 2x + b.$$

The buoy is at $(\frac{3}{4}, 0)$ and so

$$0 = 2\left(\frac{3}{4}\right) + b$$

giving $b = -\dfrac{3}{2}$ and making the equation of the line

$$y = 2x - \frac{3}{2}$$

Your position is at the intersection of these two lines, which can be found by solving

$$\begin{cases} y = -5x + 2 \\ y = 2x - \dfrac{3}{2} \end{cases}$$

simultaneously, giving

$$-5x + 2 = 2x - \frac{3}{2}$$

$$7x = \frac{7}{2}$$

so

$$x = \frac{1}{2}$$

Substituting $x = \frac{1}{2}$ into $y = -5x + 2$ gives

$$y = -5\left(\frac{1}{2}\right) + 2 = -\frac{1}{2}$$

so you are at the point $(\frac{1}{2}, -\frac{1}{2})$.

PROBLEM SET   6.1

Solve the following sets of equations graphically. Be sure to check your answers.

1. $\begin{cases} x + y = 11 \\ 3x - y = 5 \end{cases}$

2. $\begin{cases} 5x + 3y = 7 \\ 3x - 5y = -23 \end{cases}$

3. $\begin{cases} 5x - 2y = 1 \\ x + 3y = 7 \end{cases}$

4. $\begin{cases} x + y = 1 \\ x - y = 3 \end{cases}$

5. $\begin{cases} 2x + 3y = 1 \\ 4x + 6y = 2 \end{cases}$

6. $\begin{cases} x + 2y = 5 \\ 3x - y - 1 = 0 \end{cases}$

7. $\begin{cases} x + y - 5 = 0 \\ 2x - y - 1 = 0 \end{cases}$

8. $\begin{cases} x - 3y = 1 \\ x = 2y \end{cases}$

9. $\begin{cases} x + y + 1 = 0 \\ 0 = 3x - y - 3 \end{cases}$

10. $\begin{cases} 3x + 2y = 2 \\ x - y = 9 \end{cases}$

11. $\begin{cases} 3x - 5y = 5 \\ 7x + y = 75 \end{cases}$

12. $\begin{cases} 2x - 3y = 2 \\ 4x + 7y = -9 \end{cases}$

13. Your surveying equipment tells you that the line joining your tripod to a landmark at $(0, 3)$ has slope $-\frac{4}{3}$, and the line joining your tripod to another landmark at $(1, 4)$ has slope $-\frac{5}{2}$. Where are you?

14. One set of railroad tracks runs along the straight line through the point $(1, 0)$ with slope $-\frac{1}{4}$. Another set of railroad tracks runs along the straight line through $(4, 0)$ with slope $-\frac{1}{7}$. Where do the tracks cross?

15. A straight highway passes through a town located at $(0, 2)$ with slope $-\frac{5}{3}$. Another straight highway passes through a town located at $(-1, -4)$ with slope $\frac{1}{4}$. Where do the highways intersect?

16. An airplane flying along a straight-line path passes over $(2, 1)$ and $(3, 5)$. A second plane, also flying along a straight-line path, passes over $(0, -5)$ and $(5, 11)$. At what point do their paths intersect?

17. Two ships at sea are proceeding along on straight-line courses that will cause them to collide. Using standard navigation maps, one ship passes through the points $(1, -\frac{1}{2})$ and $(3, 3)$, while the other passes through $(2, 0)$ and $(5, 9)$. Where do they collide?

18. A person A leaves the point $(-1, 4)$ and walks along a line with slope $-\frac{1}{2}$ at 2 mph. If another person B leaves $(0, -4)$ at the same time, and walks along a line with slope 2, where will he cross A's path? How fast must B walk in order to meet A?

19. Show that the altitudes of a triangle are concurrent (meet in one point). [*Hint:* A tringle can be represented by three points in the plane. For simplicity, they could be $(0, 0)$, $(a, 0)$, (b, c) and still represent any arbitrary triangle. An altitude of a triangle is a line perpendicular to one side of the triangle that passes through the opposite vertex.]

20. If the equations of two lines are

$$y = Ax + B$$

$$y = Bx + A \qquad A \neq 0, B \neq 0$$

and if these lines are perpendicular, where do they intersect?

6.2   NONLINEAR SIMULTANEOUS EQUATIONS AND THEIR GRAPHS

Suppose you draw the graphs of $y = x$ and $y = x^2$ on the same axes, as in Figure 6.4. It is clear the graphs cross at the origin and at one other point, but how do we find the coordinates of that other point? Realize that we are again looking for a point that lies on both graphs, namely, the solution to the simultaneous equations

$$\begin{cases} y = x \\ y = x^2 \end{cases}$$

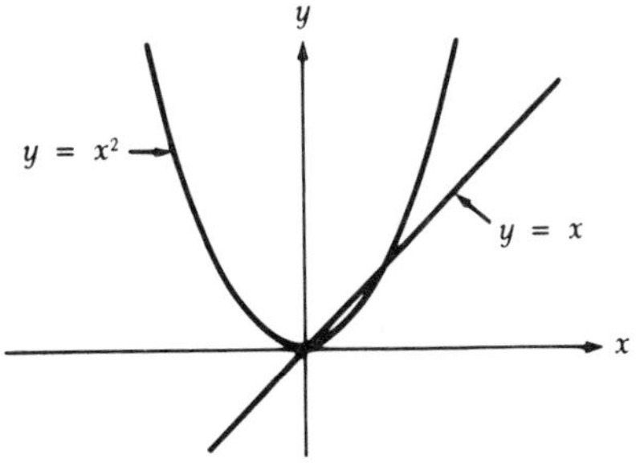

FIG. 6.4

To solve these equations, notice that if $y = x$ and $y = x^2$, then

$$x = x^2.$$

Therefore,

$$x^2 - x = 0$$

or

$$x(x - 1) = 0$$

so

$$x = 0 \quad \text{or} \quad x = 1$$

Substituting back into either equation, if $x = 0$, then $y = 0$; and if $x = 1$, then $y = 1$. Therefore the points of intersection are $(0, 0)$ and $(1, 1)$.

Check: These points at least look reasonable, since we knew the curves cut at the origin, and were expecting a second point in the first quadrant.
Also, substituting $x = 0$ and $y = 0$

into $y = x$ and $y = x^2$

gives $0 = 0$ and $0 = 0^2$, so $(0, 0)$ lies on both curves.

Similarly, substituting $x = 1$ and $y = 1$

into $y = x$ and $y = x^2$

gives $1 = 1$ and $1 = 1^2$, so $(1, 1)$ lies on both curves.

Please notice that it is *very important* which x goes with which y. For example, $x = 0$, $y = 1$ is *not* a solution to either equation, and the point $(0, 1)$ does not lie on either graph.

EXAMPLE: *Find the points of intersection of*

$$y = 2x \quad and \quad y = 1 + \frac{1}{x}$$

It is a good idea to draw a rough graph of these equations first, so that you will know how many intersections you are looking for, and whether the answers you get are at least reasonable.

$y = 2x$ is a line of slope 2 through the origin.

$y = 1 + \dfrac{1}{x}$ is the graph of $y = \dfrac{1}{x}$ moved up by one.

From Figure 6.5 you can see that we should be looking for two points of intersection, one in the first and one in the third quadrant.

Solving

$$\begin{cases} y = 2x \\ y = 1 + \dfrac{1}{x} \end{cases}$$

simultaneously gives

$$2x = 1 + \frac{1}{x} \quad \text{(since both sides are equal to } y)$$

This is a fractional equation, so we should multiply through by x to clear of fractions:

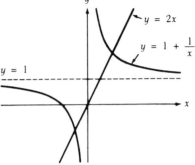

$$2x^2 = x + 1$$

$$2x^2 - x - 1 = 0$$

Therefore

$$(2x + 1)(x - 1) = 0$$

so either

$$x = -\frac{1}{2} \quad \text{or} \quad x = 1$$

FIG. 6.5

Substituting into either equation (but $y = 2x$ is easiest) shows that if $x = -\frac{1}{2}$, $y = -1$; and if $x = 1$, $y = 2$. *Therefore the points of intersection are $(-\frac{1}{2}, -1)$ and $(1, 2)$.*

These points certainly look reasonable from the graph; you should check them by substituting into the equations.

EXAMPLE: *Find the point(s) of intersection of $y = x^2$ and $x = y^2$.*

The graphs are shown in Figure 6.6.

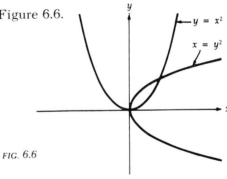

FIG. 6.6

If $y = x^2$ and $x = y^2$, then substituting for y into the second equation gives

$$x = (x^2)^2 \quad \text{or} \quad x = x^4$$

Therefore

$$x^4 - x = 0$$

To solve, we must factor this. x is clearly a factor, so we have

$$x(x^3 - 1) = 0$$

Now $(x - 1)$ is a factor of $x^3 - 1$. We can get the other factor by division:

$$\frac{x^3 - 1}{x - 1} = x^2 + x + 1$$

so

$$x^3 - 1 = (x - 1)(x^2 + x + 1)$$

Therefore the equation factors to

$$x(x - 1)(x^2 + x + 1) = 0$$

and so either

$$x = 0 \text{ or } x - 1 = 0 \text{ or } x^2 + x + 1 = 0$$

Now, $x^2 + x + 1 = 0$ has no (real) roots because its discriminant $(= -3)$ is negative. So either $x = 0$ or $x = 1$ are the only solutions.

Substituting into the first equation, when $x = 0$, $y = 0$; and when $x = 1$, $y = 1$. *Therefore the points of intersection are (0, 0) and (1, 1).*

CHECK: Substituting $x = 0$, $y = 0$ into

$$y = x^2 \quad \text{and} \quad x = y^2$$

gives

$$0 = 0^2 \quad \text{and} \quad 0 = 0^2$$

so (0, 0) is a point of intersection. Similarly,

$$1 = 1^2 \quad \text{and} \quad 1 = 1^2$$

so (1, 1) is also a point of intersection.

   Note: It is extremly important that you check your solutions in *both* the original equations. In any set of nonlinear simultaneous equations, extraneous roots can come up and the only way you'll detect them is by checking. So check!

For example, suppose we had calculated y from the second equation $(x = y^2)$ instead of the first. Substituting $x = 0$ gives $0 = y^2$, so $y = 0$ as before, but when $x = 1$, then $1 = y^2$ so $y = \pm 1$. We had all of these solutions before except $x = 1$, $y = -1$. However, substituting this into the first equation, $y = x^2$, gives $-1 = 1^2$, which is not true. So $x = 1$, $y = -1$ is an extraneous solution (and should be thrown out).

The graph confirms that $x = 1$, $y = -1$ is not a solution to this system because it shows that the point $(1, -1)$ could not be a point of intersection of $y = x^2$ and $x = y^2$.

As you may have noticed, we got an extraneous root when we used the equation that is quadratic in y to calculate y. The fact that it is quadratic in y means that most x values lead to two y values. So, given the x coordinate of the point of intersection, we end up finding *two* points with that x value, rather than just the single point of intersection that we are looking for.

PROBLEM SET   6.2

Find the points of intersection of the following sets of equations.   Sketch graphs to check the reasonableness of your answers.

1. $\begin{cases} y = x^2 \\ y = x^3 \end{cases}$

2. $\begin{cases} y = x^3 \\ y = x \end{cases}$

3. $\begin{cases} y = \dfrac{1}{x} \\ y = x^2 \end{cases}$

4. $\begin{cases} y = \dfrac{1}{x} \\ y = x^3 \end{cases}$

5. $\begin{cases} y = x^2 \\ y = \sqrt{x} \end{cases}$

6. $\begin{cases} y = x^2 \\ y = -\sqrt{x} \end{cases}$

7. $\begin{cases} y^2 = 4x \\ y = 4x \end{cases}$

8. $\begin{cases} y^2 = 4x \\ y + 4x = 0 \end{cases}$

9. $\begin{cases} y^2 + 4x = 0 \\ y^2 = 4x \end{cases}$

10. $\begin{cases} y = 4x^2 \\ y^2 = 2x \end{cases}$

11. $\begin{cases} y = 3x \\ xy = 1 \end{cases}$

12. $\begin{cases} y = \dfrac{x}{x - 1} \\ y = 2x \end{cases}$

13. $\begin{cases} y = x - 1 \\ y = \dfrac{2(x - 1)}{(x - 2)^2} \end{cases}$

14. $\begin{cases} 4y^2 = 6x \\ 4y - x - 6 = 0 \end{cases}$

15. $\begin{cases} y^2 = 8 - 4x \\ y^2 = 6x - 32 \end{cases}$

16. $\begin{cases} x^2 + y^2 = 25 \\ xy = 12 \end{cases}$

17. $\begin{cases} \dfrac{2}{x} + \dfrac{2}{y} = 10 \\ \dfrac{2}{x} - \dfrac{2}{y} = 2 \end{cases}$

18. $\begin{cases} \dfrac{3}{x} + \dfrac{2}{y} = \dfrac{9}{2} \\ \dfrac{12}{x} - \dfrac{2}{y} = 3 \end{cases}$

19. $\begin{cases} y = |x| \\ y = x^2 \end{cases}$

20. $\begin{cases} y = |x| \\ y = x^3 \end{cases}$

21. $\begin{cases} y = |x| \\ y = \dfrac{1}{x} \end{cases}$

22. $\begin{cases} y = |x - 1| \\ y = (x - 1)^2 \end{cases}$

23. $\begin{cases} y = |x - 1| \\ y = -(x - 1)^2 \end{cases}$

24. $\begin{cases} y = |x + 1| \\ y = (x + 1)^3 \end{cases}$

25. $\begin{cases} y = |x + 1| \\ y = -(x + 1)^3 \end{cases}$

6.3   SPECIAL CASE: THE CONICS

Drawing the graphs of

$$x^2 + y^2 = 5$$

which is a circle of radius $\sqrt{5}$, and of

$$y = x - 1$$

which is a line of slope 1, leads you to suspect that they have two points of intersection, as shown in Figure 6.7. The easiest way to find these points of intersection is to solve

$$x^2 + y^2 = 5$$

$$y = x - 1$$

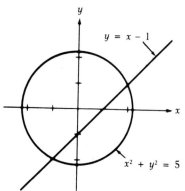

FIG. 6.7

simultaneously.

Substituting for y into the first equation gives

$$x^2 + (x - 1)^2 = 5$$

or

$$2x^2 - 2x - 4 = 0$$

so

$$2(x + 1)(x - 2) = 0$$

Therefore either $x = -1$ or $x = 2$.

We will find y from the equation $y = x - 1$. Substituting $x = -1$ gives $y = -2$, and $x = 2$ gives $y = 1$, so it seems that the points of intersection are

$$(-1, -2) \quad \text{and} \quad (2, 1)$$

These certainly look reasonable, but you should check both these points in both equations (they do work!). Notice what would have happened if we had tried to find y using the other equation, $x^2 + y^2 = 5$. Because it is quadratic in y, we would have got extraneous roots as in the last example in the previous section (try it!).

EXAMPLE: *Find the points of intersection of the circle $x^2 + y^2 = 25$ and the line $4y = 3x + 25$.*

Solve these equations simultaneously. First solve the line for y:

$$y = \frac{3}{4}x + \frac{25}{4}$$

Then substitute into the circle:

$$x^2 + \left(\frac{3}{4}x + \frac{25}{4}\right)^2 = 25$$

which leads to

$$\frac{25}{16}x^2 + \frac{75}{8}x + \frac{(25)^2}{16} - 25 = 0$$

Cancelling 25, multiplying through by 16, and cleaning up eventually gives you:

$$x^2 + 6x + 9 = 0$$

or

$$(x + 3)^2 = 0$$

so

$$x = -3$$

Substituting $x = -3$ into the equation for the line gives $y = 4$. *Therefore the only point of intersection seems to be* $(-3, 4)$

[Checking: $x^2 + y^2 = (-3)^2 + 4^2 = 9 + 16 = 25 \ldots$ O.K.!]

What does it mean that there is only one point of intersection? The only lines that cut a circle once are tangents to the circle, and so that's what we must have here. See Figure 6.8.

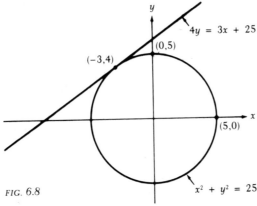

FIG. 6.8

EXAMPLE: *Find the points of intersection of the circle $x^2 + y^2 = 1$ with the line $x + y = 2$.*

Solving $x + y = 2$ for y gives

$$y = 2 - x$$

Substituting into the circle gives

$$x^2 + (2 - x)^2 = 1$$

or

$$2x^2 - 4x + 3 = 0$$

Unfortunately, the discriminant of this equation is −8, so there are no (real) roots. This means that the line does not intersect the equation at all—a fact that is confirmed by the graph in Figure 6.9.

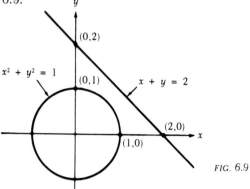

FIG. 6.9

The point of these last three examples is to show how the different numbers of solutions to the simultaneous equations are reflected in the different numbers of points of intersection of the circle and the line. Let's look at an apparently more complicated example, but one that shows the same thing.

EXAMPLE: *Find the points of intersection of the hyperbola*

$$\frac{x^2}{9} - y^2 = 1$$

with each of the ellipses

$$\frac{x^2}{16} + y^2 = 1 \qquad \frac{x^2}{9} + y^2 = 1 \qquad \frac{x^2}{4} + y^2 = 1$$

This time we will draw the graphs first. To begin,

$$\frac{x^2}{9} - y^2 = 1$$

is a hyperbola that opens horizontally and has x intercepts of ±3. On the other hand

$$\frac{x^2}{16} + y^2 = 1$$

is an ellipse that has x intercepts of ±4 and y intercepts of ±1. The other two ellipses have the same y intercepts but have x intercepts of ±3 and ±2, respectively. So what we have are the graphs in Figure 6.10.

Figure 6.10 makes it pretty clear that the hyperbola and the smallest ellipse have no points of intersection. To show this algebraically, solve:

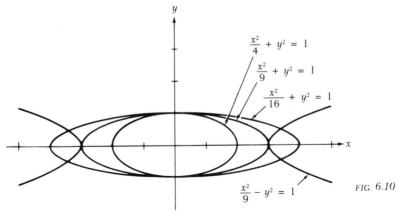

FIG. 6.10

$$\begin{cases} \dfrac{x^2}{4} + y^2 = 1 \\[2mm] \dfrac{x^2}{9} - y^2 = 1 \end{cases}$$

Adding gives

$$\frac{13}{36}x^2 = 2$$

so

$$x^2 = \frac{72}{13} \qquad \left(\text{or } x = \pm\sqrt{\frac{72}{13}} \right)$$

Substituting into the first equation to find y:

$$\frac{1}{4}\left(\frac{72}{13}\right) + y^2 = 1$$

gives

$$y^2 = -\frac{5}{13}$$

This negative value for y^2 tells us that there is no point at which these two graphs cross.

As you might expect from their graphs.

$$\begin{cases} \dfrac{x^2}{9} + y^2 = 1 \\[2mm] \dfrac{x^2}{9} - y^2 = 1 \end{cases}$$

have two points of intersection, namely $(3, 0)$ and $(-3, 0)$, which you can get by solving this system of equations.

The last two,

$$\begin{cases} \dfrac{x^2}{16} + y^2 = 1 \\ \dfrac{x^2}{9} - y^2 = 1 \end{cases}$$

can be solved by adding:

$$\frac{25x^2}{144} = 2$$

$$x^2 = \frac{288}{25}$$

so

$$x = \pm\frac{12\sqrt{2}}{5}$$

Substituting into the ellipse to find y gives

$$y^2 = 1 - \frac{x^2}{16} = 1 - \frac{1}{16}\left(\frac{288}{25}\right) = \frac{7}{25}$$

so

$$y = \pm\frac{\sqrt{7}}{5}$$

Therefore the points of intersection are

$$\left(\frac{12\sqrt{2}}{5}, \frac{\sqrt{7}}{5}\right); \quad \left(\frac{12\sqrt{2}}{5}, \frac{-\sqrt{7}}{5}\right)$$

$$\left(\frac{-12\sqrt{2}}{5}, \frac{\sqrt{7}}{5}\right); \quad \left(\frac{-12\sqrt{2}}{5}, \frac{-\sqrt{7}}{5}\right)$$

Note: This example shows two conics intersectiong in zero, two, and four points. It is also possible for two conics to intersect in one or three points, as you will see in the problem set following this section.

PROBLEM SET   6.3

Find the points of intersection of the following sets of equations. Sketch graphs to check the reasonableness of your answers.

1. $\begin{cases} y = x + 1 \\ y^2 = 2x - 1 \end{cases}$

2. $\begin{cases} x^2 + 3y^2 = 24 \\ x - 2y = 8 \end{cases}$

3. $\begin{cases} x^2 + y^2 = 25 \\ 3x - 4y = 0 \end{cases}$

4. $\begin{cases} y^2 - 4y = 3x - 1 \\ 3y - 7 = 4x \end{cases}$

5. $\begin{cases} 3x^2 = 27 + y^2 \\ x^2 + 45 = y^2 \end{cases}$

6. $\begin{cases} 10x^2 + 6y^2 = 184 \\ x^2 + \dfrac{5y^2}{2} = 26 \end{cases}$

7. $\begin{cases} 9x^2 + y^2 = 80 \\ x^2 + 9y^2 = 80 \end{cases}$

8. $\begin{cases} x^2 + y^2 = 16 \\ y^2 = x + 4 \end{cases}$

9. $\begin{cases} x^2 + y^2 = 16 \\ \dfrac{x^2}{16} - \dfrac{y^2}{16} = 1 \end{cases}$

10. $\begin{cases} x^2 + 4y^2 = 20 \\ x^2 + y^2 = 14 \end{cases}$

11. $\begin{cases} y^2 = 2x^2 - 8 \\ \dfrac{x^2}{4} + \dfrac{y^2}{9} = 1 \end{cases}$

12. $\begin{cases} \dfrac{x^2}{16} + \dfrac{y^2}{36} = 1 \\ \dfrac{x^2}{36} + \dfrac{y^2}{16} = 1 \end{cases}$

13. $\begin{cases} 4x^2 + 9y^2 = 72 \\ 3x^2 - 2y^2 = 19 \end{cases}$

14. $\begin{cases} y = x^2 - x - 1 \\ y = 2(x + \frac{3}{2}) \end{cases}$

15. $\begin{cases} 6x^2 = 14y^2 + 24 \\ 2x - 6y + 4 = 0 \end{cases}$

16. $\begin{cases} 2x^2 - 22 = -5y^2 \\ 3x^2 + 1 = y^2 \end{cases}$

17. $\begin{cases} x^2 + y^2 = 16 \\ 4x^2 - 36y^2 = 144 \end{cases}$

18. $\begin{cases} y^2 - 9x^2 = 9 \\ 9x^2 - y^2 = 9 \end{cases}$

19. Write equations for any two conics that intersect in one point.

20. Write equations for any two conics that intersect in two points.

21. Write equations for two conics that intersect in three points.

22. Write equations for two conics that intersect in four points.

23. Write equations for two circles that touch at only one point.

24. Write equations for an ellipse and a parabola that touch only at one vertex of the ellipse.

25. Write equations for a hyperbola and a parabola that touch at only one point.

6.4   GRAPHING INEQUALITIES IN TWO VARIABLES

You know how to graph an inequality in one variable by using the number line. For example, $x > 7$ is shown in Figure 6.11.

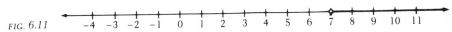

FIG. 6.11

Suppose we are given an inequality in two variables, such as $y > x$. Then a point in the plane satisfies this inequality if its y coordinate is greater than its x coordinate. For example, $(3, 5)$ satisfies the inequality because $5 > 3$, whereas (2, 1) does not, because 1 is not greater than 2. Graphing the inequality means marking all the points in the plane that satisfy it.

Let us look at the line $y = x$. This acts as a divider between those points whose y coordinates are larger than their x coordinates and those points whose x coordinates are larger than their y coordinates. Checking a few points on either side of the line, you can easily find out which side is which. For example, the point (3, 5) is above the line, and any other point that satisfies the inequality will be above the line, too. Therefore the solution to the inequality is the shaded region in Figure 6.12, excluding the line $y = x$.

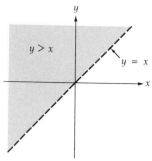

FIG. 6.12

EXAMPLE: *Graph $y - x \leq 3(1 - x)$.*

We must first find the points that lie on the boundary of this region. These are the points that satisfy the equation $y - x = 3(1 - x)$, which we can rewrite as $y = -2x + 3$ (a line with slope -2, y intercept 3, x intercept $\frac{3}{2}$). Our inequality is equivalent to $y \leq -2x + 3$, which is satisfied by points such as (0, 0) and (−1, 2). Therefore we want the region below the line $y = -2x + 3$. See Figure 6.13. This time the points on the line $y = -2x + 3$ are included in the solution of the inequality because we do want the points with $y - x = 3(1 - x)$ as well as those with $y - x < 3(1 - x)$.

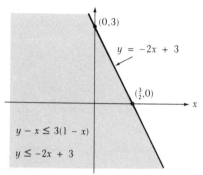

FIG. 6.13

EXAMPLE: *Graph the solution to $y < 1 - x$ and $4 - y \leq 2 - x$.*

Notice the "and," which means that we are looking for the points (x, y) whose coordinates satisfy *both* inequalities at once. To find these, we will graph the solutions to each inequality and then look for the overlap.

For $y < 1 - x$,   first   graph   the   line $y = 1 - x$ or $y = -x + 1$ (slope -1, y intercept 1, x intercept 1) and shade the side containing points such as $(0, 0), (-1, 0)$, which satisfy the inequality.

For $4 - y \leq 2 - x$,   first   graph   the   line $4 - y = 2 - x$ or $y = x + 2$ (slope 1, y intercept 2, x intercept -2).   The inequality $4 - y \leq 2 - x$ is equivalent to $y \geq x + 2$, so we want the side of the line containing points such as $(0, 4), (-2, 3)$, and we want to include the line $y = x + 2$. The graphs are shown in Figure 6.14.

The doubly shaded region is the solution to the problem.

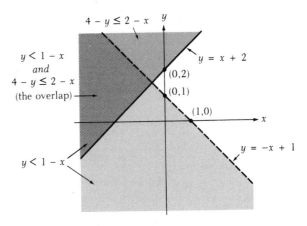

FIG. 6.14

EXAMPLE: *Graph $x^2 + y^2 \leq 1$, $x^2 + y^2 \geq 4$ on the same axes.*

If $x^2 + y^2 \leq 1$, then $\sqrt{x^2 + y^2} \leq 1$, so the distance from the point (x, y) to the origin is less than or equal to 1. This means that the point is inside the circle $x^2 + y^2 = 1$. The points that satisfy the inequality are therefore those inside or on this circle.

If $x^2 + y^2 \geq 4$, then $\sqrt{x^2 + y^2} \geq 2$, so the point (x, y) is outside or on the circle $x^2 + y^2 = 4$. These regions are shown in Figure 6.15.

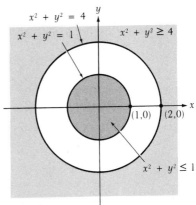

FIG. 6.15

PROBLEM SET 6.4

Graph the solutions to the following inequalities:

1. $y + x \geqslant 2(2 + x)$
2. $2y - 4 \leqslant x$
3. $x^2 + y^2 \leqslant 9$
4. $y^2 \geqslant (x - 3)^2$
5. $\dfrac{x^2}{4} + \dfrac{y^2}{9} \leqslant 1$
6. $\dfrac{x^2}{16} - \dfrac{y^2}{25} \geqslant 1$

7. $(y - 3) \geqslant 2(x - 4)^2$
8. $-2y \leqslant (x + 1)^2$
9. $3y^2 \leqslant 9x^2$
10. $(x - 2)^2 + (y - 3)^2 \geqslant 1$
11. $2y + 2 \leqslant 3x + 3$
12. $y \leqslant 2x + 3$ and $y \geqslant -x + 4$
13. $y \leqslant 3 - x$ or $y \leqslant 2x$

14. $x^2 + y^2 \leqslant 1$ or $(x - 3)^2 + (y - 4)^2 \geqslant 25$
15. $(x - 2)^2 + (y - 2)^2 > 4$ and $y > x + 1$
16. $\dfrac{x^2}{4} - \dfrac{y^2}{9} \leqslant 1$ and $\dfrac{y^2}{4} - \dfrac{x^2}{9} \geqslant 1$
17. $2x^2 + 2y^2 \leqslant 5$ and $x^2 + y^2 \geqslant 3$
18. $(x - 3)^2 + (y - 4)^2 \leqslant 16$ and $(x - 4)^2 + (y - 3)^2 \leqslant 16$
19. $\dfrac{x}{2} + \dfrac{y}{3} \geqslant 1$ and $\dfrac{x^2}{4} + \dfrac{y^2}{9} \geqslant 1$
20. $\dfrac{(x - 3)^2}{16} - \dfrac{(y - 2)^2}{9} \leqslant 1$ and $y \geqslant 2x$
21. $(x + 5)^2 + (y - 1)^2 \geqslant 75$ and $5x^2 + 7y^2 \leqslant 35$
22. $9(x + 1)^2 + 4(y - 3)^2 \leqslant 36$ and $x^2 + y^2 \geqslant 49$
23. $\dfrac{(x + 1)^2}{\frac{9}{4}} + \dfrac{(y - 2)^2}{\frac{25}{4}} \geqslant 1$ and $\dfrac{(x - 3)^2}{12} + \dfrac{(y + 4)^2}{36} \leqslant 1$
24. $16x^2 + 32x + 9y^2 - 36y \leqslant 92$ and $\dfrac{x^2}{16} - \dfrac{y^2}{9} = 1$
25. $\dfrac{x^2}{25} - \dfrac{y^2}{36} \leqslant 1$ and $\dfrac{y^2}{36} - \dfrac{x^2}{25} \leqslant 1$

CHAPTER 6 REVIEW

Solve the following systems of equations, using graphs to make sure your answers are reasonable:

1. $\begin{cases} x + y = 9 \\ 2x - y = 3 \end{cases}$
2. $\begin{cases} 2x + 3y = 7 \\ 5x + 2y = 12 \end{cases}$
3. $\begin{cases} 3x - 4y = 10 \\ 2x + 3y = 18 \end{cases}$

4. $\begin{cases} 5x + 3y = 29 \\ x + 2y = 3 \end{cases}$
5. $\begin{cases} 2x + 11y = \frac{13}{2} \\ x - y = 0 \end{cases}$
6. $\begin{cases} 4x - 7y = 3 \\ 3x + \frac{9}{2}y = -6 \end{cases}$

7. $\begin{cases} 13x - 15y = -49 \\ 2x + 3y = 19 \end{cases}$
8. $\begin{cases} \dfrac{2}{x} - 2 = \dfrac{3}{y} \\ \dfrac{4}{x} + 9 + \dfrac{7}{y} = 0 \end{cases}$

9. $\begin{cases} \dfrac{1}{x+y} - \dfrac{1}{2} = 0 \\[2mm] \dfrac{1}{x-y} - \dfrac{1}{4} = 0 \end{cases}$

10. $\begin{cases} y = \dfrac{x+1}{x+2} \\[2mm] y = 2x \end{cases}$

11. $\begin{cases} y = 2x^2 - 1 \\[2mm] y = \dfrac{1}{x} \end{cases}$

12. $\begin{cases} y = x^3 - 1 \\ y = 1 - x^3 \end{cases}$

13. $\begin{cases} 2y^2 = x - 3 \\ y = 2x - 6 \end{cases}$

14. $\begin{cases} y = \dfrac{1}{(x-3)^2} \\[2mm] y = (x-3)^2 \end{cases}$

15. $\begin{cases} (x-1)^2 + (y-2)^2 = 4 \\ (x+1)^2 + (y+2)^2 = 4 \end{cases}$

16. $\begin{cases} 3(x+2)^2 + 12(y+4)^2 = 48 \\ 3(x+2)^2 - 12(y+4)^2 = 48 \end{cases}$

17. $\begin{cases} y^2 - 64x^2 = 64 \\ 64x^2 + y^2 = 17 \end{cases}$

18. $\begin{cases} \dfrac{(x+1)^2}{9} - \dfrac{(y+2)^2}{4} = 1 \\[2mm] \dfrac{(y+2)^2}{4} - \dfrac{(x+1)^2}{9} = 1 \end{cases}$

19. $\begin{cases} x - 3 = (y-2)^2 \\ (x-3)^2 + (y-2)^2 = 9 \end{cases}$

20. $\begin{cases} (x-4) = (y-4)^2 \\ (x-4)^2 = (y-4) \end{cases}$

21. $\begin{cases} y = 2(x - \frac{1}{2})^2 \\ y = 2(x + \frac{1}{2})^2 \end{cases}$

22. $x + y \geq 3$ and $2x - 2y \leq 5$

23. $3x - 7y \leq 8$ or $5x + 4y \geq 6$

24. $x^2 + y^2 \leq 4$ and $y \geq 2x^2$

25. $4(x-1)^2 + 9(y-7)^2 \leq 36$ and $(x+1)^2 + (y-6)^2 \geq 16$

26. $\dfrac{(x-3)^2}{49} - \dfrac{(y-3)^2}{64} \leq 1$ and $\dfrac{(x-4)^2}{49} + \dfrac{(y-4)^2}{64} \leq 1$

27. $x^2 + 2y^2 \leq 4$ and $6x^2 + 10x \leq 4$

28. $x - 3y \geq 2$ and $x + 3y \geq 2$ and $y \leq 6$

29. Solve simultaneously:

$$\frac{x^2}{a^2} - \frac{y^2}{b^2} = 1$$

$$y = \frac{b}{a}x$$

What does this tell you about a hyperbola and its asymptotes?

30. A ship at sea can determine its position by measuring its distance from radio beacons at known locations. If the ship is 4 miles from a beacon located at $(1, 2)$ and 3 miles from a beacon at $(4, 5)$, what are its possible positions?

31. On some navigation maps the United States is laid out on a Cartesian plane, and cities and towns are known by their coordinates rather than their names. Last year, all the cities above the line joining New York $(15, 11)$ and San Francisco $(-14, 7)$ had miserable winter weather, while the part of the country west of the line joining Minneapolis $(-1, 10)$ and Houston $(-2, -9)$ had a bad summer. What part of the country had bad weather all year round?

32. It is possible to find the equation of a circle containing three non-collinear points. Use the following hints and the suggested procedure to find the equation of the circle containing the points $(9, 7)$, $(3, 9)$ and $(-5, -7)$.

Hint 1: If a point is equidistant from A and B, and is equidistant from B

and C, then it is equidistant from A, B, and C; that is, it is the center of the circle containing A, B, and C.

Hint 2: The set of points equidistant from A and B lies on the line that bisects the line segment AB and is perpendicular to the segment.

(a) Find the line containing the points equidistant from (9, 7) and (3, 9).

(b) Find the line containing the points equidistant from (9, 7) and (−5, −7).

(c) Find the intersection of the lines found in (a) and (b) above.

(d) How far is the point you found in (c) from (9, 7)?

(e) Use the information in (c) and (d) to find the equation of the circle.

7   FUNCTIONS

7.1   DEFINITION OF A FUNCTION

The word "function" is used in everyday speech in statements like "the amount a person earns is a function of how many years of schooling he or she has had" or "the number of votes cast for a candidate is a function of the amount of money spent in his or her campaign." You might also say to a newcomer who asked you how much a keg of beer cost in your part of the country that the price was a function of where you bought it.

In each case we mean that one quantity in some way depends on or is determined by the other. For example, the amount a person earns depends on how much education he or she has had because the longer we stay in school the higher our salary will be (or so they keep telling us!). Similarly, the number of votes a candidate gets is definitely affected by the amount spent in his or her election campaign, which is why the government decided to limit campaign spending. And in the third example, the price of a keg of beer depends on whether you buy it at a discount store or a regular store.

In mathematics, the word "function" is used in much the same way, except that it is restricted to the cases in which one quantity is completely and unambiguously determined by the other. This is not so in the first two examples. A person's level of education does not by itself determine how much he or she will earn, nor does knowing how much was spent in a politician's election campaign tell you exactly how many votes he or she will get. In the third example, however, the place where the keg is bought does in fact determine the price. Therefore, mathematicians would say that the first two examples are not functions, but they would agree that the price of a keg of beer is a function of where you buy it. In the same way mathematicians would agree that the area of a circle is a function of its radius, meaning that the area depends on and is completely determined by its radius. In other words, once the radius of a circle is known, the area of the circle is completely determined (by using the formula $A = \pi r^2$). And (since we were bound to get back to x's and y's sometime!) suppose x and y are related by the formula $y = x^2$. Then y is a function of x because the value of y is completely and unambiguously determined by the value of x. For instance, if $x = 1$, than y has to be 1; if $x = 3$, then y must be 9; and so on for every other value of x. In general, we say that:

> *y is a function of* x if the value of y depends on and is completely determined by the value of x. Then x is called the *independent variable; y* is the *dependent variable.*

This should start to remind you of something. Many times in the last few chapters we have looked at equations that gave us y in terms of x. We used such an equation to find the values of y given by particular values of x, and wrote them down in a table of values. We then drew a graph to show just how y depended on x. Therefore, in all the examples where y was given by an equation of the form

$$y = \text{(some expression involving } x\text{'s but no } y\text{'s)}$$

y was a function of x. So even though you may not have realized it, you've actually been working with functions for quite a while.

What For?

The reason that functions were studied in the first place is simply that they're very useful. As you can see, we're frequently concerned with the relationship between two quantities. But besides that, calculus is nothing but the study of functions—how they behave and how they vary—so it is essential to have an idea of what functions are before getting to calculus.

How Are Functions Represented?

Given that functions are useful, the next thing we need is a way of writing them down so that we can work with them easily. We know that if y is a function of x, then the value of y is determined by the value of x. So the question is: How should we represent the relationship between x and y? How should we show which values of y come from which values of x?

There are three main ways.

A List or Table   In the case of the beer drinker, a list of all the distributors and the prices they charge would make very clear the relation between the price of a keg and where it was bought. A table of values is also a list. Unfortunately, it can't include all the possible values of x, and so it doesn't tell you everything about how x and y are related.

A Graph   The reason we spent all that time drawing graphs is that they are a particularly vivid way of showing how y depends on x. For example, at a moment's glance you can tell whether large values of x lead to large or small values of y; what y is when $x = 0$; and so on. However, a graph cannot always tell you accurately the value of y coming from a given value of x, because when you're reading a graph it's difficult to be sure whether the y coordinate of a certain point is 2, or 2.1 or 2.01, or what.

A Formula   The easiest way of representing most functions is by a formula. For example,

$$y = x^2$$

tells you exactly how the y value is related to the x value—it's the square of the x value. The formula is both accurate and easy to write down. This is the way we'll represent functions.

Examples of Functions

EXAMPLE: *If $p - q = 5$, is p a function of q? Is q a function of p?*

You can rewrite the equation as

$$p = q + 5$$

which tells you that whatever value q has, p is five greater (for example, if $q = 1$, $p = 6$; if $q = 3$, $p = 8$). Certainly once we have decided on a value for q, the value of p is completely determined, so *p is a function of q.*

On the other hand, you can write $p - q = 5$ as

$$q = p - 5$$

This tells you how to find the value of q if you know the value of p. A value for p will determine the value of q (because q has to be five less than p), and so *q is a function of p* also.

EXAMPLE: *If $|m| = t$, is m a function of t? Is t a function of m?*

Here m and t are certainly related to one another by the equation $|m| = t$. For m to be a function of t, the only question is whether or not the value of t completely determines the value of m. Suppose we pick a particular value of t, say 3. Then $|m| = 3$, so $m = 3$ or -3. Since we have got two values of m from one value of t, we have to conclude that the value of m is not completely determined by the value of t. Therefore *m is not a function of t.*

On the other hand, if we start with a value for m, say 5, then $t = |5| = 5$, so we get just one value for t. Whatever value we take for m, we will get only one value for t. Therefore, *t is a function of m*.

Definitions of Domain and Range

Domain   Suppose we are considering the area A of a circle as a function of its radius r. Then r can have any positive value, or zero, but it cannot be negative (because a negative radius makes no sense). We say that the domain of this function is all the positive real numbers and zero.

In the function represented by the formula

$$y = \frac{1}{x}$$

the indepent variable, x, can be anything except zero (because $\frac{1}{0}$ is undefined). In this case we say that the domain of this function is all the real numbers except zero.

In general:

> *The domain of a function* is all the possible values of the independent variable.

Range   Now consider the function represented by $y = x^2$. No matter what value you choose for x, this formula will only give you positive values or zero for y. We say that the range of this function is all the positive real numbers and zero.

In general:

> *The range of a function* is all the possible values of the dependent variable, got by substituting all possible values of the independent variable.

EXAMPLE: *Find the domain and range of $y = -\sqrt{x}$.*

Here x must be positive or zero (assuming we want only real values of y). So

domain = all positive numbers and zero

The $\sqrt{}$ always gives the positive square root of any number, so if $y = -\sqrt{x}$, y will always come out to be negative or zero. Hence,

range = all negative numbers and zero

PROBLEM SET 7.1

For each of the following, state (a) whether y is a function of x; (b) whether x is a function of y.

1. $y = 2x$
2. $y - 1 = x + 5$
3. $y = x^2$
4. $3y = 2x^2 - 1$
5. $y = |x - 1|$
6. $|y| = x + 1$
7. $y = |x^2|$
8. $y = x^3 - x^2$
9. $x + y = 7$
10. $y^2 = x - 1$
11. $y = \sqrt{x + 4}$
12. $\sqrt{y} = x - 3$

Find the domain and range of the following (x is the independent variable):

13. $y = 2x - 4$
14. $y = 3 - x$
15. $y = \sqrt{x - 7}$
16. $y = \sqrt{7 - x}$
17. $y = -x^2 + 1$
18. $y = \dfrac{1}{x}$
19. $y = \dfrac{1}{2x - 3}$
20. $y = \dfrac{1}{\sqrt{x + 3}}$
21. $y = \dfrac{1}{\sqrt{x - 3}}$
22. $y = \dfrac{1}{(x + 1)(x - 1)}$
23. $y = -\sqrt{x + 30}$
24. $y = \sqrt{x^2 + 3x - 2}$
25. $y = |x - 4|$
26. $y = |x^2 + 1|$
27. $y = \dfrac{x}{x + 2}$
28. $y = \dfrac{x - 3}{x - 4}$
29. $y = -\sqrt{2x - 8}$
30. $y = \dfrac{\sqrt{3x - 12}}{\sqrt{3x - 11}}$

For each of the following, state (a) whether v is a function of w; (b) whether w is a function of v.

31. $vw = 4$
32. $vw^2 = 4 - v$
33. $\sqrt{v + 1} - \sqrt{2w} = 8$
34. $|v - 3| = w + 8$
35. $v^2 + w^2 = 81$
36. $w = |\sqrt{v^2 - v}|$
37. $w = \begin{cases} 1 \text{ if } v \text{ is an integer} \\ -1 \text{ if } v \text{ is not an integer} \end{cases}$
38. $v = \begin{cases} w - 1 \text{ if } w \le 3 \\ 2 \quad\;\; \text{ if } w > 3 \end{cases}$
39. $w = \begin{cases} v + 1 \text{ if } v \ge 2 \\ v - 1 \text{ if } v < 2 \end{cases}$
40. $w = 0.08 \cdot 10^2$
41. $\dfrac{v}{\sqrt{w^3 + 1}} = 2$

Find the domain and range of each of the following (t is the independent variable):

42. $u = \dfrac{t}{(t + 2)(t - 2)}$
43. $u = \sqrt{\tfrac{1}{9} - t^2}$
44. $u = \sqrt{(t - 2)(t - 4)}$
45. $ut = 4$
46. $u = \dfrac{t^2 - 3t + 5}{t}$
47. $t^2 u = 7 - t$
48. $u = \sqrt{t^3 + 1}$
49. $u = \sqrt[3]{t^3 + 8}$
50. $u = |t^2 - 1|$

7.2   FUNCTIONAL NOTATION: $f(x)$

Consider the function given by the formula

$$y = x^2 + 1$$

Then

$$x = 0 \quad \text{leads to} \quad y = 1$$
$$x = 2 \quad \text{leads to} \quad y = 5$$
$$x = -3 \quad \text{leads to} \quad y = 10$$

and so on.

The question now arises as to the best way of communicating what x a given y came from. The x value could be written in brackets after the y value, or it could be written as a subscript to the y value, or in any one of a million other ways. What is actually done is this. Suppose the function is called f, meaning that f assigns to every number, x, its square plus one, or y. Then the y value that you get from $x = 0$ will be written as $f(0)$, read "f of zero." Since $x = 0$ leads to $y = 1$,

$$f(0) = \{y \text{ value corresponding to } x = 0\} = 1$$

so

$$f(0) = 1$$

Similarly, $x = 1$ leads to $y = 2$.

$$f(1) = \{\text{the } y \text{ value corresponding to } x = 1\} = 2$$

so

$$f(1) = 2$$

and

$$f(-3) = \{\text{the } y \text{ value corresponding to } x = -3\} = 10$$

so

$$f(-3) = 10$$

What about $f(7)$? Well, $f(7)$ is the y that goes with 7, which is $7^2 + 1 = 50$, so

$$f(7) = 7^2 + 1 = 50$$

By the same reasoning,

$$f(-5) = \{\text{the } y \text{ corresponding to } x = -5\}$$

so

$$f(-5) = (-5)^2 + 1 = 26$$

In general,

$$f(a) = \{\text{the } y \text{ corresponding to } x = a\}$$

so

$$f(a) = a^2 + 1$$

It is usual to use x instead of a, so the above equation is often written

$$f(x) = x^2 + 1$$

meaning that $f(x)$, the y value corresponding to x, is given by the formula $x^2 + 1$.

$f(x)$ is called *functional notation*.

To remind you that $f(x)$ is a y value, you may also come across

$$y = f(x)$$

This equation tells you that y is a variable that is a function of x.

You may also see the equation

$$y = f(x) = x^2 + 1$$

which tells you that y is a function of x *and* how to find y given any x.

In summary:

$$y = f(x)$$

means that $f(x)$ is the y value corresponding to x under the function f.

Notes:

1. $f(x)$ *does* not *mean* f *times* x. The expression $f(x)$, or "f of x," means the y value that goes with x, and has nothing to do with multiplication.

For example, if $f(x) = x^2 + 1$ then

$$f(\text{any number}) = (\text{that number})^2 + 1$$

so

$$f(0) = \{\text{the } y \text{ corresponding to } x = 0\} = 1$$

But f times 0, or $f \cdot 0$, if it meant anything, would have to be 0.

So f of zero, $f(0)$, and f times zero, $f \cdot 0$, are definitely different.

2. f *as opposed to* $f(x)$. The distinction between these two is that f is the name for the whole function, i.e. the relationship itself, whereas $f(x)$ is the name of a particular y value, namely that arising from x.

Similarly, g and h denote functions and $g(x)$ and $h(x)$ [or $h(s)$, $h(t)$, etc.] denote values of the dependent variables of g and h.

3. *Dummy variables.* The particular letter used to denote the independent variable in a function matters very little. For example, the functions

$$f(x) = x^2 + 1$$

and

$$f(t) = t^2 + 1$$

both mean

$$f(\text{any number}) = (\text{that number})^2 + 1$$

The x or the t is called a *dummy variable* because all it is trying to tell you is that you must put in the same number on the left as on the right, but the letter you use for that number is irrelevant. Certainly the name "dummy variable" leaves you in little doubt as to its status!

4. *Evaluating a function at x means finding $f(x)$.*

EXAMPLE: *If $f(x) = x^2 - 2$, evaluate the function at the values shown.*

$$f(0) = 0^2 - 2 = -2$$
$$f(-2) = (-2)^2 - 2 = 4 - 2 = 2$$
$$f(\sqrt{2}) = (\sqrt{2})^2 - 2 = 2 - 2 = 0$$

Now $x = \sqrt{2}$ is called *a zero of the function $f(x) = x^2 - 2$* because the function evaluated at $\sqrt{2}$ is zero. Also

$$f(p) = p^2 - 2$$
$$f(p^2) = (p^2)^2 - 2$$

[Here we replaced x by p^2 because the function says: $f(\text{any number}) = (\text{that number})^2 - 2$ and that number can be p^2]. Also

$$f(a + 4) = (a + 4)^2 - 2$$

[Here we substituted $(a + 4)$ for x on both sides of $f(x) = x^2 - 2$.]

$$f(a) + 4 = \underbrace{(a^2 - 2)}_{f(a)} + 4 = a^2 + 2$$

Please look at the difference between $f(a + 4)$ and $f(a) + 4$. Whatever is in the parentheses after the f is what has been substituted for x, so in $f(a + 4)$ you must substitute $(a + 4)$ for x in the formula for $f(x)$. For $f(a) + 4$, the only thing in the brackets after the f is a, so you substitute a in the formula to find $f(a)$. Then, after you have found $f(a)$, you add 4 to it to get $f(a) + 4$.

EXAMPLE: *If $f(x) = \dfrac{1}{x^2 - 4}$ evaluate the function at the values shown.*

$$f(-1) = \frac{1}{(-1)^2 - 4} = \frac{1}{1 - 4} = -\frac{1}{3}$$

$$f(s) = \frac{1}{s^2 - 4}$$

$$f(s^2) = \frac{1}{(s^2)^2 - 4}$$

In this case we replaced x by s^2, since the function says:

$$f(\text{any number}) = \frac{1}{(\text{that number})^2 - 4}$$

On the other hand,

$$[f(s)]^2 = \left[\frac{1}{s^2 - 4}\right]^2 = \frac{1}{(s^2 - 4)^2}$$

Notice that there is a difference between $f(s^2)$ and $[f(s)]^2$. The same thing happens with $f\left(\frac{1}{s}\right)$ and $\frac{1}{f(s)}$:

$$f\left(\frac{1}{s}\right) = \frac{1}{\left(\frac{1}{s}\right)^2 - 4} \qquad \left(\text{replacing } x \text{ by } \frac{1}{s}\right)$$

$$= \frac{1}{\dfrac{1 - 4s^2}{s^2}}$$

$$= \frac{s^2}{1 - 4s^2}$$

But

$$\frac{1}{f(s)} = \frac{1}{\dfrac{1}{s^2 - 4}} \qquad \left[\text{writing } \frac{1}{s^2 - 4} \text{ for } f(s)\right]$$

$$= s^2 - 4$$

so $f\left(\frac{1}{s}\right)$ and $\frac{1}{f(s)}$ are certainly different.

More complicated x's work the same way:

$$f(a + h) = \frac{1}{(a + h)^2 - 4}$$

$$f(\sqrt{a + 4}) = \frac{1}{(\sqrt{a + 4})^2 - 4} = \frac{1}{(a + 4) - 4} = \frac{1}{a}$$

You should also look at

$$f(2) = \frac{1}{2^2 - 4} = \frac{1}{4 - 4} = \frac{1}{0} \qquad \text{which is undefined}$$

So $f(2)$ is undefined, and, for the same reason, so is $f(-2)$. These are the values of x for which the function simply does not make sense. The domain of a function is all the possible x values, and so *the domain of this function is all real numbers except 2 and −2* (these being the only two numbers that make the denominator zero).

EXAMPLE: *If $f(x) = x^2$, find and simplify*

$\qquad$ (a) $f(a + 1) - f(a)$ $\qquad$ (b) $f(a + h) - f(a)$ $\qquad$ (c) $\dfrac{f(a + h) - f(a)}{h}$

$\qquad$ (a) $f(a + 1) - f(a) = (a + 1)^2 - a^2$

$$= a^2 + 2a + 1 - a^2$$

$$= 2a + 1$$

$\qquad$ (b) $f(a + h) - f(a) = (a + h)^2 - a^2$

$$= a^2 + 2ah + h^2 - a^2$$

$$= 2ah + h^2$$

$\qquad$ (c) $\dfrac{f(a + h) - f(a)}{h} = \dfrac{2ah + h^2}{h} \qquad$ [using answer to (b)]

$$= \frac{\cancel{h}(2a + h)}{\cancel{h}}$$

$$= 2a + h$$

EXAMPLE: *If $g(x) = \dfrac{x + 2}{x + 1}$, find the domain of g and solve the equation $g(x) = 3$.*

The domain of g is all the x values for which the function makes sense, which is all the real numbers except −1, because −1 makes the denominator zero.

Solving the equation $g(x) = 3$ means finding what values of x make $y = 3$. Since

$$g(x) = \frac{x + 2}{x + 1}$$

$g(x) = 3$ means

$$\frac{x + 2}{x + 1} = 3$$

and this is an ordinary equation that we can solve the ordinary way. We get

$$x + 2 = 3(x + 1)$$

or

$$x + 2 = 3x + 3$$

so

$$-1 = 2x$$

or

$$x = -\frac{1}{2}$$

CHECK:

$$g\left(-\frac{1}{2}\right) = \frac{-\frac{1}{2} + 2}{-\frac{1}{2} + 1} = \frac{\frac{3}{2}}{\frac{1}{2}} = \frac{3}{2} \cdot \frac{2}{1} = 3$$

So $x = -\frac{1}{2}$ does correspond to $y = 3$, that is,

$$x = -\frac{1}{2} \quad \text{does satisfy} \quad g(x) = 3.$$

EXAMPLE: *If $h(x) = x^2 - 26$ and $k(x) = 1 - 2x^2$, for what values of x is $h(x) < k(x)$?*

Here we are taking any value of x and comparing the two y values that h and k associate with that x. We are looking for the x's that make the y value of k larger than the y value of h.

Using the formulas for $h(x)$ and $k(x)$, we see that if

$$h(x) < k(x)$$

then

$$x^2 - 26 < 1 - 2x^2$$

so

$$3x^2 < 27 \qquad \text{or} \qquad x^2 < 9$$

which means

$$-3 < x < 3$$

that is, $-3 < x$ and $x < 3$

PROBLEM SET   7.2

1. Let $y = \dfrac{x}{x+1}$.
   (a) If $x = 2$, what is y?
   (b) If $x = 3$, what is y?
   (c) If $y = 2$, what is x?
   (d) If $y = 3$, what is x?

2. Let $f(x) = \dfrac{x}{x+1}$.
   (a) If $x = 2$, what is $f(x)$?
   (b) If $x = 3$, what is $f(x)$?
   (c) Find x such that $f(x) = 2$.
   (d) Solve for x: $f(x) = 3$.

3. Let $y = \sqrt{25 - x^2}$.
   (a) If $x = 3$, what is y?
   (b) If $x = 5$, what is y?
   (c) If $y = 3$, what is x?
   (d) If $y = 5$, what is x?

4. Let $f(x) = \sqrt{25 - x^2}$.
   (a) If $x = 3$, what is $f(x)$?
   (b) What is $f(5)$?
   (c) Find x such that $f(x) = 3$.
   (d) Solve for x: $f(x) = 5$.

Let $f(x) = 2x + 1$. Find:

5. $f(0)$

6. $f(-4)$

7. $f(100)$

8. $f(a)$

9. $f(a^2)$

10. $[f(a)]^2$

11. $f\left(\dfrac{1}{a}\right)$

12. $\dfrac{1}{f(a)}$

13. x such that $f(x) = 5$

14. x such that $f(x) = \frac{1}{5}$

15. a such that $f(a^2) = 33$

16. a such that $f\left(\dfrac{1}{a}\right) = 3$

17. If $f(x) = 2x + 1$ and $g(x) = 3x - 2$, find x such that $g(x) = f(x)$.

Let $g(x) = x^2 - 4x + 3$. Find:

18. $g(0)$

19. $g(1)$

20. x such that $g(x) = 0$

21. x making $g(x) = 15$

22. x making $g(x) = 3$

23. $g(p^2)$

24. $[g(p)]^2$ (Do not simplify.)

Let $h(b) = \dfrac{1}{b^2 + 9}$. Find:

25. $h(2)$

26. $h(a)$

27. $h(\sqrt{a})$

28. $h(a^2)$

29. $[h(a)]^2$

30. $h\left(\dfrac{1}{a}\right)$

31. $\dfrac{1}{h(a)}$

32. x if $h(x) = \frac{1}{13}$

33. x if $h(x) = \frac{1}{25}$

Let $f(x) = x^3$. Find:

34. $f(p) + 1$

35. $f(p + 1)$

36. $f(p + 1) - f(p)$

37. $f(a + h)$

38. $\dfrac{f(a + h) - f(a)}{h}$

Let $f(x) = \dfrac{1}{x^2 + 1}$. Find:

39. $f(\sqrt{x})$

40. $f(x^2)$

41. $[f(x)]^2$

42. $f(x + 1)$

43. $f(x^2 + 1)$

44. $f(x^2) + f(1)$

For any function $f(x)$, define a new function $\hat{f}(x)$ by $\hat{f}(x) = f(x + 1) - f(x)$. What is $\hat{f}(x)$ if:

45. $f(x) = 2x$

46. $f(x) = x^2$

47. $f(t) = \dfrac{1}{t}$

48. $f(t) = \sqrt{t}$

49. A square-bottomed box has depth twice the size of its base. Write the volume, v, as a function of x, the side of the base.

50. A star is made by cutting quarter circles from a square sheet of cardboard of side x, as in Figure 7.1. Find $A(x)$, the area of the star as a function of x.

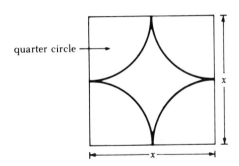

quarter circle

FIG. 7.1

If $f(x) = 0.1x + x^2$, evaluate:

51. $f(0.1)$

52. $f(1.2 \cdot 10^{-1})$

53. $f(10^{-2})$

54. $f(-0.2)$

55. $f(-3 \cdot 10^{-1})$

If $g(x) = \dfrac{(a - x)(a + x)}{1 - b}$, evaluate and simplify:

56. $g(0)$

57. $g(ab)$

58. $g(a\sqrt{b})$

CHAPTER 7 REVIEW

Let $f(x) = x^2 - 8x + 7$. Find the value(s) of x for which:

1. $f(x) = 0$

2. $f(x) = 7$

3. $f(x) = -5$

4. $f(x) = -9$

5. $f(x) = -13$

For any function $f(x)$, define a new function $\hat{f}(x)$ by

$$\hat{f}(x) = \frac{f(x+a) - f(x)}{a}$$

where a is a constant. What is $\hat{f}(x)$ if:

6. $f(x) = 2x + 1$

7. $f(x) = x^2 + 2$

8. $f(x) = \frac{x}{3} - 1$

9. $f(x) = 2x^3$

10. $f(x) = \frac{x+1}{x}$

11. If $f(x) = 2x + 3$, find $f(0)$ and solve $f(x) = 0$.

12. If $h(x) = (x+1)^2$, solve $h(a+1) = 25$ for a.

For each of the following functions, solve $f(x) + f(0) = 0$ for x.

13. $f(x) = 2x$

14. $f(x) = x^2 - 1$

15. $f(x) = \frac{x}{3}$

16. $f(x) = \frac{x^2}{4} - 12$

17. $f(x) = \sqrt{x}$

18. $f(x) = \sqrt{x^2 + 1} - 2$

A rectangular box with a lid has width 2 feet more than its depth, and length 5 feet more than its depth. Let $d =$ depth, and $s =$ total surface area of box.

19. Write s as a function of d.

20. Find $s(1)$.

21. Find the depth of the box when the total surface area is 100 square feet.

22. Write the volume, v, of the box as a function of d.

23. Find $v(2)$.

24. Can you find an x such that $v(x) = 120$?

Let $g(t) = \frac{t+1}{t-2}$.   Find:

25. $g(t+1)$

26. $g(t) + g(1)$

27. $g\left(\frac{t+1}{2}\right)$

28. $\frac{g(t+1)}{g(2)}$

29. $g\left(\frac{t^2+2}{t-1}\right)$

30. $g\left(\frac{a^3}{a+4}\right)$

Let $h(v) = \frac{(v+1)^2}{2v}$.   Find:

31. $h(x^2 - 3x)$

32. $h(\sqrt{v})$

33. $h\left(\frac{v-3}{v+2}\right)$

34. $\frac{h(v-3)}{h(v+2)}$

35. $h[(a+1)^2]$ (Do not simplify.)

36. v such that $h(v-1) = 0$

37. If $f(x) = (x - 1)^2$, simplify

$$\frac{[f(x)] \cdot [f(0)]}{(x - 1)}$$

38. The profit from selling a car is one-third of the cost, minus a share of overhead expense. This share is computed by dividing the total overhead by the number of cars sold. If cars cost the dealer $3600 and a dealer's total overhead is $16,800, determine the profit per car as a function of the number of cars sold.

39. How many cars must the dealer in Problem 38 sell to make a profit of $150 per car?

40. If first-class postage costs 13¢ for the first ounce or fraction thereof, and 11¢ for each additional ounce or fraction thereof, write postage needed as a function of weight.

41. Show that if $f(x) = 2x - 4$, then

$$f\left(\frac{a + c}{2}\right) = \frac{f(a) + f(c)}{2}$$

42. Is the result in Problem 41 true for any function of the form $f(x) = mx + b$? Show why or why not.

43. Is the result in Problem 41 true for functions of the form $f(x) = x^2$? Show why or why not.

44. If $f(x) = \frac{1}{x}$, show that $f\left(\frac{a + c}{2}\right) = 2f(a + c)$.

45. If $f(x) = \frac{1}{x^2}$, show that $f(b \cdot d) = f(b) \cdot f(d)$.

8   FUNCTIONS AND GRAPHS

8.1   PUTTING EVERYTHING INTO FUNCTIONAL NOTATION

Although you may not realize it, you already know a great deal about functions. The chapters on graphing were all about how y could depend on x in various ways, and so you saw a great many functions, though we never called them that. In particular, you may recognize the following functions.

The *linear function* which has the form

$$f(x) = mx + b \qquad (m, b \text{ constants and } m \neq 0)$$

for example,

$$f(x) = 2x - 3$$

All linear functions have straight-line graphs.

The *quadratic function*, which has the form

$$f(x) = ax^2 + bx + c \qquad (a, b, c \text{ constants and } a \neq 0)$$

for example,

$$f(x) = -x^2 + 2x + 3$$

All quadratic functions have parabolic graphs.

The *cubic function* which has the form

$$f(x) = ax^3 + bx^2 + cx + d \qquad (a, b, c, d \text{ constants and } a \neq 0)$$

for example,

$$f(x) = 2x^3 - 3x^2 + 4x - 5$$

In a previous chapter you also met the *reciprocal function,*

$$f(x) = \frac{1}{x}$$

If we feel like being more general, we can say that the first three functions above are examples of *polynomial functions,*

$$f(x) = a_n x^n + a_{n-1} x^{n-1} + \cdots + a_1 x + a_0$$

where n is the degree of the polynomial and the coefficients $a_n, a_{n-1}, \ldots, a_0$ are constants. For example,

$$f(x) = 12x^4 + x^3 - 5x - 2$$

Here $n = 4$, $a_4 = 12$, $a_3 = 1$, $a_2 = 0$, $a_1 = -5$, and $a_0 = -2$.

All four functions above are examples of *rational functions,* which are functions of the form

$$f(x) = \frac{p(x)}{q(x)}$$

where $p(x)$ and $q(x)$ are polynomials. For example,

$$f(x) = \frac{3x^2 - 16x + 5}{2x^3 + 5x^2 + 2x}$$

Here $p(x) = 3x^2 - 16x + 5$ and $q(x) = 2x^3 + 5x^2 + 2x.$

In previous chapters, functions were written in the form $y = mx + b$ rather than $f(x) = mx + b$ because we didn't have the $f(x)$ notation at that point, but both forms represent the same function.

We have already spent a great deal of time drawing graphs of every imaginable equation, because a graph is a particularly good way of displaying y's dependence on x. In exactly the same way, you can draw a graph of any function expressed using the $f(x)$ notation.

For example, if $f(x) = 2x + 1$, you can make a table of corresponding x and y values, that is, a table of x and $f(x)$ values.

x	$y = f(x)$
-1	$-1 = f(-1)$
0	$1 = (0)$
1	$3 = f(1)$
2	$5 = f(2)$
3	$7 = f(3)$

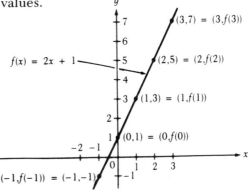

FIG. 8.1

Figure 8.1 shows the graph.

EXAMPLE: *The graph of* $f(x) = x^2 + 2$ *is shown in Figure 8.2.*

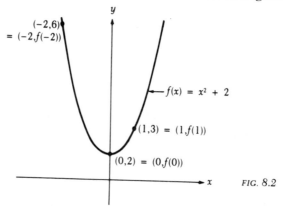

FIG. 8.2

You know that in order to graph an equation you plot points (x, y) whose coordinates satisfy that equation—in other words, whose y coordinate is derived from its x coordinate by using the equation.

If a function is written in the form $y = f(x)$, then the y corresponding to some x is $f(x)$, and so we must plot points of the form $(x, f(x))$. Thus, *every function* $y = f(x)$ *can be represented by a graph, which is made up of all the points of the form* $(x, f(x))$.

As we have seen in Chapter 3, the geometric properties of a graph and the algebraic properties of an equation are closely related. In the same way, the graph of a function reflects many of the properties of that function, and looking at the graph can tell you a good deal about the function.

The next sections show exactly what you can learn about a function from its graph.

PROBLEM SET   8.1

1. The graph of a linear function passes through the points $(a, f(a))$, $(c, f(c))$. What is the slope of this graph?

Suppose g is a function given by the graph in Figure 8.3.

2. What is $g(0)$?
3. What is $g(3)$?
4. What is $g(1)$?
5. What is x such that $g(x) = -5$?
6. For what values of x is $g(x) > 0$?
7. For what values of x is $g(x) < -5$?
8. What kind of function might g be?

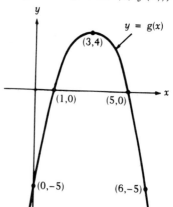

FIG. 8.3

Suppose h is a function given by the
graph in Figure 8.4.

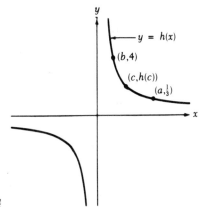

   9. What kind of function might h be?
   10. For what values of x is $h(x) > 0$?
   11. What is $h(a)$?
   12. What is $h(b)$?
   13. What is the distance from the origin
to the point $(c, h(c))$?

FIG. 8.4

8.2   ZEROS AND INTERCEPTS

By a *zero of a function* we mean a value of x that makes the function zero—in
other words, an x that makes y zero.
So *if x is a zero of a function, then $f(x) = 0$.*

   For example, if $f(x) = x^3 - x$, then the values of x that make $f(x) = 0$ satisfy

$$x^3 - x = 0$$

or

$$x(x - 1)(x + 1) = 0$$

giving $x = 0$, 1, or -1. Therefore, 0, 1, and -1 are the zeros of the function
$f(x) = x^3 - x$.

   Now look at the graph of $f(x) = x^3 - x$ in Figure 8.5 (this graph was drawn
in detail in Section 4.1). Obviously, *the zeros of the function are just the
same as the x intercepts*—which is not surprising since the zeros make $y = 0$,
and the x intercepts are where the graph cuts the x axis, which is where
$y = 0$.

   Now you might wonder what the y intercept means in functional notation.
The y intercept is the y coordinate of the point at which the function crosses
the y axis. On the y axis $x = 0$, and the y coordinate of the point at which $x = 0$
is $y = f(0)$.

   Therefore *the y intercept is $f(0)$,* whereas *the x intercepts are the x values
that make $f(x) = 0$.*

   Note: Realize that there is a big difference
between $f(0)$ and $f(x) = 0$.

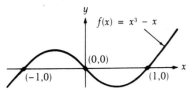

FIG. 8.5

EXAMPLE: *If* $f(x) = -2x + 8$, *find* $f(0)$ *and solve* $f(x) = 0$; *find* $f(1)$ *and solve*
$f(x) = 1$.

Substituting $x = 0$ gives

$$f(0) = -2 \cdot 0 + 8 = 8$$

But

$$f(x) = 0$$

means

$$-2x + 8 = 0$$

giving $x = 4$

(see Figure 8.6).

Similarly, substituting $x = 1$ gives

$$f(1) = -2 \cdot 1 + 8 = 6$$

But

$$f(x) = 1$$

means $-2x + 8 = 1$

so $x = \dfrac{7}{2}$

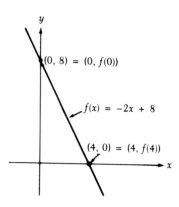

FIG. 8.6

PROBLEM SET   8.2

For each of the following functions, find $f(0)$ and solve $f(x) = 0$. Then graph
the function, labeling all intercepts.

1. $f(x) = x + 3$
2. $f(x) = 2x - 10$
3. $f(x) = -\frac{1}{2}x$
4. $f(x) = x^2 + 1$
5. $f(x) = x^2 + 7x + 12$
6. $f(x) = x^2 - 3x - 4$
7. $f(x) = 2x^2 - 11x - 6$
8. $f(x) = x^3 - 2$
9. $f(x) = x^4 - x^2$
10. $f(x) = x^3 + 4x^2 - 10x + 5$
11. $f(x) = \dfrac{1}{x - 1}$
12. $f(x) = \dfrac{1}{x^2 - 4}$

13. $f(x) = |x - 2|$
14. $f(x) = \dfrac{x + 2}{x - 3}$
15. $f(x) = \dfrac{(x - \frac{1}{4})(x + \sqrt{\frac{3}{4}})}{x^2}$
16. $f(x) = \dfrac{x^3 - 1}{x + 2}$
17. $f(x) = \dfrac{x - 3.7 \cdot 10^{-1}}{(x + 1.4)(x + 2.7)}$
18. $f(x) = \dfrac{2x^2 - 4}{x^2 - 8x + 12}$
19. $f(x) = \dfrac{|x + 2|}{|x - 3|}$
20. $f(x) = \sqrt{x^2 - 6x + 9}$

8.3   ODD AND EVEN FUNCTIONS

We already know what makes an equation have a graph with odd or even symmetry (Section 3.2). We'll now see what has to be true of a function for its
graph to have these properties.

Consider the function $f(x) = x^2$. If you substitute 2 and -2 for x, the function takes on the same value both times, since

$$f(2) = 2^2 = 4 \quad \text{and} \quad f(-2) = (-2)^2 = 4$$

Similarly,

$$f(5) = 25 \quad \text{and} \quad f(-5) = 25$$

so

$$f(5) = f(-5)$$

As you might guess, this is true for any x, since

$$f(-x) = (-x)^2 = x^2 = f(x)$$

so

$$f(x) = f(-x) \quad \text{for any } x$$

A function that has the property that $f(x) = f(-x)$ for all x is called an even
function. Now $f(-x) = f(x)$ means that you get the same value of y from x as
from $-x$. But this is exactly what you need for a graph with even symmetry
(symmetry about the y axis). So an even function has even symmetry.

Now consider the function $g(x) = x^3$. Substituting 2 and -2 for x gives opposite values, since

$$g(2) = 2^3 = 8 \quad \text{and} \quad g(-2) = (-2)^3 = -8$$

so

$$g(-2) = -g(2)$$

Similarly,

$$g(3) = 3^3 = 27 \quad \text{and} \quad g(-3) = (-3)^3 = -27$$

so

$$g(-3) = -g(3)$$

In general,

$$g(-x) = (-x)^3 = -x^3 = -g(x)$$

so

$$g(-x) = -g(x)$$

A function that has the property that $g(-x) = -g(x)$ for all x is called an *odd
function.* If $g(-x) = -g(x)$, then you get opposite values of y from x and $-x$.
This ensures that the graph has odd symmetry (symmetry about the origin).
So an odd function has odd symmetry.

You should notice that all even powers of x (including x^0) are even functions, while odd powers are odd functions. For example, if $f(x) = x^{10}$,

$$f(-x) = (-x)^{10} = x^{10} = f(x)$$

and if $g(x) = x^7$,

$$g(-x) = (-x)^7 = -x^7 = -g(x)$$

The sum (or difference) of even functions is even, and of odd functions is odd, so

$$f(x) = 10x^8 - 7x^4 + 3 \qquad \text{is even}$$

and

$$g(x) = -3x^5 + 16x^3 + x \qquad \text{is odd}$$

For an example of an even function that is *not* a polynomial in x, look at the absolute value function:

$$f(x) = |x|$$

This function is even because

$$f(-x) = |-x| = |x| = f(x)$$

When we get to trigonometry we will see many more examples of odd and even functions that are not polynomials or rational functions of x.

PROBLEM SET   8.3

Determine whether the following functions are odd or even (or neither) and sketch their graphs.

1. $f(x) = x^2 + 2$
2. $f(x) = \dfrac{1}{x^2 - 9}$
3. $f(x) = (x - 1)^2$
4. $f(x) = x^4 - 16$
5. $f(x) = 2x^3 + x$
6. $f(x) = 10x(x^5 + 3x^3 - 4x)$
7. $f(x) = 5$
8. $f(x) = (x^2 + 1)^2$
9. $f(x) = 3x^6 + 5x^4 + 7x^2 + 1$

10. $f(x) = \dfrac{x^2 - 1}{x}$
11. $f(x) = \dfrac{x^4 - x^2 + 1}{x^3}$
12. $f(x) = \dfrac{x^3 + x}{x - 1}$
13. $f(x) = \dfrac{2x^3 - x}{2x}$
14. $f(x) = \dfrac{x^4 + 1}{x}$

8.4   THE VERTICAL LINE TEST

If x and y are connected by the equation $y = x^2$, then y is a function of x; if they are related by $y^2 = x$, then y is not a function of x (because each value of x gives rise to two values of y). Let us compare the graphs of these two equations shown in Figure 8.7.

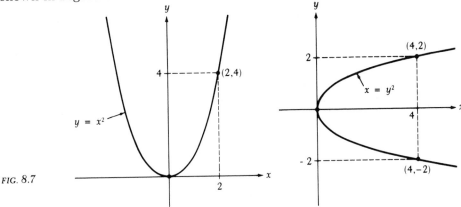

FIG. 8.7

On the graph of $y = x^2$ there is only one point above each x value, corresponding to the one y value given by the equation. On the graph of $y^2 = x$ there is a point above and a point below each positive x value, corresponding to the two y values given by the equation.

All the points with a given x value lie on the vertical line passing through that value on the x axis. So you can always tell how many y values correspond to an x value by looking at how many times the vertical line cuts the graph. This gives us:

The Vertical Line Test

If every vertical line cuts the graph at most once, then every x gives rise to at most one y and we are looking at the graph of a function. If some vertical line cuts the graph more than once, then that x value gives rise to more than one y, and we do not have a function.

EXAMPLE: *Test the circle $x^2 + y^2 = 1$.*

This has the graph shown in Figure 8.8.

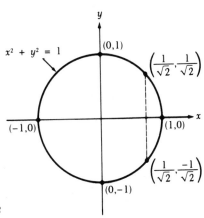

FIG. 8.8

Some vertical lines cut the circle twice, corresponding to the fact that for a given x the equation gives two values of y:

$$y = \sqrt{1 - x^2} \quad \text{and} \quad y = -\sqrt{1 - x^2} \quad \text{(where } \sqrt{} \text{ means the positive square root)}$$

Therefore the circle is not the graph of a function.

If, however, we restrict ourselves to the top half of the circle, it will represent a function because any vertical line cuts it only once (or not at all), as you can see in Figure 8.9.

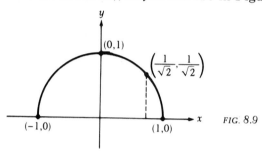

FIG. 8.9

So the semicircle is the graph of some function—let's find out what. We get this semicircle by throwing out all the points on the circle with negative y, and keeping those with positive y. Since the y coordinate of a point on the circle is given either by $y = \sqrt{1 - x^2}$ or by $y = -\sqrt{1 - x^2}$, the upper half of the circle must satisfy

$$y = \sqrt{1 - x^2}$$

This confirms the fact that the semicircle is the graph of a function, because here is its equation, giving a unique y for every x.

As you might guess, the equation

$$y = -\sqrt{1 - x^2}$$

is that of the lower half of the circle, which is also the graph of a function.

PROBLEM SET   8.4

In Problems 1–10, determine which graphs represent functions.

1.

2.

3.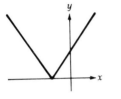

4.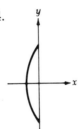

5.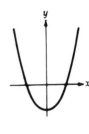

6.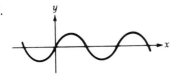

7.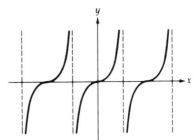

8.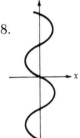

9.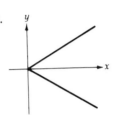

10.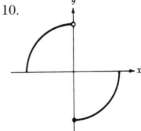

Graph the following, and use the vertical line test to determine which are functions, taking x as the independent variable.

11. $y = \sqrt{9 - x^2}$

12. $x^2 + y^2 = 16$

13. $\dfrac{x^2}{4} + \dfrac{y^2}{25} = 1$

14. $y = \begin{cases} x, & x < 0 \\ 0, & x = 0 \\ -x, & x > 0 \end{cases}$

15. $y = \begin{cases} 0, & x > 0 \\ -x, & x \le 0 \end{cases}$

16. $x = \sqrt{4 - y^2}$

17. $\sqrt{x} = \sqrt{y}$

8.5   DOMAINS AND RANGES

As you may remember, the *domain* of a function is all the possible values that the independent variable may take on; the *range* is all the possible values of the dependent variable.

The domain of any polynomial function is all the reals, because you can substitute any number into a polynomial and have it make sense. However, in the function $f(x) = \dfrac{1}{x}$, x cannot be zero (though any other real number is all right); therefore the domain of $f(x) = \dfrac{1}{x}$ is all the reals except zero. For $f(x) = \sqrt{x}$, x cannot be negative, and so its domain is all positive numbers and zero.

The graph of a function shows very clearly what its domain and range are. Let us look at the examples in Figure 8.10.

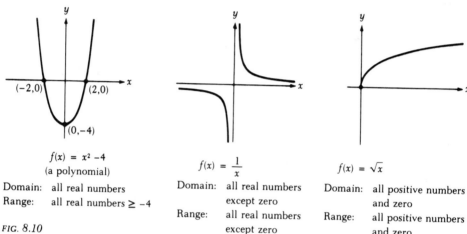

$$f(x) = x^2 - 4$$
(a polynomial)

Domain: all real numbers
Range: all real numbers ≥ -4

$$f(x) = \frac{1}{x}$$

Domain: all real numbers
           except zero
Range: all real numbers
           except zero

$$f(x) = \sqrt{x}$$

Domain: all positive numbers
           and zero
Range: all positive numbers
           and zero

FIG. 8.10

To see what the graph has to do with domains and ranges of a function, think how you can use the graph to find the value of a function. Given some value of x, you use the graph to "read off" the corresponding value of y, as shown in Figure 8.11.

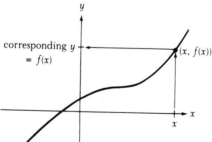

FIG. 8.11

Now the domain of f is all the possible values of x that you could start with, and still get a y value. Therefore, x must be at such a point that a vertical line through it will cut the graph—otherwise the graph can't possibly give you a y value. Therefore, *the domain is those x values that give rise to vertical lines which cut the graph.*

From Figure 8.10 you can see immediately that any vertical line cuts the graph of $f(x) = x^2 - 4$, and so its domain is all the real numbers. In the case of $f(x) = \dfrac{1}{x}$ the only vertical line that does not cut the graph is the y axis (the line $x = 0$), and so the domain is all reals except zero. For $f(x) = \sqrt{x}$, a vertical line will cut the graph if it is on or to the right of the y axis, and therefore the domain of $f(x) = \sqrt{x}$ is all positive real numbers and zero.

The range of a function can be found by exactly the same method, using horizontal lines instead of vertical. This is because the range of a function is all the possible y values that may be "read off" as corresponding to some x value. Consequently, if y is to be in the range, a horizontal line through y must cut the graph, and so *the range is all the possible y's that give rise to horizontal lines which cut the graph.*

Notice that a horizontal line cuts $f(x) = x^2 - 4$ only if it is at or above $y = -4$, and so the range is all real numbers greater than or equal to -4. Any horizontal line except the x axis (the line $y = 0$) cuts the graph of $f(x) = \dfrac{1}{x}$, so the range of this function is all the reals except zero. In the case of $f(x) = \sqrt{x}$, any horizontal line on or above the x axis cuts the graph; the range is therefore all positive real numbers and zero.

EXAMPLE: *Find the domain and range of* $f(x) = \dfrac{1}{(x-1)^2}.$

The graph of this function is shown in Figure 8.12 (see Section 4.2 if you aren't sure how to graph this).

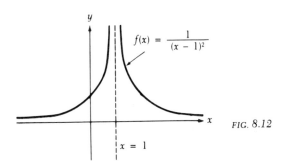

FIG. 8.12

Domain: all real real numbers except 1, that is, all x, except $x = 1$, since all vertical lines except $x = 1$ cut the graph.

Range: all positive real numbers (but not zero), that is, all $y > 0$, since any horizontal line above (but not on) the x axis cuts the graph, but no lines below the x axis do.

EXAMPLE: *Find the domain and range of* $f(x) = \sqrt{1 - x^2}$.

This is the equation of the
upper half of the circle
$x^2 + y^2 = 1$.
Therefore its graph looks
like that in Figure 8.13.

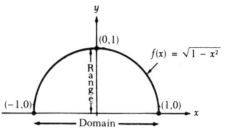

FIG. 8.13

$Domain$ = all real numbers between -1 and 1 (inclusive), that is,

$$-1 \le x \le 1$$

$Range$ = all real numbers between 0 and 1 (inclusive), that is,

$$0 \le y \le 1$$

PROBLEM SET   8.5

Find the domain and range of the following functions:

1.

2.

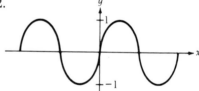

3.

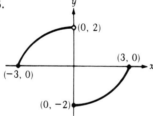

4.

5.

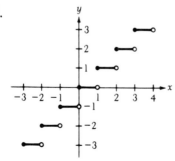

6.

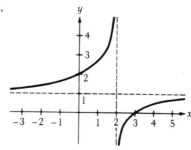

7.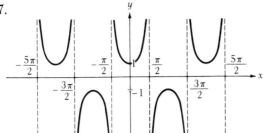

8.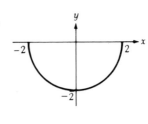

9.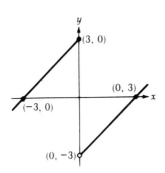

10.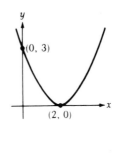

11. What function could have a graph like that in Problem 1?
12. What function could have the graph represented in Problem 4?
13. What function could have a graph like that in Problem 6?
14. What function could have a graph like that in Problem 8?
15. What function could be represented by the graph in Problem 9?
16. What function could have the graph represented in Problem 10?

CHAPTER 8 REVIEW

Suppose $y = f(x)$ has the graph in Figure 8.14.

1. Evaluate $f(6)$.
2. Solve $f(x) = 2$.
3. For what values of x is $f(x) > 0$?
4. For what values of x is $|f(x)| < 2$?

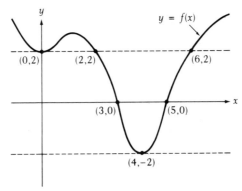

FIG. 8.14

Suppose g and h are functions given by the graphs in Figure 8.15. In terms of letters on the graph, what is:

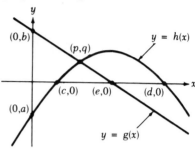

5. $g(0)$
6. $h(0)$
7. x such that $g(x) = 0$
8. x such that $h(x) = 0$
9. x such that $g(x) = h(x)$

FIG. 8.15

Suppose $y = k(x)$ is given by the graph in Figure 8.16.

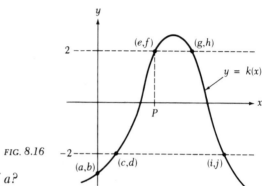

FIG. 8.16

10. What is the value of f? Of a?
11. What are the coordinates of P in terms of the other letters?
12. Is b positive or negative? What about c?
13. Solve for x: $k(x) = -2$.
14. For what values of x is $k(x) > 2$?

A direct-dialed long-distance call from Boston to New Orleans costs 52¢ for the first minute and 36¢ for each additional minute or fraction thereof.

15. Express the cost, C, as a function of the time, t.
16. Graph the function $C(t)$ you found in Problem 15.
17. What is the domain of the function C?
18. What is the range of the function?
19. For what values of t is $C(t) = \$1.60$?

For each of the following functions: (a) find $f(0)$; (b) solve $f(x) = 0$; (c) state whether the function is even, odd, or neither; (d) find the domain; (e) find the range; and (f) sketch the graph of the function.

20. $f(x) = \frac{1}{3}x - 7$

21. $f(x) = 2x^2 + 5$

22. $f(x) = |x - \frac{7}{2}|$

23. $f(x) = |x| - \frac{7}{2}$

24. $f(x) = (\sqrt{x - \frac{7}{2}})^2$

25. $f(x) = \dfrac{1}{x^2 - 3}$

26. $f(x) = \dfrac{(x + \sqrt{2})(x + 2)}{(x - 10^{-1})}$

27. $f(x) = \dfrac{(x^2 + \sqrt{3})(x^2 - 3)}{x + \sqrt{3}}$

28. $f(x) = 3x^2 - x - 2$

29. $f(x) = -(x - 1)^2$

30. $f(x) = \dfrac{1}{x^3 + x}$

On an operator-assisted long-distance call from Boston to Chicago at night, the charge is $2.05 for the first 3 minutes and 13¢ for each additional minute or fraction thereof. A direct-dialed call from Boston to Chicago made during the day costs 53¢ for the first minute and 34¢ per minute thereafter.

31. Find the function, A, relating the cost of an operator-assisted night call to the time spoken, t. Graph $A(t)$.

32. Find the function, D, relating the cost of a direct-dialed daytime call to the time spoken, t. Graph $D(t)$. [Do this on the same axes used in Problem 31.]

33. Use your graphs to determine if there is a t such that $D(t) = A(t)$. If so, what is the value of t at this point?

34. What is the domain of $A(t)$?

35. What is the range of $D(t)$?

36. For what values of t is $A(t)$ less than half of $D(t)$?

The builders of the bullring in Madrid weren't sure whether to make the ring rectangular or circular. They were not sure just how much fencing they had to build the walls of the bullring, but they did know that they wanted the ring to have as large an area as possible for the amount of fencing eventually to be used.

37. Express the area of circle, C, as a function of the amount of fencing, p, to be used in building the perimeter of the ring. Graph $C(p)$.

38. Express the area of a square, S, as a function of the amount of fencing, p, to be used in building the perimeter of the ring. Graph $S(p)$.

39. If the promoters know that they have only 400 feet of fencing, what is the domain of C? Of S?

40. Use your graphs to determine the values of t such that $S(t) < 8000$.

41. For what values of t is $C(t) > 7000$?

42. For what values of t is $C(t) < S(t)$?

43. Given that the bullring must be at least 2000 square feet in area and that no more than 300 feet of fencing are available, how should the bullring be built so as to maximize the area?

You may have noticed in Problems 37–43 that it was assumed that a square shape was the best shape to use if the bullring was to be rectangular. Let's see if this is always the case.

44. For a given perimeter of 12, relate the area of a rectangle, A, to the size of the height of the rectangle, x. Graph $A(x)$.

45. What is the domain of A? The range?

46. Repeat Problems 44 and 45, this time using a given perimeter of 4.

47. Repeat Problems 44 and 45, this time using a given perimeter of 1000.

48. Now what do you think about the wisdom of looking only at square bullrings?

For each of the following functions: (a) find $f(0)$; (b) solve $f(x) = 0$; (c) state whether the function is even, odd, or neither; (d) find the domain; (e) find the range; and (f) sketch the graph of the function.

49. $f(x) = (2x^3 - x)^2$

50. $f(x) = \dfrac{x^3}{x^4 - 4}$

51. $f(x) = \dfrac{(2x + \sqrt{12})(x - \sqrt{15})}{100\left(\dfrac{x}{10} - 2\right)(x - 15)}$

52. $f(x) = \dfrac{(x - 1)^2}{|x + 2|}$

53. $f(x) = \dfrac{1}{x^2 + 4}$

54. Two lines with the same slope,

$$y(x) = Ax + 1$$

$$y(x) = Ax + 2$$

must be parallel, but the distance between them will depend on the value of A. Using the two lines above, find a formula for the distance between them as a function of their slope. [*Hint*: See Problem 43 in Section 2.5] Sketch a graph of this function. (Be sure to consider negative slopes.)

9 COMPOSITE AND INVERSE FUNCTIONS

9.1   DEFINITION OF COMPOSITE FUNCTIONS

There's a particular way of combining two functions to make a new one that turns out to be very useful. It goes like this.

We start with two functions, say

$$f(x) = x + 2 \quad \text{and} \quad g(x) = x^2$$

To create the new function, we take any x and calculate $f(x)$, then we substitute the result into $g(x)$. The number that we end up with is determined by x. In fact it is a function of x which we will call $h(x)$.

Let's see what happens to 1. We first calculate $f(1)$:

$$f(1) = 1 + 2 = 3$$

then substitute the 3 into g:

$$g(3) = 3^2 = 9$$

So, starting with $x = 1$ and applying f and then g gives 9. So

$$h(1) = 9$$

The way we got $h(1)$ can be pictured like this:

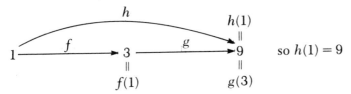

Let's see what happens with $x = 3$:

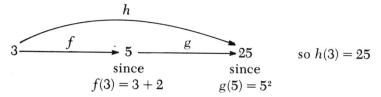

so $h(3) = 25$

since
$f(3) = 3 + 2$

since
$g(5) = 5^2$

Look at $x = 4$ also:

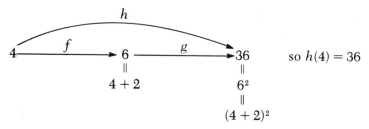

so $h(4) = 36$

$4 + 2$

6^2
$\|$
$(4 + 2)^2$

What is the general formula for h? To find out, let's try the process for $x = a$:

$$a \xrightarrow{\ \ f\ \ } a + 2 \xrightarrow{\ \ g\ \ } (a + 2)^2 \qquad \text{so } h(a) = (a + 2)^2$$

$f(a)$ ... $g(a + 2)$

Since x is the usual name for the independent variable, this formula is generally written:

$$h(x) = (x + 2)^2$$

$h(x)$ is called *the composition of $f(x)$ and $g(x)$*, written $g(f(x))$ and read "g of f of x." So, in this example,

$$g(f(x)) = (x + 2)^2$$

In general, $g(f(x))$ *means that the function f is applied first and then g:*

$$x \xrightarrow{\ \ f\ \ } f(x) \xrightarrow{\ \ g\ \ } g\,(f\,(x))$$

Notice that $g(f(x))$ works just like the order of operations—you start at the inside and work out; $f(x)$ is called the *inside* or *first function* and $g(x)$ is the *outside* or *second function*. $g(f(x))$ is sometimes called *a function of a function* because g is now a function of $f(x)$, which is itself a function. $g(f(x))$ is also called a *composite function*.

Alternate Notation: The function $g(f(x))$ is also sometimes written as

$$(g \cdot f)\,(x)$$

where $g \cdot f$ means that the two functions g and f have been composed, with g as the outside and f as the inside function. We will stick to the $g(f(x))$ notation.

Compositions Versus Products

There's a big difference between a composition and a product of two functions. For example, if $f(x)$ and $g(x)$ are as before, their composition is

$$g(f(x)) = (x + 2)^2 = x^2 + 4x + 4$$

whereas their product is

$$g(x) \cdot f(x) = x^2(x + 2) = x^3 + 2x^2$$

EXAMPLE: *If $f(x) = x + 1$ and $g(x) = x^2 + 3$, find $g(f(2))$, $g(f(x))$, $f(g(x))$.*

Finding $g(f(2))$ means finding $f(2)$ and then finding g of that:

$$g(f(2)) = g(\underbrace{2 + 1}) = g(3) = 3^2 + 3 = 12$$

here $f(2)$ is
replacing x in $g(x)$

Similarly, finding $g(f(x))$ means replacing x by $f(x)$ in the formula for $g(x)$:

$$g(f(x)) = g(\underbrace{x + 1}_{f(x)}) = \underbrace{(x + 1)^2 + 3}_{g(x) \text{ with } (x + 1) \text{ replacing } x}$$

Finding $f(g(x))$ means making $g(x)$ the inside function and replacing x by it in the formula for $f(x)$:

$$f(g(x)) = f(\underbrace{x^2 + 3}_{\text{here is } g(x)}) = \underbrace{(x^2 + 3) + 1}_{f(x) \text{ with } (x^2 + 3) \text{ replacing } x}$$

Note: Since

$$g(f(x)) = (x + 1)^2 + 3 = x^2 + 2x + 4$$

and

$$f(g(x)) = x^2 + 4$$

you can see that $g(f(x))$ and $f(g(x))$ are different functions. This is perhaps a bit surprising, since for many of the things we do the order doesn't matter. For example, $2 \cdot 3$ is the same as $3 \cdot 2$, and $5 + 7$ is the same as $7 + 5$. However, just as it matters which you put on first, your shoes or your socks, so it makes a difference which is on the inside, f or g.

EXAMPLE: *If $h(x) = \dfrac{1}{x}$ and $k(x) = \sqrt{x + 2}$, find $k(h(x))$, $h(k(x))$, $k(k(x))$.*

Finding $k(h(x))$ means replacing x in $k(x)$ by $h(x)$:

$$k(h(x)) = k\left(\underbrace{\frac{1}{x}}_{h(x)}\right) = \underbrace{\sqrt{\frac{1}{x} + 2}}_{\text{replaces } x}$$

Finding $h(k(x))$ means replacing x in $h(x)$ by $k(x)$:

$$h(k(x)) = h(\sqrt{x+2}\,) = \underbrace{\frac{1}{\underbrace{\sqrt{x+2}}}}$$

$\underbrace{\phantom{h(\sqrt{x+2}\,)}}_{k(x)}$ $\underbrace{\phantom{\sqrt{x+2}}}_{\text{replaces } x}$

Finding $k(k(x))$ means replacing x in $k(x)$ by $k(x)$ itself.

$$k(k(x)) = k(\underbrace{\sqrt{x+2}}_{k(x)}\,) = \sqrt{\underbrace{\sqrt{x+2}}_{\text{replaces } x} + 2}$$

How a Composite Function Might Come Up in the "Real World"

When a stone is dropped in a pond, a series of waves spreads outward in a circle from the place at which the stone hits the water (See Figure 9.1). Suppose you wanted to calculate the area A of the water that has been disturbed at any given time. A is a function of r, the radius of the disturbance, and

$$A = \pi r^2$$

But r is itself a function of the time t (usually measured from the moment the stone hit the water). So a value of t determines the radius at that moment, which in turn determines the area. Hence, A is a function of t.

To be specific, suppose you are told that r is given by

$$r = 10t \qquad (r \text{ in feet, } t \text{ in minutes})$$

Now substituting for r into $A = \pi r^2$ gives

$$A = \pi(10t)^2$$

or

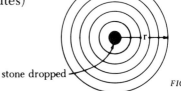

stone dropped

FIG. 9.1

$$A = 100\pi t^2$$

This shows that A is a "function of a function." A is a function of r, and r is a function of t, and the formula shows you exactly how A depends on t.

"Decomposing" a Function

It is often helpful to be able to look at a complicated function as the composition of two simpler ones. Here's how this can be done.

EXAMPLE: *Write $h(x) = (x^2 + 1)^3$ as the composition of two functions.*

Just looking at this makes you think that $x^2 + 1$ is the inside function and the cube is the outside function.

Let us say $f(x) = x^2 + 1$ and $g(x) = x^3$ and check our guess:

$$g(f(x)) = g(x^2 + 1) = (x^2 + 1)^3 \qquad \text{O.K.!}$$

Note: This kind of question always has a "trivial" answer: Suppose you let

$f(x) = x$ and $g(x) = (x^2 + 1)^3$,   the original function. Then

$g(f(x)) = g(x) = (x^2 + 1)^3$

But such an answer doesn't really tell you anything you didn't know before, and is certainly no use in practice!

EXAMPLE: *Express* $\dfrac{1}{x^2 + 4}$ *as a (nontrivial) composition.*

What the outside function should be is less obvious here until you think what you have to do to calculate

$$\frac{1}{x^2 + 4}$$

for any particular x. First you must square the x and add 4, and then take the reciprocal. Therefore you might take the outside function to be the reciprocal function, because taking the reciprocal is the last thing you do.

So let

$f(x) = x^2 + 4$ and $g(x) = \dfrac{1}{x}$

Then

$g(f(x)) = g(x^2 + 4) = \dfrac{1}{x^2 + 4}$ O.K.

Note: There are lots of other ways of expressing $\dfrac{1}{x^2 + 4}$

as a composition. For example, you could let $f(x) = x^2$ because squaring is the first thing you do. Then

$g(x) = \dfrac{1}{x + 4}$

so

$g(f(x)) = g(x^2) = \dfrac{1}{x^2 + 4}$

EXAMPLE: *Find g (x) if*

$g(f(x)) = \dfrac{1}{(x^2 + 2)^2}$ *and* $f(x) = x^2 + 2$

Here you are given the inside function and have to find the outside. To do that, ask yourself what has to be done to $f(x)$, that is, to $x^2 + 2$, to get the composite function. To get $g(f(x))$ you have to take one over the square of $x^2 + 2$, so g must be the function that takes one over the square of things. In other words:

$$g(x) = \frac{1}{x^2}$$

Check: $g(f(x)) = g(x^2 + 2) = \dfrac{1}{(x^2 + 2)^2}$ O.K.

EXAMPLE: *Find $f(x)$ if*

$$g(f(x)) = \sqrt{x + 5} \quad and \quad g(x) = \sqrt{x}$$

Notice that here we are asked to find the inside function. We have to find what function must be substituted for x in $\sqrt{x}$ to get $\sqrt{x + 5}$

Clearly, $f(x) = x + 5$ will do just fine, because

$$g(f(x)) = g(x + 5) = \sqrt{x + 5}$$

PROBLEM SET 9.1

Let $f(x) = x^2 + 1$ and $g(x) = \dfrac{1}{x}$. Find:

1. $f(2)$
2. $f(3)$
3. $f(v)$
4. $f(t)$
5. $f\left(\dfrac{1}{x}\right)$
6. $f(g(x))$

7. $g(1)$
8. $g(-2)$
9. $g(v)$
10. $g(t)$
11. $g(x^2 + 1)$
12. $g(f(x))$

Let $h(x) = x + 1$ and $j(x) = \dfrac{1}{x^2 + 1}$. Find:

13. $h(4)$
14. $h(-3)$
15. $h(w)$
16. $h\left(\dfrac{1}{x^2 + 1}\right)$
17. $h(j(x))$

18. $j(1)$
19. $j(2)$
20. $j(u)$
21. $j(x + 1)$
22. $j(h(x))$

Let $k(x) = x^2$, $m(x) = \dfrac{1}{x - 1}$, and $n(x) = \dfrac{2x^2}{x + 1}$. Find:

23. $k(m(x))$
24. $m(k(x))$
25. $k(x) \cdot m(x)$
26. $m(x) \cdot k(x)$
27. $k(n(x))$
28. $k(x) \cdot n(x)$

29. $n(k(x))$
30. $m(n(x))$
31. $n(m(x))$
32. $m(m(x))$
33. $[m(x)]^2$
34. $m(x^2)$

35. An air balloon is being blown up in such a way that its radius, r, at time t is given by the function $r = t^2$.

   (a) Write the volume of the balloon, v, as a function of its radius, r.

   (b) Use this to write the volume as a function of time.

36. A circular ink blot is created at time $t = 0$ (i.e., its radius is 0 when $t = 0$).

   (a) Express the area of the blot, A, as a function of its radius.

   (b) If the radius of the blot increases at a rate of 2 centimeters per minute, express the radius as a function of time.

   (c) Use (a) and (b) to express the area of the blot as a function of time.

37. Suppose a person's income tax, I, is 15% of his gross income, G, and the surtax, S, is 10% of the income tax.

   (a) Express the income tax as a function of G.

   (b) Express the surtax as a function of I.

   (c) Express S as a function of G.

Let $f(x) = x^2$, $g(x) = \dfrac{1}{x}$, $h(x) = x + 1$. Find:

38. $f(2)$

39. $g(f(2))$

40. $h(g(f(2)))$

41. $f(3)$

42. $g(f(3))$

43. $h(g(f(3)))$

44. $f(a)$

45. $g(f(a))$

46. $h(g(f(a)))$

47. $h(g(f(x)))$

48. $g(h(f(x)))$

49. $g(f(h(x)))$

50. $f(g(h(x)))$

51. $f(h(g(x)))$

52. $f(x) \cdot g(h(x))$

53. $f(x) \cdot g(x) \cdot h(x)$

Let $k(x) = \sqrt{x-1}$, $m(x) = (x+1)^2$, $n(x) = \dfrac{1}{x}$. Find:

54. $n(k(m(x)))$

55. $n(m(k(x)))$

56. $k(m(n(x)))$

57. $k(n(m(x)))$

58. $m(k(n(x)))$

59. $m(n(k(x)))$

Let $k(x) = x^2 + 2$. Find:

60. $h(x)$ if $h(k(x)) = (x^2 + 2)^3$

61. $g(x)$ if $g(k(x)) = \dfrac{1}{x^2 + 2}$

62. $f(x)$ if $k(f(x)) = \left(\dfrac{1}{x}\right)^2 + 2$

63. $m(x)$ if $m(k(x)) = \dfrac{1}{\sqrt{x^2 + 2}}$

9.2   DOMAINS AND RANGES OF COMPOSITE FUNCTIONS

EXAMPLE: *If*

$$f(x) = x^2 \quad and \quad g(x) = \frac{1}{x-1}$$

find the domain and range of $f(g(x))$ and $g(f(x))$.

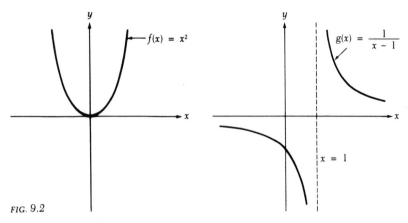

FIG. 9.2

Before we start, look at the graphs of f and g in Figure 9.2 to see that:

Domain of f is all real numbers.

Range of f is all positive reals and zero.

Domain of g is all reals except 1.

Range of g is all reals except 0.

Now to the composite functions:

$$f(g(x)) = f\left(\frac{1}{x-1}\right) = \frac{1}{(x-1)^2}$$

To find domain and range it helps to draw a graph, so look at the one in Figure 9.3.

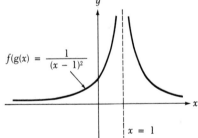

FIG. 9.3

Domain of $f(g(x))$ is all real numbers except 1. (Note: This is the same as the domain of $g(x)$.)

Range of $f(g(x))$ is all real numbers greater than but excluding zero.

On the other hand, consider

$$g(f(x)) = g(x^2) = \frac{1}{x^2 - 1}$$

Again a graph is helpful; so look at Figure 9.4.

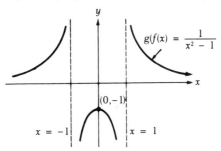

$$g(f(x)) = \frac{1}{x^2 - 1}$$

FIG. 9.4

Domain of g(f(x)) is all reals except 1 and −1. (Note: This is *not* the same as the domain of *f* or *g*.)

Range of g(f(x)) is all reals except those between −1 and 0; −1 is in the range, 0 is not. (Note: This is not the same as the range of *f* or *g*.)

The moral of this problem is that the domain and range of a composite function do not have to be the domain or range of either of the constituent functions. The domain of a composite function, however, cannot be larger than the domain of the inside function (because you can't put anything into the composite function that you can't put into the inside one), and the range of a composite function cannot be larger than the range of the outside function (because you can't get something out of the composite function that you can't get out of the outside function).

PROBLEM SET   9.2

Let $f(x) = x^2 + 1$, $g(x) = \sqrt{x}$, $h(x) = \dfrac{1}{x + 1}$, $k(x) = \dfrac{1}{x^2 - 4}$.   Find the domain and range of:

1. $f(g(x))$
2. $g(f(x))$
3. $f(h(x))$
4. $h(f(x))$
5. $f(k(x))$
6. $k(f(x))$

7. $g(h(x))$
8. $h(g(x))$
9. $g(k(x))$
10. $k(g(x))$
11. $h(k(x))$
12. $k(h(x))$

13. $f(g(h(x)))$
14. $f(g(k(x)))$
15. $f(h(k(x)))$
16. $g(h(k(x)))$
17. $g(f(k(x)))$
18. $k(h(f(x)))$

Using the functions defined for Problems 1–18, find the domain and range of:

19. $f(g(h(k(x))))$
20. $g(f(k(h(x))))$
21. $f(h(k(g(x))))$
22. $k(g(f(h(x))))$

23. $g(f(g(f(x))))$
24. $f(g(f(g(x))))$
25. $f\left(\dfrac{1}{h(x)}\right)$

9.3   DEFINITION OF INVERSE FUNCTIONS

As you may know, crickets chirp more as the weather gets warmer. In fact, there is a formula that gives n, the number of times a cricket chirps per minute, as a function of the temperature, t, measured in degrees Fahrenheit:

$$n = f(t) = 4t - 40$$

So, given the temperature, we can determine how many times a minute a cricket should chirp.

Now suppose that one summer night you are without a thermometer and decide to use the crickets to tell the temperature. That means you will need to go backward from n to find t. Fortunately, the equation

$$n = 4t - 40$$

can be solved for t, giving

$$t = \frac{n}{4} + 10$$

If we define

$$g(n) = \frac{n}{4} + 10$$

then g is the function giving t as a function of n.

The functions f and g are "opposites" of one another in the sense that f starts with a temperature and gives you the number of chirps, and g takes the number of chirps and gives you the temperature.

Schematically:

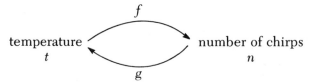

The functions f and g are actually opposites in an even more precise way. Suppose that you started with a certain temperature, and using f, calculated the number of chirps you'd get at that temperature. Then suppose you used g to calculate the temperature required to produce that number of chirps. But g was invented to tell you what temperature had given rise to a certain number of chirps, and so, unless your arithmetic is horrible, you should end up with the temperature you started with.

This means that if you apply the function f to t, and then apply g to that, you get back to the t with which you started:

$$t \xrightarrow{\quad f \quad} n \xrightarrow{\quad g \quad} t$$

But applying f and then g in succession is composing them, and so what we have show is that

$$g(f(t)) = t$$

Similarly, if we start with the number of chirps, use g to calculate the temperature, and then use f to find how many chirps this temperature generates, we'll get the number we started with—we are clearly going in circles. In functional notation:

$$n \xrightarrow{\quad g \quad} t \xrightarrow{\quad f \quad} n \quad \text{ or } \quad f(g(n)) = n$$

It is useful to check algebraically that the functions f and g do indeed satisfy these relationships. Now

$$f(t) = 4t - 40 \quad \text{and} \quad g(n) = \frac{n}{4} + 10$$

so

$$g(f(t)) = g(4t - 40) = \frac{(4t - 40)}{4} + 10 = t - 10 + 10 = t$$

so

$$g(f(t)) = t$$

Similarly

$$f(g(n)) = f\left(\frac{n}{4} + 10\right) = 4\left(\frac{n}{4} + 10\right) - 40 = n + 40 - 40 = n$$

so

$$f(g(n)) = n$$

Since the actual letter used for the independent variable is unimportant, we could equally well have written x instead of t and n, giving:

$$g(f(x)) = x \quad \text{and} \quad f(g(x)) = x$$

These equations tell us that applying f and g in turn always takes us back to where we started, namely x. Therefore, f and g each "undo" what the other one does, and so are called *inverse functions* to one another. In summary:

---

f and g are inverse functions if they satisfy

$$g(f(x)) = x$$

and

$$f(g(x)) = x$$

---

g is said to be the *inverse of* f and f is the *inverse of* g.

If we want to point out the fact that some function is the *inverse function to f, we call it f⁻¹.* The equations connecting f^{-1} and f read:

$$f^{-1}(f(x)) = x$$

and

$$f(f^{-1}(x)) = x$$

An objection to the $f^{-1}(x)$ notation   In many ways $f^{-1}(x)$ is horrible notation because it might make you think we are talking about $(f(x))^{-1}$ or $\dfrac{1}{f(x)}$, which we most definitely are not.

$$f^{-1}(x) \text{ is the inverse function of } f \text{ applied to } x$$

and

$$(f(x))^{-1} \text{ is } \frac{1}{f(x)}, \text{ the reciprocal of } f(x)$$

These are different!
   For example, if $f(x) = 4x - 40$, then

$$f^{-1}(x) = \frac{x}{4} + 10 \qquad (f^{-1} \text{ is what we were calling } g \text{ earlier})$$

whereas

$$(f(x))^{-1} = \frac{1}{4x - 40}$$

Unfortunately, the f^{-1} notation is so commonly used that we will have to put up with it.

PROBLEM SET   9.3

Show that the following pairs of functions are inverses.
1. $f(x) = x + 2$ and $g(t) = t - 2$
2. $h(u) = 3u - 4$ and $k(v) = \dfrac{v + 4}{3}$
3. $f(x) = x^2 + 1$ and $g(s) = \sqrt{s - 1}$
4. $h(x) = \sqrt{2x}$ and $k(t) = \dfrac{t^2}{2}$
5. $g(s) = \sqrt[3]{s + 1}$ and $m(u) = u^3 - 1$
6. $f(x) = \dfrac{x}{4} - \dfrac{3}{2}$ and $g(t) = 4\left(t + \dfrac{3}{2}\right)$
7. $h(t) = \sqrt{t} + 1$ and $k(s) = (s - 1)^2$

8. $f(u) = u^3 + 2$ and $g(x) = \sqrt[3]{x - 2}$

9. $f(x) = \dfrac{1}{x}$ and $g(t) = \dfrac{1}{t}$

10. $f(s) = \dfrac{1}{s + 3}$ and $h(x) = \dfrac{1}{x} - 3$

11. $f(x) = 2x^3 - 4$ and $g(t) = \sqrt[3]{\dfrac{t + 4}{2}}$

12. $h(t) = t^3 - 6$ and $m(u) = \sqrt[3]{u^2 + 6}$

13. $f(t) = (t^2 - 1)^2 + 2$ and $g(x) = \sqrt{\sqrt{x - 2} + 1}$

14. $h(x) = (\sqrt{x} + 3)^2 - 4$ and $f(s) = (\sqrt{s + 4} - 3)^2$

15. $f(x) = \dfrac{x + 2}{x + 1}$ and $g(t) = \dfrac{t - 2}{1 - t}$

16. $h(t) = \dfrac{t}{t + 3}$ and $m(u) = \dfrac{3u}{1 - u}$

9.4   CONSTRUCTING INVERSE FUNCTIONS

Suppose we are given a function like

$$f(x) = \frac{x + 2}{x + 1}$$

and want to find its inverse, $f^{-1}(x)$. The best way is to remember that f and f^{-1} must satisfy

$$f(f^{-1}(x)) = x$$

Using the formula for $f(x)$ and replacing x by $f^{-1}(x)$, we see that

$$f(f^{-1}(x)) = \frac{f^{-1}(x) + 2}{f^{-1}(x) + 1} = x$$

This is a literal equation which can be solved for $f^{-1}(x)$ as follows:

$$\frac{f^{-1}(x) + 2}{f^{-1}(x) + 1} = x$$

so

$$f^{-1}(x) + 2 = x(f^{-1}(x) + 1)$$
$$f^{-1}(x) + 2 = xf^{-1}(x) + x$$
$$f^{-1}(x) - xf^{-1}(x) = x - 2$$
$$f^{-1}(x)(1 - x) = x - 2$$
$$f^{-1}(x) = \frac{x - 2}{1 - x}$$

[*Note:* **This method doesn't work if you use** $f^{-1}(f(x)) = x$ **instead of** $f(f^{-1}(x)) = x$. **Using** $f^{-1}(f(x)) = x$ **tells you that** $f^{-1}\left(\dfrac{x+2}{x+1}\right) = x$ **which is not much help because it doesn't contain** $f^{-1}(x)$. **The moral is that you need to have the unknown function on the inside.**]

Check: Since we got $f^{-1}(x)$ by assuming that it satisfied $f(f^{-1}(x)) = x$, we will only check that it satisfies the other equation, namely $f^{-1}(f(x)) = x$.

$$f^{-1}(f(x)) = f^{-1}\left(\frac{x+2}{x+1}\right) \qquad \text{since } f(x) = \frac{x+2}{x+1}$$

$$= \frac{\dfrac{x+2}{x+1} - 2}{1 - \dfrac{x+2}{x+1}} \qquad \text{since } f^{-1}(x) = \frac{x-2}{1-x}$$

$$= \frac{\dfrac{(x+2) - 2(x+1)}{x+1}}{\dfrac{(x+1) - (x+2)}{x+1}} \qquad \begin{array}{l}\text{(getting an L.C.D. in top and} \\ \text{bottom)}\end{array}$$

$$= \frac{x + 2 - 2x - 2}{x + 1 - x - 2}$$

$$= \frac{-x}{-1}$$

$$= x \qquad\qquad\qquad \text{O.K.!}$$

Alternative Method for Finding f^{-1}

If you think of the function f as giving you y in terms of x by the formula

$$y = \frac{x+2}{x+1}$$

then finding f^{-1} means finding a formula that will give you x in terms of y—in other words, solving the formula for x:

$$y(x + 1) = x + 2$$

$$y - 2 = x - yx$$

so

$$x = \frac{y-2}{1-y}$$

Therefore the function of y that gives you x is

$$f^{-1}(y) = \frac{y-2}{1-y}$$

Notation: $f^{-1}(x)$ and $f^{-1}(y)$

Comparing the formulas that we got for f^{-1} by the two methods,

$$f^{-1}(x) = \frac{x-2}{1-x} \quad \text{and} \quad f^{-1}(y) = \frac{y-2}{1-y}$$

you see that they represent exactly the same function, only in one the independent variable is called x, and in the second, y.

Since the name of the independent variable does not matter, both of these answers are equally right. However, each formula arises from a different way of looking at inverse functions, and so one may be useful in one place and one another.

$f^{-1}(x)$ puts f^{-1} on a footing with every other function, in which the independent variable is usually called x. This way of looking at f^{-1} is necessary if we are to draw a graph of f^{-1}, in which case the independent variable must be along the horizontal axis and is usually called x.

$f^{-1}(y)$ emphasizes the fact that f^{-1} is the inverse function of $y = f(x)$, meaning that it starts with a y and gives you back the x from which it came; the independent variable is therefore y. Writing f^{-1} this way is useful if we want to study f^{-1} via the graph of f. There f will take you from the variables along the x axis to those along the y axis, and f^{-1} will take you from the y axis back to the x axis. See Figure 9.5.

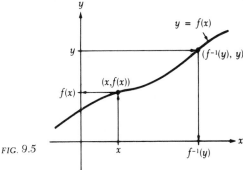

FIG. 9.5

EXAMPLE: *Find the inverse function of $f(x) = x^2$.*

In this function we get y from x by the formula:

$$y = x^2$$

We want the inverse function, that is, a formula giving us x in terms of y. So we solve for x in terms of y, getting:

$$x = \pm\sqrt{y}$$

Therefore the inverse function seems to be

$$f^{-1}(y) = \pm\sqrt{y}$$

But this is very confusing, because we seem to have come up with *two* different values for $f^{-1}(y)$, namely $+\sqrt{y}$ and $-\sqrt{y}$.

Looking again at the original function, $f(x) = x^2$, will give you an idea of what has happened. Suppose $y = f(x) = x^2$; then in looking for an inverse function we are asking for a function that will tell us what x any y came from. But any positive y came from two *different* x's (namely, $\sqrt{y}$ and $-\sqrt{y}$) and so there can be no such thing as an inverse function. The "double function"

$$f^{-1}(y) = \pm\sqrt{y}$$

is the best we can do. See Figure 9.6.

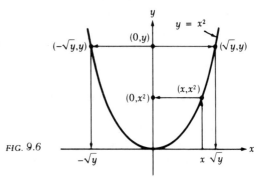

FIG. 9.6

However, it is possible to construct a real and proper single-valued inverse function. Suppose you say that you are interested just in positive x's: Then there is no ambiguity as to which x any y comes from—it must be $\sqrt{y}$ (the positive square root). Under these conditions, the inverse function of $f(x) = x^2$ can be defined, and is

$$f^{-1}(y) = \sqrt{y}$$

Throwing out all negative x's amounts to restricting the domain of the original function f. The graph is then as shown in Figure 9.7.

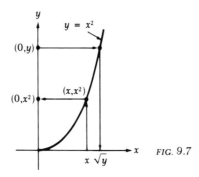

FIG. 9.7

From the graph you can see that each y value now gives rise to only one x value, because each horizontal line cuts the graph at most once.

Therefore, restricting the domain of the original function has made it possible to define an inverse function.

PROBLEM SET   9.4

Find inverses of the following functions.   Don't worry about restricting domains.

1. $f(x) = 3x$
2. $f(x) = -2x - 7$
3. $f(t) = 2t^2$
4. $h(x) = 12x^2$
5. $f(x) = 3(x^2 + 12)$
6. $g(x) = \dfrac{1}{x - 3}$
7. $k(x) = \dfrac{1}{x^2}$
8. $f(x) = \dfrac{1}{x - 4}$
9. $f(x) = \dfrac{x + 2}{x}$
10. $g(t) = \dfrac{t + 2}{t - 2}$
11. $h(u) = \dfrac{2u + 1}{u}$
12. $f(x) = \dfrac{-2x - 3}{x + 6}$
13. $f(x) = \sqrt{x + 2}$
14. $f(x) = \sqrt{x} - 2$
15. $g(t) = (t + 1)^2$
16. $h(x) = (\sqrt{x} - 3)^2$
17. $f(t) = \sqrt{t - 1} - 2$
18. $f(x) = \sqrt[3]{3x} + 2$
19. $g(u) = \dfrac{1}{\sqrt{u}}$
20. $h(u) = \dfrac{1}{\sqrt{2u + 2}}$
21. $f(s) = \dfrac{1}{\sqrt{s}} + 2$

22. $g(s) = \dfrac{1}{\sqrt{s} + 2}$
23. $h(x) = \left(\dfrac{1}{x}\right)^2$
24. $h(x) = \left(\dfrac{1}{x} - 2\right)^2$
25. $f(x) = \left(\dfrac{x}{x + 4}\right)^2$
26. $g(x) = \left(\dfrac{x - 1}{x - 6}\right)^3$
27. $k(x) = \sqrt{\dfrac{x}{x - 1}}$
28. $m(x) = \sqrt{\dfrac{x + 1}{x}}$
29. $f(x) = \sqrt{\dfrac{x - 1}{x + 1}}$
30. $f(x) = \dfrac{1}{\sqrt{\dfrac{x - 1}{x}}}$

31. $f(x) = \sqrt{\sqrt{x + 2} - 2}$
32. $f(x) = [(x + 1)^2 - 3]^2$
33. $f(x) = \dfrac{1}{\left(\dfrac{x + 1}{x - 3}\right)^2}$
34. $g(x) = \left(\sqrt{\dfrac{x + 1}{x - 3}}\right)^3$
35. $h(x) = \left(\sqrt{\dfrac{x}{x + 4}} - 2\right)^2$

Construct the inverses of the following functions, making any necessary restrictions on the domain of the original function:

36. $f(x) = x^2 + 1$
37. $f(x) = \dfrac{1}{x^4 - 2}$
38. $g(t) = \sqrt{t^2 + 9}$
39. $h(u) = (u^2 + 3)^2$

40. What do you notice about all of the original functions in Problems 36–39? Are these the only kinds of functions for which the domain must be restricted, or are there others?

Restrict the domain of the following functions so that an inverse function can be defined, but do not construct the inverse function.

41. $f(x) = x^2 - 2x - 3$
42. $f(x) = x^2 + 4x + 4$
43. $f(x) = x^2 - 1$
44. $f(x) = x^2 - 9x + 20$
45. $f(x) = x^3 + 3x^2 + 2x$

9.5   GRAPHS, DOMAINS, AND RANGES OF INVERSE FUNCTIONS

The domain and range of an inverse function, f^{-1}, can easily be found by looking at the graph of f, and without drawing a graph of f^{-1} itself.

For example, consider $f(x) = x^2$. Then if x is restricted to be positive, the inverse function is $f^{-1}(y) = \sqrt{y}$, and the graph is as shown in Figure 9.8.

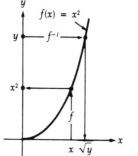

FIG. 9.8

The Domain of f^{-1} is all the possible values of its independent variable, namely y. The only possible y values you can put into the formula $f^{-1}(y) = \sqrt{y}$ are positive real numbers or zero, and those therefore constitute the domain.

The Range of f^{-1} is all possible values of its dependent variable, x. Clearly only positive real numbers or zero can be produced by the formula $x = f^{-1}(y) = \sqrt{y}$, and so these are the domain.

You should notice that you can also get the domain of f^{-1} by seeing which horizontal lines cut the graph of $y = f(x)$ (namely, those of positive or zero y), and the range by looking at which vertical lines cut the graph (namely, those of positive or zero x). This is the opposite way round from usual (when we look at vertical lines to find the domain, and horizontal lines to find the range), because the independent variable for f^{-1} is arranged up the y axis and the dependent variable along the x axis.

EXAMPLE: *Find the domain and range of the inverse function of*

$$f(x) = 1 + \frac{1}{x-2}$$

Notice first that we only have to find the domain and range of f^{-1}; we aren't asked to find f^{-1} itself. Since, in the examples above, the way we were finding the domains and ranges of f^{-1} was by looking at the graphs of f, there is actually no need to find f^{-1}. But we do have to draw a graph of

$$f(x) = 1 + \frac{1}{x-2}$$

This we can get by moving the graph of $y = \dfrac{1}{x}$ over by 2, so that it "blows up" at $x = 2$, and then up by 1, so that the horizontal asymptote is $y = 1$. The result is shown in Figure 9.9.

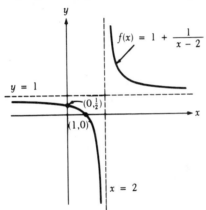

FIG. 9.9

domain of f^{-1} = all y values for which you can read off an x

= all real numbers except 1

range of f^{-1} = all x values that can be read off

= all real numbers except 2

You should notice that

domain of f = all reals except 2

range of f = all reals except 1

so that, in this case at least:

$$\boxed{\begin{aligned} \text{domain of } f^{-1} &= \text{range of } f \\ \text{range of } f^{-1} &= \text{domain of } f \end{aligned}}$$

The domains and ranges of f and f^{-1} will always be related in this way (unless you have to restrict the domain of f to create f^{-1}) because f^{-1} was invented to do "the opposite" of f, and so f^{-1}'s input (domain) is f's output (range) and vice versa.

Up until now we have been studying f^{-1} by looking at the graph of f. It would be interesting to draw f^{-1} a graph of its own—that is, one where the domain is down the x axis and the range is up the y axis in the usual fashion, instead of this funny business of having f^{-1}'s domain displayed on the y axis. This means thinking of f^{-1} as a function of x rather than y, and drawing a graph of $y = f^{-1}(x)$.

For example, consider $f(x) = x^2$. Then the inverse function is $f^{-1}(x) = \sqrt{x}$ (for $x \geq 0$) and its graph is shown in Figure 9.10. The graph of f is drawn in for comparison. You will see that it is in the same relation to the y axis as the graph of f^{-1} is to the x axis—which is not surprising since the function f^{-1} can be thought of as the function f working backwards from the y axis.

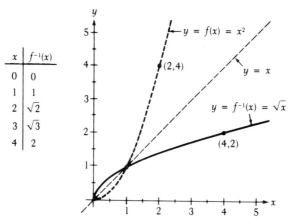

FIG. 9.10

More precisely, what you should notice is that the graphs of f and f^{-1} are symmetric about the diagonal line through the origin (the line $y = x$). In other words, the graph of $f^{-1}(x) = \sqrt{x}$ is the reflection of the graph of $f(x) = x^2$ in the line $y = x$.

The reason for this depends on the fact that when you reflect in the line $y = x$, the x and y coordinates change places, so the point $(1, 3)$ goes into the point $(3, 1)$ and the point (a, b) into (b, a), as shown in Figure 9.11. Now points on the graph of $y = x^2$ have coordinates like $(1, 1), (2, 4), (3, 9)$, in which the second coordinate is the square of the first. Under reflection these points go into $(1, 1), (4, 2), (9, 3)$, in which the first coordinate is the square of the second, or, equivalently, the second coordinate is the (positive) square root of the first. Hence all the reflected points are on the graph of $y = \sqrt{x}$. Therefore reflection takes points on $y = x^2$ into points on $y = \sqrt{x}$.

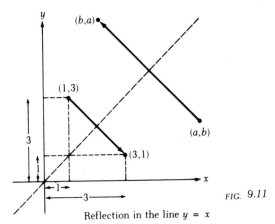

Reflection in the line $y = x$

FIG. *9.11*

Exactly the same arguments apply to the graphs of $y = f(x)$ and $y = f^{-1}(x)$. So it is always true that:

> The graphs of $y = f(x)$ and $y = f^{-1}(x)$ are reflections of one another in the line $y = x$.

PROBLEM SET   9.5

Determine the domain and range of the inverses of the following functions by examining the graph of the original function. You need not find the inverse function.

1. $f(x) = \dfrac{1}{x + 2}$

2. $f(x) = \dfrac{1}{x} + 2$

3. $f(x) = \dfrac{1}{x + 3} - 3$

4. $f(x) = \sqrt{x - 4}$

5. $f(x) = \sqrt{x + 2} - 3$

6. $f(x) = x^3 + 1$

7. $f(x) = (\sqrt{x + 1})^2$

8. $f(x) = \dfrac{1}{3x}$

9. $f(x) = 3(x + 2)$

10. $f(x) = \dfrac{1}{x^2 - x}, \quad x \ge 0$

Graph the following functions to determine the domains and ranges of their inverses. Place restrictions on the domain of the original function where appropriate.

11. $f(x) = x^2 - 2x - 3$ (*Hint:* Complete the square.)

12. $f(x) = x^2 + 6x + 9$

13. $f(x) = x^2 - 8$

14. $f(x) = x^4 + 2$

15. $f(x) = (x - 1)^4$

16. $f(x) = (x + 1)^4 - 2$

For each of the following functions: (a) graph the function; (b) construct the inverse; (c) give the domain and range of the inverse; (d) graph the inverse function on the same set of axes as the original function; and (e) check to see that the two graphs are reflections of one another in the line $y = x$.

17. $f(x) = 2x$
18. $f(x) = x - 3$
19. $f(x) = -2x + 2$
20. $f(x) = -3x - 4$
21. $f(x) = -x$
22. $f(x) = x$
23. $f(x) = x^3$

24. $f(x) = x^3 + 1$
25. $f(x) = (x - 1)^3$
26. $f(x) = (x + 2)^3 - 3$
27. $f(x) = (2x + 1)^3 - 2$
28. $f(x) = x^5$
29. $f(x) = x^5 - 1$
30. $f(x) = (x + 1)^5$

For each of the following functions: (a) graph the function; (b) construct the inverse, restricting the domain where necessary; (c) give the domain and range of the inverse; (d) graph the inverse function on the same set of axes as the original function; and (e) check to see that the two graphs are reflections of one another in the line $y = x$.

31. $f(x) = 3 + \dfrac{1}{x}$

32. $f(x) = -2 - \dfrac{1}{x - 1}$

33. $f(x) = \dfrac{x - 6}{x + 4}$

34. $f(x) = \sqrt{x^2 - 16}$

35. $f(x) = 1 - \dfrac{x - 1}{x + 1}$

36. $f(x) = x^2 - 2x - 3$
    (*Hint:* Complete the square.)

37. $f(x) = 2x^2 - 5x + 2$

38. $f(x) = x^2 - 3x + 2$

39. $f(x) = \sqrt{x^2 - 3x + 2}$

40. Explain in your own words why the graphs of $y = f(x)$ and $y = f^{-1}(x)$ are reflections of one another in the line $y = x$.

CHAPTER 9 REVIEW

1. Suppose

$$f(x) = \frac{x}{x + 1} \quad \text{and} \quad g(x) = \frac{1}{x + 1}$$

(a) Show that $f(x) + g(x) = (x + 2)f(x + 1)g(x)$.
(b) Find $f^{-1}(x)$ and $f(g(x))$.
(c) Decompose $g(x)$ into two functions [that is, find $P(x)$ and $Q(x)$ such that $P(Q(x)) = g(x)$].

2. Write $h(x) = 10x$ as a composite function [that is, write $h(x) = f(g(x))$, where neither $f(x) = x$ nor $g(x) = x$].

Let $D(x) = \dfrac{2x}{x - 1}$ and $F(x) = \dfrac{2x + 3}{3}$.

3. Find $D(F(x))$.
4. Find $F^{-1}(x)$.

5. Sketch a graph of $D^{-1}(x)$.

6. Given that $f(x) = \dfrac{2x + 1}{x + 3}$

    (a) Find $f^{-1}(x)$.

    (b) Graph $f^{-1}(x)$.

7. Find nontrivial functions g, h, and j making

$$g(h(j\,(x))) = \frac{x(x + 2)}{5}$$

8. $f(x) = 2x + 1$; $g(x) = \left|\dfrac{x + 1}{x - 3}\right|$

What value(s) of x make $f(g(x)) = 10$?

Let $h(x) = \dfrac{x + 2}{x - 1}$ and $g(x) = x^2 + 6x + 6$

Find:

  9. $h^{-1}(x)$

 10. $h^{-1}(g(x))$

Let $f(x) = 2x$ and $k(x) = \dfrac{2}{x^2 - 4}$

Find:

 11. The domain of k

 12. The composition $k(f(x))$. Simplify your answer.

 13. Find nontrivial functions $k(x)$ and $h(x)$ such that

$$k(h(x)) = \frac{3x^4 + 4}{3x^4 - 6}$$

 14. If $f(x) = \dfrac{1}{x^2}$, what is $f(f(x))$?

 15. Find a pair of nontrivial functions, f and g, such that

$$f(g(x)) = \sqrt{x^2 + 1}$$

Let $f(x) = x^2 + 9$ and $g(x) = 2x + 5$

Find:

 16. $f(g(x))$

 17. $g(f(x))$

Let $f(x) = x - 1$ and $g(s) = \dfrac{s^2 + 2s}{s + 1}$

 18. What is $g(f(x))$?

 19. What is $g^{-1}(x)$?

 20. Graph $g(f(x))$.

 21. If $f(x) = (x - 1)^2$, simplify $\dfrac{f(x^2)}{f(x)}$

Barbara is worth twice her weight in gold. In other words, she is worth twice what she would be worth were she made of gold.

22. Express Barbara's weight in ounces, N, as a function of her weight in pounds, P.

23. Express Barbara's worth, W, as a function of her weight in ounces, N, if gold sells for \$35 per ounce.

24. Express Barbara's worth as a function of her weight in pounds; that is, find $W(N(P))$.

25. Find the inverse to the function you found in Problem 24.

26. How much is Barbara worth if she weighs 110 pounds?

27. How much does Barbara weigh if she is worth \$58,800?

If the number of cricket chirps per minute, n, is related to the temperature, t (in degrees Fahrenheit) by the function $n = 4t - 40$:

28. Find the function relating the number of chirps to the temperature in degrees Celsius, C. [Remember that $C = \frac{5}{9}(t - 32)$.]

29. Find the inverse of the function found in Problem 28.

30. If there are 70 chirps per minute, what is the Celsius temperature?

31. Let $f(x) = \dfrac{x + 4}{x}$ and $f(g(x)) = \dfrac{x + 14}{x + 2}$

(a) Find $g(x)$.
(b) Graph $f^{-1}(x)$.

32. If $f(x) = (x - 1)^2$, find the simplest expression for $\dfrac{f(f(x))}{x^2}$

33. Suppose we define a function $F(x)$ by the "functional equation"

$$F(x + 1) = (x + 1) \cdot F(x)$$

and we know furthermore that $F(0) = 1$. Evaluate $F(1)$, $F(2)$, $F(3)$, and give an explicit formula, not involving the function F itself, for $F(n)$, where n is a positive integer.

Let $f(x) = \dfrac{2x + 1}{x - 3}$ and $g(x) = \dfrac{x - 1}{2x - 1}$

34. Find and simplify $g(g(x))$.
35. Find and simplify $f^{-1}(x)$.
36. Plot the graph of $f^{-1}(x)$ using the methods we have discussed.

Let $f(x) = \dfrac{x - 9}{x - 4}$ and $g(x) = x^2$

37. Find $f(g(x))$.
38. Draw the graph of $f(g(x))$.
39. Find $f^{-1}(x)$.

Let $f(x) = \dfrac{x+1}{x-3}$ and $g(x) = 2x - 5$

40. Find $f^{-1}(x)$, $g^{-1}(x)$ and $f(g(x))$
41. Verify that $(fg)^{-1}(x) = g^{-1}(f^{-1}(x))$

42. Find $G\left(\dfrac{1}{x}\right)$ if $G^{-1}(\theta) = \left(\dfrac{1+\theta}{1-\theta}\right)^2$

The Town of Functionville has its own train system, pictured in Figure 9.12. Trains there run only clockwise, and their speed is controlled by the three tunnels, A, B, and C. If a train enters tunnel A at x mph, it leaves at $3x - 8$ mph, while a train entering tunnel B at x mph leaves at $\dfrac{2x + 3}{5}$ mph.

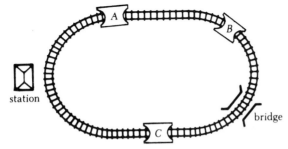

FIG. 9.12

43. If a train leaves the station at 7 mph, how fast is it traveling when it gets to the bridge?

44. If a train leaves the station at a speed of x mph, how fast is it traveling when it gets to the bridge?

45. If a train enters tunnel C at x mph, how fast must it be moving when it leaves so that the train will always be passing the station at the same speed?

10   EXPONENTS AND RADICALS

10.1   EXPONENTS AND THE EXPONENT RULES

Just as you can think of multiplication as an abbreviated way of writing addition, since for example

$$5x = x + x + x + x + x$$

so you can think of exponents as an abbreviated way of writing multiplication. For example

$$x^7 = x \cdot x \cdot x \cdot x \cdot x \cdot x \cdot x$$

x^7 is said to be in *exponential form*, where x is the *base*, and 7 the *exponent* or the *power*.

Scientific notation should convince you that exponents can be a real convenience, expecially for expressing extremely large or extremely small numbers. For example, it would take 3,000,000,000,000,000,000,000,000,000 candles to produce as much light as the sun. In scientific notation this number is written much more compactly as $3 \cdot 10^{27}$. The weight of an atom of gold is 0.000,000,000,000,000,000,000,115 ounces, which becomes $1.15 \cdot 10^{-22}$ ounces in scientific notation.

Exponents are equally useful in dealing with algebraic expressions, and so we will look into them in considerable detail. I assume that you are happy with the definition of x^n (where n is a positive integer), namely:

$$x^n = \underbrace{x \cdot x \cdot \cdots \cdot x}_{n \text{ factors}}$$

Using this definition, you can see that if we multiply x^n and x^m, we get

$$x^n \cdot x^m = \underbrace{(x \cdot x \cdots x)}_{n \text{ factors}} \cdot \underbrace{(x \cdot x \cdot x \cdots x)}_{m \text{ factors}} = x^{n+m}$$
$$\underbrace{}_{(n+m) \text{ factors in all}}$$

This gives us

> ### *The First Exponent Rule:* $x^n \cdot x^m = x^{n+m}$

Thus, $2^3 \cdot 2^5 = (2 \cdot 2 \cdot 2) \cdot (2 \cdot 2 \cdot 2 \cdot 2 \cdot 2) = 2^8$. Therefore when you multiply numbers in exponential form, you add the exponents, but remember to leave the base alone. Unfortunately this rule cannot be used to multiply something like $2^3 \cdot 3^5$, in which the bases are different.

If now we divide x^m by x^n, assuming $m > n$ (in other words, that the fraction is top heavy) we get:

$$\frac{x^m}{x^n} = \frac{\overbrace{x \cdot x \cdot x \cdots x \cdot x}^{m \text{ factors}}}{\underbrace{x \cdot x \cdot x \cdots x}_{n \text{ factors}}} = \frac{\overbrace{x \cdot x \cdots x}^{(m-n) \text{ factors}}}{1} = x^{m-n}$$

cancel n factors out
of top and bottom, leaving
$(m - n)$ in the top

So we get

> ### *The Second Exponent Rule:* $\dfrac{x^m}{x^n} = x^{m-n}$ for $m > n$

For example, $\dfrac{3^7}{3^4} = \dfrac{3 \cdot 3 \cdot 3 \cdot 3 \cdot 3 \cdot 3 \cdot 3}{3 \cdot 3 \cdot 3 \cdot 3} = 3^3$. Therefore to divide numbers in exponential form, you subtract the exponents — but again please leave the base alone. As before, this is of no help for things like $\dfrac{2^7}{5^4}$, which just have to be left the way they are.

Lastly, if we take the n-th power of x^m, we get

$$(x^m)^n = (x^m) \cdot (x^m) \cdots (x^m) = \underbrace{(x \cdots x)}_{m \text{ factors of } x} (x \cdots x) \cdots (x \cdots x) = x^{mn}$$

n factors of $(x \cdots x)$

giving mn factors of x altogether

This gives us

> ### *The Third Exponent Rule:* $(x^m)^n = x^{mn}$

For example, $(5^2)^3 = (5 \cdot 5) \cdot (5 \cdot 5) \cdot (5 \cdot 5) = 5^6$. Therefore to find the "power of a power", you multiply exponents.

So far we have defined x^n where n is a positive integer and we have three rules that expressions of this form must obey. Now comes the question of negative and fractional exponents—that is, what is meant by x^{-3} or $x^{5/2}$. But before we go through the definitions, you presumably want to know why we're defining them at all. What gap do they fill?

To answer that we have to look at a specific value of x—suppose we take $x = 10$. Then we have

$$10^1 = 10$$

$$10^2 = 100$$

$$10^3 = 1000$$

$$10^4 = 10{,}000 \text{ etc.}$$

So we have a series of numbers starting at 10 and going up from there. Given that writing numbers in exponential form is useful (and it really is!), it would be helpful if we could write *all* numbers in this form—including both those between the various powers of ten and those less than ten. As you might guess from the fact that $10^1 = 10$ and $10^2 = 100$, things like $10^{3/2}$ will be between 10 and 100, and things like $10^{5/2}$ will be between 100 and 1000. It is even possible to define 10^π, which will then be between 1000 and 10,000. As you may remember, 10^0 turns out to be 1, and the numbers between one and zero are given by negative exponents.

Now a word about how the definitions are made. In theory, we're free to define x^{-3}, $x^{5/2}$, and so on, in any way we please—obviously some definitions are more convenient than others, but we're free to choose whichever we want. In practice, of course, we have to figure out what it is that we want x^{-3} or $x^{5/2}$ to do, and then pick the definition that does that. In the case of $x = 10$, we know that we want $10^{5/2}$ to be between $10^2 = 100$ and $10^3 = 1000$ and 10^{-3} to be less than 1, but that doesn't help much for the purposes of a definition. More useful is the fact that for positive integral exponents we have those three rules, and those rules are simple and extremely helpful. It would be a major tragedy if things like x^{-3} and $x^{5/2}$ did not satisfy those rules, and therefore we will define them so that they do. We assume that they do satisfy the rules and then see what the rules tell us about how they have to be defined.

Definition of x^0

The Second Exponent Rule tells us that if $m > n$

$$\frac{x^m}{x^n} = x^{m-n}$$

Suppose we assume that this rule holds even when m and n are equal, say $m = n = 2$. Then

$$\frac{x^2}{x^2} = x^{2-2} = x^0$$

But the left-hand side of this must be 1 (anything divided by itself must be 1). So we will have to define

$$x^0 = 1 \qquad \text{for any value of } x \text{ except zero.}$$

For example, $5^0 = 1$, $\left(\dfrac{1}{17}\right)^0 = 1$, $(44.3)^0 = 1$.

Definition of x^{-n}: n a Positive Integer.

First we define x^{-1}, which we again do by assuming that the exponent rules hold in cases for which we didn't prove them. In particular, suppose we can use the second exponent rule to evaluate $\dfrac{x^2}{x^3}$, giving

$$\frac{x^2}{x^3} = x^{2-3} = x^{-1}.$$

But if you cancel an x^2 top and bottom, the left-hand side reduces to $\dfrac{1}{x}$, so we had better define

$$x^{-1} = \frac{1}{x}.$$

Now for x^{-2}. Presuming upon the second rule again

$$\frac{x^2}{x^4} = x^{2-4} = x^{-2}.$$

Cancelling x^2 top and bottom on the left gives $\dfrac{1}{x^2}$ so we define

$$x^{-2} = \frac{1}{x^2}$$

Similar arguments give

$$x^{-3} = \frac{1}{x^3} \text{ and } x^{-4} = \frac{1}{x^4}$$

and in general

$$x^{-n} = \frac{1}{x^n} \qquad \text{for any value of } x \text{ except zero and any positive integer } n.$$

For example,

$$10^0 = 1, \quad 10^{-1} = \frac{1}{10} = 0.1, \quad 10^{-2} = \frac{1}{10^2} = \frac{1}{100} = 0.01, \text{ etc.}$$

Notice that

$$10^{-3} < 10^{-2} < 10^{-1} < 10^0 < 10^1 < 10^2 \ldots \text{ etc.}$$

as we had hoped. Similarly,

$$2^{-1} = \frac{1}{2} \qquad 2^{-n} = \frac{1}{2^n}$$

The definition works just the same for fractions and decimals:

EXAMPLES:
$$\left(\frac{1}{3}\right)^{-1} = \frac{1}{\left(\frac{1}{3}\right)} = 3$$

$$\left(\frac{4}{5}\right)^{-1} = \frac{1}{\left(\frac{4}{5}\right)} = \frac{5}{4}$$

$$\left(\frac{4}{5}\right)^{-2} = \frac{1}{\left(\frac{4}{5}\right)^2} = \frac{1}{\frac{16}{25}} = \frac{25}{16}$$

$$(0.01)^{-1} = \frac{1}{0.01} = 100$$

$$(0.2)^{-3} = \frac{1}{(0.2)^3} = \frac{1}{0.008} = 125$$

Definition of $x^{1/n}$: n a Positive Integer

We will start by defining $x^{1/2}$, again by assuming the truth of the exponent rules for cases in which they were not proved. Suppose we use the third rule on $(x^{1/2})^2$. Then

$$(x^{1/2})^2 = x^{(1/2)(2)} = x^1 = x$$

So $x^{1/2}$ is a number that when squared gives x, which means $x^{1/2}$ is plus or minus the square root of x. Since we are in the business of *defining* $x^{1/2}$ (as opposed to solving for it), we can decide whether we want to use the $+$ or the $-$ sign. Needless to say, we would prefer to have $x^{1/2}$ positive, so define

$$x^{1/2} = \sqrt{x} \qquad \text{Note: This is a real number only if } x \text{ is positive}$$

The definition of $x^{1/3}$ goes along the same lines. Suppose we use the third exponent rule on $(x^{1/3})^3$:

$$(x^{1/3})^3 = x^{(1/3)(3)} = x^1 = x$$

So $x^{1/3}$ cubed is x, which means $x^{1/3}$ has to be the cube root of x. There is no choice about sign here, since the cube root of a number must have the same sign as the number (for example, $2^3 = 8$, giving $\sqrt[3]{8} = 2$; $(-2)^3 = -8$, giving $\sqrt[3]{-8} = -2$). So we have to define

$$x^{1/3} = \sqrt[3]{x}$$

As you might suspect,

$$x^{1/4} = \sqrt[4]{x} \qquad \text{(Assume } x \text{ positive to make } x^{1/4} \text{ real)}$$

where $\sqrt[4]{x}$ is the positive fourth root of x, or the positive number whose fourth power is x.

And, in general:

$$\boxed{\; x^{1/n} = \sqrt[n]{x} \qquad \begin{array}{l} \text{(For even } n\text{, assume } x \text{ is} \\ \text{positive to make } x^{1/n} \text{ real)} \end{array} \;}$$

where $\sqrt[n]{x}$ is the nth root of x, or the number whose nth power is x. If there is any choice, $\sqrt[n]{x}$ is taken to be positive. For example, $4^{1/2} = \sqrt{4} = 2$; $27^{1/3} = \sqrt[3]{27} = 3$; $32^{1/5} = \sqrt[5]{32} = 2$; and $4^{1/3} = \sqrt[3]{4}$—which can't be simplified.

Definition of $x^{m/n}$: m, n Positive Integers

Now the question is, what is meant by things like $x^{2/3}$ or $x^{5/4}$? Again, we assume that the exponent rules hold, giving:

$$x^{2/3} = x^{(1/3)(2)} = (x^{1/3})^2 \qquad \text{(by third exponent rule)}$$
$$= (\sqrt[3]{x})^2 \qquad \text{(by previous definition)}$$

Alternatively, exactly the same argument gives us:

$$x^{2/3} = x^{(2)(1/3)} = (x^2)^{1/3} \qquad \text{(by third exponent rule)}$$
$$= \sqrt[3]{x^2} \qquad \text{(by previous definition)}$$

So it looks as though the exponent rules have led to two different definitions of $x^{2/3}$, namely $x^{2/3} = (\sqrt[3]{x})^2$ and $x^{2/3} = \sqrt[3]{x^2}$. Fortunately, these two turn out to be the same. Remember that one of the things you *can* do with square roots is to say that the square root of a product is the product of the square roots; that is,

$$\sqrt{ab} = \sqrt{a} \cdot \sqrt{b}$$

The same thing is true with cube roots, fourth roots, nth roots:

$$\sqrt[3]{ab} = \sqrt[3]{a}\,\sqrt[3]{b}$$
$$\sqrt[n]{ab} = \sqrt[n]{a}\,\sqrt[n]{b}$$

This means that

$$\sqrt[3]{x^2} = \sqrt[3]{x \cdot x} = \sqrt[3]{x} \cdot \sqrt[3]{x} = (\sqrt[3]{x})^2$$

cube root of a   product of
product   cube roots

so we can define $x^{2/3}$ either way:

$$x^{2/3} = (\sqrt[3]{x})^2 = \sqrt[3]{x^2}$$

Similarly,

$$x^{5/4} = x^{(1/4)(5)} = (x^{1/4})^5 \qquad \text{(by third exponent rule)}$$
$$= (\sqrt[4]{x})^5 \qquad \text{(definition of } x^{1/4})$$

or

$$x^{5/4} = x^{(5)(1/4)} = (x^5)^{1/4} \qquad \text{(by third exponent rule)}$$
$$= \sqrt[4]{x^5} \qquad \text{(definition)}$$

Now, $(\sqrt[4]{x})^5$ and $\sqrt[4]{x^5}$ are equal, so $x^{5/4}$ can be defined either way

$$x^{5/4} = (\sqrt[4]{x})^5 = \sqrt[4]{x^5}$$

In general:

$$x^{m/n} = (\sqrt[n]{x})^m = \sqrt[n]{x^m} \qquad \textit{Note: For even } n, \text{ this}$$
is a real number only
if x is positive.

Definition of $x^{-m/n}$: *m, n* Positive Integers

Assuming the third exponent rule:

$$x^{-m/n} = x^{(m/n)(-1)} = (x^{m/n})^{-1} \qquad \text{(by third rule)}$$
$$= ((\sqrt[n]{x})^m)^{-1} \qquad \text{(by definition of } x^{m/n})$$
$$= \frac{1}{(\sqrt[n]{x})^m} \qquad \text{(by definition of } x^{-1})$$

So

$$x^{-m/n} = \frac{1}{(\sqrt[n]{x})^m} \qquad \text{Again, } x \text{ must be positive for even } n.$$

For example,

$$(4)^{3/2} = (\sqrt{4})^3 = 2^3 = 8$$

If we had worked this out using the other definition, we would have found:

$$(4)^{3/2} = \sqrt{4^3} = \sqrt{64} = 8$$

So the answer comes out the same (as it most certainly is supposed to), but you have to deal with larger numbers if you take the power first and the root second. Consequently, the first way is usually easier. Evaluating $(27)^{4/3}$ both ways will dramatize this.

The first way:

$$(27)^{4/3} = (\sqrt[3]{27})^4 = 3^4 = 81$$

The second way:

$$(27)^{4/3} = \sqrt[3]{27^4} = ???$$

Now 27^4 is a huge number which I would hate to have to work out, and then we would have to take its fourth root, which would be worse. The first way is definitely better!

Notice that

$$1^{1/2} = \sqrt{1} = 1, \qquad 1^{2/3} = (\sqrt[3]{1})^2 = 1^2 = 1,$$

$$1^0 = 1, \qquad 1^{-1} = \frac{1}{1} = 1, \qquad 1^{-2} = \frac{1}{1^2} = 1$$

In short, all powers of 1 are 1.

EXAMPLES:

$$8^{-1/3} = \frac{1}{8^{1/3}} = \frac{1}{\sqrt[3]{8}} = \frac{1}{2}$$

$$16^{-5/4} = \frac{1}{16^{5/4}} = \frac{1}{(\sqrt[4]{16})^5} = \frac{1}{2^5} = \frac{1}{32}$$

$$(0.01)^{1/2} = \sqrt{0.01} = 0.1$$

$$(0.01)^{-1/2} = \frac{1}{(0.01)^{1/2}} = \frac{1}{0.1} = 10$$

$$(0.027)^{-2/3} = \frac{1}{(\sqrt[3]{0.027})^2} = \frac{1}{(0.3)^2} = \frac{1}{0.09} = \frac{100}{9}$$

Definition of x^r: r Irrational

If r is irrational, then it cannot be written as a fraction in the form $\frac{m}{n}$ (for m and n integers), and so we can't use any of the definitions so far. However, any irrational number can always be approximated as closely as you want by a fraction. Therefore, to find x^r, use the above definitions to find $x^{m/n}$, where m/n is approximately equal to r. The closer $\dfrac{m}{n}$ is to r, the closer $x^{m/n}$ is to x^r, so you can find x^r as accurately as you want by taking $\dfrac{m}{n}$ close enough to r.

There is one problem concerning the exponent rules which I will bring up and then ignore. You originally saw them proved just for the case in which the exponents were positive integers. We then assumed they were true for other exponents and used them to determine how expressions with other exponents must be defined to satisfy them. But in each case only *one* rule was used to make the definition, and so although we know that the quantity was defined to satisfy that one rule, what about the other two? We really should be checking that they are satisfied too—but we won't, because it takes time, and I presume that you believe it.

To summarize before moving on:

Definitions

$$x^0 = 1$$

$$x^{-n} = \frac{1}{x^n}$$

$$x^{1/n} = \sqrt[n]{x}$$

$$x^{m/n} = (\sqrt[n]{x})^m$$

$$x^{-m/n} = \frac{1}{(\sqrt[n]{x})^m}$$

The Exponent Rules

1. $x^m \cdot x^n = x^{m+n}$

2. $\dfrac{x^m}{x^n} = x^{m-n}$

3. $(x^m)^n = x^{mn}$

PROBLEM SET   10.1

Evaluate:

1. $9^{1/2}$
2. $2^4 \cdot 2^3$
3. $4^2 \cdot 3^2$
4. $16^{3/4}$
5. $16^{-1/4}$
6. $(\frac{1}{4})^{1/2}$
7. $(\frac{1}{5})^{-1}$
8. $(0.16)^{3/2}$

9. $\dfrac{2^{25}}{2^{22}}$
10. 4^{-2}
11. $(3^{-1})^2$
12. $(\frac{1}{64})^{-1/2}$
13. $(\frac{2}{3})^{-2}$
14. $(10^{1.2})(10^{2.8})$
15. $(-8)^{-1/3}$

Simplify without leaving any negative exponents or radical signs.

16. $5^{1+n}\, 5^{1-n}$

17. $\dfrac{3^{n+2}}{3^n}$

18. $x^{2n}x^{n-2}$

19. $(2x)^3 2^{-4}$

20. $(-8x)^{-1/3}$

21. $\dfrac{(6a)^0}{6a^0}$

22. $(4b^3)^2$

23. $\sqrt{\dfrac{9}{x^6}}$

24. $\dfrac{8a^3}{(-2a)^3}$

25. $(a^{-1})^{-1}$

26. $(2x^{-1/2})^{-2}$

27. $\dfrac{2x^{-2} - x^{-1}}{2x^{-2}}$

28. $\dfrac{x^{2a+b}\, x^{2a-b}}{x^{a+b}\, x^{a-b}}$

29. $(a^{1/2}b^{-2}c^{1/4})^{-2}\,(\tfrac{1}{2}a^{-1}b^2\sqrt{c})^{-3}$

30. $(a^{1/2} + b^{1/2})^2$

Evaluate and write the answer in scientific notation.

31. $\left(\dfrac{10^8}{3}\right)^{-1}$

32. $\dfrac{2(2 \cdot 10^{-3})^3}{(4 \cdot 10^{-4})^{1/2}}$

33. $\dfrac{4 \cdot 10^2}{4 \cdot 10^{-2}}$.

34. $\dfrac{5 \cdot 3^2 \cdot 10^{-2}}{0.2}$

35. $\dfrac{5.9 \cdot 10^{99}}{10^{-100}}$

36. $\sqrt{4 \cdot 10^{-4}}$

37. $\dfrac{(2.1 \cdot 10^{-2})\,(3 \cdot 10^{-5})}{7 \cdot 10^{-4}}$

38. $2^{-2}(4 \cdot 10^{-3})10^5$

39. $\dfrac{\frac{1}{9}(4.5 \cdot 10^5)}{(1.5 \cdot 10^{-2})}$

40. $(4 \cdot 10^3)\,(12 \cdot 10^5)\,(48^{-1} \cdot 10^2)\,\dfrac{1}{300{,}000}$

Evaluate:

41. $\dfrac{1}{3^{-2}}$

42. $\dfrac{10^{1.5}}{10^{0.5}}$

43. $\left(\dfrac{2^{-2}}{2^{-3}}\right)^2$

44. $(2^{-2} - 2^{-3})^2$

45. $(\tfrac{4}{9})^{3/2}(\tfrac{1}{8})^{-1/3}(4)^0$

46. $[(2^3)^{1/2}]^{1/3}$

47. $[144(0.0001)^{1/2}]]^{1/2}$

48. $\dfrac{0.9}{10^{-3}} \cdot 3^{-1}$

49. $(0.125)^{-2/3} + \dfrac{3}{2 + 2^{-1}}$

50. $\left(\dfrac{10^{2.65}}{10^{2.15}}\right)^4$

Simplify, leaving no radical signs or negative exponents:

51. $\dfrac{15x^7y^2}{3x^6y^3}$

52. $\dfrac{(\sqrt{ac} + \sqrt{b})^2}{ac + b}$

53. $(ab)\,(ab^2)^{1/2}\,(ab^3)^{-1/3}$

54. $\left(\dfrac{a^{1/2}b^{2/3}}{c^{3/4}}\right)^6 \left(\dfrac{\sqrt{c}}{a^{1/4}b^{1/3}}\right)^9$

55. $\dfrac{36^{x+3}}{6^{x-1}}$

56. $\dfrac{1}{\sqrt{\dfrac{4a^4}{(\frac{1}{2}a)^2}}}$

57. $\dfrac{(m^x + n^x)^2}{m^x - b^x} \cdot \dfrac{m^x}{m^{2x} + 2(mn)^x + n^{2x}}$

58. $\dfrac{(4n^2)^{2/3}}{(4n^2)^{1/6}}$

59. $(a^{-1} + b^{-1} + c^{-1})^{-1}$

60. $\left(\dfrac{a^x + a^{-x}}{2}\right)^2 - \left(\dfrac{a^x - a^{-x}}{2}\right)^2$

Evaluate and write the answer in scientific notation:
61. $0.00036 + 4.2 \cdot 10^{-4}$
62. $3.1 \cdot 10^{-9} + 1.2 \cdot 10^{-10}$
63. $3 \cdot 10^{-6} + 0.4 \cdot 10^{-4} - 0.02 \cdot 10^{-6}$
64. $(3 \cdot 10^2) + (0.02 \cdot 10^4)$
65. $(2.5 \cdot 10^{-2}) (3 \cdot 10^{-6} - 2 \cdot 10^{-5})$
66. $\dfrac{4.2}{10^{13}} + 0.1 \cdot 10^{-11}$
67. $(47 \cdot 10^{-9} - 1.7 \cdot 10^{-8})^{-1}$

10.2   THINGS YOU CAN AND CAN'T DO WITH EXPONENTS

You will, I hope, remember that

$$(x + y)^2 \quad \text{is not equal to} \quad x^2 + y^2$$

because if you multiply out $(x + y)^2$ you get

$$(x + y)^2 = x^2 + 2xy + y^2$$

You may also remember that

$$\sqrt{x + y} \quad \text{is not equal to} \quad \sqrt{x} + \sqrt{y}$$

(try values and see—for example, $x = 9$, $y = 16$). This can be written

$$(x + y)^{1/2} \neq x^{1/2} + y^{1/2}$$

In general, it is true that

$$\boxed{(x + y)^n \neq x^n + y^n \quad \text{(unless } n = 1)}$$

Similarly

$$\boxed{(x - y)^n \neq x^n - y^n}$$

On the other hand,

$$\sqrt{xy} = \sqrt{x} \cdot \sqrt{y}$$

or

$$(xy)^{1/2} = x^{1/2} y^{1/2}$$

Also,

$$(xy)^2 = (xy) \cdot (xy) = x^2 y^2$$

So I hope that you can believe that

$$\boxed{(xy)^n = x^n y^n}$$

Using this relationship with y replaced by y^{-1}, we see that

$$(xy^{-1})^n = x^n (y^{-1})^n = x^n y^{-n}$$

which can be written as

$$\boxed{\left(\frac{x}{y}\right)^n = \frac{x^n}{y^n}}$$

EXAMPLE: *Rewrite*

$$\left(\frac{x^4 y^{-2}}{z^{-6}}\right)^{1/2} \quad and \quad \frac{\pi L^{1/3} q^0}{L^{-1/5}}$$

without zero, negative, or fractional exponents.

$$\left(\frac{x^4 y^{-2}}{z^{-6}}\right)^{1/2} = \frac{x^{4/2} \cdot y^{-2/2}}{z^{-6/2}} = \frac{x^2 y^{-1}}{z^{-3}} = \frac{x^2 \dfrac{1}{y}}{\dfrac{1}{z^3}} = \frac{x^2 z^3}{y}$$

Using $(xy)^n = x^n y^n$

and $\left(\dfrac{x}{y}\right)^n = \dfrac{x^n}{y^n}$

using definition
of negative exponents

$$\frac{\pi L^{1/3} q^0}{L^{-1/5}} = \pi L^{(1/3)-(-1/5)} \cdot 1 = \pi L^{8/15} = \pi \left(\sqrt[15]{L}\right)^8$$

using second
exponent rule
and definition of x^0

EXAMPLE: *Simplify $(a^{-1} + b^{-1})^{-1}$.*

Note: This is *not* equal to $a^1 + b^1$

$$(a^{-1} + b^{-1})^{-1} = \frac{1}{\dfrac{1}{a} + \dfrac{1}{b}} = \frac{1}{\dfrac{b+a}{ab}} = \frac{ab}{b+a}$$

EXAMPLE: *Show that*

$$\left(\frac{e^x + e^{-x}}{2}\right)^2 - \left(\frac{e^x - e^{-x}}{2}\right)^2 = 1$$

where e is a positive constant that comes up in calculus and problems about the growth of a population, and whose value is about 2.7.

To show that this is true means showing that the left side and the right side are really different forms of the same expression—it is an identity, not an equation.

Simplifying the left-hand side:

$$\left(\frac{e^x + e^{-x}}{2}\right)^2 - \left(\frac{e^x - e^{-x}}{2}\right)^2$$

$$= \frac{1}{4}\left[(e^x)^2 + 2e^x \cdot e^{-x} + (e^{-x})^2\right] - \frac{1}{4}\left[(e^x)^2 - 2e^x \cdot e^{-x} + (e^{-x})^2\right]$$

$$= \frac{1}{4}\left[e^{2x} + 2e^{x-x} + e^{-2x} - (e^{2x} - 2e^{x-x} + e^{-2x})\right]$$

$$= \frac{1}{4}\left(e^{2x} + 2 + e^{-2x} - e^{2x} + 2 - e^{-2x}\right)$$

$$= \frac{1}{4}\left(4\right)$$

$$= 1$$

So simplifying the left-hand side leads you to the right-hand side.

PROBLEM SET   10.2

Express the following expressions in the form $a \cdot 2^p$:

1. 16	3. $32 + 16$	5. 4^3	7. $\frac{1}{16}$
2. 32	4. $32 \cdot 16$	6. $(\frac{1}{2})^{24}$	

Solve the following equations for x:

8. $2^x = 8$	16. $27^x = 9$	24. $(0.04)^x = \frac{1}{5}$
9. $(0.01)^x = 0.1$	17. $(\sqrt{2})^x = 4$	25. $(\frac{1}{4})^x = 0.25$
10. $8^x = \sqrt{8}$	18. $2^x = 1$	26. $(0.04)^x = \frac{1}{125}$
11. $10^x = 0.1$	19. $2x^{-2} = 18$	27. $(0.36)^x = 0.216$
12. $125 = 5^x$	20. $10^x = 10$	28. $(0.001)^x = 10$
13. $8^x = \frac{1}{2}$	21. $8^x = 0.5$	29. $27^{-x} = \frac{1}{81}$
14. $0.0001 = 10^x$	22. $(0.04)^x = 25$	30. $(2x)(2^2) = \frac{1}{64}$
15. $8^x = 2$	23. $3^{2x} = 81$	

True or false?

31. $x^{3/2} = x^3 \cdot x^{1/2}$

32. $x^a + x^b = x^{a+b}$

33. $ax^m - ay^m = a(x^m - y^m)$

34. $x^a \cdot x^a = x^{2a}$

35. $y^3 \cdot y^3 = y^9$

Find n:

36. $y^3 \cdot x^3 = (xy)^n$

37. $(vx^2)^{-6} = \dfrac{x^n}{v^6}$

38. $\sqrt[3]{r^2 s^2} = (rs)^n$

39. $\dfrac{(z+y)^2}{(z+y)^{1/2}} = (z+y)^{n/2}$

40. $(32)^n = 2$

41. $\dfrac{\sqrt[6]{c^6 d^3}}{cd^7} = \left(\dfrac{1}{d}\right)^n$

42. $x^2 y^3 z^4 - 2x^3 y^4 z^5 = 2x^2 y^3 z^4 (n - xyz)$

43. Show that

$$\frac{1}{e^x + e^{-x}} = \frac{e^x}{e^{2x} + 1}$$

44. Solve for y:

$$\left[e^{(y-1)}\right]^{2(y+2)} = 1$$

45. Simplify $(1 + 5x)^7 - x^7(x^{-1} + 5)^7$.

46. Evaluate $(5.63)^2 + (2.63)^2 - 2(5.63)(2.63)$.

47. Solve for x and y:

$$\frac{x}{y} = 3$$

$$(e^{x/3})(e^{3y})(e^{-2}) = 1$$

48. Notice that $4^x - 5 \cdot 2^x + 6$ can be written as $(2^x)^2 - 5(2^x) + 6$. Comparing this with $y^2 - 5y + 6$, factor

$$4^x - 5 \cdot 2^x + 6$$

49. Solve for x:

$$100^x - 101 \cdot 10^x + 100 = 0$$

10.3   THINGS YOU CAN AND CAN'T DO WITH RADICALS

A radical is a root, for example $\sqrt{a}$ or $\sqrt[5]{x}$. Since a root can always be written in the form $a^{1/n}$, what you can and cannot do with radicals is just a special case of what you can and cannot do with exponents.

Therefore the exponent rules,

$$(x + y)^p \neq x^p + y^p$$

$$(xy)^p = x^p y^p$$

$$\left(\frac{x}{y}\right)^p = \frac{x^p}{y^p}$$

with p replaced by $\frac{1}{n}$ give you the following rules for radicals:

$$\sqrt[n]{x + y} \neq \sqrt[n]{x} + \sqrt[n]{y}$$

$$\sqrt[n]{xy} = \sqrt[n]{x}\sqrt[n]{y}$$

$$\sqrt[n]{\frac{x}{y}} = \frac{\sqrt[n]{x}}{\sqrt[n]{y}}$$

In order to be able to simplify expressions involving radicals it is helpful to notice that

$$\boxed{\sqrt{a^2} = |a|}$$

The justification for this result is that the square root of a^2 has the same magnitude as the original number a, but $\sqrt{a^2}$ will always be positive, even if a is negative, so $\sqrt{a^2}$ must be the absolute value of a.

Simplifying Radicals

An nth root is said to be *in simplest form* if, when the expression under the radical sign is fully factored, it contains no factors to powers greater than or equal to n.

EXAMPLES: $\sqrt{4a^2b^3} = \sqrt{2^2a^2b^2b} = \sqrt{2^2}\sqrt{a^2}\sqrt{b^2}\sqrt{b} = 2ab\sqrt{b}$ This is in simplest form, assuming that a, b are positive so that you don't need absolute values and that $\sqrt{b}$ is real.

$$\sqrt[3]{\frac{x^7z^{-4}}{16}} = \sqrt[3]{\frac{x^3x^3x}{2^32z^3z}} = \frac{\sqrt[3]{x^3}\sqrt[3]{x^3}\sqrt[3]{x}}{\sqrt[3]{2^3}\sqrt[3]{2}\sqrt[3]{z^3}\sqrt[3]{z}} = \frac{x^2\sqrt[3]{x}}{2\sqrt[3]{2}z\sqrt[3]{z}}$$

$$= \frac{x^2}{2z}\sqrt[3]{\frac{x}{2z}}$$ This is in simplest form.

$\sqrt{z^2p^4} = \sqrt{z^2(p^2)^2} = \sqrt{z^2}\sqrt{(p^2)^2} = |z|p^2$. This is in simplest form. You don't need absolute values on p^2 because it is positive anyhow.

Rationalizing Denominators

It sometimes helps in simplification to be able to write all the fractions containing radicals with the radicals in the numerator. This is called rationalizing the denominator. It is usually done by multiplying top and bottom by some carefully chosen quantity. Here are some examples in which the denominator is a multiple of a radical.

EXAMPLE: *Rationalize the denominator of* $\dfrac{1}{\sqrt{3}}$.

Multiply top and bottom by $\sqrt{3}$

$$\frac{1}{\sqrt{3}} = \frac{1}{\sqrt{3}} \cdot \frac{\sqrt{3}}{\sqrt{3}} = \frac{\sqrt{3}}{3}$$

EXAMPLE: *Rationalize the denominator of* $\dfrac{1}{5q^{2/3}}$

Multiply top and bottom by $q^{1/3}$

$$\frac{1}{5q^{2/3}} = \frac{1}{5q^{2/3}} \cdot \frac{q^{1/3}}{q^{1/3}} = \frac{q^{1/3}}{5q}$$

EXAMPLE: *Rationalize the denominator of* $\dfrac{1}{c\sqrt[5]{ab}}$

Multiply top and bottom by $(\sqrt[5]{ab})^4$:

$$\frac{1}{c\sqrt[5]{ab}} = \frac{1}{c\sqrt[5]{ab}} \cdot \frac{(\sqrt[5]{ab})^4}{(\sqrt[5]{ab})^4} = \frac{(\sqrt[5]{ab})^4}{c(\sqrt[5]{ab})^5} = \frac{(\sqrt[5]{ab})^4}{cab}$$

If the denominator is the sum or difference of square roots, try this:

EXAMPLE: *Rationalize the denominator of* $\dfrac{p}{\sqrt{p} - \sqrt{q}}$

Multiply top and bottom by $\sqrt{p} + \sqrt{q}$.

$$\frac{p}{\sqrt{p} - \sqrt{q}} = \frac{p(\sqrt{p} + \sqrt{q})}{(\sqrt{p} - \sqrt{q})(\sqrt{p} + \sqrt{q})}$$

$$= \frac{p(\sqrt{p} + \sqrt{q})}{(\sqrt{p})^2 - (\sqrt{q})^2} = \frac{p(\sqrt{p} + \sqrt{q})}{p - q}$$

Notice that the first method—used when you have the multiple of a radical in the denominator—works not only for square roots, but also for cube roots, fourth roots, and n^{th} roots.

The method of the second example makes use of the fact that

$$(a + b)(a - b) = a^2 - b^2 \qquad \text{(the difference of squares)}$$

and then arranges things so that $a^2 - b^2$ is in the denominator and that a^2 and b^2 are not radicals—which means the method works only when a and b are square roots, and not cube or nth roots.

PROBLEM SET 10.3

Simplify and write without radical signs using rational exponents:

1. $\sqrt{2^9}$

2. $\sqrt{(2n^2)(3n)^2}$

3. $\sqrt[3]{\sqrt{2^3}}$

4. $\sqrt[4]{b}\sqrt[5]{b}$

5. $\sqrt[3]{\frac{1}{4}} + \sqrt[3]{\frac{1}{32}}$

6. $\sqrt{\dfrac{4^{3n}}{4^n}}$

7. $\sqrt[2n]{9x^2y^4}$

8. $\dfrac{\sqrt{45x^3}}{\sqrt{5x}}$

9. $\sqrt[10]{x^2y^4z^6}$

10. $\sqrt{a^{32}b^4 - b^8}$

11. $\sqrt[3]{16a^{-3}b^2c^3}$

12. $\sqrt[5]{64x^{10}}$

13. $\sqrt{a^2b^2 + a^2c^2}$

14. $\sqrt[3]{\dfrac{1}{x^3}}$

15. $\dfrac{\sqrt[3]{8x^3y^6}}{\sqrt{xy^{-2}}}$

16. $\sqrt[3]{40} \cdot \dfrac{1}{\sqrt[3]{5}}$

17. $(\sqrt{x+1} + \sqrt{x})(\sqrt{x+1} - x)$

18. $\dfrac{1}{\sqrt{2y+r}} + \dfrac{2x\sqrt{2y+r}}{2y+r}$

19. $\sqrt[3]{\dfrac{x}{y^2}} + \sqrt[3]{\dfrac{y}{x^2}} - \sqrt[3]{\dfrac{1}{x^2y^2}}$

20. $\sqrt[n]{\dfrac{a^n}{b^{n-1}}}$

Solve for x:

21. $\sqrt[4]{\frac{3}{8}} = 3^{1/4}2^x$

22. $\sqrt[6]{27a^3} = (3a)^x$

23. $\sqrt[12]{81a^4t^8} = (3at^2)^x$

24. $\sqrt[6]{8} = x$

True or false?

25. $\sqrt{(x+y)^2} = x + y$

26. $\sqrt{x^2 + y^2} = x + y$

27. $\sqrt[4]{x^2y^4} = y\sqrt{x}$

28. $\sqrt{a^{-2}b^{1/2}} = \dfrac{1}{ab^2}$

29. $\dfrac{a^{-1} - b^{-1}}{a + b} = \dfrac{b - a}{ab(a + b)}$

30. $(y + z)^2 (y + z)^{1/2} (y + z)^{-3} = \dfrac{1}{\sqrt{y+z}}$

Simplify by rationalizing the denominator:

31. $\dfrac{5 + \sqrt{3}}{5 - \sqrt{3}}$

32. $\dfrac{1}{3 - \sqrt{2}}$

33. $\dfrac{\sqrt{5} - \sqrt{3}}{\sqrt{3} + \sqrt{5}}$

34. $\dfrac{1}{\sqrt{x+2} - \sqrt{x}}$

35. $\dfrac{a}{x + \sqrt{y}}$

Solve for the indicated variable:

36. $\sqrt{1 + x^2} = s^{-1}t^2$ (x)

37. $\sqrt{a + c} = \sqrt{a} + \sqrt{b}$ (b)

38. $a^{-1} + b^{-1} = c$ (a)

39. $\sqrt{x} - \sqrt{x + 8} = -2$ (x)

40. $\sqrt{3r + 1} + \sqrt{3r - 11} = 2$ (r)

Simplify and write without radical signs using rational exponents:

41. $\frac{2}{3}\sqrt[3]{\frac{3}{4}} - 5\sqrt[3]{\frac{2}{9}} + \sqrt{48}$

42. $\sqrt[6]{27a^3} - \sqrt[9]{8a^3}$

43. $a\sqrt[3]{2x} + \sqrt[3]{128a^3x} - \frac{3}{a}\sqrt[3]{250a^6x}$

44. $\sqrt{4z^6c^2 + 36z^6 - 24z^6c}$

45. $\dfrac{(\sqrt{36x^4})^3 \left[\dfrac{y}{(x^2y^4)^{1/3}}\right]^{3/2}}{12xy}$

46. $\dfrac{(p^{-2/n})\sqrt[2n]{p^3}}{p\sqrt[n]{p^{-3}}}$

47. $\left(\dfrac{x^9y^3}{xy^{-5}}\right)\left(\sqrt[5]{\dfrac{x^{11}y^{10}}{x^3\sqrt[3]{x^9}}}\right)$

48. $\dfrac{(5m^2)^2 \, (\sqrt[4]{81m^2})^4 - m^{16}}{m^4 - m^3}$

49. $\sqrt[5]{(a - b)^2x} - \sqrt[3]{(2a - 3b)^2x} + \sqrt{a^2x + 2abx + b^2x}$

50. $\sqrt[4]{\dfrac{\sqrt{\dfrac{x^9h^{12}}{x^{-7}z^4}} \cdot \sqrt{\dfrac{h^{12}y^{20}\sqrt{y^4}}{x(x^{14}y^7)}}}{\sqrt{\dfrac{z^8y^3}{x^4y}} \cdot 16h^8}}$

51. If $f(x) = \sqrt{\dfrac{x + 1}{x - 1}}$ and $g(x) = \dfrac{1 + x^2}{2x}$

     simplify $f(g(x))$.

52. If $f(x) = \sqrt{x}$, find $f(a^2x) - a^2f(x)$.

53. Show that

$$\frac{b}{a}\sqrt{x^2 - a^2} = \sqrt{\left(\frac{bx}{a}\right)^2 - b^2}$$

54. Factor and hence simplify $\sqrt{13.5}\,\sqrt{6}(2.705) - \sqrt{6}(0.705)\sqrt{13.5}$.

55. Show that

$$\sqrt{\frac{1 + p}{1 - p}}(1 + p)^{-1} = \frac{1}{\sqrt{1 - p^2}}$$

10.4   EXPONENTIAL GRAPHS

Let's draw a graph of $y = 10^x$ and of $y = 10^{-x}$. Since we have not drawn a graph of equations like this before, we must make a table of values:

x	$y = 10^x$		x	$y = 10^{-x}$
−3	0.001		−3	1000
−2	0.01		−2	100
−1	0.1		−1	10
0	1		0	1
1	10		1	0.1
2	100		2	0.01
3	1000		3	0.001

The graphs are shown in Figure 10.1.

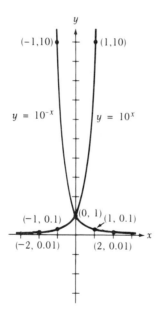

FIG. 10.1

You can see that the graph of $y = 10^x$ climbs very, very fast for large x—in fact it can be shown that it climbs faster than any power of x. Notice also that the two graphs are reflections of one another in the y axis. This is because replacing x by $-x$ changes the equation $y = 10^x$ into $y = 10^{-x}$ and vice versa.

PROBLEM SET   10.4

Sketch a graph of the following functions:

1. $f(x) = 2^x$
2. $f(x) = 10^x$
3. $f(x) = 3 \cdot 10^x$
4. $g(x) = -2 \cdot 10^x$
5. $h(x) = 10^{-x}$
6. $f(x) = 10^{2x}$
7. $l(x) = (\tfrac{1}{2})^x$
8. $p(x) = (0.1)^x$

9. $r(x) = A2^x$ if $A > 0$
10. $t(x) = A2^x$ if $A < 0$
11. $f(x) = 10^{x+1}$
12. $f(x) = 10^x + 1$
13. $f(x) = 10^{x^2}$
14. $\tfrac{1}{2}g(x) = -10^{-x}$
15. $f(x) = |5^x|$
16. $f(x) = 5^{|x|}$

CHAPTER 10 REVIEW

Evaluate:

1. $(\tfrac{2}{3})^{-2}$
2. $(-27)^{-1/3}$
3. $(\tfrac{1}{4})^0$
4. $\dfrac{5^{-2}}{4 \cdot 10^{-2}}$
5. $5^3 \cdot 2^3$

6. $(3.6 \cdot 10^4)(4.1 \cdot 10^3)$
7. $(3 \cdot 10^2) + (0.02 \cdot 10^4)$
8. $\dfrac{(4 \cdot 10^{-6})(5 \cdot 10^{-3})}{(6 \cdot 10^1)(0.15 \cdot 10^{-4})}$
9. $2^{-4} - 2^{-3} + 2^2 - 2^0 + 2^3$
10. $(2\tfrac{1}{4})^{-1/2}$

Express the following in the form $a \cdot 2^p$:

11. $2^3 \cdot 2^7 - 2^5 \cdot 2^3$
12. $(\tfrac{1}{4})^{17}$
13. $16 \cdot \left(\dfrac{64}{\frac{1}{8}}\right)$

14. $\left| 8^{2/5}(\tfrac{1}{16})^{3/2} \right.$
15. $\left| (\tfrac{1}{16})^3 \right| 4^7 \cdot 8^{-3}$

Simplify:

16. $a^n \cdot a$
17. $\left(\dfrac{1}{2}a^3\right)\left(\dfrac{1}{4a^3}\right)$
18. $(x^n)^2$
19. $\sqrt[15]{a^{30}}$

20. $\left(\dfrac{a^2}{b}\right)^7 \left(\dfrac{b^2}{a^3}\right)^6$
21. $(\sqrt{a} + \sqrt{b})^2$
22. $\dfrac{20^n}{5^n}$

Insert the proper inequality ($>$, $\leq$, $\geq$, $<$, or $=$):

23. $\dfrac{7}{5\sqrt{10}}$ _____ $\dfrac{6\sqrt{5}}{25\sqrt{2}}$
24. $\dfrac{2}{\sqrt{3} + \sqrt{12}}$ _____ $\dfrac{\sqrt{2}\sqrt{3}}{\sqrt{18}}$
25. $\dfrac{2^{4/3}}{\sqrt[3]{2}}$ _____ 2

Solve for x:

26. $8^x = 4^x(\tfrac{1}{2})^{-1}$
27. $12(10^{2x}) - 4(10^x) - 1 = 0$

28. $8(100^x) - 2(10^x) = 1$

29. $2^{2x+3} - 2^{x+1} - 1 = 0$ [*Hint:* $2x + 3 = 2(x + 1) + 1$]

30. $\dfrac{a^2\left(\dfrac{1}{a^3}\right)}{a^x} = 1$

31. Show that

$$\frac{1}{2^x + 2^{-x}} = \frac{2^x}{4^x + 1}$$

32. Show that

$$\sqrt{1 + \frac{x^{-1} - x}{2}} = \frac{x^{-1} + x}{2}$$

33. Show that

$$\frac{8}{27a^2}\left(1 + \frac{9}{4}a^2x\right)^{3/2} = \frac{1}{a^2}\left(\frac{4}{9} + a^2x\right)^{3/2}$$

34. Show that

$$\frac{b}{a}[x - \sqrt{x^2 - a^2}] = \frac{ab}{x + \sqrt{x^2 - a^2}}$$

35. Simplify

$$\sqrt{(a_1 - a_2)^2} - \sqrt{(a_2 - a_1)^2}$$

36. If

$$f(x) = \frac{(a - x)(a + x)}{1 - b}$$

then simplify $f(a\sqrt{b})$.

37. Simplify

$$\frac{2\sqrt{ax + y}}{\sqrt{4ax + 4y}}$$

38. Simplify

$$\sqrt{a} + \sqrt{b} + \frac{b}{\sqrt{a} - \sqrt{b}}$$

39. Simplify

$$\frac{1}{\sqrt{1 - \left(\dfrac{x}{a}\right)^2}} \cdot \frac{1}{a}$$

40. Multiply and simplify

$$\sqrt{ab}\left(\frac{A\sqrt{b}}{\sqrt{a}} + \frac{B\sqrt{a}}{\sqrt{b}}\right)$$

41. Simplify

$$(1 + p)^{1/2}\,\frac{1}{\sqrt{p^2 + p^3}}$$

42. Simplify

$$\frac{\sqrt{a - b}}{\sqrt{a^2 - b^2}}$$

43. Combine into a single fraction

$$(x^2 + a^2)^{1/2} - x(x^2 + a^2)^{-3/2} \cdot x$$

44. Simplify
$$\frac{\sqrt{u^2 + x^2} - x^2(u^2 + x^2)^{-1/2}}{(u^2 + x^2)^{3/2}}$$

45. Simplify
$$\frac{(u^2 + x^2)^{1/2} - 2x^2(u^2 + x^2)^{-1/2}}{(u^2 - x^2)^{1/2}}$$

46. Simplify
$$\sqrt[4]{4(x^2 + 2ax + a^2)} - \sqrt{2x + 2a}$$

47. Find the value of α for which
$$(8 + x)^{1/3} = \alpha\left(1 + \frac{x}{8}\right)^{1/3}$$

48. Define $f(x) = 3^{1-x}$.
    (a) Find $f(0)$.
    (b) For what x does $f(x) = 1$?
    (c) Does $f(x)$ increase or decrease as x increases?

49. Solve for p:
$$\frac{2p^2 + 19p + 17}{\sqrt{p^2 + 2p + 1}} = -p$$

50. Graph $y = |3^x|$ and $y = 3^{|x|}$ on the same axes.

51. (a) Match the equations to the graphs:
    (I) $\frac{1}{2}y = 10^x$
    (II) $\frac{1}{2}y = 3^x$
    (III) $\frac{1}{2}y = 10^{-x}$
    (IV) $\frac{1}{2}y = 3^{-x}$
    (b) Label the axes.

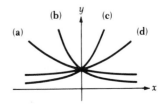

52. The number of bacteria, N, present after an amount of time t in a particular culture can be found using the equation
$$N = 5e^t$$
where e is a constant that will be introduced properly in Chapter 12, but whose value is about 2.7.
    (a) Graph the function.
    (b) What is the y intercept?
    (c) Looking at the graph, approximately how long does it take the culture to double in size?

11 LOGARITHMS IN EXPRESSIONS

11.1   SHORTCUTS TO MULTIPLICATION AND DIVISION: DEFINITION OF LOGS

The widespread use of scientific notation shows how convenient the exponential form can be for writing numbers. Besides being convenient for writing very, very large or very, very small numbers, scientific notation makes the multiplication and division of numbers much easier. For example, to multiply $2 \cdot 10^7$ by $3 \cdot 10^2$ you multiply the numbers in front of the powers of ten, and add the exponents:

$$(2 \cdot 10^7) \cdot (3 \cdot 10^2) = (2 \cdot 3) \cdot (10^7 \cdot 10^2)$$

$$= 6 \cdot 10^{7+2}$$

$$= 6 \cdot 10^9$$

Doing this multiplication out the long way would be much more tiresome because you would do nothing but count zeros and decimal places.

Suppose, however, we have to multiply something like

$$(2.71 \cdot 10^{13}) \cdot (3.02 \cdot 10^{-7})$$

We have to work out $2.71 \cdot 3.02$ and $10^{13} \cdot 10^{-7}$, giving:

$$(2.71 \cdot 10^{13}) \cdot (3.02 \cdot 10^{-7}) = (2.71 \cdot 3.02) \cdot (10^{13} \cdot 10^{-7})$$

$$= 8.1842 \cdot 10^6$$

Multiplying the powers of ten together is really easy, but multiplying $2.71 \cdot 3.02$ is almost (although not quite!) as much nuisance as multiplying the numbers $2.71 \cdot 10^{13}$ and $3.02 \cdot 10^{-7}$ without using scientific notation. The thing that makes multiplying the powers of ten so easy is that the only calculation we have to do is to add the exponents—and adding is so much easier than multiplying. This gives you the idea that if only you could write a number entirely in exponential form (in other words, in the form $10^{(\text{something})}$), then multiplication would become much simpler.

For example, suppose someone tells you that

$$2.71 = 10^{0.433}$$

and

$$3.02 = 10^{0.480}$$

then

$$(2.71 \cdot 10^{13}) \cdot (3.02 \cdot 10^{-7}) = 10^{0.433} \cdot 10^{13} \cdot 10^{0.480} \cdot 10^{-7}$$
$$= 10^{0.433+13+0.480-7}$$
$$= 10^{6.913}$$

You can write $10^{6.913}$ as $10^{0.913+6} = 10^{0.913} \cdot 10^6$, so suppose you were also told that $10^{0.913} = 8.1842$. Then your calculation reads

$$(2.71 \cdot 10^{13}) \cdot (3.02 \cdot 10^{-7}) = 10^{6.913}$$
$$= 10^{0.913} \cdot 10^6$$
$$= 8.1842 \cdot 10^6$$

This time absolutely *no* multiplication was required—all we had to do was add exponents. Of course, there is one big problem: In order to be able to do this shortened method, you had to be told three things—namely, that $2.71 = 10^{0.433}$, $3.02 = 10^{0.480}$, and $10^{0.913} = 8.1842$. Knowing how to write numbers in exponential form is obviously the clue to the whole business, and unfortunately we have no idea how to do it.

There is, however, a method of calculating the power of ten needed to give any number, and people have made up tables in which you can look up these powers. Therefore to multiply

$$(2.71 \cdot 10^{13}) \cdot (3.02 \cdot 10^{-7})$$

you look up 2.71 and 3.02 in these tables to find out what powers of ten are needed (0.433 and 0.480, respectively). Then you do the multiplication by adding the exponents, and finally use the table again to find out what number has an exponent of 0.913 (it turns out to be 8.1842).

A division problem can be done much the same way except that you subtract exponents instead of adding them.

EXAMPLE: *Divide* $\dfrac{263{,}000}{0.41}$

given that the power of ten needed to give 2.63 is 0.42, that the power of ten needed to give 4.1 is 0.6128, and that the power of ten needed to give 6.42 is 0.8072.

The information we are given means that

$$2.63 = 10^{0.42}, \qquad 4.1 = 10^{0.6128}, \qquad 6.42 = 10^{0.8072}$$

Notice that we haven't been given the powers of ten needed to give 263,000 or 0.41, but those for 2.63 and 4.1 instead. To use these we will first have to put 263,000 and 0.41 into scientific notation:

$$\frac{263{,}000}{0.41} = \frac{2.63 \cdot 10^5}{4.1 \cdot 10^{-1}} = \frac{10^{0.42} \cdot 10^5}{10^{0.6128} \cdot 10^{-1}}$$

$$= 10^{0.42 + 5 - 0.6128 - (-1)} \qquad \text{(using the exponent rules)}$$

$$= 10^{5.8072}$$

But

$$10^{5.8072} = 10^{0.8072} \cdot 10^5 = 6.42 \cdot 10^5 = 642{,}000$$

so

$$\frac{263{,}000}{0.41} = 642{,}000$$

Multiplication and division are clearly made much easier by writing all the numbers in exponential form, for the multiplication problems become addition and the division problems subtraction. If you first write all the numbers in scientific notation, then the only numbers whose powers of ten you have to look up in tables will be between one and ten. Therefore, with a set of tables giving powers from numbers from one to ten, you can do any multiplication or division whatsoever just by using the tables, addition, and subtraction.

This "shortcut" method of doing multiplication and division was discovered in the seventeenth century, to the great relief of the astronomers living at the time. The tables of the powers of ten needed to give any number were drawn up then, and until the invention of computers were used for the majority of scientific calculations.

The idea of "the power of ten needed to give a number" is obviously an important one, and so is given a special name: the *logarithm* of the number. In other words:

The *logarithm of a number is the power to which ten must be raised to give that number.* That is,

if $M = 10^a$, then a is the logarithm of M, written $a = \log M$.

For example,

$$1000 = 10^3 \qquad \text{so} \qquad 3 = \log 1000$$

$$0.01 = 10^{-2} \qquad \text{so} \qquad -2 = \log 0.01$$

$$6.4 = 10^{0.8062} \qquad \text{so} \quad 0.8062 = \log 6.4$$

$$\sqrt{10} = 10^{1/2} \qquad \text{so} \qquad \frac{1}{2} = \log(\sqrt{10})$$

Please note: A logarithm is just an exponent!

The antilogarithm of something means the number that has that thing as its logarithm. In other words

if antilog $a = M$, then $a = \log M$, so $10^a = M$.

For example,

$$10^2 = 100 \quad \text{so} \quad \text{antilog} \quad 2 = 100$$

$$10^{-3} = 0.001 \quad \text{so} \quad \text{antilog} \ -3 = 0.001$$

$$10^{2/3} = 4.64 \quad \text{so} \quad \text{antilog} \ \frac{2}{3} = 4.64$$

PROBLEM SET   11.1

Evaluate:

1. $\log 1$
2. $\log(0.1)$
3. $\log(10^0)$
4. $\log \sqrt{10}$
5. $\log 10^5$
6. $\log 10^2$
7. $\log \dfrac{1}{\sqrt{10}}$
8. $10^{\log 100}$
9. $10^{\log 1}$
10. $10^{\log(0.01)}$

11. If $\log 252 = 2.4014$, then what is $10^{1.4014}$?
12. If $\log 408 = 2.6107$, then what is $10^{3.6107}$?
13. If $\log 692 = 2.8401$, then what is $10^{(0.8401-1)}$?
14. If $\log 0.341 = 0.5328 - 1$, then what is $10^{4.5328}$?
15. If $5 = 10^{0.7}$, then find $\log 25$.

Assuming that $2 = 10^{0.3010}$ and $3 = 10^{0.4771}$, write the following in the form 10^x:

16. 6
17. 20
18. $\frac{27}{4}$
19. 5
20. 0.24

11.2   FINDING LOGS FROM TABLES

To Find the Logarithm of Numbers between 1 and 10

The table in the appendix give you the logarithms of numbers between 1 and 10. Most logarithms are irrational numbers and so the values printed in tables are only approximate. Since $1 = 10^0$ (so log $1 = 0$) and $10 = 10^1$ (so log $10 = 1$), we would expect the logarithms of numbers between 1 and 10 to fall between 0 and 1—which you can see from the table to be the case. (Table 11.1 on the facing page is a part of that table.)

To find the logarithm of 2.35 from the table, look down the left-hand column until you find the first two digits (2.3) and then move over until you are under the 5 column, and then read off the logarithm—it is 0.3711. See Table 11.1. Therefore

$$\log 2.35 = 0.3711 \qquad \text{(approximately)}$$

To Find the Logarithm of Numbers Less than 1 or Greater than 10

To find the logarithm of 374, you first write it in scientific notation: $374 = 3.74 \cdot 10^2$. Looking up 3.74 in the table tells you that log $3.74 = 0.5729$, or $3.74 = 10^{0.5729}$. Therefore

$$374 = 3.74 \cdot 10^2 = 10^{0.5729} \cdot 10^2 = 10^{0.5729+2} = 10^{2.5729}$$

so

$$\log 374 = 2.5729$$

Generalizing this:

To Find the Log of a Number N:

1. Write N in scientific notation; the power of ten occurring is called the *characteristic* (e.g., $374 = 3.74 \cdot 10^{2}$ ←characteristic).
2. Look up the number in front of the power of ten in the table; its logarithm is called the *mantissa* (e.g., log $3.74 = 0.5729$ ← mantissa).
3. Add the characteristic and the mantissa to get log N (e.g., log $374 = 2 + 0.5729 = 2.5729$).

Table 11.1   Logarithms

N	0	1	2	3	4	5	6	7	8	9
1.0	.0000	.0043	.0086	.0128	.0170	.0212	.0253	.0294	.0334	.0374
1.1	.0414	.0453	.0492	.0531	.0569	.0607	.0645	.0682	.0719	.0755
1.2	.0792	.0828	.0864	.0899	.0934	.0969	.1004	.1038	.1072	.1106
1.3	.1139	.1173	.1206	.1239	.1271	.1303	.1335	.1367	.1399	.1430
1.4	.1461	.1492	.1523	.1553	.1584	.1614	.1644	.1673	.1703	.1732
1.5	.1761	.1790	.1818	.1847	.1875	.1903	.1931	.1959	.1987	.2014
1.6	.2041	.2068	.2095	.2122	.2148	.2175	.2201	.2227	.2253	.2279
1.7	.2304	.2330	.2355	.2380	.2405	.2430	.2455	.2480	.2504	.2529
1.8	.2553	.2577	.2601	.2625	.2648	.2672	.2695	.2718	.2742	.2765
1.9	.2788	.2810	.2833	.2856	.2878	.2900	.2923	.2945	.2967	.2989
2.0	.3010	.3032	.3054	.3075	.3096	.3118	.3139	.3160	.3181	.3201
2.1	.3222	.3243	.3263	.3284	.3304	.3324	.3345	.3365	.3385	.3404
2.2	.3424	.3444	.3464	.3483	.3502	.3522	.3541	.3560	.3579	.3598
2.3	.3617	.3636	.3655	.3674	.3692	.3711	.3792	.3747	.3766	.3784
2.4	.3802	.3820	.3838	.3856	.3874	.3892	.3909	.3927	.3945	.3962
2.5	.3979	.3997	.4014	.4031	.4048	.4065	.4082	.4099	.4116	.4133
2.6	.4150	.4166	.4183	.4200	.4216	.4232	.4249	.4265	.4281	.4298
2.7	.4314	.4330	.4346	.4362	.4378	.4393	.4409	.4425	.4440	.4456
2.8	.4472	.4487	.4502	.4518	.4533	.4548	.4564	.4579	.4594	.4609
2.9	.4624	.4639	.4654	.4669	.4683	.4698	.4713	.4728	.4742	.4757
3.0	.4771	.4786	.4800	.4814	.4829	.4843	.4857	.4871	.4886	.4900
3.1	.4914	.4928	.4942	.4955	.4969	.4983	.4997	.5011	.5024	.5038
3.2	.5051	.5065	.5079	.5092	.5105	.5119	.5132	.5145	.5159	.5172
3.3	.5185	.5198	.5211	.5224	.5237	.5250	.5263	.5276	.5289	.5302
3.4	.5315	.5328	.5340	.5353	.5366	.5378	.5391	.5403	.5416	.5428
3.5	.5441	.5453	.5465	.5478	.5490	.5502	.5514	.5527	.5539	.5551
3.6	.5563	.5575	.5587	.5599	.5611	.5623	.5635	.5647	.5658	.5670
3.7	.5682	.5694	.5705	.5717	.5729	.5740	.5752	.5763	.5775	.5786
3.8	.5798	.5809	.5821	.5832	.5843	.5855	.5866	.5877	.5888	.5899
3.9	.5911	.5922	.5933	.5944	.5955	.5966	.5977	.5988	.5999	.6010
4.0	.6021	.6031	.6042	.6053	.6064	.6075	.6085	.6096	.6107	.6117
4.1	.6128	.6138	.6149	.6160	.6170	.6180	.6191	.6201	.6212	.6222
4.2	.6232	.6243	.6253	.6263	.6274	.6283	.6294	.6304	.6314	.6325
4.3	.6335	.6345	.6355	.6365	.6375	.6385	.6395	.6405	.6415	.6425
4.4	.6435	.6444	.6454	.6464	.6474	.6484	.6493	.6503	.6513	.6522
4.5	.6532	.6542	.6551	.6561	.6571	.6580	.6590	.6599	.6609	.6618
4.6	.6628	.6637	.6646	.6656	.6665	.6675	.6684	.6693	.6702	.6712
4.7	.6721	.6730	.6739	.6749	.6758	.6767	.6776	.6785	.6794	.6803
4.8	.6812	.6821	.6830	.6839	.6848	.6857	.6866	.6875	.6884	.6893
4.9	.6902	.6911	.6920	.6928	.6937	.6946	.6955	.6964	.6972	.6981

EXAMPLE: *Find the logarithms of 96,700.*

$96,700 = 9.67 \cdot 10^4$

From the table in the appendix,

$\log 9.67 = 0.9854$

Therefore;

$\log(96,700) = 4 + 0.9854 = 4.9854.$

In other words,

$96,700 = 9.67 \cdot 10^4 = 10^{0.9854} \cdot 10^4 = 10^{4.9854}$

EXAMPLE: *Find the logarithm of 0.017.*

$0.017 = 1.7 \cdot 10^{-2}$

From the table,

$\log 1.7 = 0.2304$

Therefore

$\log 0.017 = -2 - 0.2304 = -1.7696$

or

$0.017 = 1.7 \cdot 10^{-2} = 10^{0.2304} \cdot 10^{-2} = 10^{-1.7696}$

Danger: Concerning Logarithms of Numbers Less than 1

It turns out (for reasons that are not apparent yet) that it is much more convenient to leave log 0.017 in the form $-2 + 0.2304$ than to convert it to -1.7696. Unfortunately, it is all too easy to let $-2 + 0.2304$ turn into -2.2304 (which is *quite* different, since in -2.2304 the minus sign applies to *both* the 2 and the 0.2304, whereas in $-2 + 0.2304$ it applies only to the 2), and so we will write $-2 + 0.2304$ as $0.2304 - 2$.

EXAMPLE: *Find log 0.00521, log 5.21, log 52.1, and log 52,100.*

Writing each of these numbers in scientific notation:

$0.00521 = 5.21 \cdot 10^{-3}$

$5.21 = 5.21 \cdot 10^{0}$

$52.1 = 5.21 \cdot 10^{1}$

$52,100 = 5.21 \cdot 10^{4}$

From the table,

log 5.21 = 0.7168

Therefore

$\log(0.00521) = 0.7168 + -3 = 0.7168 - 3$

$\log(5.21) = 0.7168 + 0 = 0.7168$

$\log(52.1) = 0.7168 + 1 = 1.7168$

$\log(52{,}100) = 0.7168 + 4 = 4.7168$

Therefore, looking at the logarithm of a number: *The characteristic tells you the power of ten, and the mantissa is the logarithm of the number in front of the power of ten.* That is:

If $b \cdot 10^c$ is in scientific notation,

$$\log(b \cdot 10^c) = \underbrace{c}_{} + \underbrace{\log b}_{\text{mantissa}}$$

characteristic

To Find Antilogarithms

Suppose we want to find antilog 2.8756, that is, the number, N, whose logarithm is 2.8756. The 2 is the characteristic of log N, so N must be (some number) $\cdot 10^2$. If you look for 0.8756 in the tables, you find it on the same line as 7.5 and under the 1 column, so $0.8756 = \log 7.51$. So 7.51 must be the number in front of the power of 10, and N must be $7.51 \cdot 10^2$.

Generalizing,

To Find the Antilogarithm of a Number

(that is, to find a number N
whose logarithm is given)

1. Make sure the logarithm is written with the decimal part of the number positive, because the tables contain only positive mantissas (e.g., $2.8756 = 2 + 0.8756$).
2. The whole-number part of the logarithm is the characteristic c, and N contains a factor 10^c (e.g., N contains a factor of 10^2).
3. The decimal part of the logarithm is the mantissa. Look it up in the body of the table to find that it is the logarithm of some number. Call this number b (e.g. $0.8756 = \log 7.51$).
4. Multiply b by 10^c to get N, that is, $N = b \cdot 10^c$ (e.g., $N = 7.51 \cdot 10^2$).

EXAMPLE: *Find antilog(3.6702), antilog(0.6702−2), and antilog(−4.3298).*

Now, 3.6702 and .6702−2 are in the correct form for finding their antilogarithms; −4.3298 is not, because the minus sign refers to both the 4 and the 0.3298. To get it into the right form, we use the following trick. Since −4.3298 is between −5 and −4, and is to the right of −5, it can be written in the form

$$-4 - 0.3298 = -5 + (1 - 0.3298) = -5 + 0.6702$$

So the trick amounts to adding one to the mantissa and subtracting one from the characteristic.

Therefore we want to find the antilogarithms of 3.6702, 0.6702 − 2, and 0.6702 − 5.

Since the mantissa is the same in each case, all that changes is the characteristic; the numbers we are looking for will all be the same except for their powers of ten.
Looking for 0.6702 in the table gives

$$0.6702 = \log (4.68)$$

so

$$3.6702 = \log(4.68 \cdot 10^3) \quad \text{or}$$
$$\text{antilog } 3.6702 = 4.68 \cdot 10^3 = 4680$$

$$0.6702 - 2 = \log(4.68 \cdot 10^{-2}) \quad \text{or}$$
$$\text{antilog}(0.6702 - 2) = 4.68 \cdot 10^{-2} = 0.0468$$

$$0.6702 - 5 = \log(4.68 \cdot 10^{-5}) \quad \text{or}$$
$$\text{antilog}(0.6702 - 5) = 4.68 \cdot 10^{-5} = 0.0000468$$

Note: We have to make a decision about what to do in cases where we are looking for an antilog and find that the exact mantissa we need is not in the tables. For the purposes of this book we will pick the closest value and use that. The answer that results is not absolutely accurate, but that is all right since the values in the table are only approximate anyway.

PROBLEM SET   11.2

Evaluate:

1. log 1.04	6. log 0.233	11. log 7.08
2. log 95	7. log 0.00018	12. log 263,000
3. log 4600	8. $\log(5.4 \cdot 10^6)$	13. log 0.00263
4. log 0.253	9. log 507	14. $\log(1.883 \cdot 10^{-2})$
5. log 3070	10. log 0.00507	15. log 13.72

16. antilog 0.3010
17. antilog 1.9138
18. antilog 6.1959
19. antilog 3.9243
20. antilog 2.0719
21. antilog(0.9509 − 1)
22. antilog(0.9284 − 4)
23. antilog(12.7649)
24. antilog 0.6767
25. antilog(3.4639 − 5)

26. antilog(−7.8996)
27. antilog(−3.7832)
28. antilog(−2.4881)
29. antilog(−4.4034)
30. antilog(−3.6990)
31. antilog 2.5011
32. antilog 3.6010
33. antilog $\frac{1}{2}$
34. antilog $\frac{9}{8}$
35. antilog$(-\sqrt{\frac{9}{4}})$

Write x in scientific notation:

36. $x = 10^{-5.6073}$
37. $x = 10^{0.9263}$
38. $x = 10^{-2.2161}$

39. $10^{3.6417} = x$
40. $10^{-0.0259} = x$

Evaluate:

41. antilog(log 100)
42. antilog(log 0.1)
43. antilog[log (0)]
44. antilog[log(32.3)]
45. antilog[log ($\frac{4}{7}$)]

46. log(antilog 20)
47. log[antilog(12.1)]
48. log(antilog 1)
49. log(antilog 670.32)
50. log[antilog (−7.8881)]

11.3   THE LOG RULES

Multiplying Using Logarithms

Suppose we want to multiply 418 by 85.3 using logarithms. We must first write each of the numbers in exponential form, which means looking up their logarithms.

$$\log 418 = \log(4.18 \cdot 10^2) = 2.6212 \quad \text{so} \quad 418 = 10^{2.6212}$$

$$\log 85.3 = \log(8.53 \cdot 10^1) = 1.9309 \quad \text{so} \quad 85.3 = 10^{1.9309}$$

Therefore

$$418 \cdot 85.3 = 10^{2.6212} \cdot 10^{1.9309}$$
$$= 10^{2.6212+1.9309} \quad \text{(by the first exponent rule)}$$
$$= 10^{4.5521}$$

Therefore the logarithm of the product is 4.5521, that is,

$$\log(418 \cdot 85.3) = 4.5521$$

You see that in order to get $\log(418 \cdot 85.3)$, we had to add $\log(418)$ and $\log(85.3)$. In other words,

$$\log(418 \cdot 85.3) = \log(418) + \log(85.3)$$

Generally, *to find the log of a product, add the logs of the factors.* This amounts to:

The First Logarithm Rule

$$\log MN = \log M + \log N$$

You have already seen this justified for the particular case when $M = 418$ and $N = 85.3$.

The general proof goes like this:

$$\text{Call } \log M = a, \quad \text{so} \quad M = 10^a$$
$$\text{and } \log N = b, \quad \text{so} \quad N = 10^b.$$

Then

$$MN = 10^a \cdot 10^b$$
$$= 10^{a+b} \quad \text{(by the first exponent rule)}$$

But $MN = 10^{a+b}$ means that the power of ten that you need to get MN is $a + b$. Therefore,

$$\log MN = a + b$$

so

$$\log MN = \log M + \log N \quad \begin{array}{l} \text{(since } a = \log M \\ \quad b = \log N) \end{array}$$

Dividing Using Logarithms

Now suppose you want to divide $\dfrac{418}{85.3}$ using logarithms. Since $418 = 10^{2.6212}$ and $85.3 = 10^{1.9309}$,

$$\frac{418}{85.3} = \frac{10^{2.6212}}{10^{1.9309}}$$
$$= 10^{2.6212 - 1.9309} \quad \text{(by the second exponent rule)}$$
$$= 10^{0.6903}$$

Therefore the logarithm of the fraction is 0.6903, that is,

$$\log\left(\frac{418}{85.3}\right) = 0.6903$$

Here in order to find $\log\left(\dfrac{418}{85.3}\right)$ we had to subtract log 85.3 from log 418. In other words,

$$\log\left(\frac{418}{85.3}\right) = \log 418 - \log 85.3$$

More generally, *to find the log of a fraction, subtract the log of the denominator from the log of the numerator.* This amounts to:

The Second Logarithm Rule

$$\log \frac{M}{N} = \log M - \log N$$

Again, you have seen this justified for the special case $M = 418$ and $N = 85.3$.

The general proof looks like this:

Call $\log M = a$ and $\log N = b$ as before, so

$$M = 10^a \quad \text{and} \quad N = 10^b$$

Then

$$\frac{M}{N} = \frac{10^a}{10^b}$$

$$= 10^{a-b} \qquad \text{(by the second exponent rule)}$$

But $\dfrac{M}{N} = 10^{a-b}$ means that the power of ten needed to give $\dfrac{M}{N}$ is $a - b$.

Therefore

$$\log \frac{M}{N} = a - b$$

so

$$\log \frac{M}{N} = \log M - \log N \qquad \text{(since } a = \log M$$
$$b = \log N)$$

Taking Powers and Roots Using Logarithms

What happens if we want to calculate $(85.3)^2$ or $\sqrt[5]{418}$? The first of these could obviously be done by straightforward multiplication, though logarithms save a lot of work. But $\sqrt[5]{418}$ can only be found using logarithms, or a calculator.

First, let's do $(85.3)^2$: We've seen that

$$85.3 = 10^{1.9309}$$

so

$$(85.3)^2 = (10^{1.9309})^2$$

$$= 10^{2(1.9309)} \qquad \text{(by the third exponent rule)}$$

$$= 10^{3.8618}$$

Therefore

$$\log[(85.3)^2] = 3.8618$$

Notice that to find $\log[(85.3)^2]$, you have to multiply $\log 85.3$ by 2.

In general, *to find the log of the square of a number, multiply the log of the number by 2.*

As you might guess, you find the log of the cube of a number by multiplying its log by 3.

In general, *to find the log of the nth power of a number, multiply the log of the number by n.*

What about roots? Remember that $\sqrt[5]{418} = (418)^{1/5}$, and we've already seen that

$$418 = 10^{2.6212}$$

so

$$\sqrt[5]{418} = (418)^{1/5} = (10^{2.6212})^{1/5}$$

$$= 10^{(1/5)(2.6212)}$$

$$= 10^{0.5242}$$

[Actually, $\frac{1}{5}(1.6212) = 0.52424$, but we are only interested in four decimal places since our tables only give logarithms to four places.] Hence

$$\log \sqrt[5]{418} = 0.5242$$

so to find $\log \sqrt[5]{418}$, we have to divide $\log 418$ by 5 (or multiply by $\frac{1}{5}$), and therefore it should seem reasonable that *to find the log of the nth root of a number, you divide the log of the number by n (or multiply by 1/n).*

The results about powers and roots can be summarized:

The Third Logarithm Rule

$$\log(M^p) = p \log M$$

If $p = n$, an integer, this gives you the result about powers. If $p = \dfrac{1}{n}$, this gives you the result about roots.

You have seen the justification for this rule in the case where $M = 85.3$ and $p = 2$, and where $M = 418$ and $p = \frac{1}{5}$.

The general proof is this: Suppose $\log M = a$ so $M = 10^a$

Then

$$M^p = (10^a)^p$$

$$= 10^{pa} \qquad \text{(by the third exponent rule)}$$

But $M^p = 10^{pa}$ means that

$$\log M^p = pa$$

or

$$\log M^p = p \log M \qquad \text{(since } a = \log M)$$

PROBLEM SET   11.3

If $\log 2 = 0.3010$ and $\log 3 = 0.4771$, evaluate the following expressions:

1. $\log 6$
2. $\log 3000$
3. $\log 12$
4. $\log \frac{9}{8}$
5. $\log \frac{8}{9}$
6. $\log \frac{20}{3}$
7. $\log \frac{4}{27}$
8. $\log \sqrt{\frac{20}{3}}$
9. $\log 5$
10. $\log 150$

True or false? Justify your answers, and correct the statement if false.

11. $\log \dfrac{a+b}{c} = \log a + \log b - \log c$

12. $\dfrac{x}{y} = \log x - \log y$

13. $(\log A)^4 = 4 \log A$

14. $\log \left(\dfrac{P}{QR}\right) = \dfrac{\log P}{\log Q + \log R}$

15. $\log [\log(xy)] = \log(\log x) + \log(\log y)$

16. $\dfrac{\log A}{\log B} = \log A - \log B$

17. $\log(a \cdot 2B) = \log a + 2 \log B$

18. $\log \dfrac{P^2}{Q} = \dfrac{\log P - 2 \log Q}{2}$

19. $\log a\left(b + \dfrac{c}{d}\right) = \log ab + \log(ac - d)$

20. $\log a\left(b + \dfrac{c}{d}\right) = \log a + \log(bd + c) - \log d$

Evaluate the following:

21. $\log\left(\dfrac{1}{507}\right)$

22. $\log\left(\dfrac{5.07}{7.08}\right)$

23. $\log\left[(21)\,(0.061)\right]$
24. $\log\sqrt{51}$
25. $\log(2.5^3)$

Write the following expressions in the form log x:

26. $2\log 9$
27. $3\log 4$
28. $-\frac{1}{2}\log 10^6$
29. $-\log 9^2$
30. $\log 28 - \log 20$

31. $3 - \log 2$
32. $\log 9x - \log x$
33. $1 - \log 5$
34. $2(\log 5 + \log 20)$
35. $\dfrac{\log 9}{2} - \log 30$

Evaluate:

36. antilog(2 log 3)
37. antilog(2 log x)
38. antilog(1 + log 5)

39. antilog(1 + log x)
40. antilog(log 5 − log 0.2)

41. Explain what is wrong with the following reasoning:

$$a^x + b = 0$$

$$\log a^x + \log b = 0$$

$$x\,\log a + \log b = 0$$

$$x\log a = -\log b$$

$$x = \frac{-\log b}{\log a}$$

$$x = -\log\left(\frac{b}{a}\right)$$

42. Prove that $\log \dfrac{1}{a} = -\log a$

43. Prove the second log rule using the first and third log rules.

44. Solve for x:

$$\log x = \log a - \log b$$

45. Solve for x:

$$\log x = 2\log A + \log \frac{B}{3}$$

46. Show that $\dfrac{\log x^k}{kx}$ is independent of the value of k.

47. Simplify

$$\frac{\sqrt{\log 10{,}000}}{\log \sqrt{10}}$$

48. Simplify the following without using tables:

$$2 \log(9) - \log(8.1)$$

(combine this into one term and then evaluate.)

Let $\log n = 0.6$.

49. Find $\log(10\ n)$
50. Find $\log \sqrt{n}$
51. In combining $\log 1 + \log 2 + \log 3$, a student argued that

$$\log 1 + \log 2 + \log 3 = \log(1 + 2 + 3) = \log 6$$

Is the procedure correct? Is the result correct? If it is required that

$$\log(a + b + c) = \log a + \log b + \log c$$

express c in terms of a and b.

Find the logarithm of the variable in parentheses in terms of the logarithms of the other variables:

52. $v = \frac{4}{3}\pi r^3$ (v)
53. $d = \frac{1}{2}at^2$ (d)
54. $m = m_0 \sqrt{1 - \dfrac{u^2}{c^2}}$ (m)
55. $\frac{1}{2}mv^2 = \frac{3}{2}kT$ (v)

11.4   THE RELATION BETWEEN TAKING LOGS AND TAKING EXPONENTS

There are two equations that point out the extremely close connection between powers of ten and logarithms. They are derived from examples like the following:

$$\log(10^2) = 2$$ (because the power of ten you need to get 10^2 is 2)

or

$$\log(10^{0.8756}) = 0.8756$$ (because the power of ten you need to get $10^{0.8756}$ is 0.8756)

In general,

$$\boxed{\log (10^a) = a}$$

because the power of ten needed to get 10^a is a.

Now consider:

$$10^{\log 100} = 10^2 = 100 \qquad \text{(since } \log 100 = 2)$$

Also,

$$10^{\log 3} = 10^{0.4771}$$

but $$10^{0.4771} = 3$$

because 0.4771 was specially constructed to be the power of ten you needed to get 3, so

$$10^{\log 3} = 10^{0.4771} = 3$$

Similarly,

$$10^{\log 85.3} = 10^{1.9309} = 85.3$$

because 1.9309 is called log 85.3 precisely because it makes $10^{1.9309} = 85.3$. These results,

$$10^{\log 100} = 100$$

$$10^{\log 3} = 3$$

$$10^{\log 85.3} = 85.3$$

are enough to make you want to guess that

$$10^{\log M} = M$$

To see why this is true, remember that log M is the power of ten required to give M. This means that log M is specially chosen to fill the equation

$$10^{(?)} = M$$

And so filling log M in for the (?) gives

$$10^{(\log M)} = M$$

Summarizing:

---

log M = a if and only if M = 10ᵃ

The Three Logarithm Rules

1. $\log MN = \log M + \log N$

2. $\log \dfrac{M}{N} = \log M - \log N$

3. $\log M^p = p \log M$

The Three Exponent Rules

1. $10^a \cdot 10^b = 10^{a+b}$

2. $\dfrac{10^a}{10^b} = 10^{a-b}$

3. $(10^a)^p = 10^{pa}$

Also:

$$\log 10^a = a \quad \text{and} \quad 10^{\log M} = M$$

PROBLEM SET   11.4

Evaluate or simplify the following expressions:

1. $10^{0.3010}$
2. $10^{\log 2}$
3. $10^{1.3979}$
4. $10^{\log 25}$
5. $10^{\log a}$
6. $10^{\log (a+b)}$
7. $10^{\log[pq/(p+q)]}$
8. $10^{-\log 100}$
9. $10^{-\log 510}$
10. $10^{2 \log 100}$
11. $10^{-\log 5}$

12. $10^{-\log x}$
13. $10^{-2 \log 510}$
14. $10^{\log(1/10)}$
15. $10^{\log N}$
16. $\log \dfrac{1}{\sqrt{10}}$
17. $\log \dfrac{1}{b} + \log b$
18. $\log(100 \cdot \sqrt{10})$
19. $\log 10 + \log 10^{-2}$

20. $\log 10^a + \log 10^b$
21. $10^{-\log 8}$
22. $10^{-\log 0.3}$
23. $10^{\log 100} \cdot 10^{\log 10}$
24. $10^{-\log 25.3}$
25. $10^{-\log 117}$
26. $10^{-\log x}$
27. $10^{\log(10^{ax+b})}$
28. $10^{\log(\log x)}$
29. $\log 10^{\log 5}$
30. $\log 10^{\log y}$

31. Show that $(\sqrt{10})^{\log 5} = \sqrt{5}$.
32. Show that $10^{1+\log x} = 10x$.

Solve for the unknown:

33. $x^{\log 3} = 3$
34. $x^{\log b} = b$ (solve for x)
35. $5(10^{\log y}) = 1$
36. $10^{\log y} = 17$
37. $x^{\log 31} = 31$
38. $\dfrac{1}{13} = \dfrac{1}{10^{\log x}}$
39. $\dfrac{1}{a} = 10^{-\log x}$ (solve for x)
40. $\dfrac{1}{p} - 10^{-\log x} = 0$ (solve for x)

Let $f(x) = 10^x$ and $g(x) = \log x$.

41. Evaluate $g(10^{0.7})$.
42. Evaluate $g(10^a)$.
43. Evaluate $f(\log 7)$.
44. Evaluate $f(\log x)$.
45. Evaluate $f(a \log x)$.
46. Evaluate $f(a + \log x)$.
47. Evaluate $g(10^a 10^x)$.
48. Evaluate $g[f(2) \cdot f(4)]$.
49. Simplify $f[g(x)]$.
50. Simplify $g[f(x)]$.

11.5   LOGS TO DIFFERENT BASES

Everything we have done in the last three sections has been concerned with expressing numbers as powers of ten. But we might be interested in expressing numbers as powers of something else—for example, of 2. Then:

$$\frac{1}{2} = 2^{-1},\ 1 = 2^0,\ 2 = 2^1,\ 4 = 2^2,\ 8 = 2^3,\ 16 = 2^4,\ 32 = 2^5, \text{ etc.}$$

The exponents are still called logarithms, but this time they are *logarithms to base 2*, written $\log_2$.

EXAMPLE: $16 = 2^4$, so the power of two needed to give 16 is 4. In other words, the logarithm to base 2 of 16 is 4, or:

$$\log_2 16 = 4$$

If $M = 2^a$, the power of two needed to give M is a. Therefore the logarithm to base 2 of M is a, or

$$\log_2 M = a$$

EXAMPLE: $81 = 27^{4/3}$, so the power of 27 needed to give 81 is $\frac{4}{3}$, and

$$\log_{27} 81 = \frac{4}{3}$$

The most general possible *definition of a logarithm* is this:

If the power of k needed to give M is a, that is,

$$\text{if } M = k^a$$

then the logarithm to base k of M is a, that is,

$$\log_k M = a$$

(k must be positive).

In other words, *a logarithm to base k is just an exponent of k.*

The equation $M = k^a$ is said to be in *exponential form*, while $\log_k M = a$ is in *logarithmic form*.

You can see that what we were calling just "logarithms" earlier were really "logarithms to base 10," and that if $10^a = b$, we should have written $\log_{10} b = a$, rather than $\log b = a$. However, logs to base 10 are used so often that they are called *common logarithms*, and the subscript 10 is left out. Hence $\log b = a$ is understood to mean $\log_{10} b = a$.

Just to make life confusing, however, logarithms to base e become very important in calculus (e is an irrational number whose value is about 2.7). These are called *natural logarithms* and in a calculus book, no subscript usually means a natural logarithm—so there $\log b$ means $\log_e b$. In this book, however, no subscript will always mean log to base 10.

EXAMPLE: *Find $\log_2 \frac{1}{8}$.*

Suppose $\log_2 \frac{1}{8} = a$. Then a is the power of two needed to give $\frac{1}{8}$, or $2^a = \frac{1}{8}$. Now

$$2^{-3} = \frac{1}{2^3} = \frac{1}{8}$$

so

$$\log_2 \frac{1}{8} = -3$$

EXAMPLE: *Find* $\log_8 \frac{1}{2}$.

If $\log_8 \frac{1}{2} = b$ *then* $8^b = \frac{1}{2}$ and

$$8^{-1/3} = \frac{1}{8^{1/3}} = \frac{1}{\sqrt[3]{8}} = \frac{1}{2}$$

so

$$\log_8 \frac{1}{2} = -\frac{1}{3}$$

EXAMPLE: *Find* $\log_5 1$.

$\log_5 1 = 0$ because $5^0 = 1$

For the same reason,

$$\boxed{\log_k 1 = 0 \quad \text{for any } k > 0}$$

Also notice that $k^1 = k$, so

$$\boxed{\log_k k = 1 \quad \text{for any } k > 0}$$

The Log Rules

The log rules are all true for logarithms to any base. The proofs for a general base are precisely the same as for base 10—just erase all the 10's and put in k's:

The Three Logarithm Rules	The Three Exponent Rules
1. $\log_k MN = \log_k M + \log_k N$	1. $k^a \cdot k^b = k^{a+b}$
2. $\log_k \dfrac{M}{N} = \log_k M - \log_k N$	2. $\dfrac{k^a}{k^b} = k^{a-b}$
3. $\log_k M^p = p \log_k M$	3. $(k^a)^p = k^{pa}$

Analogues of $\log 10^a = a$ and $10^{\log M} = M$ also hold for general bases. $\log_k k^a$ means the power of k you need to give k^a, which is obviously a, so:

$$\log_k k^a = a$$

And as for the other result, remember that $\log_k M$ is the power of k needed to give M; that is, $\log_k M$ fits the equation

$$k^{(?)} = M$$

so

$$k^{(\log_k M)} = M$$

Examples of the Use of the Log Rules

The log rules are very important in everything from simplification to radiocarbon dating. You should therefore work with them until using them seems completely natural. The main thing to keep in mind is that you have to use them *exactly* as they are. The reason for such an apparently obvious remark is that the log rules often strike people as a little strange, and so they tend to rewrite them in another form. But this usually changes the meaning of the rule, so it is safer to use them exactly as they are.

EXAMPLE: *Find* $\log_k \left(\dfrac{x^2 \sqrt{z}}{y} \right)$ *given* $\log_k x = 1.2$
$\log_k y = 0.3$
$\log_k z = 0.8$

We will expand

$$\log_k \frac{x^2 \sqrt{z}}{y}$$

in terms of $\log_k x$, $\log_k y$, $\log_k z$, so that we can use the given information.

$$\log_k \left(\frac{x^2 \sqrt{z}}{y} \right) = \log_k (x^2 \sqrt{z}) - \log_k y \qquad \text{(by Rule 2)}$$

$$= \log_k x^2 + \log_k \sqrt{z} - \log_k y \quad \text{(by Rule 1)}$$

$$= 2 \log_k x + \tfrac{1}{2} \log_k z - \log_k y \quad \begin{array}{l}\text{(by Rule 3 and} \\ \sqrt{z} = z^{1/2}\text{)}\end{array}$$

$$= 2\,(1.2) + \tfrac{1}{2}\,(0.8) - (0.3)$$

$$= 2.5$$

EXAMPLE: *Given log 3 = 0.4771 and log 5 = 0.6990, find log 45 without using tables.*

In order to find log 45 without using tables, we must express 45 as the product or quotient of numbers whose logs we know, and then we can use the log rules. Now,

$$45 = 3^2 \cdot 5$$

so

$$\log 45 = \log (3^2 \cdot 5)$$
$$= \log 3^2 + \log 5 \qquad \text{(by Rule 1)}$$
$$= 2 \log 3 + \log 5 \qquad \text{(by Rule 3)}$$
$$= 2(0.4771) + 0.6990$$
$$= 1.6532$$

Therefore

$$\log 45 = 1.6532$$

EXAMPLE: *Write*

$$log_k(T - 1) - \frac{log_k s}{2} + n \, log_k(Q + 1)$$

as the log of one quantity.

$$log_k(T - 1) - \frac{log_k s}{2} + n \, log_k(Q + 1)$$
$$= \log_k(T - 1) - \log_k(s^{1/2}) + \log_k[(Q + 1)^n] \qquad \text{(by Rule 3)}$$
$$= \log_k\left(\frac{T - 1}{s^{1/2}}\right) + \log_k[(Q + 1)^n] \qquad \text{(by Rule 2)}$$
$$= \log_k\left[\frac{(T - 1)(Q + 1)^n}{\sqrt{s}}\right] \qquad \text{(by Rule 1)}$$

EXAMPLE: *Write the equation $N = Ae^{ct}$ in logarithmic form.*

This can be done in many ways, depending on what base we decide to use for the logs. An obvious choice is e, since we have powers of e in the equation. Taking logarithms to base e of both sides of the equation

$$N = Ae^{ct}$$

gives

$$\log_e N = \log_e(Ae^{ct})$$
$$\log_e N = \log_e A + \log_e e^{ct} \qquad \text{(by Rule 1)}$$
$$\log_e N = \log_e A + ct \qquad \text{(since $\log_e e^x = x$)}$$

If we had chosen a different base, say 10, we could have proceeded the same way: If

$$N = Ae^{ct}$$

then

$\log_{10} N = \log_{10}(Ae^{ct})$ (taking logs of both sides)

$\log_{10} N = \log_{10} A + \log_{10} e^{ct}$ (by Rule 1)

$\log_{10} N = \log_{10} A + ct \log_{10} e$ (by Rule 3)

This is not the same equation as before, but it is in logarithmic form and so is equally correct.

$N = Ae^{ct}$ is an *exponential growth equation*, and equations of this form come up frequently in biology and other sciences. It will be looked at in detail in Section 12.1 and graphed in Section 12.3.

EXAMPLE: *Write $\log_k M = N + 2 \log_k P$ in exponential form.*

If

$$\log_k M = N + 2 \log_k P$$

then

$$\log_k M - 2 \log_k P = N$$

so

$\log_k M - \log_k P^2 = N$ (by Rule 3)

or

$\log_k\left(\dfrac{M}{P^2}\right) = N$ (by Rule 2)

so

$\dfrac{M}{P^2} = k^N$ (by definition of $\log_k$)

How
Logs to Different Bases
are Related

One of the things it would be nice to know is how the logs of the same number to different bases are related. For example, what is the relation between $\log_2 33$ and $\log_{10} 33$? Or $\log_{10} M$ and $\log_e M$?

Suppose we look for a relationship between $\log_{10} M$ and $\log_e M$. Let

$$\log_{10} M = a$$

so

$$M = 10^a$$

Suppose we now take logs to base e of both sides, giving

$$\log_e M = \log_e 10^a$$

Therefore

$$\log_e M = a \log_e 10 \qquad \text{(by Rule 3)}$$

$$\log_e M = (\log_{10} M)(\log_e 10) \qquad \text{(since } a = \log_{10} M)$$

or

$$\frac{\log_e M}{\log_e 10} = \log_{10} M$$

This tells us that for any number M, its log to base 10 is its log to base e, divided by a constant—the constant being $\log_e 10$.

Therefore, in theory (though I'd never want to do it in practice), you could find common logs (to base 10) from a table of natural logs (to base e) by dividing by $\log_e 10$ (which you could get from the table).

In the same way as we derived

$$\log_e M = (\log_{10} M)(\log_e 10)$$

We can show that the relationship between logarithms to bases other than 10 and e is

$$\log_k M = (\log_L M)(\log_k L).$$

This equation is usually easier to remember when it is written in the form

$$\boxed{\log_k M = (\log_k L)(\log_L M)}$$

PROBLEM SET   11.5

Write the following equations in logarithmic form:

1. $3^3 = 27$
2. $8^{2/3} = 4$
3. $4^{-1/2} = \frac{1}{2}$
4. $16^{1/4} = 2$
5. $9^{-5/2} = \frac{1}{243}$

Write the following equations in exponential form:

6. $\log_9 81 = 2$
7. $\log_{16} 8 = \frac{3}{4}$
8. $\log_a 1 = 0$
9. $\log_x a = b$
10. $\log_{10} p = \log_{10} q + \log_{10} r$
11. $\log_x y + \log_x z = a$
12. $2 \log_x 3 + \log_x 1 - \log_x y = 1$
13. $\log_x y - \log_x p = a$
14. $\log_b m = 21 \log_b y + \log_b z + 1$
15. $\log_a 5 - 2 \log_a 5 + 3 \log_a 5 - 4 \log_a 5 = 2$

Evaluate the following expressions:

16. $2^{\log_2 8}$
17. $2^{\log_2 3}$
18. $2^{\log_2 x}$
19. $7^{\log_7 (a+b)}$
20. $7^{\log_7 a + \log_7 b}$

Solve for the unknown in the following equations:

21. $\log_2 32 = y$ 25. $\log_r 32 = \frac{5}{3}$ 29. $-\frac{1}{3} = \log_t 2$
22. $\log_5 125 = r$ 26. $5^{2x} = 125$ 30. $\log_{\sqrt{3}} 27 = y$
23. $\log_4 z = \frac{1}{2}$ 27. $\log_3 x = 3$
24. $\log_3 \frac{1}{81} = a$ 28. $3 = \log_m 125$

Simplify the following expressions:

31. $\log_2(2^{-1/2})^2$
32. $\log_2(2^{x/2})^2$
33. $e^{-\ln x^2}$ (Note: $\ln x = \log_e x$)
34. $\log_5 1 + \log_5 \frac{1}{25}$
35. $\log_2\left(\dfrac{\sqrt{2}}{2}\right)^{-2/5}$

36. $\log_2 \sqrt{32}$
37. $\log_{\sqrt{2}} 4$
38. If $\log_3 (2) = 0.635$, find $\log_3 \frac{1}{6}$
39. If $\log_a 6 = 0.8$, find $\log_a 36$.
40. If $\log_b 81 = 0.9084$, find $\log_b 9$.

Evaluate:

$9^{(1/2) \log_3 7}$

$\dfrac{\log_7 10}{\log_3 10}$

$\log_3 12$

45. $\log_2 15$
46. If $\log_x 4 = 5$, evaluate $\log_x 2$.
47. If $a^2 = b^3$, find $\log_a b$.

48. Simplify

$$\frac{\log_a a^{36}}{\log_a a^9}$$

49. Write the following expression as a single log to the base 10:

$$\frac{\log_{10} 44}{\log_{10} 10} - \log_{10} 2$$

50. Why is $2 < \log_2 5 < \frac{5}{2}$?

51. Consider the statement $\log_p 7^5 = 5$. Rewrite this in exponential form, and thus find p.

52. Given $\ln 10 = \log_e 10 = 2.3$ and $\log_{10} e = \dfrac{1}{2.3}$. Calculate

$$x = \frac{\log_e 10 + 10}{e}$$

53. If $f(x) = \log_3 x$ and $g(x) = x^2$,
   (a) Find $f[g(9)]$.
   (b) Find $f[g(x)]$.

Solve for x:

54. $\log_6(x^2 - 5x) = 1$
55. $\log_2\left(\dfrac{1}{4}\right) + x = \log_e\left(\dfrac{1}{e}\right)$

56. If $a = b^n$, evaluate the following:

   (a) $\log_n\left(\dfrac{\log_a b}{\log_b a}\right)$

   (b) $\log_b(\log_a b) + \log_b(\log_b a)$

11.6   CALCULATIONS USING LOGS

The original reason for the study of logarithms was their importance in calculations. Using them, the worst multiplication and division problems become straightforward addition and subtraction, and even finding roots becomes quite painless (something it is definitely not without logs or computers). For several hundred years before calculators took over, logarithms were used all day and every day by scientists and engineers, and those who did not use tables of logarithms instead used a slide rule, which is based on logarithms.

Today calculators obviously make calculating with logarithms much less important, but learning to do calculations is probably the best way to gain a good understanding of the log rules, which are still as important as they ever were. This section therefore will include examples of all the different types of calculations that can be done using logarithms.

EXAMPLE: *Find* $\dfrac{3.14 \cdot 370}{15}$

Let $x = \dfrac{3.14 \cdot 370}{15}$

Then taking logs (to base 10) of both sides:

$\log x = \log\left(\dfrac{3.14 \cdot 370}{15}\right)$

$\qquad = \log(3.14 \cdot 370) - \log 15$ (by Rule 2)

$\qquad = \log 3.14 + \log 370 - \log 15$ (by Rule 1)

$\qquad = 0.4969 + 2.5682 - 1.1761$ (from the tables)

$\qquad = 1.8890$

Therefore

$\log x =$ the log of the answer we want $= 1.8890$

so

$x =$ the number whose log is 1.8890

or

$x = $ antilog $1.8890 = 7.75 \cdot 10^1$

so

$x = \dfrac{3.14 \cdot 370}{15} = 77.5$

I think you will agree that doing this by long multiplication and division would have been worse!

Checks: Approximate Calculations

It is important to know roughly how big the answer to a calculation should be so that you can catch any glaring errors (even a small error in the characteristic can make a considerable difference in the size of the answer!). To find an approximate answer, replace each factor by a number with only one nonzero digit. For example,

$$\frac{3.14 \cdot 370}{15} \cong \frac{3 \cdot 400}{20} = 60 \qquad \text{which is close enough to the real answer of } 77.45$$

(is approximately equal to)

EXAMPLE: *Evaluate*

$$\frac{(1.95 \cdot 10^{-7}) \cdot (0.263)}{0.00981}$$

Let

$$x = \frac{(1.95 \cdot 10^{-7}) \cdot (0.263)}{0.00981}$$

Then taking logs of both sides and using the log rules:

$$\log x = \log\left(\frac{1.95 \cdot 10^{-7} \cdot 0.263}{0.00981}\right)$$

$$= \log(1.95 \cdot 10^{-7}) + \log(0.263) - \log(0.00981)$$

$$= 0.2900 - 7 + 0.4200 - 1 - (0.9917 - 3)$$

[*Note:* The parentheses around $(0.9917 - 3)$ are absolutely essential because we are meant to be subtracting the whole of $\log(0.00981)$. Without the parentheses the minus refers only to the 0.9917, so we would get $0.2900 - 7 + 0.4200 - 1 - 0.9917 - 3$.]

So

$$\log x = 0.2900 - 7 + 0.4200 - 1 - (0.9917 - 3)$$

$$= 0.2900 - 7 + 0.4200 - 1 - 0.9917 + 3$$

$$= 0.2900 + 0.4200 - 0.9917 - 7 - 1 + 3$$

$$= 0.7100 - 0.9917 - 5$$

At this point we run into a problem because $0.7100 - 0.9917 = -0.2817$, so

$$\log x = -0.2817 - 5$$

This is a problem because the mantissa (decimal part) of $\log x$ has come out negative and so we can't look it up in the tables to find x. However, we add and subtract one to make the mantissa positive.

$$\log x = -0.2817 - 5$$
$$= (-0.2817 + 1) - 1 - 5$$
$$= 0.7183 - 6$$

Now we can look up 0.7183 to find x:

$$x = \text{antilog}\,(0.7183 - 6) = 5.23 \cdot 10^{-6}$$

Therefore

$$\frac{(1.95 \cdot 10^{-7}) \cdot (0.263)}{(0.00981)} = 5.23 \cdot 10^{-6}$$

CHECK:
$$\frac{(1.95 \cdot 10^{-7}) \cdot (0.263)}{(0.00981)} \simeq \frac{2 \cdot 10^{-7} \cdot (0.3)}{0.01}$$
$$= \frac{0.6 \cdot 10^{-7}}{10^{-2}}$$
$$= 6 \cdot 10^{-8 - (-2)}$$
$$= 6 \cdot 10^{-6} \quad \text{O.K.}$$

Note—Useful Trick: In the calculation above we had to rewrite $\log x$ in a different form by adding one to the mantissa and subtracting one from the characteristic. This kind of trick is going to turn up again, so you should feel comfortable with the fact that

$$-0.2817 - 5 = 0.7183 - 6 \quad \text{(the example above)}$$

and

$$0.7183 - 6 = 1.7183 - 7$$
$$= 2.7183 - 8$$
$$= 3.7183 - 9$$
$$= 25.7183 - 31 \quad \text{etc.}$$

EXAMPLE: *Find $\sqrt[3]{79.5^2}$.*

Let $x = \sqrt[3]{(79.5)^2} = (79.5)^{2/3}$.   Then

$\log x = \log(79.5)^{2/3}$

$\qquad = \dfrac{2}{3} \log(79.5)$ (by Rule 3)

$\qquad = \dfrac{2}{3}(1.9004)$

$\qquad = 1.2669$

So

$x = \text{antilog } 1.2669 = 1.85 \cdot 10^1$

Therefore

$\sqrt[3]{(79.5)^2} = 18.5$

CHECK: $4^3 = 64$

so

$\sqrt[3]{(79.5)^2} = (\sqrt[3]{79.5})^2 \simeq (\sqrt[3]{64})^2 = 4^2 = 16$ which is O.K.

EXAMPLE: *Find $\sqrt{0.00335}$.*

Let $x = \sqrt{0.00335} = (0.00335)^{1/2}$.   Then

$\log x = \log(0.00335)^{1/2}$

$\qquad = \dfrac{1}{2} \log(0.00335)$ (by Rule 3)

$\qquad = \dfrac{1}{2}(0.5250 - 3)$ (note: parentheses are necessary)

Here we run into trouble again because dividing through by 2 gives $0.2625 - 1.5$, which is no good because we've got a negative decimal. This is where the "useful trick" comes in again: we have to rewrite $\log 0.00335$ in such a way that 2 goes into the characteristic exactly. This can be done by saying

$0.5250 - 3 = 1.5250 - 4$

so

$$\log x = \frac{1}{2}(0.5250 - 3)$$

$$= \frac{1}{2}(1.5250 - 4)$$

$$= 0.7625 - 2$$

so

$$x = \text{antilog } (0.7625 - 2) = 5.79 \cdot 10^{-2}$$

or

$$\sqrt{0.00335} = 0.0579$$

CHECK: It is easier to square numbers than to square root them, so we'll check that $(0.0579)^2 = 0.00335$. Approximately:

$$(0.0579)^2 = (0.06)^2 = 0.0036 \qquad \text{which is O.K.}$$

EXAMPLE: *There is a legend about a King of Persia who was very fond of playing games. He grew tired of all the games in the kingdom and offered a reward for anyone who would bring him a new game. One day an old man brought the King a chess board and showed him how to play. The King was so delighted that he offered the old man fantastic treasures. The old man replied that he was poor and could never accustom himself to a life of wealth. He said that while riches were useless to him, if the King could supply him with some rice he could feed his family better. The King agreed. The old man asked that one grain of rice be placed on the first square of the chess board he had brought, two on the second, four on the third, eight on the fourth, and so on. The King, still shocked by the old man's refusal of wealth, immediately consented to the strange request and ordered his guards to fill it. But the King's officers became alarmed, because the country's entire rice supply had soon been exhausted. The old man laughed with glee, for he knew that the amount of rice he had asked for was worth more than diamonds and gold. The King, however, was not very amused by the old man's ruse, and chopped off his head. [The moral is clear: Math is dangerous stuff and should be handled with care!] The question is, how many grains of rice were on the last (64th) square of the chess board?*

To work this out, you must realize that the first square contains $1 = 2^0$ grains, the second contains $2 = 2^1$ grains, the third $4 = 2^2$, the fourth $8 = 2^3$, and so on.

Therefore the 64th square contains 2^{63} grains of rice. To find $x = 2^{63}$, take logs, giving:

$\log x = \log 2^{63}$

$\qquad = 63 \log 2 \qquad$ (by Rule 3)

$\qquad = 63(0.3010)$

$\qquad = 18.9630$

so

$\qquad x = $ antilog $18.9630 = 9.18 \cdot 10^{18}$ grains of rice.

As it turns out, the total amount of rice the old man had asked for would be enough to cover the whole of Persia with a layer several inches deep!

EXAMPLE: *Suppose that Rip van Winkle fell asleep having just deposited his $175 pay check in a savings bank offering 5% interest, compounded annually. How much money would he have when he woke up 20 years later?*

A 5% interest rate, compounded annually, means that at the end of each year the bank adds to your account 5% of what is in it. Thus at the end of his first year asleep, Rip van Winkle would have:

$175 + 5\%$ interest on $175

$\qquad = \$175 + \dfrac{5}{100} \cdot 175$

$\qquad = \$175 \left(1 + \dfrac{5}{100}\right)$

$\qquad = \$175(1.05)$

At the end of two years he would have:

$175(1.05) + 5\%$ interest on $175(1.05)

$\qquad = \$175(1.05) + \dfrac{5}{100} \cdot 175(1.05)$

$\qquad = \$175(1.05) \left(1 + \dfrac{5}{100}\right)$

$\qquad = \$175(1.05)^2$

At the end of three years he would have:

$175(1.05)^2 + 5\%$ interest on $175(1.05)^2$

$$= \$175(1.05)^2 + \frac{5}{100} \cdot 175(1.05)^2$$

$$= \$175(1.05)^2\left(1 + \frac{5}{100}\right)$$

$$= \$175(1.05)^3$$

and so on.

Therefore, at the end of 20 years Rip van Winkle would have $175(1.05)^{20}$. To find this number, use logarithms.

Let

$$x = 175(1.05)^{20}$$

so

$$\log x = \log[175(1.05)^{20}]$$

$$= \log 175 + \log(1.05)^{20} \qquad \text{(by Rule 1)}$$

$$= \log 175 + 20 \log 1.05 \qquad \text{(by Rule 3)}$$

$$= 2.2430 + 20(0.0212)$$

$$= 2.6670$$

so

$$x = \text{antilog } 2.6670 = 4.65 \cdot 10^2$$

So Rip van Winkle would have $465, or much more than twice what he had deposited.

Interest Rate Problems

The Rip van Winkle problem is an example of a general type of problem in which P, called the principal, is invested at an interest rate of $r\%$ per year compounded annually. After a time of t years you have a yield of y, where

$$\boxed{y = P\left(1 + \frac{r}{100}\right)^t}$$

(For Rip Van Winkle, $p = 175$, $r = 5$, $t = 20$, and $y = 465$.)

PROBLEM SET   11.6

Evaluate using logs:
1. $325 \cdot 12.7$
2. $0.256 \cdot 0.38$
3. $\dfrac{16.4}{2.31}$
4. $(3.11)^6$
5. $\sqrt{10}$
6. $\sqrt[3]{6.27}$
7. $\sqrt{5\sqrt{5}}$
8. $\sqrt{0.0248}$
9. $\dfrac{89.3}{\sqrt[3]{57.1}}$
10. $(0.0342)^{-3/2}(0.1461)$
11. $\dfrac{33\sqrt{65}}{(2.1)^2}$
12. $\sqrt[3]{\dfrac{(425)^2}{0.48\sqrt{104}}}$

13. $\dfrac{5\sqrt{36.5}}{3\sqrt{15.2}}$
14. $\dfrac{29 \cdot 10^3(0.7)^2}{\sqrt{1841}\,(3.732)^3}$
15. $[(1.1)^{1.1}]^{1.1}$
16. $\dfrac{(0.0073)\,(5190)}{23}$
17. $\sqrt[4]{\left(\dfrac{1}{0.0025}\right)^2}$
18. $\dfrac{3\sqrt{5.0 \cdot 10^3}}{(57)\,(2)\,(24.7)}$
19. $\sqrt[3]{\left(\dfrac{1.6 \cdot 10^2 \cdot (3.05 \cdot 10^1)}{21500} + 10^0\right)^{6.6}}$
20. $\dfrac{5.8 \cdot 10^{-2} - 0.002}{(1.4 \cdot 10^{-3})200}$

Express as a single log in simplest terms:
21. $\frac{2}{3}\log_{10} 4 + \frac{1}{3}\log_{10} 54 - \frac{2}{3}\log_{10} 2$
22. $\frac{1}{2}\log_3 24 + \log_3 5\sqrt{2} - \frac{2}{3}\log_3 6 - \frac{4}{5}\log_3 \frac{1}{2}$
23. $2\log_2 6 + \log_2 \frac{1}{3} - 2\log_2 3 + \log_2 24$
24. $\frac{1}{3}\log_2 13 + \log_2 \sqrt[3]{\dfrac{1}{39}} + \log_2 6$
25. $\log_c x + \log_c 3x - 2\log_c 2x$
26. $3\log_b y - \log_b 2x - \log_b 2z$
27. $\dfrac{\log_2 x + \log_2 3x - 2\log_2 2x}{\log_2 \frac{4}{3}}$
28. $\log_2 15 - \log_2 6 - \log_2 20$
29. $\log_b 2 + 2\log_b x - \frac{1}{2}\log_b L - \frac{1}{2}\log_b z$
30. $\dfrac{\log_m 5 - \log_m 2}{\log_m \left(\frac{4}{25}\right)}$
31. In trying to compute $\sqrt{110} + \sqrt[3]{53}$, a student wrote:

$$x = \sqrt{110} + \sqrt[3]{53}$$

$$\log x = \frac{1}{2}\log(110) + \frac{1}{3}\log(53)$$

$$= \frac{1}{2}(2.0414) + \frac{1}{3}(1.7243)$$

$$= 1.0207 + 0.5748$$

$$= 1.5955$$

so

$$x = 3.94 \cdot 10^1 = 39.4$$

But 39.4 is much too large a number to be correct.
  (a) What did the student do wrong
  (b) Work out the correct solution.

Express $\log_{10} A$ in terms of $\log_{10} x$ and $\log_{10} y$:
  32. $A = x^3 \sqrt{y}$

  33. $A = \dfrac{y^n}{(x^p)^{1/q}}$

Use logs to compute the following:
  34. The radius of the circle $x^2 + y^2 = 89$.
  35. The volume of a room of dimensions 26.7 feet by 12.3 feet by 9.9 feet.
  36. The period, T, of a single pendulum of length 1.35 centimeters, where T is given by the formula

$$T = \pi \sqrt{\frac{l}{g}}$$

where

$$l = \text{length in centimeters}$$
$$g = 980$$
$$\pi = 3.14$$

  37. The weight of a gold ingot with the dimensions shown in Figure 11.1, if the density of gold = 19.3 grams per cubic centimeter

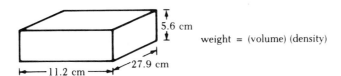

FIG. 11.1

weight = (volume) (density)

  38. The radius of a spherical gold ball of weight 40.9 milligrams. $\left(\text{Volume} = \dfrac{\text{weight}}{\text{density}}\right)$
  39. How long it would take a photon (a light particle) to travel once around the earth (radius of earth = 6.37×10^8 centimeters, velocity of light = 3×10^{10} centimeters per second).

  40. Given a pizza of radius 10 inches, what would the area of one slice be if 13 people shared the pizza? Let $\pi = 3.142$.

  41. The earth travels (around the sun) in a circular path of radius 93 million miles. How far does it travel in one year? In one hour?

  42. Verify the following identity:

$$\log\left(\frac{1}{a} + \frac{1}{b}\right) - \log\left(\frac{1}{a}\right) - \log\left(\frac{1}{b}\right) = \log(a + b)$$

43. Factor: $(\log xy)^2 + \log(x^3) + 3 \log y + 2$.

44. For a cylinder of height $h = 50$ inches, and radius r inches, the surface area $S(r)$ is given by the formula

$$S(r) = 2\pi(50)r + 2\pi r^2$$

Evaluate the ratio $\dfrac{S(3.25)}{S(2.25)}$

45. The weight of a man d miles above the surface of the earth is given by the formula

$$W = \frac{W_0(4.1)^2}{(4.1 + d)^2}$$

where W_0 is the man's weight at the surface of the earth. Determine the approximate weight of a 160-pound man in a plane 5 miles above the earth.

46. Assuming that Indians really did sell Manhattan for the legendary $24, and further assuming that the sale took place exactly 300 years ago, compute the amount of money the Indians would now have if they had placed the money in a bank that paid 5% interest per year, compounded annually, and kept it there until today.

47. Stirling's formula is used to find the approximate values of $n!$ (read "n factorial") when n is very large $(n > 100)$. Now

$$n! = n(n-1)\,(n-2) \cdots 2 \cdot 1$$

for example,

$$5! = 5 \cdot 4 \cdot 3 \cdot 3 \cdot 1 = 60$$

Stirling's formula is as follows:

$$\log_e n! = n \log_e n - n$$

Use the formula to calculate 365!

48. Calculate 1010! using Stirling's Formula.

49. John Maggio deposited his life savings at the Texas National Bank in September of his freshman year. In March he realized that while Texas National was paying only 5% interest, Houston First Federal Bank would pay 6%. So, just six months after his initial deposit, John transferred all of his money to the Houston First Federal Bank. Ten years later (when he finally graduated from college after repeated leaves-of-absence), he withdrew all of his money and found that he had $1000. How much was his initial investment if the interest was compounded quarterly?

$$\left[\text{total money} = (\text{money invested}) \left[1 + \left(\frac{1}{4}\right)\left(\frac{r}{100}\right)\right]^{4n}\right.$$

where n = number of years and r = % interest per annum compounded quarterly.]

50. A particularly prolific microorganism has baffled all of modern science by dividing into three (rather than the usual two) every hour. If there are ten of these little bugs in a petri dish at 9:00 one morning, how many will be there by quitting time at 5:00?

CHAPTER 11 REVIEW

Find:
1. log 97
2. log 0.0907
3. antilog(1.0907)
4. antilog(0.0907 − 7)

5. log(437) − log(0.437)
6. log($\frac{1}{35}$)
7. log$\left(\frac{2.7}{10^3}\right)$

8. antilog(−0.3010)
9. antilog(−0.7)
10. antilog(log 5.05)

Simplify:

11. $\dfrac{\log x^2}{\log x}$

12. log xy − log y
13. $\log_3 3^3$

14. $p^{\log_p q^2}$

If log $Q = 0.4$, find:

15. $\log\dfrac{1}{Q^2}$

16. log 10Q

17. $\dfrac{1}{\log Q}$

Evaluate:

18. $\dfrac{\sqrt{2}}{\sqrt[5]{3}}$

19. $\sqrt{2} - \sqrt[5]{3}$

20. $\dfrac{-(0.291)^3}{0.00185}$

21. $\dfrac{58.25}{0.0631} + \dfrac{29037}{9.56}$

22. $\dfrac{\sqrt{0.045}}{(1.3 + \sqrt{0.810})^2}$

23. $\dfrac{\sqrt[3]{0.073}}{1.2 + 31.7}$

24. $\dfrac{\sqrt[4]{(0.043)^2}}{(65,300)\,(0.00187)}$

Solve for x:

25. $\left(\dfrac{4.09}{0.637}x\right)^2 = 0.462$

26. $x = \dfrac{(1.3 \cdot 10^5)^2}{\sqrt{428}}$

27. $\left(\sqrt[3]{\sqrt{(\log_p 10)^{15}}}\right)^4 = (\log_p 10)^x$

28. Find $\log_z \dfrac{1}{\sqrt{z}}$

29. If $\log_B 4 = 1.2$, find $\log_B 2$.
30. Evaluate $3^{\log_3 0.6}$

Simplify:
31. log $x^2 y$ − log y
32. log$(AB)^2$ − log B^2

33. Solve for A in terms of B, C, and D:

$$\frac{\log A}{B} = \log C + \log D$$

34. If log x − log $y = $ log z, express y in terms of x and z.
35. Simplify $a^{k \log_a x}$.

36. What value of x makes

$$\frac{a^3 \cdot a^x}{a^{3x}} = 1$$

37. A certain species of bacteria defies all laws of biology when it reproduces by splitting into three separate bacteria. It does this splitting every hour. If we start with ten of these weird bacteria, how many will we have after 11 hours? (*Hint:* How many do you have after 1 hour? After 2 hours? After 3 hours? After n hours?) Express your answer to three significant digits.

38. The Richter scale for earthquakes is logarithmic (i.e., a rating of 4 is ten times as strong as a rating of 3, which is ten times as strong as a rating of 2, etc.)
    (a) What would a tremor that was twice as strong as 3 be rated?
    (b) How many times stronger is an earthquake that is rated $5\frac{1}{2}$ than one that is rated 5?

39. One-third of log B equals one-fifth of log C, and half of log C equals log D. If $D = 32$, what is B?

40. If

$$\frac{1}{\sqrt{5}}\left(\frac{1 + \sqrt{5}}{2}\right)^n$$

(where n is a positive integer) is approximated to the nearest integer, then this integer is the nth Fibonacci number. A Fibonacci number is the sum of the two preceding Fibonacci numbers, where 1, 1, 2, 3, 5, 8, 13, 21 are the first eight numbers in the sequence. Find the 20th Fibonacci number using logs. (Fibonacci numbers play an interesting role in describing various kinds of biological growth, especially with regard to flowers.)

41. You know how it is with homework: The further you get the more tired you get and the slower it goes. Suppose you spend x minutes on the first question and each succeeding question takes as much time as all previous questions put together (that is, question 2 takes x minutes; question 3 takes $x + x = 2x$; question 4 takes $x + x + 2x = 4x$, etc.). If eight questions take 3 hours, how long did the first question take? Use logs to find the answer to the nearest second.

42. Find the area of the shape shown in Figure 11.2.

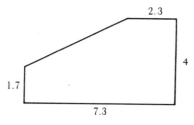

FIG. 11.2

43. One quiet night in Alsace-Lorraine during World War I, Reggie Cramer won \$500 playing poker with the other G.I.'s in his outfit. Not knowing what to do with the money, he deposited it in a local bank. Two weeks later, that area was bombed heavily, and Reggie assumed that his money was gone. Miraculously, however, the bank records survived, and last month, 50 years after the deposit was made, Reggie (now 74) received word that his account was still open. If Reggie's money had been earning 6% interest compounded quarterly ever since it was deposited, how much is the account now worth?

44. Given that

$$1 + \frac{1}{2} + \frac{1}{4} + \frac{1}{8} + \cdots + \frac{1}{2^n} + \cdots = 2$$

(where this infinite sum means that the more terms you add up, the closer you get to 2), show that

$$100 = 10\sqrt{10}\ \sqrt[4]{10}\ \sqrt[8]{10}\ \sqrt[16]{10}\ \ldots$$

45. Compute $10\ \sqrt{10}\ \sqrt[4]{10}\ \sqrt[8]{10}$ and see how close the number is to the 100 you would obtain if you carried out the process indefinitely.

12 LOGARITHMS IN EQUATIONS AND FUNCTIONS

12.1   EXPONENTIAL EQUATIONS

Suppose we want to find x, where

$$5 = 2^x$$

This is called an *exponential equation* because the unknown is contained in the exponent. We have never before had to solve for a variable that was "in the air" like this, and it is not obvious how to begin. It is certainly true that

$$x = \log_2 5$$

but assuming we do not have any tables of logarithms to base 2, this does not tell us the numerical value of x. However, it does give us the idea that the way to get the x "out of the air" (i.e., out of the exponent) and on to the line, where we are used to our unknowns being, is to take logarithms. Since we have tables to base 10, we'd better take logs to base 10. Therefore

$$5 = 2^x$$

means

$$\log 5 = \log 2^x$$

or

$$\log 5 = x \log 2 \qquad \text{(by rule 3)}$$

Now remember that log 5 and log 2 are just ordinary numbers, and so this equation can be solved in exactly the same way that you would solve $7 = 3x$.

Replacing log 5 and log 2 by their numerical values gives:

$$0.6990 = x(0.3010)$$

so

$$x = \frac{0.6990}{0.3010}$$

The value of x must now be found by division. This can either be done the long way or using logarithms. If you do it the second way, you should forget that 0.6990 and 0.3010 themselves came up as logarithms. As far as we are concerned at the moment, they are just two numbers we are interested in dividing.

$$x = \frac{0.6990}{0.3010}$$

$$\log x = \log \frac{0.6990}{0.3010}$$

$$= \log 0.6990 - \log 0.3010 \qquad \text{(by rule 2)}$$

$$= (0.8445 - 1) - (0.4786 - 1)$$

$$= 0.8445 - 1 - 0.4786 + 1$$

$$= 0.3659$$

So

$$x = \text{antilog } 0.3659 = 2.32$$

$$x = 2.32$$

Check: $2^2 = 4$ and $2^3 = 8$ so if $2^x = 5$, we would expect x to be between 2 and 3, and closer to 2.

EXAMPLE: *Solve* $(0.725)^t = 0.031$ *for t.*

Taking logs of both sides:

$\log(0.725)^t = \log 0.031$

$t \log 0.725 = \log 0.031 \qquad$ (by rule 3)

So

$$t = \frac{\log 0.031}{\log 0.725}$$

$$= \frac{0.4914 - 2}{0.8603 - 1}$$

Again t must be found by division, and it would be helpful if you forget that $(0.4914 - 2)$ and $(0.8603 - 1)$ came up as logarithms and think of them just as ordinary numbers. In order to divide them you will have to convert each of them into one number. (This goes very much against the grain if you are still thinking of them as logarithms, because it will make the decimal parts negative. However, it's necessary!)

$$t = \frac{0.4914 - 2}{0.8603 - 1}$$

$$= \frac{-1.5086}{-0.1397}$$

$$= \frac{1.5086}{0.1397}$$

Now we do long division, or use logarithms again, and since I'm lazy, we'll do the second.

$$\log t = \log\left(\frac{1.5086}{0.1397}\right)$$

$$= \log 1.5086 - \log 0.1397$$

$$= 0.1790 - (0.1461 - 1)$$

$$= 0.1790 - 0.1461 + 1$$

$$= 1.0329$$

So

$$t = \text{antilog } 1.0329 = 1.08 \cdot 10^1 = 10.8$$

$$t = 10.8$$

EXAMPLE: *Solve* $(4.2)^{1+x} = 3.5(1.8)^{2-x}$.

Taking logs:

$$\log(4.2)^{1+x} = \log[3.5(1.8)^{2-x}]$$

$$\log(4.2)^{1+x} = \log 3.5 + \log(1.8)^{2-x} \qquad \text{(by rule 1)}$$

$$(1 + x)\log 4.2 = \log 3.5 + (2 - x) \log 1.8 \qquad \text{(by rule 3)}$$

This can be turned into a more ordinary-looking equation by replacing the logs by their numerical values:

$$(1 + x)\, 0.6232 = 0.5441 + (2 - x)\,(0.2553)$$

Solving for *x:*

$$0.6232 + 0.6232x = 0.5441 + 0.5106 - 0.2553x$$

$$0.6232x + 0.2553x = 0.5441 + 0.5106 - 0.6232$$

$$0.8785x = 0.4315$$

So

$$x = \frac{0.4315}{0.8785}$$

To find x, take logs again:

$$\log x = \log \frac{0.4315}{0.8785}$$

$$= \log(0.4315) - \log(0.8785)$$

$$= (0.6355 - 1) - (0.9440 - 1)$$

$$= 0.6355 - 1 - 0.9440 + 1$$

$$= 1.6355 - 2 - 0.9440 + 1 \qquad \text{(this is to make sure the mantissa comes out positive)}$$

$$= 0.6915 - 1$$

$$x = \text{antilog } (0.6915 - 1) = 4.91 \cdot 10^{-1} = 0.491$$

$$x = 0.491$$

EXAMPLE: *As most bacteria do, the Thyngs multiply according to the law*

$$\boxed{N = N_0 e^{kt}}$$

where

$t = time$ *(in hours) measured from some starting point*

$N_0 = number$ *of Thyngs at starting point, when* $t = 0$

$N = number$ *of Thyngs at time t*

$k = a$ *constant depending on the Thyng*

$e = 2.72$ *(approx.)*

At noon one day, two Thyngs land in Metropolis, and by 1 o'clock there are 30 of them. How long will it take for their numbers to swell to 1000?

The equation $N = N_0 e^{kt}$ has five letters in it, so to find a value for any one of them, we have to know values for the other four.

In this problem, we measure time from noon, when $t = 0$. At noon we have two Thyngs so $N_0 = 2$, and e is always 2.72. The problem wants us to find t when $N = 1000$, which means solving

$$1000 = 2(2.72)^{kt} \quad \text{for} \quad t$$

We can't solve this for t without knowing k. However, at $t = 1$ we know everything except k, and so can find k. When $t = 1$, $N = 30$, because there are 30 Thyngs at 1 o'clock. Hence:

$$30 = 2(2.72)^{k \cdot 1}$$

We will substitute the value for k we get from this equation into $1000 = 2(2.72)^{kt}$ and then solve for t.

But first we must solve for k:

$$30 = 2(2.72)^k$$

Taking logs:

$$\log 30 = \log[2(2.72)^k]$$

$$\log 30 = \log 2 + k \log 2.72$$

Therefore,

$$1.4771 = 0.3010 + k(0.4346)$$

$$k = \frac{1.1761}{0.4346}$$

Using logs again to calculate k:

$$\log k = \log 1.1761 - \log 0.4346$$

$$= 0.0719 - (0.6385 - 1)$$

$$= 0.4334$$

So

$$k = \text{antilog } 0.4334 = 2.71$$

Going back to the equation for t:

$$1000 = 2(2.72)^{kt}$$

and substituting the value we have just found for k:

$$1000 = 2(2.72)^{2.71t}$$

Taking logs:

$$\log 1000 = \log[2 \cdot (2.72)^{2.71t}]$$

$$\log 1000 = \log 2 + \log[(2.72)^{2.71t}]$$

$$\log 1000 = \log 2 + 2.71t \log 2.72$$

So

$$3 = 0.3010 + 2.71t\ (0.4346)$$

or

$$t = \frac{2.6990}{(2.71)\,(0.4346)}$$

Using logs to calculate t:

$$\log t = \log \left[\frac{2.6990}{(2.71)\,(0.4346)} \right]$$

$$= \log 2.6990 - \log 2.71 - \log 0.4346$$

$$= 0.4314 - 0.4330 - (0.6385 - 1)$$

$$= 0.3599$$

So

$$t = \text{antilog } 0.3599 = 2.29 \text{ hours.}$$

So it takes 2.29 hours, or just over two hours, for there to be 1,000 Thyngs. A remarkably short time!

EXAMPLE: *Radioactive Decay follows the equation*

$$\boxed{Q = Q_0 e^{-kt}}$$

where

$Q_0 = $ *initial amount of material*

$Q = $ *amount at time t later*

$e = 2.72$ *(approx.)*

$k = $ *constant depending on the substance*

We will show that no matter how much of the substance you begin with, it always takes the same time to decay to half that amount.

The reason for this is that when there is more material present it decays faster. The amount of time for half of the material to disintegrate is called the *half-life* of the substance.

More generally, the amount of time for a given fraction of the substance to decay does not depend on the initial amount. It is this fact that makes carbon dating possible, because it means that by seeing what fraction of the unstable carbon in a sample has decayed, you can tell the age of the sample.

Let us call the half-life T, and look for a relation between T and k.

T is the time for half the material to decay, so when $t = T$, $Q = \frac{1}{2}Q_0$. Therefore T satisfies:

$$\frac{1}{2}Q_0 = Q_0 e^{-kT}$$

or, cancelling Q_0's:

$$\frac{1}{2} = e^{-kT}$$

Since Q_0 now does not enter into the equation, its value cannot affect the value of T. Therefore T does not depend on Q_0, in other words, the amount of time for half the material to decay (T) is the same regardless of how much you start with (Q_0).

To find a relationship between k and T, take logs of the equation:

$$\frac{1}{2} = e^{-kT}$$

$$\log \frac{1}{2} = \log e^{-kT}$$

$$\log 1 - \log 2 = -kT \log e \qquad \text{(by rules 2 and 3)}$$
$$- \log 2 = -kT \log e \qquad (\log 1 = 0)$$

So

$$T = \frac{\log 2}{k \log e}$$

This gives you T in terms of k. Notice that it doesn't contain Q_0.

EXAMPLE: *The police were baffled by what seemed to be the perfect murder of a girl who had been found, apparently suffocated, in her kitchen. Finally, Sherlock Holmes was called in. With the aid of Dr. Watson's knowledge of botany, the mystery was solved and the following story told.*

The girl had been making bread in her kitchen, whose dimensions were 10 feet by 50 feet by 10 feet. She had formed the dough into a ball of volume $\frac{1}{8}$ cubic feet and turned away to wash some dishes. At that moment Holmes' enemy, Professor Moriarty, had added a particularly virulent strain of yeast to the dough. As a result, the bread immediately started to rise, tripling in volume every 4 minutes. Before long, the dough filled the room, stopping the clock at 3:48 and squashing the girl to death against the wall. By the time Inspector Lestrade of Scotland Yard reached the scene the next day, the yeast had worked itself out and the dough returned to its original size.

At what time did Professor Moriarty add the yeast?

To solve this, we must find a relation between the time, t minutes, after the yeast was added and the volume, V, of the bread.

After 4 minutes the dough has tripled, so

$$V = \frac{1}{6} \cdot 3$$

After 8 minutes, the dough has tripled again, so

$$V = \left(\frac{1}{6} \cdot 3\right) \cdot 3 = \frac{1}{6} \cdot 3^2$$

After 12 minutes,

$$V = \frac{1}{6} \cdot 3^3$$

So after t minutes:

$$V = \frac{1}{6} \cdot 3^{t/4}$$

We want to know how long it takes for the yeast to fill the room, which has volume $10 \cdot 50 \cdot 10 = 5000$ cubic feet. Therefore we want to solve the equation

$$5000 = \frac{1}{6} \cdot 3^{t/4}$$

Taking logs:

$$\log 5000 = \log\left(\frac{1}{6} \cdot 3^{t/4}\right)$$

$$= \frac{t}{4} \log 3 - \log 6$$

or

$$3.6990 = \frac{t}{4}(0.4771) - 0.7782$$

$$t = \frac{4(4.4772)}{(0.4771)}$$

This calculation can be done by logs, giving

$t = 37.5$ minutes or about 38 minutes

Therefore, Professor Moriarty added the yeast at about 3:10 p.m.

PROBLEM SET 12.1

Solve for the unknown.

1. $2^x = 2.2$
2. $2.5^y = 0.25$
3. $4^t = 0.12$
4. $99 = 30^a$
5. $4^{m+1} = 7$
6. $2^x = 3^{x+2}$
7. $2^z = 14^{2z-1}$

8. $5^{2b} = 15$
9. $0.67^r = 8$
10. $5^{p^2+4p} = 17.64$
11. $5^{2k} = 28 \cdot 2^k$
12. $15^{3x} = 85 \cdot 3^x$
13. $11^{x-1} = 1331$
14. $5^{2d-3} = 14$

15. $(\frac{1}{3})^{-c} = 12$
16. $(121)^{x+2} = 35^{x+1}$
17. $24 \cdot 2^y = 12 \cdot 8^{3y}$
18. $12^{z-1} = 45$
19. $3^{2w} = 15$
20. $0.8 = 0.9^{1+t}$

21. Use logs to compute $\dfrac{0.9031}{0.7782}$. Then use this to solve $6^x = 8$.

22. Solve for x: $x^5 = \dfrac{x^{0.0229}}{9}$.

23. Solve the following simultaneous equations:

$$5^{a+2b} = 10$$

$$a - b = 3$$

24. If $A = B \cdot 10^{kt}$, find $\log \dfrac{A}{B}$.

25. From the relation $P_1 v_1{}^x = P_2 v_2{}^x$, find:
    (a) The value of x when $P_1 = 16.7$, $P_2 = 36.8$, $v_1 = 57.3$, $v_2 = 32.6$
    (b) The general formula for x in terms of P_1, P_2, v_1, v_2

Solve the following equations for the variable in parentheses.

26. $R = b^x e^{at}$ (x)
27. $1 = AB^k$ (k)
28. $A + B = cy^{zt}$ (t)
29. $AB = m + p^{r+st}$ (s)
30. $\dfrac{N}{2} + m^{bt} = N$ (t)
31. $100^x = ab$ (x)

32. $R = \dfrac{1}{T} e^{-Rx}$ (x)

33. $\dfrac{n}{\sqrt{\dfrac{1}{n^3}}} = n^a$ (a)

34. $Pe^{-a} - R = S$ (a)
35. $A = B^{(c+Dv^2)}$ (v^2)

36. Let $f(x) = 3^x$ and $g(x) = \log_{27} x$.
    (a) Find x such that $f(x) = 10$.
    (b) Find $f[g(x)]$.
    (c) Find $f^{-1}(x)$.

37. At a roulette table in Las Vegas, you may bet on whether the number comes up odd or even. If you win, the house gives you back your bet plus a 55% return. If you started with \$10, and reinvest all your money on each subsequent bet, how many consecutive odd–even wins does it take to become a millionaire?

38. The population of a colony of ants grows with time according to:

$$N(t) = 100 \, e^{0.2t} - 11e^{0.3t}$$

Another colony grows according to:

$$N(t) = 20e^{0.3t} - e^{0.2t} \quad \text{where } e = 2.718$$

When are the two colonies of equal size?

39. A scientist has predicted that the world's maximum oil production will occur in 1980. After that, production will decline according to the formula:

$$\text{production} = P_{max} \cdot 10^{-(0.015)Y}$$

where Y = number of years after 1980. In what year will the world's oil production be one-half the 1980 production level?

40. How long will it take to get \$632.51 from an initial investment of \$150 if it receives 5% interest compounded annually?

41. Suppose you know that the volume, V, of a sphere is related to the radius (r) by a formula with the form:

$$V = ar^b$$

But you've forgotten both the coefficient, a, and the exponent, b. Suppose further that you have the following information:

Diameter of Sphere	Volume of Sphere
1.2	0.9048
7.2	195.4

What are a and b?

42. For a long time it has been known that the height, H, of a tree is related to its diameter, D, by an equation of the form

$$H = x(D)^y$$

This was determined experimentally (not theoretically) from data such as:

Diameter of Tree	Height of Tree
0.8	59.2
1	70.0
1.2	80.26

What is the value of the exponent y?

43. The *E. coli* cell is approximately a box with a volume of 1 cubic micron (1 micron = 10^{-6} meter; 1 cubic micron = 10^{-12} cubic centimeter (cc)).

   (a) What is the weight of one cell, if the density of a cell = 1 gram/cc? Suppose the cell can divide in two every 30 minutes, and that after it has divided n times, there are 2^n cells. (That is, after one division there are two cells; the next time both divide again, so there are four cells, i.e., 2^2 cells; after three divisions there are $2 \cdot 4 = 8$ cells, i.e., 2^3 cells; ... etc.)

(b) What is the total weight of the cells after n divisions (as a function of n)?

(c) How many divisions must occur for the weight of the E. *coli* to exceed the weight of the earth ($6 \cdot 10^{27}$ grams)?

(d) How long will it take for the weight of the E. *coli* to exceed the weight of the earth?

44. The cicatrization (healing) of a wound of a certain kind is described by the function

$$A = f(t) = B \cdot 10^{-0.221t}$$

where A is the area in square centimeters of the wound after t days, and B is the initial area of the wound.

(a) Calculate the area of the wound after 3 days, if the initial area was 107 square centimeters.

(b) How long will it take for a wound to be 95% healed, according to this formula?

45. In a strange country, on the farthest moon of the nearest planet of the farthest star, is a strange race of people. In this country there is no war, disease, pestilence, famine, or inflation. And the people love each other very much. So much, in fact, that the population triples every 4 years. If the population today is 100, after how many years will the population be 10^6?

46. Imagine another land where the population today is 100,000 and the population triples every 5 years. When will the population in the two countries in this and the previous problem be the same?

47. After sitting unattended all winter, the Idaville municipal swimming pool is about to be reopened. Unfortunately, the town fathers discover that the water in the pool contains an unacceptable 10^7 bacteria per gallon. If the pool's filter can process an entire pool full of water every half-hour, and if that filter removes 75% of the bacteria in the water that passes through it, how long must the town fathers run it before the pool water reaches an acceptable level of 10^5 bacteria per gallon?

48. Suppose you are the chief of the fire department of a major Midwestern city. As is well known, the more houses that are on fire at any one time, the more houses are likely to catch on fire, much the way on epidemic disease spreads. For this reason, and because of the obvious risk to life and property, you find it extremely important to extinguish these fires at their outset, before they spread.

Suppose the number of homes on fire, N, at a time t after the first report of a fire in the city is given by

$$N = N_0 e^{kt}$$

where $e = 2.72$, $N_0 = $ the number of homes on fire when $t = 0$, and k is some constant which depends on the particular type of fire.

(a) If the first report that your department receives is that two houses are burning, what is N_0?

(b) If your fire engines leave the fire house immediately after the first report and arrive at the scene of the fire 1 hour later to find 10 houses ablaze, what is k?

(c) If the city consists of 100,000 homes, how soon would it be consumed by fire if your department had been on strike at the time of this fire?

49. A ball is dropped from a height of 25 feet and rises up two-thirds of that distance after its first bounce. Each succeeding height is two-thirds of the previous height. How many bounces must occur in order for the ball to bounce less than 1 foot from the ground?

50. In Figure 12.1 are shown the exponential decay curves to the base e of two poisoned bacteria cultures F_1 and F_2. Assume that both cultures have the same decay rate constant p.

(a) What is the equation of F_1? Of F_2?

(b) What is the y value at arrow one in terms of p?

(c) (i) Express the half-life of F_2 in terms of p. (The answer can contain logs.)

(ii) If the half life of F_2 is 4 minutes, what is the numerical y value at arrow two?

(iii) Will F_1 have the same half-life as F_2? Why?

(d) How many minutes after t will there be only 10 bacteria in F_2?

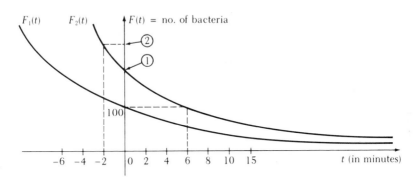

FIG. *12.1*

12.2   LOGARITHMIC EQUATIONS

Suppose we wanted to solve the equation

$$2 \log\!\left(\frac{x}{3}\right) + 5 = 9$$

which is called a *logarithmic equation*, because the unknown is contained in the logarithm. To solve it, notice that

$$2 \log\left(\frac{x}{3}\right) + 5 = 9$$

is a linear equation for $\log\left(\frac{x}{3}\right)$, giving

$$2 \log\left(\frac{x}{3}\right) = 4$$

or

$$\log\left(\frac{x}{3}\right) = 2$$

So

$$\frac{x}{3} = \text{antilog } 2 = 10^2 = 100$$

Therefore,

$$x = 300$$

EXAMPLE: *Solve for q:* $\log q = 6 - \log \frac{q}{6}$

We will treat this as an equation for $\log q$, by first using the fact that $\log \frac{q}{6} = \log q - \log 6$ so the equation becomes

$$\log q = 6 - (\log q - \log 6)$$
$$2 \log q = 6 - \log 6$$
$$2 \log q = 6 - 0.7782$$
$$2 \log q = 5.2218$$
$$\log q = 2.6109$$

So

$$q = \text{antilog } 2.6109 = 4.08 \cdot 10^2$$
$$q = 408$$

EXAMPLE: *Solve* $\log_a x - \log_a(x - 2) = \log_a 10 - \log_a(x + 1)$ *for x.*

Note that this equation is not to base 10, but it turns out not to matter because every term has a $\log_a$ in it. First let us get all the terms containing x on the left:

$$\log_a x - \log_a(x - 2) + \log_a(x + 1) = \log_a 10$$

Using the first and second log rules to rewrite the left side as the

$\log_a$ of one quantity, we see that

$$\log_a\left[\frac{x(x+1)}{x-2}\right] = \log_a 10$$

Now two different numbers cannot have the same logarithm to base a, so

$$\frac{x(x+1)}{x-2} \quad \text{and} \quad 10$$

must be the same number. In other words,

$$\frac{x(x+1)}{x-2} = 10$$

This is now a perfectly ordinary quardratic equation, which can be solved in the usual way:

$$x(x+1) = 10(x-2)$$

$$x^2 - 9x + 20 = 0$$

$$(x-4)(x-5) = 0$$

So $x = 4$ and $x = 5$ are the solutions.

EXAMPLE: *An earthquake is rated on the Richter scale, which assigns a rating to every earthquake based on its intensity relative to a "standard" earthquake. If R_1 and R_2 are the ratings of earthquakes of intensity P_1 and P_2, then the Richter scale tells you that*

$$R_1 - R_2 = \log\left(\frac{P_1}{P_2}\right)$$

If the Tehachapi earthquake of 1952 was rated 7.5, and the San Francisco earthquake of 1906 was rated 8.25, how much stronger was the San Francisco earthquake than the Tehachapi?

We have $R_1 = 8.25$, $R_2 = 7.5$, and we want to find $\dfrac{P_1}{P_2}$, the relative strengths. Substituting into the equation:

$$8.25 - 7.5 = \log\left(\frac{P_1}{P_2}\right)$$

$$0.75 = \log\left(\frac{P_1}{P_2}\right)$$

$$\frac{P_1}{P_2} = \text{antilog } 0.75 = 5.62 \cong 5\tfrac{1}{2}$$

The San Francisco earthquake was about $5\tfrac{1}{2}$ times as strong as the Tehachapi earthquake.

PROBLEM SET   12.2

Solve the following equations for the unknown.

1. $\log(x - 2) = 7$
2. $\log(3x - 20) = 5$
3. $\log(2x + 1) = 0.3$
4. $\log a^2 - \log(a - 1) = 1$
5. $\log_2(x + 1) = 1 + \log_2 x$
6. $\log(y + 1) + \log(y - 4) = 1$
7. $2 = \log 3y + \log\left(4y + \dfrac{40}{3}\right)$
8. $\log(r + 1) = \log(2r^2 + 3) - \log(2r - 5) - 1$
9. $\log x^2 + \log 2x = 7.2$
10. $\dfrac{\log 27}{3} + \log(9 - 3) = \log x$
11. $3 + \log_2 x + \log_2 3 = \log_2 96$
12. $(\log b)^2 - 3 \log b + 2 = 0$
13. $\log(u + 1)^2 = 2.3$
14. $\log(r - 1) - \log(r + 3) = \log 0.01$
15. $\log 2z - 2 \log z = -1$

Solve the following equations for the variable in parentheses in terms of the other variables.

16. $\dfrac{\log a}{b} = \log c - \log d$ $\qquad$ (a)
17. $\log x + \log a + 3 \log b = 0$ $\qquad$ (b)
18. $a = \log \sqrt[n]{x}$ $\qquad$ (x)
19. $\dfrac{\log x}{m} = \log n + \log p$ $\qquad$ (x)
20. $\log_N b = \log_{1/N} y$ $\qquad$ (y)

Verify the following identities:

21. $\log\sqrt{\dfrac{1 - x}{1 + x}} = -\dfrac{1}{2} \log\left(\dfrac{1 + x}{1 - x}\right)$
22. $\dfrac{\log(x + h) - \log x}{h} = \log\left(1 + \dfrac{h}{x}\right)^{1/h}$

Solve the following equations for the variable in parentheses.

23. $\dfrac{\log_3 a + \log_3 x}{\log_3 x} = \log_3 9$ $\qquad$ (x)
24. $\dfrac{1}{\log x - 1} = \dfrac{2}{\log x + 1}$ $\qquad$ (x)
25. $(\log \sqrt{10}) \log(y - 1) + \log 0.1 = 0$ $\qquad$ (y)
26. $\dfrac{3}{\log k + 4} + \dfrac{1}{\log b} = 1$ $\qquad$ (b)
27. $\dfrac{k^{\log_{10} t}}{k^2} = k^{1 - \log_{10} t}$ $\qquad$ (t)
28. $[\log(2w - 1)]^2 - 3 \log(2w - 1) - 10 = 0$ $\qquad$ (w)
29. $2 \log_2\left(\dfrac{1 + x}{1 - x}\right)^2 - 5 \log_2\left(\dfrac{1 + x}{1 - x}\right) + 2 = 0$ $\qquad$ (x)
30. $ax = \log_b \sqrt{b + ry^2}$ $\qquad$ (y)
31. If
$$f(x) = \log_2 \dfrac{x - 1}{2}$$
find $f^{-1}(x)$.

32. Consider $\log(-3x + 1) = x^2 + 2$.
    (a) Convert this to the equivalent statement using an exponent.
    (b) Can $\frac{1}{3} > x > 0$?
Hint: Argue (b) by considering the range of values on the left and right sides of the equality.

Solve the following inequalities for the unknown.

33. $\log p < 2$
34. $(\log r)^2 > 4$
35. $\log|x| < 3$
36. $(\log|x|)^2 > 4$
37. If $\log x$ differs from 2 by less than 0.1, where may x lie? Graph your result on a number line.
38. If $\log x$ and $\log 1$ differ by 0.01, by how much can x differ from 1?
39. A measure of the strength of earthquakes is given by

$$\frac{\log\left[\frac{1}{10}(\text{number of broken windows}) + 1\right]}{\log 2}$$

   (a) A minor tremor broke 20 windows. What is its rating?
   (b) How many windows must be broken in a quake with a rating twice as high as the rating in part (a)?
   (c) Why do you think 1 is added to one tenth the number of windows broken?

40. The Henderson-Hasselbach equation relates the pH (or acidity) of a patient's blood to the concentrations of bicarbonate (b) and of carbonic acid (c) in the blood. The equation is

$$\text{pH} = 6.1 + \log_{10}\left(\frac{b}{c}\right)$$

   (a) If in a healthy person the concentration of bicarbonate is 24 and that of carbonic acid is 1.2, what is the pH of the blood of a healthy person?
   (b) In a patient with lung disease the carbonic acid concentration increases, and if it reaches 2 the patient is considered to have severe lung failure. What would be your diagnosis for a lung patient whose bicarbonate concentration is 24 and whose blood has a pH of 7.1?

12.3   THE EXPONENTIAL AND LOGARITHMIC FUNCTIONS

The Exponential Function

You will remember that 10^x can be defined for any real number x. When x is a rational number (an integer or a fraction) we have given an explicit definition of 10^x (namely, $10^{-n} = \dfrac{1}{10^n}$, $10^{m/n} = \sqrt[n]{10^m}$, $10^0 = 1$, etc.), and when x is irrational we said that you can find 10^x as accurately as you want by looking at $10^{m/n}$, where m/n is a rational number close to x. Therefore, to any real number x there corresponds a number 10^x, and we can consider the function

$$y = f(x) = 10^x$$

This is an *exponential function*, and we drew its graph along with that of $y = 10^{-x}$ in Chapter 10. We will now look at some other exponential functions.

Change of Base   Ten is called the base of the exponential function $f(x) = 10^x$. If we change the base, we get other exponential functions, such as

$$g(x) = \left(\frac{1}{10}\right)^x \qquad h(x) = 2^x \qquad k(x) = \left(\frac{1}{2}\right)^x$$

(*Note:* we will not make the base a negative number because a^x is then not defined for all x. For example, if a is negative we cannot take its square root, so $a^{1/2}$ is not defined.)

To graph $g(x) = (\frac{1}{10})^x$ Use a table of values, or notice that

$$g(x) = \left(\frac{1}{10}\right)^x = \frac{1}{10^x} = 10^{-x} = f(-x)$$

So the value of g at $x = 2$ will be the same as the value of f at $x = -2$, and so on. Therefore the graph of $g(x) = (\frac{1}{10})^x$ is the graph of $f(x) = 10^x$ reflected in the y axis. See Figure 12.2.

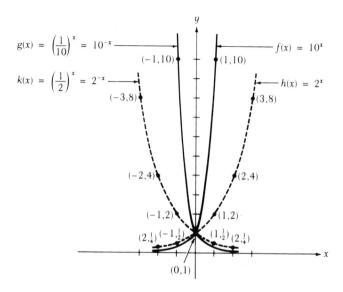

FIG. 12.2

To Graph $h(x) = 2^x$ **and** $k(x) = (\frac{1}{2})^x$ Use a table of values, and notice that if x is positive,

$$2^x < 10^x \quad \text{(for example, } 2^2 < 10^2)$$

If x is negative,

$$2^x > 10^x \quad \text{(for example, } 2^{-1} > 10^{-1})$$

Therefore the graph of $h(x) = 2^x$ is below that of $f(x) = 10^x$ for positive x, and above it for negative x; the graph of $k(x) = (\frac{1}{2})^x$ is the reflection of the graph of $h(x)$ in the y axis. See Figure 12.2.

From Figure 12.2 you can see that *the domain of an exponential function* (the x values for which there are y values) *is all the real numbers. The range of an exponential function* (all the y values that the function can produce) is all real positive numbers (but not zero).

EXAMPLE: *Plot the growth of a bacteria colony against time.*

From Section 12.1 you know that the number of bacteria in a colony grows according to the law

$$N = N_0 e^{kt}$$

where N = number at time t and N_0, e, k, t are positive constants, with $e = 2.72$ (approx.).

We have to plot a graph of N against t. This can be done using a table of values, which will show that $N_0 e^{kt}$ is much like 2^t (since k is positive), except that it cuts the vertical axis at $(0, N_0)$. See Figure 12.3. (*Note:* The equation does not apply for negative t, so we only graph it for $t > 0$.)

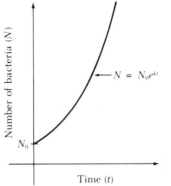

FIG. *12.3*

If N and t are related in this way, the number of bacteria is said to grow exponentially with time. You can see that as time goes on, the number of bacteria increases more and more quickly. Most populations that multiply freely increase in much the same way—hence the term "population explosion" as applied to the population of the world.

The Logarithmic Function

If x is a positive number, you can always find its logarithm; therefore we can consider the *logarithmic function:*

$$y = g(x) = \log_{10}x$$

This function is not defined for x negative or zero. Its graph is shown in Figure 12.4.

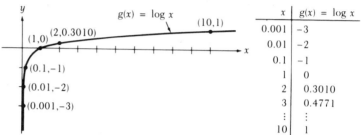

x	$g(x) = \log x$
0.001	-3
0.01	-2
0.1	-1
1	0
2	0.3010
3	0.4771
$\vdots$	$\vdots$
10	1

FIG. 12.4

The logarithmic function grows very slowly, in fact, slower than any power of x. The *domain* of the log function (all possible x's) is the positive real numbers, and its *range* (all possible y's) is all real numbers.

The Relation Between the Logarithmic and Exponential Functions

If you compare the graphs of $y = 10^x$ and $y = \log_{10}x$ in Figure 12.5, you will see that they are the same shape only oriented differently. From the labeled points it should be clear that if a point (a, b) is on one graph, then the point (b, a) is on the other. This means that the graphs are reflections of one another in the line $y = x$.

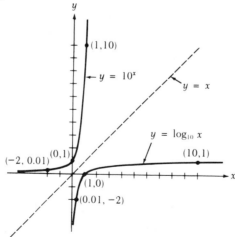

FIG. 12.5

This should be enough to make you suspicious: The only other time this has come up was in dealing with inverse functions. In fact: *The exponential and logarithmic functions are inverses of one another.*

To show this properly, we must prove that:

If
$$f(x) = k^x \quad \text{and} \quad g(x) = \log_k x$$
then
$$g[f(x)] = x \quad \text{and} \quad f[g(x)] = x$$

Proof:
$$g[f(x)] = \log_k (kx)$$

and
$$\log_k k^x = x \qquad \text{(This is one of the two results about the relationship between exponents and logarithms in Section 11.5)}$$
so
$$g[f(x)] = x$$

Also
$$f[g(x)] = k^{\log_k x}$$

and
$$k^{\log_k x} = x \qquad \text{(This is another result from Section 11.5.)}$$
so
$$f[g(x)] = x$$

Therefore the two identities of Section 11.5 ensure that $f(x) = k^x$ and $g(x) = \log_k x$ are indeed inverse functions.

This tells us that the exponential and logarithmic functions "undo" one another. For example, taking $\log_{10} 2$ gives you 0.3010; taking $10^{0.3010}$ gives you 2 back again. Conversely, if you start with 3 and take 10^3 you get 1000; taking $\log_{10} 1000$ gives you the 3 back again.

PROBLEM SET   12.3

Sketch on the same axes:

1. (a) $f(x) = \log(x)$ (b) $f(x) = -\log(x)$ (c) $f(x) = 10^x$

Sketch the following graphs.

2. $f(x) = \log_{10}(x - 1)$ 5. $g(z) = \log_2(z)$ 8. $g(x) = \log_3 x^2$
3. $f(x) = \log_{10}(x) - 1$ 6. $f(x) = -\log_{10}(x - 4)$ 9. $k(x) = \log_3 (-x)$
4. $f(x) = \log_{10}(x - 5)$ 7. $h(x) = |\log_2 x|$ 10. $Q(x) = \log_{10} 5x$

Let $f(x) = \log_{10} x$ and $g(x) = 10^x$. Find the following:

11. $f(10)$ 16. $g[f(100)]$
12. $f(1)$ 17. $f[f(10)]$
13. $g(2)$ 18. $g[g(0)]$
14. $g(0)$ 19. x if $g(x) = 1$
15. $f[g(1)]$ 20. x if $f(x) = 3$

21. Consider the statement $y = 5^x$.
    (a) Is y a function of x? Why or why not?
    (b) Find the value of y when $x = -2$.
    (c) Evaluate y when $x = \log_5 M$.
    (d) Find the y intercept of the graph of $y = 5^x$.
    (e) Are there any values of x such that y is a negative number?
    (f) What happens to y as $x \to \infty$? And as $x \to -\infty$? (Note: ∞ is a symbol for infinity.)
    (g) Graph $y = 5^x$.

22. If $f(x) = \log_{10}(1 + x)$:
    (a) Find $f(0)$.
    (b) Find x such that $f(x) = 1$.
    (c) Graph $f(x)$.

For each of the following functions $h(x)$, find $f(x)$ and $g(x)$ such that $g[f(x)] = h(x)$ [and neither $f(x) = x$ nor $g(x) = x$].

23. $h(x) = \log(2x + 7)$ 28. $h(x) = \log(x + 1) + 1$
24. $h(x) = (\log x)^2$ 29. $h(x) = 2^{\log x}$
25. $h(x) = 10^{x^2}$ 30. $h(x) = e^{1/x^2}$
26. $h(x) = 5^{-x}$ 31. Graph $y = e^x + A$ $(A > 0)$
27. $h(x) = 2^{3x}$ 32. Graph antilog$_4 x = y$.

33. If $f(x) = \log x$, for what k is

$$f\left(\frac{1}{x}\right) = k\, f(x)$$

34. If $f(x) = \log(x)$ and $g(x) = x^2$:
    (a) Find $g[f(x)]$.
    (b) Simplify

$$\frac{f[g(x)]}{2\, f(x)}$$

35. A radioactive substance decays according to

$$Q = Q_0 e^{-ct} \quad \text{where} \quad \begin{aligned} &t = \text{time} \\ &Q_0 = \text{initial amount of substance} \\ &Q = \text{amount at time } t \\ &e \cong 2.7 \\ &c = \text{positive constant} \end{aligned}$$

Graph Q against t.

36. Loudness is measured in *decibels*, where a just-audible sound is defined to be zero decibels. The loudness, D, of a sound depends on the intensity, I (the energy required to produce the sound) by the following formula:

$$D = 10 \log\left(\frac{I}{I_0}\right)$$

where $I_0 =$ intensity of the the just-audible sound. Graph loudness against relative intensity, $\left(\frac{I}{I_0}\right)$.

37. Let $f(x) = \log\left(\dfrac{1+x}{1-x}\right)$ for $-1 < x < 1$

    (a) What is

$$f\left(\frac{3x + x^3}{1 + 3x^2}\right)$$

    Simplify your answer as far as possible.

    (b) Your answer to part (a) can be expressed in one of the following ways, where k is some constant:

$$[f(x)]^k \qquad k \cdot f(x) \qquad \frac{f(x)}{k} \qquad f(x) + k$$

    Choose the correct expression and proper value for k.

38. Given the equation $y = 2^{wx}$:
    (a) Graph the equation if $w > 0$.
    (b) Graph the equation if $w < 0$.
    (c) Graph the equation if $w = 0$.
    (d) Graph the inverse function in each case.

39. Whales are known to be exceptionally noisy singers. They carry tunes that can be heard miles away under ocean waters. The intensity of the sound I of one whale that is heard by another a distance D away is found using the equation

$$\log I = 2 - \frac{D}{3}$$

    (a) Graph this function, plotting D versus I.
    (b) How close must a neighboring whale be to hear a song 5 intensity units loud?
    (c) Using the graph, over how many miles does the sound in (b) decrease its intensity by a factor of 2?

40. A very rare, perfectly symmetrical tree (shown in Figure 12.6) was found on a botanical field trip. Each of the branches of the tree forks into two branches after every foot of growth. The base of the branching contains one fork.

FIG. *12.6*

    (a) The number of branches B at a height H above the initial fork can be described by what function $B(H)$?
    (b) Graph this function.
    (c) Use the graph to approximate the number of branches 10 feet in branch height above the fork.
    (d) A second function $H(B)$ can be used to give H in terms of B. What is this function? Graph it.

12.4   SEMILOG AND LOG-LOG GRAPH PAPER

The numbers on most rulers are equally spaced, and the distance from the zero mark to a number is that number of units (centimeters, inches, etc.). This is called a *linear scale.* See Figure 12.7.

FIG. 12.7

It is sometimes useful to have a scale in which the distance from the end of the scale to a number is the logarithm of that number (or proportional to it). This is called a *logarithmic scale.* Since log 1 = 0, the point from which all the numbers are measured is marked 1. See Figure 12.8.

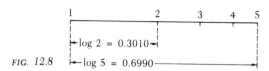

FIG. 12.8

Note that the numbers on a logarithmic scale are not evenly spaced, because logarithms do not increase linearly. However, 10, 100, 1000, 10,000 are evenly spaced, as shown in Figure 12.9.

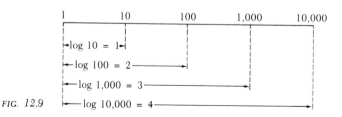

FIG. 12.9

You can also see that 2, 4, 8, 16 are also evenly spaced, since

$$\log 4 = 2 \log 2$$

$$\log 8 = 3 \log 2$$

$$\log 16 = 4 \log 2$$

so these numbers are each a distance of log 2 apart (see Figure 12.10).

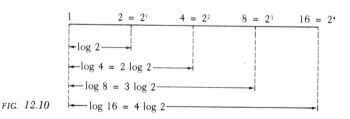

FIG. 12.10

Therefore, on a logarithmic scale, powers of integers are equally spaced. This is as you might expect, since the numbers were placed on the scale according to their logarithms, and logs *are* just the powers.

The logarithm function grows increasingly slowly: as x goes from 1 to 10, $\log x$ increases from 0 to 1, but in order for $\log x$ to climb from 1 to 2, x must go all the way from 10 to 100. This means that on a logarithmic scale, the numbers are bunched together as they get larger. See Figure 12.11. Also, note that zero is not on the scale, since $\log 0$ is undefined.

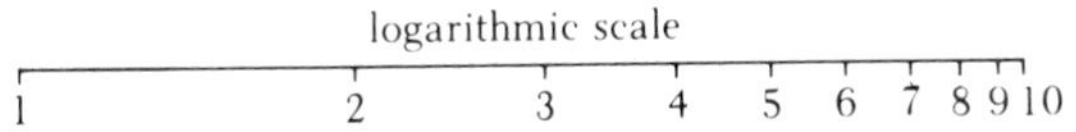

logarithmic scale

FIG. *12.11*   1   2   3   4   5   6   7 8 9 10

Semilog graph paper has a logarithmic scale up the vertical axis, and a linear (ordinary) scale along the horizontal axis; log-log graph paper has logarithmic scales on both axes. See Figure 12.12.

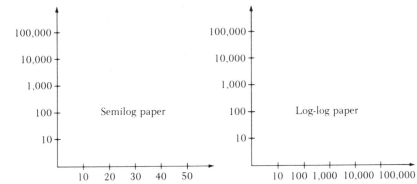

FIG. *12.12*

On semilog paper the point (x, y) is at a distance $\log y$ up the vertical axis and a distance x along the horizontal axis. In other words, plotting the point (x, y) on semilog paper is equivalent to plotting the point $(x, \log y)$ on ordinary paper. Therefore plotting y against x on semilog paper gives you just the same graph as plotting $\log y$ against x on ordinary paper.

Now plotting $\log y$ against x involves making a table of values of x and y, looking up the values of $\log y$, and then plotting the points $(x, \log y)$; whereas, to plot y against x on semilog paper, you only need a table of x and y values because the paper "looks up the logs" for you, as it were.

On log-log paper, the point (x, y) is at a distance $\log y$ up the vertical axis and at a distance $\log x$ along the horizontal axis. Plotting y against x on log-log paper therefore gives the same graph as $\log y$ against $\log x$ on regular paper.

The reason that semilog and log-log scales are so useful is that they can be used to make the graphs of powers of x and of exponential functions into straight lines.

Use of Semilog Paper

Suppose that you were interested in some quantity y that depends on some other quantity x according to the equation

$$y = Ae^{kx} \quad (e, k, A \text{ constant})$$

(For example, N, the number of bacteria, depends on t, time, this way.) Then

$$\log y = \log (Ae^{kx})$$

$$\log y = \log A + \log e^{kx}$$

$$\log y = \log A + kx \log e$$

Since e, k, and A are constants, we can write

$$\log y = a + mx \quad (\text{where } a = \log A \text{ and } m = k \log e)$$

And if we write Y instead of $\log y$, we have

$$Y = a + mx$$

This is a linear equation, so if we plotted Y against x we would get a line. But $Y = \log y$, so we get the same line by plotting y against x on semilog paper. This gives us an experimental way of determining the constants A and k in the original equation—and these constants are often important (in the bacteria example, A is the number we started with, k tells us how fast they multiply). Here's how it works:

Think of the line

$$Y = \log A + (k \log e)x \quad (\text{where } Y = \log y)$$

on regular paper. This has a Y intercept of $\log A$, meaning that it crosses the vertical axis at a height of $\log A$ above the origin. This is the point labeled A on the y scale (the log scale). So.

$$\boxed{y \text{ intercept} = A}$$

Similarly,

$$\boxed{\text{slope of line} = k \log e}$$

To calculate the slope you have to find the change in Y between two points on the line and divide by the change in x. Since the vertical scale on the semilog paper is marked in y's, not Y's, in reading off such paper you will have to take the logs of the two y values you use (because $Y = \log y$). So:

$$\text{slope} = \frac{Y_2 - Y_1}{x_2 - x_1} = \frac{\log y_2 - \log y_1}{x_2 - x_1}$$

(see Figure 12.13). Having found the slope of the line, you can find k.

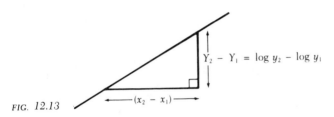

FIG. 12.13

$$Y_2 - Y_1 = \log y_2 - \log y_1$$

$$(x_2 - x_1)$$

Therefore, to find A and k experimentally, plot your data on semilog paper. You will get a line whose y intercept is A and whose slope is k times $\log e$. See Figure 12.14.

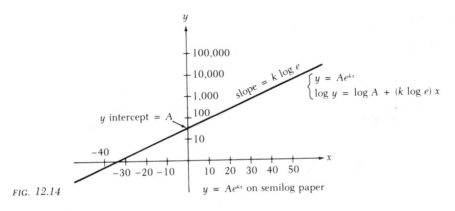

FIG. 12.14

$y = Ae^{kx}$ on semilog paper

Use of Log-Log Paper

Now suppose that y depends on x as follows:

$$y = Ax^m$$

Then

$$\log y = \log (Ax^m)$$
$$\log y = \log A + \log x^m$$
$$\log y = \log A + m \log x$$

or

$$\log y = a + m \log x \qquad (a = \log A, \text{ constant})$$

Letting $Y = \log y$ and $X = \log x$, this means that

$$Y = a + mX$$

so the graph of Y against X will be a straight line.

To plot $\log y$ against $\log x$, you can either look up the logs of both x and y and plot the result on regular graph paper, or plot x and y directly onto log-log paper, which does the logarithms for you. Either way, you will get the straight line in Figure 12.15.

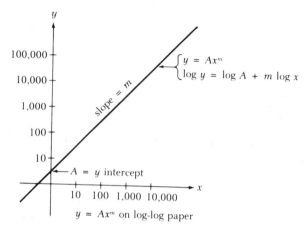

FIG. 12.15 $y = Ax^m$ on log-log paper

Again, the line can be used to find values of A and m. Thinking of the graph of

$$Y = \log A + mX$$

on regular paper, you can see its Y intercept is $\log A$. In other words, the line cuts the y axis at the point labeled A, so

$$\boxed{y \text{ intercept} = A}$$

Similarly,

$$\boxed{\text{slope of line} = m}$$ (You have to take logs of the x and y values you use to get the slope.)

Summary

If you have some data showing corresponding values of y and x, plot them on ordinary, semilog, and log-log graph paper, and see which (if any) come out a line or something close.

If the graph on *ordinary* paper is a line, then y is a linear function of x:

$$y = a + mx$$

If the graph on *semilog* paper is a line, then y is an exponential function of x:

$$y = Ae^{kx} \quad \text{where } A = y \text{ intercept}$$

$$k = \frac{\text{slope}}{\log e}$$

If the graph on *log-log* paper is a line, then y is a power function of x:

$$y = Ax^m \quad \text{where } A = y \text{ intercept}$$

$$m = \text{slope}$$

You can then use whichever graph is a line to find the values of the relevant constants.

PROBLEM SET   12.4

Plot on semilog graph paper. Mark the vertical axis in powers of 10 or e, as appropriate.

1. $y = 10 \cdot 10^x$
2. $y = 2 \cdot 10^{-x}$
3. $y = (0.1)10^{2x}$
4. $y = (0.03) \cdot 10^{-5x}$

5. $y = 5e^x$
6. $y = \dfrac{e^{-2x}}{2}$

Plot on log-log graph paper:

7. $y = 10x^2$
8. $y - 5x^3 = 0$

9. $y^2 = (0.04)x^3$
10. $10y^2x = 1$

The following are graphs of equations of the form $y = Ae^{ct}$ on semilog paper. What are A and c in each case?

11.
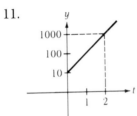

12.
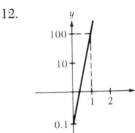

The following are graphs of equations of the form $y = Ax^m$. What are A and m in each case?

13.
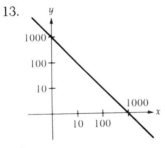

14.
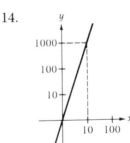

15.   Do Problem 41 in Problem Set 12.1 by graphing on log-log paper.
16.   Do Problem 42 in Problem Set 12.1 by graphing on log-log paper.
17.   Do Problem 55 in the review problems for this chapter by plotting the data given by the graph on semilog paper.
18. A very active enzyme is able to catalyze the reactions of thousands of molecules. When the rate of the enzyme's action was carefully studied, it was found that the number of enzymes, E, are related to the number of molecules, M, that they affect by the equation: $M = A \cdot 10^{KE}$, where A and K are constants. If this equation is rewritten as

$$\log M = \log A + EK \log 10$$

its graph on semilog paper looks like the one in Figure 12.16.

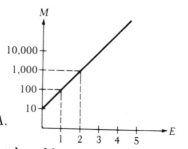

FIG. 12.16

(a) Looking at the graph, find A.
(b) Find K.
(c) How many molecules are catalyzed by 4 enzymes?

19. The growth of many viruses is incredibly fast. The growth rate of one bug in particular is shown in the graph on semilog paper in Figure 12.17; this describes the relationship between the number of bugs, B, and the time of growth in hours, t. Given $\log B = Q + Kt$:

FIG. 12.17

(a) What is K if $Q = 1$ and 1000 bugs are present after 4 hours growth?
(b) Label and put scales on both axes in Figure 12.17
(c) How many bugs will be present after 3 hours?
(d) What does the graph of B against t look like on regular graph paper?

20. The following three equations represent the three lines in Figure 12.18, on semilog graph paper:

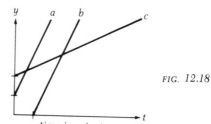

FIG. 12.18

(i) $\log Y = \log A + Wt \log e$
(ii) $\log Y = \log B + Zt \log e$
(iii) $\log Y = \log C + Wt \log e$

(a) If $A > C$, match up equations i, ii, and iii with lines a, b, and c.
(b) What would the graph of $\log Y = \log A + Zt \log e$ look like?
(c) What would the graph of $\log Y = \log B - Zt \log e$ look like?

21. During a recent election, there was an enormous amount of literature distributed throughout the country. In fact, the relationship between the number of leaflets, L, and the number of voters, V, in any given town can be described as:

$$\log L = 2 + 2 \log V$$

(a) What does the graph look like on log-log paper? (Graph $\log L$ on the vertical axis; $\log V$ on the horizontal.)
(b) Approximately how many leaflets were handed out in a town of 5000 people? 10,000 people?

(c) What would the graph look like on semilog graph paper, that is, if V were graphed along the horizontal axis and $\log L$ on the vertical?

22. Graph the following equations on either semilog or log-log paper, choosing whichever paper makes the graph come out a line:

(i) $\log Y = 20 + \frac{1}{2}\log X$

(ii) $\log Y = \log e + X \log 20$

(a) Find all X and Y intercepts.

(b) What is the slope of each line?

CHAPTER 12 REVIEW

Solve for a:

1. $3^a = 2^{a+1}$

2. $3^a = 0.5^{a+1}$

3. $\log(a - 2) = 0.37$

4. $\log(a + 2) - \log 2a = 0.78$

5. $\log_3(1 + 2a) = 1 + \log_3 a$

Solve for the variable indicated in parentheses.

6. $\log_a x - \log_a(x - 2) = \log_a 10 - \log_a(x + 1)$ (x)

7. $\log p = \log q - \log r$ (p)

8. $\log p = \dfrac{\log q}{\log r}$ (p)

9. $\log x = 2 \log A + \dfrac{\log B}{3}$ (x)

10. $\log T^2 - \log \dfrac{3T}{7} = 8.43$ (T)

Solve for x:

11. $\log[\log(x - 4)] = 0$

12. $\log(2 \log x + 3) + 1 = 0$

13. $6 \cdot 11^x = 2.07$

14. $\log(x + 4) = \log x + 0.96$

15. $3^{x+1} = 8^{x-2}$

16. A certain colony of bacteria grows according to the formula $N = 3e^{2t}$, where N is the number of bacteria, t is the time in hours, and $e = 2.72$. How many bacteria are present initially (i.e., when $t = 0$)? How long does this initial colony take to double in size?

17. A certain colony of bacteria grows according to the formula $P = P_0 2^{\alpha t}$, where P_0 is the population at time $t = 0$, and t is time in minutes. After a colony of these bacteria has been growing for 10 minutes, its population is 800. Five minutes later, its population is 1600. What was the initial population? What is α?

18. One strain of bacteria is growing at a rate such that the number of bacteria at any time t is $N = 5(10^{2t})$. A second strain is dying off at a rate such that its number at time t is $N = 500(10^{-2t})$.

(a) For both strains, graph (on the same set of axes) the number of bacteria as a function of time.

(b) At what time do both strains have the same number of bacteria?

19. (a) Calculate the following, using log tables:

$$N = \frac{3.659}{0.903}$$

(b) A monster that grows exponentially can be most terrifying. For example, consider the Blubber Beast, whose weight after birth increases exponentially according to the following equation:

$W = 2^{kt}$ where $\begin{cases} W \text{ gives the monster's weight} \\ \quad \text{in pounds at time } t \text{ minutes} \\ \quad \text{after its birth;} \\ \\ t \text{ is the time in minutes after} \\ \quad \text{birth; and} \\ \\ k \text{ is a constant.} \end{cases}$

Now suppose we know that the Blubber Beast weighs 4560 pounds just 3 minutes after its birth. Find the value of k.

20. A certain local stock brokerage uses the following formula to evaluate the stability of a stock:

$$\text{rating} = \log\left(\frac{\text{high price} - \text{low price}}{\text{time}}\right) + 1$$

(a) What is the rating of a stock that fell from a high of 30 to a low of 4 in 2 years?
(b) What is the rating of the Hoaxem Co., which fell from 30 to 17 in 1 year?

True or false? If true, justify the statement; if false, correct it.
21. $\log(-8) = -\log 8$
22. $\log \log\left(\frac{A}{B}\right) = \log \log A - \log \log B$
23. $\log_5 \log_{25} 25 = 0$
24. $(\sqrt{10})^{\log 6} = \sqrt{6}$
25. $\log 4 = 2 - \log 5$

If $f(x) = \log(x - 3)$, $g(x) = x^2$:
26. Find $f[g(x)]$
27. Find $g[f(x)]$.
28. For what values of x is $f(x)$ defined?
29. What is the domain of $f[g(x)]$?
30. Graph $y = 3^x$ and $y = \log_3 x$ on the same axes.
31. Graph $2y = \log x$.
32. Graph $y + \dfrac{\log x}{3} = 0$.

33. Graph $y = \log(x - 5)$.
34. Graph $y + 1 = \log(-x)$.

35. Let $f(x) = \log(10 + x)$.
    (a) Find the y intercept of this function.
    (b) Solve for x such that $f(x) = 0$.
    (c) Graph $f(x)$.
    (d) If $h(x) = 3x$, find $f[h(30)]$.

36. If $f(x) = 3 \cdot 10^x$, find $f^{-1}(x)$.

37. Find $f^{-1}(t)$ if $f(t) = \log_5 t$.

38. If $f(x) = 2 + \log_3 x$, find $f^{-1}(x)$.

Solve for x:

39. $(\log x^2) \cdot (\log 2x) = \log (2x^{\log x^2})$

40. $\dfrac{\log x}{m} = \log N - \log P$

Graph:

41. $y = \log x^3$

42. $y = \log|x|$

43. $y = |\log x|$

44. If $f(x) = \dfrac{e^x + e^{-x}}{e^x - e^{-x}}$ find $f^{-1}(x)$ and verify that $f[f^{-1}(x)] = x$.

45. Let $f(x) = \log\left(\dfrac{x+1}{x-1}\right)$ for $-1 < x < 1$

    (a) What is $f\left(\dfrac{1+x^2}{2x}\right)$? Simplify your answer as far as possible.

    (b) Your answer to part (a) can be expressed in one of the following ways, where k is some constant:

    $$[f(x)]^k \qquad kf(x) \qquad \dfrac{f(x)}{k} \qquad f(x) + k$$

    Choose the correct expression and proper value for k.

46. Biologists have observed experimentally that populations of certain strains of bacteria double in size every few hours, so long as there is food available. This is described by the equation:

$$P(T) = N \cdot 2^{T/R}$$

where N is the size of the original population, T is the time since the start of the experiment, R is the length of the doubling period, and $P(T)$ is the size of the population at time T. For example, a population that began with 50 individuals and doubled in size every 3 hours would be described by $P(T) = 50 \cdot 2^{T/3}$. [*Note:* T can be in any units, but R must be in the same units. Therefore in $P(T) = 50 \cdot 2^{T/3}$, T is in hours.]

    (a) What is the equation describing an experiment beginning with a population of 10 that doubles every 4 hours?
    (b) How many bacteria are there after 4 hours? A day? Two days?
    (c) In a further experiment, there is only enough food provided to keep approximately 10,000 bacteria alive. About how long will it take before the population stablizes at that level?

47. A certain bacterium, at full size, weighs $2.3 \cdot 10^{-15}$ pounds and splits in two every hour. The two new bacteria rapidly grow to full size. If the bacterium and its offspring reproduce at this rate without limit, how long will it take before the mass of bacteria outweighs the earth? The earth weighs $3.95 \cdot 10^{22}$ pounds. (*Hint:* How many bacteria are there after 1 hour? After 2 hours? After 3 hours? After N hours?)

48. A businessman took $2000 to the Hoaxem Banking Company of Fairfield, Illinois, to deposit it at an interest rate of 5%, compounded yearly. The Bank told him that his money would triple in 10 years, but the skeptical businessman knows that the Banking Manager was obviously misinformed, for the estimate errs on the short side. How long would it really take the businessman's money to triple?

49. A brawl broke out in Burger Queen between a newcomer and one of the regular customers. Eventually it became clear that the fight had started when the newcomer tried to give the regular some advice on how to invest his money. The regular wanted to invest in the shares of the Burger Queen Corporation. The newcomer snorted at that, saying that the shares were worth only $1.00 each, whereas shares of Taco Tent were worth $10.00 and doubling in value every year. The regular retorted that this might be true, but since the value of a Burger Queen share was tripling every year, he wouldn't have to wait very long before a share was worth the same as a Taco Tent share. At this point the chairs started flying and the waitresses hid behind the counter.... How long will we have to wait before a Burger Queen share will be worth the same as a Taco Tent share?

50. Solve for x and y:

$$\begin{cases} 4^x = 3y \\ 2.4^x = 5y \end{cases}$$

51. Solve for t in terms of s:

$$(10^s)^{\frac{(s+t)}{(s-t)}} = 100$$

52. Solve for x:

$$\log_x 4 = 0.3010$$

53. Not wanting to work on her math problems, a certain student stalled for time by graphing the hours of work, W, that she had to do as a function of the number of days, d, since the beginning of the semester. The function took the form: $W = 10e^{\frac{1}{4}d}$. How long did it take her to accumulate 1000 hours of work? See Figure 12.19.

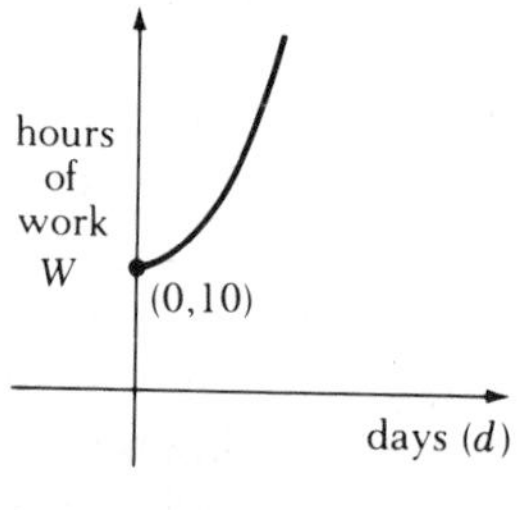

FIG. *12.19*

54. In a dance marathon, 225 people start dancing at midnight, and at each hour on the hour one-third of those remaining drop out. At what time will there be less than one person (i.e., no one) left dancing? (To make this easier, you may ignore the problem of fractional people; that is, it's O.K. to say that there are 12.7 people left after x hours or 6.35 after y hours, and so on.)

55. A huge rock located on an ocean shore wears away because of the pressure of the water against it at an exponentially decreasing rate. When the amount of rock worn away each year was studied as a function of the number of years that have passed since the observation began, the relation graphed in Figure 12.20 was found. Find a function $R(Y)$ that describes this relation.

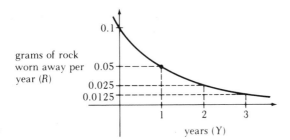

FIG. 12.20

56. A chemist wanted to measure the amount of a new purple compound she had synthesized. This quantity is related directly to the amount of purple light able to pass through the solution, and the scientist therefore set out to determine the emission, E, of light through her solution. She first measured the absorption, A, of purple light by the liquid and then calculated E using the equation: $E = -3 \log A$.

   (a) Graph E as a function of A.

   (b) If her solution had an absorption of 0.002, what was the corresponding emission?

   (c) To check the quality of her light meter, she had measured the light emission in a different machine. What is the equation from which one can calculate A by knowing E?

   (d) Graph the equation you found in (c).

57. As a culture of bacteria grows, the bacteria increase in number, B, with time, t, according to the equation $B = 3e^t$. At the same time, the amount of a necessary nutrient in the culture medium decreases, and its amount, N, as a function of time is given by $N = 300e^{-t}$.

   (a) Graph both functions on the same set of axes.

   (b) How many bacteria and nutrient particles are present originally?

   (c) At a certain point in time, there are the same number of bacteria as nutrients in the medium. Approximate this time from your graphs.

   (d) Approximately how many bacteria are present at this time?

58. Given the graphs in Figure 12.21, find A, B, C, and D.

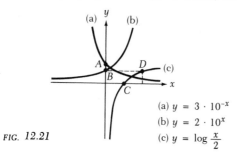

(a) $y = 3 \cdot 10^{-x}$

(b) $y = 2 \cdot 10^{x}$

(c) $y = \log \frac{x}{2}$

FIG. *12.21*

59. The graph in Figure 12.22 describes the equation $y = A \log x^{B}$, where A and B are positive. Graph the following curves, marking intercepts and asymptotes:

(a) $y = -A \log x^{B}$

(b) $y = -A \log x^{-B}$

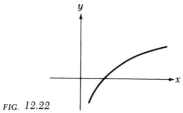

FIG. *12.22*

13 TRIGONOMETRIC FUNCTIONS IN A RIGHT TRIANGLE

13.1 THE ROAD TO TRIGONOMETRY

Trigonometry can be approached two ways. Historically it grew up as a method of studying triangles: The word trigonometry means the measurement (-metry) of three-sided figures (tri-gon). Studying triangles leads to looking at certain functions, and it turns out that these functions in fact represent a great many things in nature and in physics. Trigonometry can also be approached simply as the study of these functions, which can be defined using circles and without any mention of triangles.

We will start with the triangle approach, as it is much the most straightforward and practical, and then show how the circle definitions are just generalizations of those from triangles.

13.2 DEFINING THE TRIGONOMETRIC FUNCTIONS OF ANGLES BETWEEN 0° AND 90°

The first triangles that became of interest to people are those used in building, namely, those with one right angle in them. A right angle is the angle that you get when two perpendicular lines cross, that is, when two lines cross in such a way that all the angles are equal. Now two lines cross to give four angles, and the angles at a point are defined to add up to 360°, so a right angle must be $\frac{360°}{4} = 90°$. See Figure 13.1.

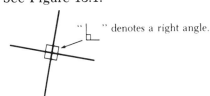

" $\llcorner$ " denotes a right angle.

FIG. 13.1

A triangle containing a right angle is called a *right triangle.* In a right triangle, the side opposite the right angle is called the *hypotenuse,* and the two sides that meet in the right angle are called the *legs.*

One of the important properties of right triangles is that given one of the angles other than the right angle, the shape (though not the size) of the triangle is completely determined. For example, since the angles of a triangle add up to 180°, in a right triangle with one angle of 52°, the third angle must be 180° − 90° − 52° = 38°. Generalizing, the two acute (less than 90°) angles of a right triangle must add up to 90°, since they plus the right angle must add up to 180°. So, *if one acute angle in a right triangle is α, the other is (90° − α).*

Notice that the right angle is always the largest angle in a right triangle (since the other two angles have only 90° to split between them), and the hypotenuse is always the largest side. See Figure 13.2.

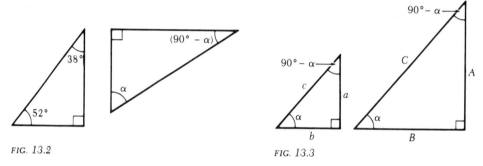

FIG. 13.2   FIG. 13.3

Suppose we now look at two right triangles, both containing the same acute angle α. This means that the second acute angle is (90° − α), and so both triangles have the same angles and are similar, as in Figure 13.3. [See appendix.]

The similarity of these two triangles means that the ratios of corresponding sides are equal:

$$\frac{a}{A} = \frac{b}{B} = \frac{c}{C}$$

Looking just at $\frac{a}{A} = \frac{c}{C}$, we see that this can be rewritten as

$$\frac{a}{c} = \frac{A}{C}$$

In other words, the ratio

$$\frac{\text{the length of the side opposite to } \alpha \ (a \text{ or } A)}{\text{the length of the hypotenuse } (c \text{ or } C)}$$

is the same in both the large and the small triangle. Therefore this ratio does not depend on the size of the triangle at all, but only on its shape. Since the shape of the right triangle is entirely determined by the angle α, this ratio depends only on the angle α. The ratio can therefore be expressed as a function of the angle α. The function is called the sine function and is written

$$\text{sine of } \alpha \quad \text{or} \quad \sin \alpha = \frac{\text{length of side opposite to } \alpha}{\text{length of hypotenuse}} = \frac{a}{c}$$

By exactly similar reasoning,

$$\frac{b}{c} = \frac{B}{C}$$

so that the ratio

$$\frac{\text{length of side adjacent to } \alpha}{\text{length of hypotenuse}}$$

is independent of the size of the triangle and depends only on the angle α. Therefore the ratio can be expressed as a function of the angle, called the cosine function:

$$\text{cosine of } \alpha \quad \text{or} \quad \cos \alpha = \frac{\text{length of side adjacent to } \alpha}{\text{length of hypotenuse}} = \frac{b}{c}$$

Similarly, $\frac{a}{b}$ is independent of the size of the triangle, and so we can define the tangent function:

$$\text{tangent of } \alpha \quad \text{or} \quad \tan \alpha = \frac{\text{length of side opposite to } \alpha}{\text{length of side adjacent to } \alpha} = \frac{a}{b}$$

We also define:

$$\text{cotangent of } \alpha \text{ or } \cot \alpha = \frac{\text{length of side adjacent to } \alpha}{\text{length of side opposite to } \alpha} = \frac{b}{a}$$

$$\text{secant of } \alpha \quad \text{or} \quad \sec \alpha = \frac{\text{length of hypotenuse}}{\text{length of side adjacent to } \alpha} = \frac{c}{b}$$

$$\text{cosecant of } \alpha \text{ or } \csc \alpha = \frac{\text{length of hypotenuse}}{\text{length of side opposite to } \alpha} = \frac{c}{a}$$

The sine, cosine, tangent, cotangent, secant, and cosecant are called the *trigonometric functions*, and these are the *triangle definitions* of these functions. Before moving on, let's summarize these definitions:

$$\sin \alpha = \frac{\text{opposite}}{\text{hypotenuse}} \qquad \cot \alpha = \frac{\text{adjacent}}{\text{opposite}}$$

$$\cos \alpha = \frac{\text{adjacent}}{\text{hypotenuse}} \qquad \sec \alpha = \frac{\text{hypotenuse}}{\text{adjacent}}$$

$$\tan \alpha = \frac{\text{opposite}}{\text{adjacent}} \qquad \csc \alpha = \frac{\text{hypotenuse}}{\text{opposite}}$$

where:

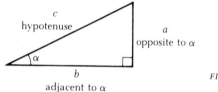

FIG. 13.4

Let's try applying these definitions to the right triangle shown in Figure 13.5. Then

$$\sin \alpha = \frac{4}{5} \qquad \cot \alpha = \frac{3}{4}$$

$$\cos \alpha = \frac{3}{5} \qquad \sec \alpha = \frac{5}{3}$$

$$\tan \alpha = \frac{4}{3} \qquad \csc \alpha = \frac{5}{4}$$

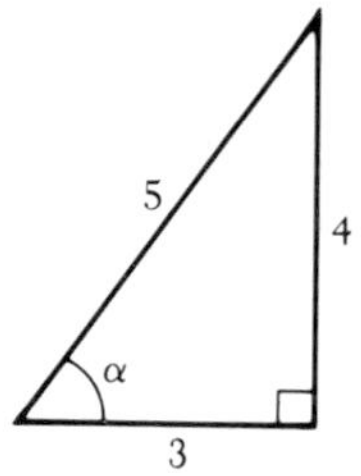

FIG. 13.5

Since the hypotenuse is always the longest side in the triangle, you can see that the sine and cosine of any angle must be less than one, and the secant and cosecant must be greater than one. Also, the tangent and the cotangent of an angle are reciprocals of one another, as are the cosine and the secant, and the sine and the cosecant.

Relation Between
Trigonometric Functions of α and $(90° - \alpha)$

Angles like α and $(90° - \alpha)$ that add up to 90° are called *complementary angles*. Remember that sin α was defined as:

$$\frac{\text{length of side opposite to } \alpha}{\text{length of hypotenuse}}$$

Therefore $\sin(90° - \alpha)$ must be defined as:

$$\frac{\text{length of side opposite to } (90° - \alpha)}{\text{length of hypotenuse}}$$

In the previous example, the side opposite to α is 4, but the side opposite to $(90° - \alpha)$ is 3, so

$$\sin \alpha = \frac{4}{5} \quad \text{while} \quad \sin(90° - \alpha) = \frac{3}{5}$$

Similarly,

$$\cos(90° - \alpha) = \frac{\text{length of side adjacent to } (90° - \alpha)}{\text{length of hypotenuse}}$$

so

$$\cos \alpha = \frac{3}{5} \quad \text{while} \quad \cos(90° - \alpha) = \frac{4}{5}$$

So you can see that,

$$\boxed{\sin \alpha = \cos(90° - \alpha) \quad \text{and} \quad \cos \alpha = \sin(90° - \alpha)}$$

because the side adjacent to α is opposite to $(90° - \alpha)$ and vice versa. So the sine of an angle is equal to the cosine of its complement. See Figure 13.6.

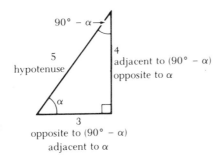

FIG. 13.6

$90° - \alpha$

5
hypotenuse

4
adjacent to $(90° - \alpha)$
opposite to α

α

3
opposite to $(90° - \alpha)$
adjacent to α

Similarly,

$$\boxed{\tan \alpha = \cot(90° - \alpha) \quad \text{and} \quad \cot \alpha = \tan(90° - \alpha)}$$

and

$$\boxed{\sec \alpha = \csc(90° - \alpha) \quad \text{and} \quad \csc \alpha = \sec(90° - \alpha)}$$

PROBLEM SET   13.2

Apply the triangle definitions to the following right triangles to determine the particular values of the trigonometric functions.

1. sin α
2. cos α
3. tan α
4. cot α
5. sec α
6. csc α

7. sin α
8. cos α
9. tan α
10. cot α
11. sec α
12. csc α

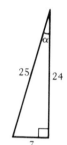

13. sin α
14. tan(90° − α)
15. cos(90° − α)
16. sec α
17. csc α
18. cot(90° − α)

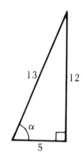

19. cos(90° − α)
20. sin(90° − α)
21. tan α
22. cot α
23. tan(90° − α)
24. csc(90° − α)

25. cos α
26. sin(90° − β)
27. cot β
28. csc β
29. sec(90° − β)
30. sec β

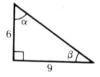

Apply the triangle definitions and the Pythagorean theorem to the following right triangles to determine the particular values of the trigonometric functions.

31. sin β
32. sin(90° − β)
33. cos α
34. tan α
35. cot β
36. sec β

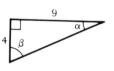

37. csc β
38. tan(90° − α)
39. sin α
40. cos(90° − β)
41. cot α
42. tan β

43. sin α
44. cos α
45. tan(90° − α)
46. cot(90° − α)
47. sec(90° − α)
48. csc(90° − α)

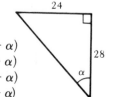

13.3   TRIGONOMETRIC TABLES

To calculate the value of a sine of a given angle from the definition you would have to draw accurately a right triangle containing that angle, measure the opposite side and the hypotenuse, and divide them. Finding a sine this way would be a great nuisance, and you will be glad to know that this has all been done for you and the results listed in tables. Tables of cosines, tangents, and all the other trigonometric functions also exist. They are sometimes called tables of natural sines or natural cosines, rather than just sines or cosines, to distinguish them from tables of logarithmic sines or logarithmic cosines (which contain the logarithms of the sines or cosines) and from hyperbolic sines or cosines, which are different again. We will use only natural sines and cosines. The table in the appendix has columns for sine, cosine, tangent, cotangent, secant, and cosecant.

To use the table you have to remember that angles are measured in degrees, and that each degree consists of 60 minutes, where 10 minutes is written as 10′. Thus $51\frac{1}{2}° = 51°30'$ and $27°59'$ is one minute short of $28°$. The word "minute" comes from the Latin for "first minute part" and, as you can see, a minute is really a very small angle. Our table has entries at intervals of 6 minutes (which is also equivalent to tenths of a degree), though other tables may go in jumps of 10 minutes or some other interval.

Using the Tables

The table below is an excerpt (covering from $12°$ to $13°12'$) from the complete table in the appendix.

To look up $\sin 13°$, for example, go down the left-hand column until you find $13°00'$. Then look in the first column to the right (headed sin), where you find 0.2250. Hence

$$\sin 13° = 0.2250$$

Angle	Sin	Cos	Tan	Cot	Sec	Csc	
12°00′	0.2079	0.9781	0.2126	4.705	1.0223	4.8097	78°00′
06	.2096	.9778	.2144	4.665	1.0227	4.7706	54
12	.2113	.9774	.2162	4.625	1.0231	4.7321	48
18	.2130	.9770	.2180	4.586	1.0235	4.6942	42
24	.2147	.9767	.2199	4.548	1.0239	4.6569	36
30	.2164	.9763	.2217	4.511	1.0243	4.6202	30
36	.2181	.9759	.2235	4.474	1.0249	4.5841	24
42	.2198	.9755	.2254	4.437	1.0251	4.5456	18
48	.2215	.9751	.2272	4.402	1.0255	4.5137	12
54	.2233	.9748	.2290	4.366	1.0259	4.4793	06
13°00′	0.2250	0.9744	0.2309	4.331	1.0263	4.4454	77°00′
06	.2267	.9740	.2327	4.297	1.0267	4.4121	54
12	.2284	.9736	.2345	4.264	1.0271	4.3792	48
	Cos	Sin	Cot	Tan	Csc	Sec	Angle

Notice that every entry in the sine and cosine table is less than or equal to 1, just as it is supposed to be, since

$$\sin \alpha = \frac{\text{opposite}}{\text{hypotenuse}}$$

and the hypotenuse is the longest side of the triangle.

The left-hand column only goes up to 45°. For an angle between 45° and 90°, the functions are listed at the *bottom* of the table, and the angle is found in the right-hand column. This column decreases from 90° to 45° as the left-hand column goes from 0° to 45°.

For example, to find tan 77°, look for 77°00′ in the right-hand column, then move left to the tangent column, remembering to use the function headings at the bottom of the table. You will see that

$$\tan 77° = 4.331$$

Notice that we get the same number, 4.331, by looking up cot 13°. This is because 77° and 13° are complementary angles (see Section 13.2), so

$$\tan 77° = \cot(90° - 77°) = \cot 13°$$

In general, if you have an angle, α, which is greater than 45°, then its complementary angle, $(90° - \alpha)$, is less than 45°. Since

$$\sin \alpha = \cos(90° - \alpha)$$

and

$$\cos \alpha = \sin(90° - \alpha)$$

and similar relations hold between tan and cot, sec and csc, any trig function of any angle greater than 45° is always equal to another trig function of an angle less than 45°. Consequently each entry in the table is a value for two different functions, and the table is only half as long as it would have been otherwise.

Use of the Minute Entries   To find cos 12°36′, look in the left-hand column for 12°00′, and then move down under 12° to 36′. Then move over to the cos column, where you find 0.9759. So

$$\cos 12°36′ = 0.9759$$

Now suppose you want to find cos 12°38′. Since the table goes in 6′ jumps, there is no entry for 38′. For the purposes of this book, we will use the closest value—here 12°36′—and say that

$$\cos 12°38′ = 0.9759$$

This tactic will introduce some small inaccuracies into our calculations, but they are seldom important. Because of these approximations, our answers will probably not be accurate to four figures, so we will round off to three figures after each series of calculations.

If more accuracy is needed, you could use a technique called interpolation, which helps you to estimate values between table entries. However, even interpolation doesn't work well everywhere (particularly in regions where the value of a function is changing rapidly, such as the tangent near 90°), so we will not cover it. If you need really accurate answers, I suggest you use a scientific calculator.

PROBLEM SET   13.3

Use the trigonometric tables to evaluate the following:

1. sin 10°	9. cot 47°	17. csc 62°42'	25. sin 3°19'
2. cos 22°	10. csc 18°	18. cot 20°48'	26. sin 86°35'
3. tan 41°	11. tan 89°	19. cos 37°18'	27. csc 9°44'
4. csc 13°	12. sec 78°	20. sin 41°24'	28. tan 78°56'
5. sin 86°	13. cot 6°	21. sec 56°6'	29. cos 14°9'
6. sec 82°	14. cos 43°	22. tan 12°36'	30. tan 17°27'
7. cos 80°	15. sin 53°24'	23. cot 18°12'	
8. tan 23°	16. cos 76°54'	24. csc 2°48'	

Order from highest to lowest in value without using the table:

31.   sin 2°   cot 40°   cos 60°

32.   sec 89°   cot 82°   tan 46°

13.4   SOLVING RIGHT TRIANGLES

One of the uses of the trigonometric functions is to "solve right triangles," which means to find the size of all the angles and sides in the triangle.

For example, if we are given the triangle in Figure 13.7, we can find a, b, and α as follows:

α is the easiest, because

$$\alpha = 90° - 15° = 75° \qquad (\alpha \text{ is the angle complementary to } 15°)$$

b can be found using the fact that

$$\sin 15° = \frac{\text{opposite}}{\text{hypotenuse}} = \frac{b}{10}$$

From the table, sin 15° = 0.2588 . So

$$\sin 15° = \frac{b}{10}$$

becomes

$$0.2588 = \frac{b}{10}$$

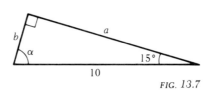

FIG. 13.7

Solving for b:

$$b = 10(0.2588) = 2.588 \cong 2.59$$ (rounding to three figures, as explained at end of last section)

a can be found using the fact that

$$\cos 15° = \frac{a}{10}$$

From the table we find

$$0.9659 = \frac{a}{10}$$

or

$$a = 10(0.9659) = 9.659 \cong 9.66$$

Therefore the triangle is as shown in Figure 13.8.

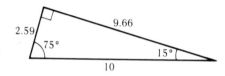

FIG. 13.8

EXAMPLE: *Suppose now that we were not given an angle, but instead were asked to solve the triangle shown in Figure 13.9.*

Then

$$\sin \alpha = \frac{5.3}{10} = 0.53$$

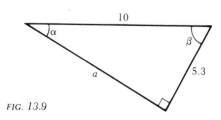

FIG. 13.9

so we know the value of $\sin \alpha$ and want to find α. This can be done by looking up 0.53 ($= 0.5300$) in the body of the table and finding the angle of which it is the sine. Since 0.5300 is not itself in the table, we will pick the closest value and use that. That closest value is 0.5299, which is opposite $32°00'$, so the angle we are looking for is $32°$. So

$$\alpha = 32°$$

Then

$$\beta = 90° - \alpha = 90° - 32° = 58°$$

To find a, we use the fact that

$$\cos \alpha = \cos 32° = \frac{a}{10}$$

From the cosine table we get

$$0.8480 = \frac{a}{10}$$

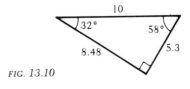

so

$$a = 10(0.8480) = 8.48$$

FIG. 13.10

So the triangle works out to be the one shown in Figure 13.10.

Note: a could also have been found from the other two sides (10 and 5.3) by using Pythagoras' theorem:

$$10^2 = (5.3)^2 + a^2$$

But this is much more nuisance to solve than an equation like

$$\cos 32° = \frac{a}{10}$$

so it is easier not to use Pythagoras here.

EXAMPLE: *Solve the right triangle in Figure 13.11.*

This can be done using tangents and secants:

$$\tan 27° = \frac{b}{2}$$

so

$$0.5095 = \frac{b}{2}$$

or

$$b = 2(0.5095) = 1.019 \cong 1.02$$

Also,

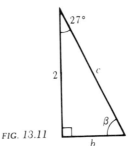

$$\sec 27° = \frac{c}{2}$$

FIG. 13.11

so

$$1.1223 = \frac{c}{2}$$

$$c = 2(1.1223) = 2.2446 \cong 2.24$$

Also,

$$\beta = 90° - 27° = 63°$$

Note: We could also have found c by saying

$$\cos 27° = \frac{2}{c}$$

so

$$0.8910 = \frac{2}{c}$$

giving

$$c = \frac{2}{0.8910}$$

However, this is a pain to work out, unless you have a calculator, so the previous method, which gave

$$c = 2(1.1223) = 2.2446$$

was much easier. The moral of this is that in solving for a side it is definitely helpful to use that trig function which puts the unknown in the numerator (in this case the secant)!

Word Problems

Right triangles occur all over the place. Buildings and maps are constructed around them, and they are vital to physics. Therefore it is not surprising that there are a mass of word problems that involve solving right triangles.

Useful Steps for Solving a Trigonometric Word Problem:

1. Draw a picture.
2. Label every length and angle given by the problem, or which can be calculated directly (without trigonometry) from the data of the problem. If you have any unused facts in the problem, wonder about them, and ask yourself if they tell you anything else about the picture.
3. Solve the triangle.

EXAMPLE: *In an experiment on gravity, Galileo is said to have dropped several balls from the top of the Leaning Tower of Pisa, and measured how long they took to reach the ground. Assuming that the tower was built to stand 180 feet high when vertical and that by Galileo's time it made an angle of 5° with the vertical, how far did each of the balls have to fall from the top to the ground?*

Absolutely the first thing to do in this kind of problem is to draw a picture, and mark on it any information that you are given. In other words you should come up with something like Figure 13.12.

Since the angle between the vertical and the ground is 90°, angle $\alpha = 90° - 5° = 85°$. Therefore,

$$\sin 85° = \frac{x}{180}$$

or

$$0.9962 = \frac{x}{180}$$

$$x = 180(0.9962) = 179.3 \cong 179 \text{ feet}$$

Therefore, each ball fell 179 feet.

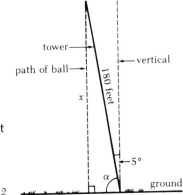

FIG. *13.12*

Just as in algebra, where certain topics always seem to come up in word problems (trains, swimming pools), so in trigonometric problems there seems to be an abundance of flagpoles and church spires. Most of these problems involve someone looking at the flagpole or church spire and determining either its height or its distance away. Frequently encountered are these:

The angle of elevation: the angle between a horizontal line and the line joining the observer to an object above the horizon. See Figure 13.13.

The angle of depression: the angle between a horizontal line and the line joining the observer to an object below the horizon. See Figure 13.14.

FIG. *13.13*

FIG. *13.14*

Bearings: These give you the direction of one point from another. For example:

N25°E means measure 25° from north in the direction of east (see Figure 13.15); and

S52°E means measure 52° from south in the direction of east (see Figure 13.16).

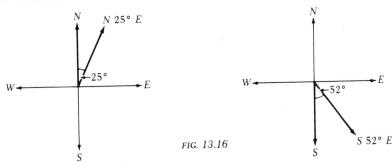

FIG. *13.15*

FIG. *13.16*

EXAMPLE: *Tenth Avenue in New York runs North–South, and 200 feet from where it crosses 46th Street is a Texaco station. If a warehouse on 46th Street, whose bearing from the gas station is N25°E, catches fire and throws sparks 230 feet, is the Texaco station in any danger of exploding?*

The gas station is in danger only if sparks could be thrown into it, so the problem here is to find the distance from the warehouse to the gas station.

First, a picture. Notice that since the warehouse on 46th is N25°E from the gas station, the gas station must be to the south of the intersection of Tenth and 46th, and the warehouse must be to the east. See Figure 13.17.

Let the distance from the warehouse to the Texaco Station be d. Then

$$\sec 25° = \frac{d}{200}$$

or

$$d = 200 \sec 25°$$

$$= 200\,(1.1034)$$

$$= 220.68 \text{ feet} \cong 221 \text{ feet}$$

Since d is less than 230 feet, the Texaco station is in trouble.

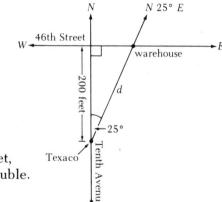

FIG. 13.17

EXAMPLE: *A German Luftwaffe pilot is flying a reconnaissance mission over Scotland in March 1941. Traveling at 400 miles per hour (mph) at an extremely low (but constant) altitude to avoid radar, he spots the Forth bridge right ahead at an angle of depression of 1°; a quarter of a minute later it is at an angle of 5°. What is the pilot's altitude? (1 mph = 88 feet per minute.)*

Since the pilot keeps a constant altitude (height), he is flying horizontally. See the picture in Figure 13.18.

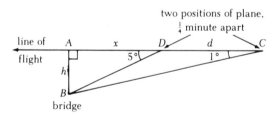

FIG. 13.18

Finding the altitude means finding h.

The plane travels from C to D in $\frac{1}{4}$ minute at 400 mph. Now 400 mph = $400 \cdot 88 = 35,200$ ft/min and distance = speed $\cdot$ time, so the distance from C to D is

$$d = 35,200 \cdot \frac{1}{4} = 8800 \text{ feet}$$

In order to use the trigonometric functions, we must have a right triangle. Looking at $\triangle ADB$, we see

$$\frac{h}{x} = \tan 5°$$

From $\triangle ACB$:

$$\frac{h}{x + d} = \tan 1°$$

or

$$\frac{h}{x + 8800} = \tan 1° \qquad (\text{since } d = 8800)$$

So we have a pair of simultaneous equations,

$$\begin{cases} \dfrac{h}{x} = \tan 5° \\[2mm] \dfrac{h}{x + 8800} = \tan 1° \end{cases}$$

that can be solved for h and x (though we only want h).

Solving the first equation for x:

$$x = \frac{h}{\tan 5°}$$

Rewriting the second as

$$h = (x + 8800) \tan 1°$$

and substituting for x:

$$h = \left(\frac{h}{\tan 5°} + 8800 \right) \tan 1°$$

Solving for h:

$$h - h\frac{\tan 1°}{\tan 5°} = 8800 \tan 1°$$

$$h\left(1 - \frac{0.0175}{0.0875} \right) = 8800 \, (0.0175)$$

$$h\left(1 - \frac{1}{5} \right) = 154$$

$$h = 154 \cdot \frac{5}{4} = 192.5 \text{ feet} \cong 193 \text{ feet}$$

So the plane is at an altitude of 193 feet.

PROBLEM SET   13.4

Let *ABC* be the right triangle shown in Figure 13.19, with $\angle C = 90°$. Find all the angles and lengths of sides of $\triangle ABC$ if:

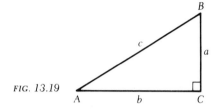

FIG. 13.19

1. $a = 24, c = 25$
2. $a = 12, c = 13$
3. $a = 1, b = 1$
4. $a = 1, b = \sqrt{3}$
5. $a = 1, c = 4$
6. $A = 62°, b = 2$
7. $a = 3.27, A = 49°30'$
8. $c = 15.6, A = 72°$
9. $b = 2.4, A = 41°10'$
10. $b = 0.76, B = 15°$
11. $a = 8, b = 15$
12. $b = 12, A = 32°$
13. $c = 25, B = 27°15'$
14. $a = b, b = 9$
15. $a = 12, A = 26°45'$
16. $b = 4, B = 34°20'$
17. $a = 20, c = 33$
18. $a = 2, b = 4$
19. $A = 13°, c = 11$
20. $b = 40, c = 80$

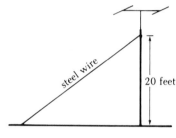

steel wire

20 feet

FIG. 13.20

21. The length of a path going diagonally across a rectangular field is 200 yards, and the path makes an angle of 25° with the fence. How much fencing is required to go around the field?

22. A staircase is to rise 10 feet 8 inches in a horizontal distance of 13 feet 4 inches. Find the inclination of the handrail to the horizontal (i.e., find the angle between the handrail and a horizontal line).

23. An antenna pole is to be braced by three steel wires which are to be attached to a point on the pole 20 feet from the ground, and which make an angle of 50° with the ground. See Figure 13.20. Find the total amount of steel wire necessary.

24. In order to find the distance to an island offshore, a man on the coast measures a baseline 200 yards long and pointing out toward the island. At each end of the baseline he measures the angle between the baseline and the top of a tall tree on the island (i.e., the angle of elevation of the top of the tree). The angles are 68°20' and 81°30'. Find the distance from the island to the nearer end of the baseline.

25. A hydrofoil going due south approaches pier *P*. Lighthouse *X* is situated 1000 yards due west of *P*, and lighthouse *Y* is 1000 yards due east of *P*. When the hydrofoil is at a point *A*, $\angle XAY = 50°$. One minute later the hydrofoil is at *B*, and it is found that $\angle XBY = 90°$. What is the speed of the hydrofoil in yards per minute?

26. A rectangle has a diagonal of length 7 and a width of 3. Find the length of the long side, and the angle formed by the diagonal and the long side.

27. A farmer owns a right triangular plot of land whose nonhypotenuse sides are x and $2x$. What angle does the hypotenuse form with the short side?

28. A tree 100 feet tall casts a shadow 120 feet long. Find the angle of elevation of the sun.

29. A tree broken in two by the wind forms a right triangle with the ground. If the broken part makes an angle of 40° with the ground and the top of the tree is now 45 feet from the base, how tall was the tree?

30. Due to a faulty compass, Daniel Boone walks to his favorite river (see Figure 13.21) along a path with bearing N30°E instead of taking the shortest path and heading due north. If the short way is 3 miles long, how much extra mileage did Daniel walk?

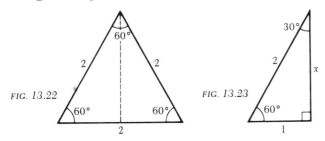

FIG. *13.21*

13.5   SPECIAL ANGLES

There are a few angles—namely, 30°, 45°, and 60°—whose sines and cosines can be found without tables and without drawing anything accurately. They come up so frequently that, much as I hate to suggest it, you should either memorize the values of their sines, cosines, and tangents, or, better, know so well where they come from that you can figure them out in no time flat.

To Find the Values of Trigonometric Functions of 30° and 60°

To do this we must draw triangles containing angles of 30° and 60°. Since 30° and 60° add up to 90°, they will both come up in the same right triangle.

Imagine you are given an equilateral triangle (all sides equal, all angles equal to 60°) with side 2. Cut this in half, as shown in Figure 13.22, to make two right triangles with angles 30° and 60°, as shown in Figure 13.23.

The third side of the triangle, x, can be found using Pythagoras' theorem, which tells us that:

$$2^2 = 1^2 + x^2$$

so

$$x^2 = 3$$

Since x is the length of a side of a triangle it must be positive, so

$$x = \sqrt{3}$$

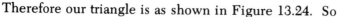

FIG. 13.24

Therefore our triangle is as shown in Figure 13.24. So

$$\sin 30° = \frac{1}{2} \qquad \sin 60° = \frac{\sqrt{3}}{2}$$

$$\cos 30° = \frac{\sqrt{3}}{2} \qquad \cos 60° = \frac{1}{2}$$

$$\tan 30° = \frac{1}{\sqrt{3}} \qquad \tan 60° = \frac{\sqrt{3}}{1} = \sqrt{3}$$

$$\cot 30° = \frac{\sqrt{3}}{1} = \sqrt{3} \qquad \cot 60° = \frac{1}{\sqrt{3}}$$

$$\sec 30° = \frac{2}{\sqrt{3}} \qquad \sec 60° = \frac{2}{1} = 2$$

$$\csc 30° = \frac{2}{1} = 2 \qquad \csc 60° = \frac{2}{\sqrt{3}}$$

To Find the Values of
the Trigonometric functions of 45°

A right triangle with one angle 45° will have its third angle 45° also (since 90° − 45° = 45°). It is therefore an isosceles triangle, that is, one with two angles and two sides equal. Let us say that each of these sides has length 1. (We are free to choose any length we like, since in the definition of the sine, the actual size of the triangle does not matter because we are concerned only with ratios.) So we have a triangle which looks like Figure 13.25.

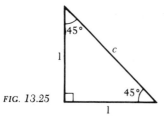

FIG. 13.25

The hypotenuse, c, can be found using Pythagoras's theorem:

$$c^2 = 1^2 + 1^2 = 2$$

Now c must be positive, since it is a length, so $c = \sqrt{2}$ and the triangle is as shown in Figure 13.26. Therefore:

$$\sin 45° = \frac{1}{\sqrt{2}} \qquad \cot 45° = \frac{1}{1} = 1$$

$$\cos 45° = \frac{1}{\sqrt{2}} \qquad \sec 45° = \frac{\sqrt{2}}{1} = \sqrt{2}$$

$$\tan 45° = \frac{1}{1} = 1 \qquad \csc 45° = \frac{\sqrt{2}}{1} = \sqrt{2}$$

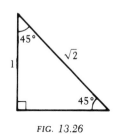

FIG. 13.26

PROBLEM SET 13.5

Let ABC be the right triangle, shown in Figure 13.27, with $\angle C = 90°$. Without using tables, find all the angles and lengths of sides of $\triangle ABC$ if:

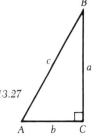

FIG. 13.27

1. $A = 30°, b = 6$
2. $A = 45°, c = \sqrt{18}$
3. $B = 30°, b = 4$
4. $B = 60°, c = 12$
5. $a = 4, b = 4\sqrt{3}$
6. $a = 13, b = 13$

7. $b = 2, c = 4$
8. $a = b, c = \dfrac{12}{\sqrt{3}}$
9. $a = 5, b = 5$
10. $A = 60°, b = 17$

Make a chart with the following row and column headings. For Problems 11–28, fill in the appropriately numbered box.

$\theta°$	sin	cos	tan	cot	sec	csc
30°	11	12	13	14	15	16
45°	17	18	19	20	21	22
60°	23	24	25	26	27	28

CHAPTER 13 REVIEW

Apply the triangle definitions to the right triangle in Figure 13.28 and evaluate the following:

1. $\sin \alpha$	7. $\sin(90° - \alpha)$
2. $\cos \alpha$	8. $\cos(90° - \alpha)$
3. $\tan \alpha$	9. $\cot(90° - \alpha)$
4. $\cot \alpha$	10. $\tan(90° - \alpha)$
5. $\sec \alpha$	11. $\csc(90° - \alpha)$
6. $\csc \alpha$	12. $\sec(90° - \alpha)$

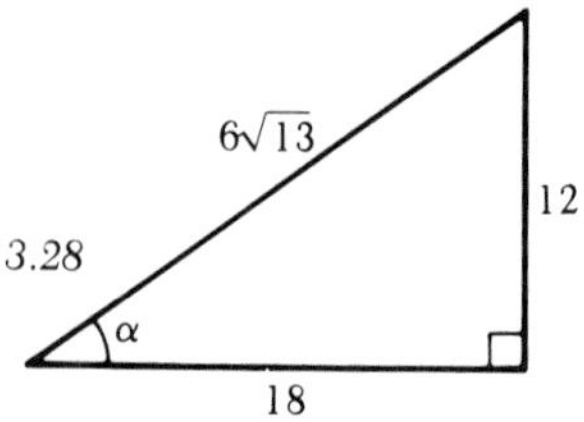

FIG. 13.28

Use the trigonometric tables to evaluate the following:

13. $\sin 21°10'$	17. $\tan 17°26'$	21. $\cot 69°27'$
14. $\sec 89°22'$	18. $\cos 40°30'$	22. $\csc 12°6'$
15. $\cot 74°41'$	19. $\tan 15°59'$	23. $\sec 11°17'$
16. $\csc 73°13'$	20. $\cos 9°24'$	24. $\sin 51°8'$

Let ABC be the right triangle shown in Figure 13.29, with $\angle C = 90°$. Find all all the angles and lengths of sides of $\triangle ABC$ if:

25. $a = 12, A = 35°$	30. $c = 8, B = 35°$
26. $c = 40, A = 23°$	31. $A = 68°, b = 4$
27. $a = 4, b = 7$	32. $a = 18, B = 12°$
28. $c = 9, a = 7$	33. $b = 6, c = 9$
29. $B = 49°, b = 3.2$	

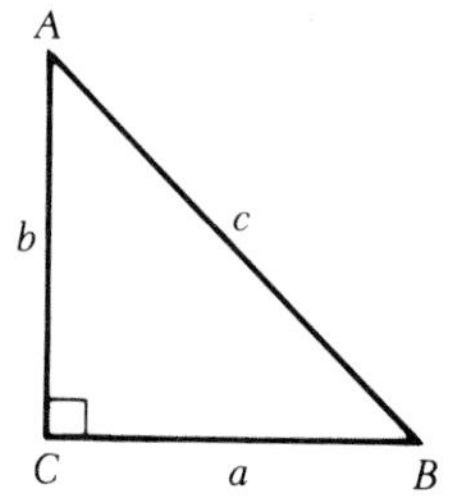

FIG. 13.29

Determine the equations of the following lines on a Cartesian plane.

34. The line through the origin inclined at an angle of 15° above the positive x axis

35. The line through the origin inclined at an angle of 30° above the positive x axis

36. The line through the origin inclined at an angle of 45° above the positive x axis

37. The line through the origin inclined at an angle of 57°13' above the positive x axis

Determine the angle formed by each of the following lines and the positive x axis:

38. $y = \dfrac{x}{3}$	41. $y = 2x$
	42. $y = 3x$
39. $y = \dfrac{x}{2}$	43. $y = 5x$
	44. $y = 10x$
40. $y = x$	

45. Find the intersection of $y = -2x + 4$ and a line through the origin inclined at an angle of 55° above the positive x axis.

46. For best results, a playground slide should be inclined at an angle of 40° from the ground. (Assume the bottom of the slide touches the ground.) In a certain playground, there is a space 13 feet long, measured along the ground. If we put a slide in this space (raised to the ideal 40° angle, of course), how long will the slide be?

47. A ladder 22.74 feet long leans against a wall. The angle that the ladder makes with the wall is 21°43′. How far is the bottom of the ladder from the bottom of the wall?

48. An observer on the ground finds that the angle of elevation of a balloon is 51°20′. How high is the balloon if the point directly under the balloon is 237 feet from the observer?

49. At noon, a bus leaves Boston and travels due north at 40 miles per hour. At 2:00 pm, a train leaves Boston and travels due west at 60 mph. When are the bus and the train 200 miles apart? (Trig is not necessary here.)

50. If the bus and train in Problem 49 stop when they are 200 miles apart, what will be the train's bearing from the bus?

51. On a field trip, a geologist sketched the hillside in Figure 13.30. How thick are the two exposed strata (i.e., find x and y)?

52. Occasionally oil gushes from a well and catches on fire. A common technique for fighting such fires is to drill a relief well and pump the oil away from the bottom of the burning well. See Figure 13.31.

    (a) For obvious reasons, the driller of the relief well wants to be as far away from the burning well as possible. If he must drill the first 1000 feet of the relief well straight down and then he can continue his hole at 30° to the vertical until he hits the other well at 6000 feet, what is the maximum distance d at which the driller can position his drilling rig from the burning well?

    (b) The driller wants to line the relief hole with steel pipe. How many feet of pipe should he order?

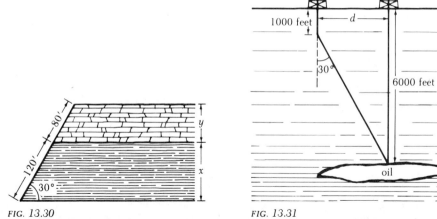

FIG. 13.30   FIG. 13.31

14 TRIGONOMETRIC FUNCTIONS OF A GENERAL ANGLE

14.1 ANGLES IN STANDARD POSITION

So far we can define the trigonometric functions of any angle between 0° and 90°. Other angles won't fit into a right triangle, and so our definitions are useless for them. We will now give a totally different definition of sine, cosine, and the other trig functions that will apply to *any* angle, and that will agree with the triangle definition in the case where the angle is between 0° and 90°.

Until now we have looked at an angle as the measure of the "width" of the "wedge" (or sector) between two lines that meet at a point. From this viewpoint, a negative angle makes no sense, since the sector can't have negative size. Similarly, no sector, and therefore no angle, could be larger than 360°. However, an angle can also be regarded as a rotation from one line (called the initial side) to another (called the terminal side), as illustrated in Figure 14.1.

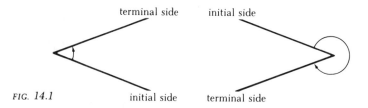

FIG. 14.1

terminal side   initial side

initial side   terminal side

The arrow is put on to show the direction of rotation so you do not, in fact, have to label the initial and terminal sides.

Now, there are two different directions of rotation possible: clockwise or counterclockwise. Therefore we can call one a positive angle, and one a negative angle. The convention is that a *counterclockwise* rotation is a *positive* angle and a *clockwise* rotation is a *negative* angle. For example, angles of

+80° and −80° are shown in Figure 14.2; angles of more than 180° look like those in Figure 14.3; and an angle of more than 360° means a rotation through more than one complete revolution, as shown in Figure 14.4.

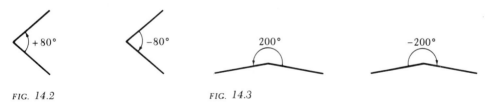

FIG. 14.2   FIG. 14.3

Suppose we now lay our angle on top of a set of axes in the plane, with the vertex (tip) of the angle at the origin, and the initial side along the positive *x* axis, as in Figure 14.5. The angle is then said to be in *standard position*.

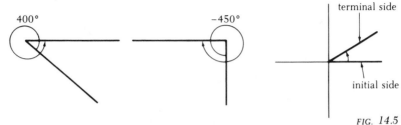

FIG. 14.4

FIG. 14.5

EXAMPLE: *Put −100°, 540°, and −330° in standard position.*

Look at Figure 14.6 for the result.

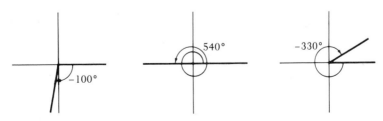

FIG. 14.6

An angle whose terminal side lies in the second quadrant (say) is called an *angle in the second quadrant,* and similarly for the other quadrants. See Figure 14.7.

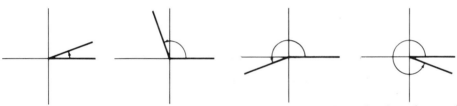

a first-quadrant angle   a second-quadrant angle   a third-quadrant angle   a fourth-quadrant angle

FIG. 14.7

PROBLEM SET   14.1

Draw a diagram showing the following angles in standard position.

1. 90°	6. 60°	11. −90°
2. 270°	7. 240°	12. −330°
3. 720°	8. −120°	13. −770°
4. 30°	9. 390°	14. 1000°
5. −30°	10. −180°	15. −900°

14.2   DEFINITIONS OF THE TRIGONOMETRIC FUNCTIONS OF A GENERAL ANGLE

Suppose we are given an angle θ, which may be positive or negative and of any size. To define the trigonometric functions:

1. Put the angle θ in standard position.
2. Find $P(\theta)$, the point at which the terminal side cuts the unit circle (a circle of radius 1 centered at the origin). See Figure 14.8.
3. Define:

$$\cos\theta = x \text{ coordinate of } P(\theta)$$

$$\sin\theta = y \text{ coordinate of } P(\theta)$$

or:

If $P(\theta) = (x, y)$,

$$\cos\theta = x \qquad \cot\theta = \frac{x}{y}$$

$$\sin\theta = y \qquad \sec\theta = \frac{1}{x}$$

$$\tan\theta = \frac{y}{x} \qquad \csc\theta = \frac{1}{y}$$

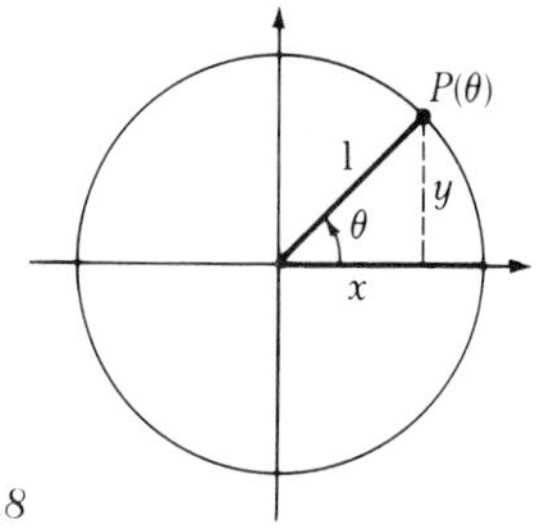

FIG. 14.8

These are the *circle definitions* of the trigonometric functions.

From these definitions you can see that there are several relationships among these functions. For example:

$$\tan\theta = \frac{y}{x} = \frac{\sin\theta}{\cos\theta} \qquad \cot\theta = \frac{x}{y} = \frac{\cos\theta}{\sin\theta}$$

$$\sec\theta = \frac{1}{x} = \frac{1}{\cos\theta} \qquad \csc\theta = \frac{1}{y} = \frac{1}{\sin\theta}$$

You can also see that the sine and cosine are always between −1 and 1 because, being sides of a right triangle with hypotenuse 1, x and y cannot have magnitudes greater than 1.

EXAMPLE: *Find cos 90° and sin 90°.*

Drawing 90° in standard position,
we get Figure 14.9.
P is the point (0, 1). So

cos 90° = 0   sin 90° = 1

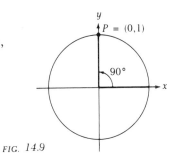

FIG. *14.9*

EXAMPLE: *Find cos 0°, sin 0°, and cot 0°.*

Putting 0° in standard position, gives Figure 14.10.

P = (1, 0)

so

cos 0° = 1   sin 0° = 0

$\cot 0° = \dfrac{1}{0} = $ undefined

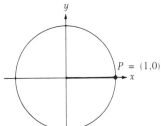

FIG. *14.10*

EXAMPLE: *Find cos 270°, sin 270°, and tan 270°.*

Drawing 270° in standard position, gives Figure 14.11.

P = (0, −1)

so

cos 270° = 0   sin 270° = −1

$\tan 270° = \dfrac{-1}{0} = $ undefined

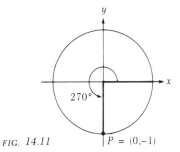

FIG. *14.11*

EXAMPLE: *Find cos(−450°).*

Putting −450° in standard position as in Figure 14.12, we see that
its terminal side is the same as that of 270°. Since the terminal
sides of 270° and −450° are the same, P will be the same for each.
Therefore all the trigonometric functions of the two angles are
exactly the same, and in particular:

cos(−450°) = cos 270° = 0

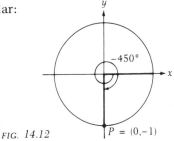

FIG. *14.12*

EXAMPLE: *Find cos 135°, sin 135°, and tan 135°.*

First put 135° in standard position, as in Figure 14.13. The trouble now is to find the coordinates of P. Suppose we drop a perpendicular from P to the x axis, meeting it at Q, as in Figure 14.13.

$$PO = 1$$

since this is the radius of the circle, and

$$\alpha = 180° - 135° = 45°$$

since α and 135° lie on a line, and

$$\beta = 90° - 45° = 45°$$

because $\beta = 90° - \alpha$. So

$\triangle PQO$ is isosceles

Therefore $a = b$.

Applying Pythagoras's theorem to $\triangle PQO$, we see that

$$a^2 + b^2 = 1^2$$

or

$$a^2 + a^2 = 1 \quad \text{(since } a = b\text{)}$$

$$2a^2 = 1$$

$$a = \pm \frac{1}{\sqrt{2}}$$

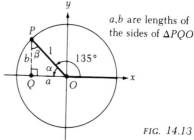

a,b are lengths of the sides of $\triangle PQO$

FIG. 14.13

But a and b are lengths and so must be positive; therefore

$$a = b = \frac{1}{\sqrt{2}}$$

P is in the second quadrant and so it has a negative x coordinate and a positive y coordinate.

Therefore, since to get to P from the origin you move to the left $\dfrac{1}{\sqrt{2}}$ and then up $\dfrac{1}{\sqrt{2}}$, its coordinates are

$$P = \left(-\frac{1}{\sqrt{2}}, \frac{1}{\sqrt{2}} \right)$$

So

$$\cos 135° = -\frac{1}{\sqrt{2}} \qquad \sin 135° = \frac{1}{\sqrt{2}} \qquad \tan 135° = \frac{\frac{1}{\sqrt{2}}}{-\frac{1}{\sqrt{2}}} = -1$$

EXAMPLE: *Find cos 205° and sin 205°.*

Putting 205° in standard position as in Figure 14.14, shows us that it is in the third quadrant.

Again, dropping a perpendicular from P to Q, we get a triangle in which $PO = 1$, since this is the radius of the circle.   Now $\alpha = 205° - 180° = 25°$, so we have the triangle in Figure 14.15 to solve.

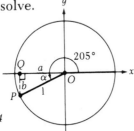

FIG. 14.14

FIG. 14.15

Since b is the side opposite the 25° angle, and a is the adjacent side:

$$\cos 25° = \frac{a}{1} \quad \text{and} \quad \sin 25° = \frac{b}{1}$$

That is,

$$0.9063 = \frac{a}{1} \quad \text{and} \quad 0.4226 = \frac{b}{1} \quad \text{(looking up 25° in the tables)}$$

Therefore

$$a = 0.9063 \quad \text{and} \quad b = 0.4226$$

But both of P's coordinates are negative, since it is in the third quadrant, so

$$P = (-0.9063, -0.4226)$$

Therefore

$$\cos 205° = -0.9063 \qquad \sin 205° = -0.4226$$

We can show that the circle definitions of the trigonometric functions reduce to the triangle definitions for angles between 0° and 90°. Here's how:

If θ is between 0° and 90°, it is in the first quadrant. If we drop a perpendicular from P to the x axis, meeting it at Q, as shown in Figure 14.16, then

$$OQ = x$$

and

$$QP = y$$

because x and y are both positive, since P is in the first quadrant.   Also, $OP = 1$, since it is the radius of the circle.

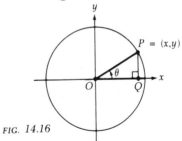

FIG. 14.16

Therefore θ is contained in the triangle OPQ shown in Figure 14.17.

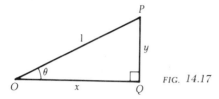

FIG. 14.17

The triangle definitions give:

$$\cos \theta = \frac{\text{adjacent}}{\text{hypotenuse}} = \frac{x}{1} = x \qquad \cot \theta = \frac{\text{adjacent}}{\text{opposite}} = \frac{x}{y}$$

$$\sin \theta = \frac{\text{opposite}}{\text{hypotenuse}} = \frac{y}{1} = y \qquad \sec \theta = \frac{\text{hypotenuse}}{\text{adjacent}} = \frac{1}{x}$$

$$\tan \theta = \frac{\text{opposite}}{\text{adjacent}} = \frac{y}{x} \qquad \csc \theta = \frac{\text{hypotenuse}}{\text{opposite}} = \frac{1}{y}$$

But these are exactly what the circle definitions give, and therefore the circle and triangle definitions agree for angles between 0° and 90°.

PROBLEM SET   14.2

By drawing the angle in standard position, find:

1. tan 0°
2. cos −90°
3. sin 180°
4. tan 90°
5. sec 0°
6. csc 90°
7. sin 150°
8. cos −150°
9. tan 120°
10. cot 210°
11. sin 540°
12. cos 450°
13. sin −35°
14. cos −35°
15. tan −35°

What are the signs of sin θ and cos θ if the terminal side of θ is in:

16. The first quadrant
17. The second quadrant
18. The third quadrant
19. The fourth quadrant

In which quadrant(s) is:

20. sin θ positive and cos θ negative
21. tan θ positive
22. sec θ positive
23. csc θ negative
24. cot θ negative
25. sin θ positive and tan θ positive
26. cos θ negative and tan θ negative
27. sin θ negative and sec θ negative
28. sin θ negative and csc θ positive
29. cot θ positive and tan θ positive
30. cot θ negative and tan θ positive

Find all values of θ between $0°$ and $360°$ satisfying:

31. $\sin \theta = 0$
32. $\cos \theta = 1$
33. $\tan \theta = 0$
34. $\sin \theta = -1$
35. $\sec \theta = 1$
36. $\csc \theta = -1$

37. $\sin \theta = \dfrac{1}{2}$

38. $\cos \theta = \dfrac{1}{\sqrt{2}}$

39. $\tan \theta = 1$
40. $\cot \theta = -1$

41. $\sec \theta = \sqrt{2}$
42. $\csc \theta = -\sqrt{2}$
43. $\sin \theta = 0.1736$
44. $\cos \theta = 0.3420$
45. $\sin \theta = -0.9848$

14.3   EVALUATING THE TRIGONOMETRIC FUNCTIONS USING RELATED ANGLES

From the examples at the end of the last section you can see that calculating the trigonometric functions from the circle definition is a lot of work. Fortunately there is a much simpler method, and one that will work for any angle. This method is based on the trick we used in the last section, that of dropping a perpendicular to the x axis.

Let us draw a picture in which, for argument's sake, θ is in the second quadrant. See Figure 14.18. The angle, α, between the terminal aide and the x axis is called the *related angle* to the original angle. α is considered positive and is between $0°$ and $90°$. (Related angles are not defined for angles whose terminal sides are on the axes. This is no problem since we can easily find the trig functions of such angles directly from the definitions.)

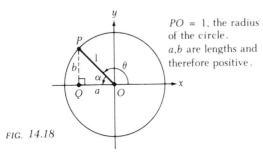

PO = 1, the radius of the circle.
a,b are lengths and therefore positive.

FIG. 14.18

The angle α is not in standard position and so we cannot use the circle definitions for $\sin \alpha$, $\cos \alpha$, and so on. However, since α is contained in a right triangle, we can use the triangle definitions to say

$$\cos \alpha = \frac{a}{1} = a \qquad \cot \alpha = \frac{a}{b}$$

$$\sin \alpha = \frac{b}{1} = b \qquad \sec \alpha = \frac{1}{a}$$

$$\tan \alpha = \frac{b}{a} \qquad \csc \alpha = \frac{1}{b}$$

Now, by the circular definitions for θ, if $P = (x, y)$,

$$\cos \theta = x \qquad \cot \theta = \frac{x}{y}$$

$$\sin \theta = y \qquad \sec \theta = \frac{1}{x}$$

$$\tan \theta = \frac{y}{x} \qquad \csc \theta = \frac{1}{y}$$

But to get to P from the origin you have to go a units horizontally and b vertically, where a and b are positive, so

$$P = (x, y) = (\pm a, \pm b)$$

where different signs apply in different quadrants. (For example, in the fourth quadrant x is positive and y is negative, so $x = a$, $y = -b$.)
  Therefore

  x and a are the same except possibly for sign

and

  y and b are the same except possibly for sign.

So if you compare the trig functions of α and of θ, you will see that they are the same except possibly for sign. In other words,

$$\cos \theta = \pm \cos \alpha$$
$$\sin \theta = \pm \sin \alpha$$
$$\tan \theta = \pm \tan \alpha$$
$$\cot \theta = \pm \cot \alpha$$
$$\sec \theta = \pm \sec \alpha$$
$$\csc \theta = \pm \csc \alpha$$

where θ is any angle, and α is its related angle and different signs apply in different quadrants

To figure out which signs apply, look at the signs of x and y in the quadrant in question, and figure out the sign of the function you are interested in. For example, in the second quadrant, x is negative and y positive, so the cosine is negative (because $\cos \theta = x$), the sine is positive ($\sin \theta = y$), and the tangent is negative ($\tan \theta = y/x$). In the third quadrant, x and y are both negative, so the cosine and sine are negative and the tangent is positive.
  Now the angle α is between $0°$ and $90°$ and so is in the tables. Therefore to find the value of, say, $\cos \theta$, we can look up $\cos \alpha$ in the tables and figure out what sign must apply from the quadrant θ is in. To find $\cot \theta$, look up $\cot \alpha$, and put the right sign in front, and so on.

To summarize:

To evaluate the trigonometric functions of a general angle, θ:

1. Find the related angle, the angle between the terminal side and the nearest part of the x axis.
2. Look up α in the tables of whichever function you are interested in.
3. Get the right sign by looking at which quadrant θ is in.

EXAMPLE: *Find sin 280°.*

Drawing 280° in standard position, we get Figure 14.19 and see that the related angle is 80°. (*Note:* The related angle is always measured to the x axis, *not* the y axis, which would give 10°.).

Now, sin 80° = 0.9848 (from the tables), and since sin θ = y, sin θ is negative in the fourth quadrant. Therefore

sin 280° = −sin 80°

that is,

sin 280° = −0.9848

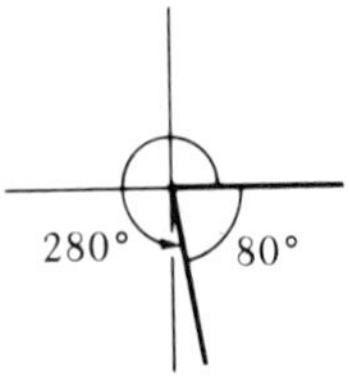

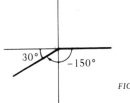

FIG. 14.19

EXAMPLE: *Find cot(−150°).*

Figure 14.20 shows that 150° has a related angle of 30°.

From the tables (or from the "special triangles"),

cot 30° = 1.7321 (= $\sqrt{3}$)

In quadrant III, x and y are both negative so cot θ is positive.

Therefore

cot(−150°) = cot 30°

So

cot(−150°) = 1.7321

FIG. 14.20

EXAMPLE: *Find cos 50° and cos (−50°).*

Figure 14.21 shows the angles 50° and −50° in standard position.

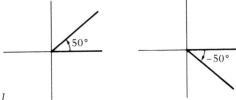

FIG. 14.21

Now, cos 50° can be looked up in the tables directly; there is no need for related angles.

cos 50° = 0.6428

Also, −50° has a related angle of 50° (because the related angle is the positive angle between the terminal side and the x-axis). In the fourth quadrant the cosine is positive, so

cos −50° = 0.6428

Therefore

cos −50° = cos 50°

Later we will show that for any angle θ: cos (−θ) = cos θ

EXAMPLE: *Find sin 130° and sin (−130°).*

The angles are shown in standard position in Figure 14.22. The related angle is 50° in both cases.

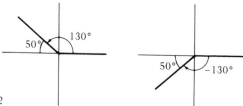

FIG. 14.22

From the tables,

sin 50° = 0.7660

Now sin θ = y, and in the second quadrant (where 130° is), y is positive, and in the third (where −130° is), y is negative. So

sin 130° = 0.7660   and   sin (−130°) = −0.7660

Therefore

sin (−130°) = −sin 130°

Later we will show that for any angle θ: sin (−θ) = −sin θ

EXAMPLE: *Find sin 20°, sin 380°, sin 740°, and sin(20° + 360°n), where n is an integer.*

First, 20° + 360°n means 20 + 360 · 1 = 380°, or 20 + 360 · 2 = 740°,   or   20 + 360 · 3 = 1100°,   or   20 + 360 · 4 = 1460°,   or 20 + 360(−2) = −700°, and so on, that is, 20° + n complete revolutions through 360°. So 20° + 360°n is the most general expression for an angle that has its terminal side at the same place as 20°.

Therefore, all the angles in Figure 14.23 have the same terminal side. So P is the same for them all, and they all have the same values for the trig functions.

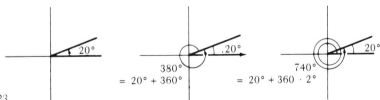

FIG. *14.23*

From the tables,

$$\sin 20° = 0.3420$$

so

$$\sin 380° = \sin 740° = \sin(20° + 360°n) = 0.3420$$

EXAMPLE: *Find cos 20°, cos 160°, cos 200°, and cos 340°.*

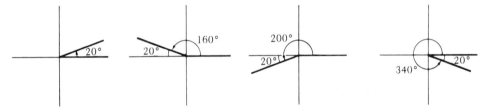

FIG. *14.24*

All these angles, shown in Figure 14.24, have related angles of 20°, and from the tables

$$\cos 20° = 0.9397$$

The cosine is positive in the first and fourth quadrants and negative in the second and third, so:

$$\cos \ 20° \ = 0.9397$$
$$\cos \ 160° = -0.9397$$
$$\cos \ 200° = -0.9397$$
$$\cos \ 340° = 0.9397$$

Since $(20° + 360°n)$ is the most general expression for an angle whose terminal side coincides with 20°:

$$\cos(20° + 360°n) = 0.9397$$

Similarly, $(160° + 360°n)$ is the most general expression for an angle whose terminal side coincides with 160°, so

$$\cos(160° + 360°n) = -0.9397$$

EXAMPLE: *Find θ if cos θ = 0.9397 and φ if cos φ = −0.9397.*

Looking at the last example you can see that:

$$\cos(20° + 360°n) = 0.9397$$

and

$$\cos(340° + 360°n) = 0.9397 \qquad [(340° + 360°n) \text{ has the same terminal side as } 340°]$$

Hence if

$$\cos θ = 0.9397$$

then

$$θ = 20° \quad \text{or} \quad (20° + 360°n)$$

or

$$θ = 340° \quad \text{or} \quad (340° + 360°n)$$

Similarly, since $\cos(160° + 360°n) = \cos(200° + 360°n) = -0.9397$,

if

$$\cos φ = -0.9397$$

then

$$φ = 160° \quad \text{or} \quad (160° + 360°n)$$

or

$$φ = 200° \quad \text{or} \quad (200° + 360°n)$$

How could we have done this without using the previous example? For both θ and φ, the related angle α must satisfy

$$\cos α = 0.9397$$

The tables will tell you that

$$\cos 20° = 0.9397$$

and so the related angle, being between 0° and 90°, must be 20°, that is,

$$α = 20°$$

Then θ could have its terminal side in any one of four positions, as shown in Figure 14.25.

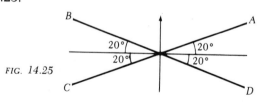

FIG. 14.25

Now the cosine is positive in the first and fourth quadrants, and negative in the second and third. Therefore all the angles with terminal sides at A and D have cosines of 0.9397, and those at B and C have cosines of -0.9397.

Since $\cos \theta$ is positive, θ must be one of the angles shown in Figure 14.26, or any angle with the same terminal side, that is,

$$\theta = 20° + 360°n \quad \text{or} \quad \theta = 340° + 360°n$$

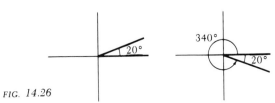

FIG. 14.26

On the other hand, ϕ must be one of the angles in Figure 14.27, or any angle with the same terminal side, that is,

$$\phi = 160° + 360°n \quad \text{or} \quad \phi = 200° + 360°n$$

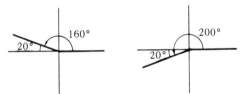

FIG. 14.27

EXAMPLE: *Find θ if $\sin \theta = -0.9511$.*

The related angle, α, must satisfy

$$\sin \alpha = 0.9511$$

so (looking in the tables)

$$\alpha = 72°$$

Sines are positive in the first and second quadrants, and negative in the third and fourth. Therefore θ must be in quadrant III or IV, and have a related angle of 72°. Therefore θ is one of the angles shown in Figure 14.28, namely

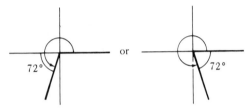

FIG. 14.28

$$180° + 72° = 252° \quad \text{or} \quad 360° - 72° = 288°$$

or any other angle with the same terminal side, that is,

$$\theta = 252° + 360°n \quad \text{or} \quad \theta = 288° + 360°n$$

EXAMPLE: *Find θ if tan θ = tan 53°.*

It is tempting to look at this equation and conclude that $\theta = 53°$. Certainly $\theta = 53°$ is one solution to the equation, but how do we know that there aren't others?

In fact, there are others, because what tan θ = tan 53° tells us is that the related angle to θ is 53°. Also, since tan 53° is positive, tan θ must be positive and therefore θ must be in quadrant I or III. So θ must be one of the angles shown in Figure 14.29, or any angle with the same terminal side, namely,

$$\theta = 53° + 360°n \quad \text{or} \quad \theta = 233° + 360°n$$

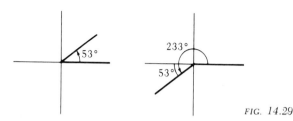

FIG. 14.29

Note: As you see, it is not necessary actually to find tan 53° using tables. If, however, the problem had been to find θ when tan θ = cos 53°, then we would have had to find cos 53° first and then find θ as in the two previous examples.

PROBLEM SET   14.3

Use related angles to evaluate:

1. sin 24°
2. cos 72°
3. tan 85°
4. tan 90°
5. sin 325°
6. cot 110°
7. sec 200°

8. cos 236°30′
9. cos (−236°30′)
10. sin 181°30′
11. sin (−181°30′)
12. sin 997°
13. cos −997°
14. cot 1012°

15. csc 170°13′
16. cot (−116°)
17. sin (−38°12′)
18. cos (−95°8′)
19. tan (−95°8′)
20. sin (95°8′)

Find all angles between 0° and 360° satisfying

21. sin θ = 0.9659
22. sin θ = −0.2588
23. cos θ + 0.4226 = 0
24. 2 cos θ = 1
25. sin θ = 0.5707
26. tan θ = 2.1445
27. tan θ = −0.3640
28. cot θ = 4.0108

29. sec θ = 2.0000
30. csc θ = −1.7434
31. sec θ = 1.5558
32. sin θ = 0.1776
33. sin θ = −0.9450
34. cos θ = 0.2792
35. tan θ + 0.3605 = 0

Solve completely (i.e. find *all* angles which satisfy the equation):

36. $\sin \theta = 1$
37. $\cos \theta = 0$
38. $\tan \theta = \dfrac{1}{\sqrt{2}}$
39. $\sin \theta = \dfrac{1}{2}$
40. $\cos \theta = -\dfrac{1}{2}$

41. $\sin \theta = \dfrac{3}{5}$
42. $\cos \theta = \dfrac{4}{5}$
43. $\tan \theta = \dfrac{3}{4}$
44. $2 \sin \theta - 3 = 0$
45. $3 \sec \theta - 2 = 0$

CHAPTER 14 REVIEW

Draw a diagram showing the following angles in standard position:

1. $10°$
2. $100°$
3. $260°$
4. $-135°$
5. $480°$

Evaluate:

6. $\sin 125°$
7. $\cos (-135°)$
8. $\tan 95°$
9. $\cot 190°$
10. $\sec (-285°)$
11. $\sec 3601°$
12. $\sec (-181°)$
13. $\cot (602°11')$
14. $\sin (-921°)$
15. $\cos (2°14')$
16. $\sec (179°)$
17. $\csc (181°)$
18. $\cos (61°13')$
19. $\tan (-217°8')$
20. $\cot (408°19')$

21. If P is the point $(\frac{3}{5}, -\frac{4}{5})$ what is $\tan \theta$?
22. If the point $(3, 4)$ lies on the terminal side of the angle θ, what is $\sin \theta$?
23. If the point $(3, 6)$ lies on the terminal side of the angle θ, what is $\cos \theta$?
24. If the point $(-2, -3)$ lies on the terminal side of the angle θ, what is $\sin \theta$? $\tan \theta$?
25. If the terminal side of a first quadrant angle is the line $y = \frac{3}{4}x$, what is $\cos \theta$?
26. If the terminal side of the second quadrant angle θ is the line $y + 2x = 0$, find $\sin \theta$.

Find all values of θ satisfying:

27. $\cos \theta = -\cos 40°$
28. $\sin \theta = -\sin 52°$
29. $\tan \theta = \tan 80°$
30. $\cot \theta = -\cot 17°$

15   THE GENERAL TRIANGLE

15.1   THE LAW OF SINES

In Section 13.4 we used the triangle definitions of the trigonometric functions to solve right triangles and to do word problems. What about solving non-right triangles? These can always be solved by the earlier methods by drawing in an altitude which divides the triangle into two right triangles, as in Figure 15.1. However, things are made much easier by the fact that this has been done in general, producing two laws, the *law of sines* and the *law of cosines*, which can be used to solve any triangle.

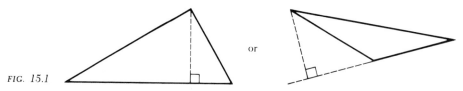

FIG. 15.1

First, let us agree on some notation. In a triangle ABC, a will denote the length of the side opposite the vertex (corner) A, and A will be used to mean both the vertex A and the angle there, and so on. See Figure 15.2. This is of course extremely sloppy notation, but it is better than having to introduce a third set of letters for the angles.

Now for the law of sines:

The Law of Sines
In any triangle,
$$\frac{\sin A}{a} = \frac{\sin B}{b} = \frac{\sin C}{c}$$

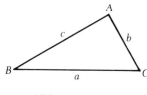

FIG. 15.2

Before we prove this law, realize that it is at least reasonable. The largest side of a triangle is always opposite the largest angle (try and draw a triangle which isn't this way). Also, from the tables you can see that up to 90°, the sine of an angle increases with the angle. Putting these two facts together, it is certainly possible that the ratio

$$\frac{\text{sine of an angle}}{\text{side opposite the angle}}$$

should be a constant.

But now we must show that the sine rule actually is true, rather than that there's no reason it shouldn't be.

Proof of the Law of Sines

We will prove this only for the case in which all the angles in the triangle are acute (less than 90°). The law of sines is just as true in an obtuse-angled triangle and the proof runs along the same lines, but the picture looks a bit different.

Draw a perpendicular (or altitude) from C cutting AB at D as shown in Figure 15.3. Let $DC = h$. Then, looking at $\triangle ADC$, which is a right triangle,

$$\sin A = \frac{h}{b}$$

or

$$h = b \sin A$$

Looking at $\triangle DCB$ gives

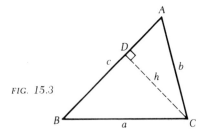

FIG. 15.3

$$\sin B = \frac{h}{a}$$

or

$$h = a \sin B$$

Therefore

$$b \sin A = h = a \sin B$$

or

$$b \sin A = a \sin B$$

Dividing through by ab:

$$\frac{\sin A}{a} = \frac{\sin B}{b}$$

which gives us the first part of the law.

By similar arguments we can get

$$\frac{\sin A}{a} = \frac{\sin C}{c}$$

and

$$\frac{\sin B}{b} = \frac{\sin C}{c}$$

So, altogether,

$$\frac{\sin A}{a} = \frac{\sin B}{b} = \frac{\sin C}{c}$$

Use of the Law of Sines

The law of sines provides three possible equations of the form

$$\frac{\sin A}{a} = \frac{\sin B}{b}$$

which we could use to solve for an unknown side or angle. To solve such an equation we must know three of the four quantities in it—for example, three of A, a, B, b. Therefore, *we must know on angle and the side opposite it, plus one other quantity to be able to use the law of sines.*

EXAMPLE: *Solve the triangle with $B = 62°$, $C = 40°$, $b = 20$.*

First draw the triangle, as shown in Figure 15.4. We need to find a, A, and c. First let's get c:

Since

$$\frac{\sin B}{b} = \frac{\sin C}{c}$$

we know that

$$\frac{\sin 62°}{20} = \frac{\sin 40°}{c}$$

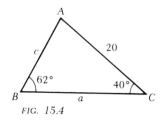

FIG. 15.4

or

$$\frac{0.8829}{20} = \frac{0.6428}{c} \qquad \text{(using tables)}$$

So

$$c = 20\left(\frac{0.6428}{0.8829}\right)$$

This can be worked out using logs, a slide rule, a calculator, or long division (preferably not this!), giving

$$c = 14.56 \cong 14.6$$

Angle A can be found from the fact that the angles of a triangle add up to 180°. Therefore,

$$\begin{aligned} A &= 180° - C - B \\ &= 180° - 40° - 62° \\ &= 78° \end{aligned}$$

The length of a can be found from

$$\frac{\sin A}{a} = \frac{\sin B}{b}$$

which gives

$$\frac{\sin 78°}{a} = \frac{\sin 62°}{20}$$

or

$$\frac{0.9781}{a} = \frac{0.8829}{20}$$

so

$$a = 20\left(\frac{0.9781}{0.8829}\right)$$

which gives

$$a = 22.15 \cong 22.2$$

EXAMPLE: *Solve the triangle with $B = 42°$, $a = 7$, $b = 10$.*

First draw a picture, as shown in Figure 15.5. Then

$$\frac{\sin A}{a} = \frac{\sin B}{b}$$

gives

$$\frac{\sin A}{7} = \frac{\sin 42°}{10}$$

or

$$\frac{\sin A}{7} = \frac{0.6691}{10}$$

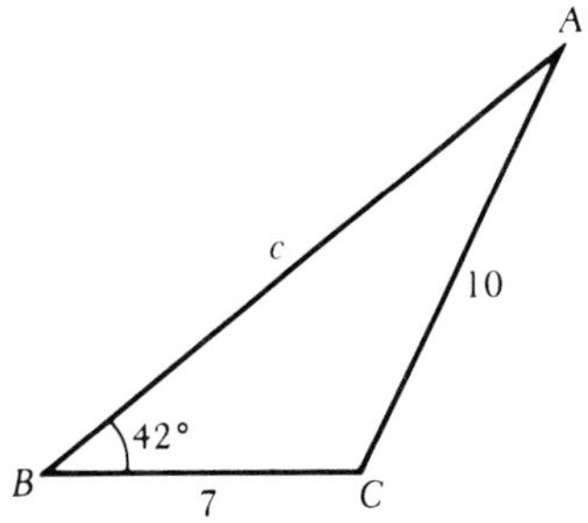

FIG. 15.5

giving

$$\sin A = \frac{7}{10}(0.6691)$$

or

$$\sin A = 0.4684 \qquad \text{(to four decimal places)}$$

We now have to find A, which is done as in the examples at the end of the last chapter. The tables tell us that

$$\sin 27°54' \cong 0.4684$$ (using 0.4679, which is the closest value in the tables)

So 27°54' is the related angle to A. The sine is positive in the first and second quadrants (see Figure 15.6), so A is 27°54' or 152°6'.

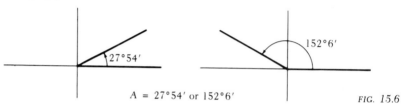

$A = 27°54'$ or $152°6'$ FIG. 15.6

Now A cannot be 152°6', because then A and B would add up to 194°6', or more than 180°, which is impossible if they are to be in a triangle. Therefore,

$$A = 27°54'$$

Since $A + B + C = 180°$,

$$C = 180° - 42° - 27°54'$$

$$C = 110°6'$$

Using

$$\frac{\sin B}{b} = \frac{\sin C}{c}$$

we can find c:

$$\frac{\sin 42°}{10} = \frac{\sin 110°6'}{c}$$

or

$$\frac{0.6691}{10} = \frac{0.9391}{c}$$ (using $\sin 110°6' = \sin 69°54'$)

so

$$c = \frac{10\,(0.9391)}{0.6691}$$

$$c = 14.0$$

PROBLEM SET   15.1

Solve the triangle ABC. See Figure 15.7
1. $B = 30°$, $C = 17°$, $c = 0.5$
2. $A = 121°$, $B = 38°$, $a = 33.4$
3. $A = 42°$, $B = 68°$, $a = 23.5$
4. $A = 56°$, $B = 87°$, $b = 40.8$
5. $C = 19°$, $a = 17.2$, $c = 20.4$
6. $A = 74°$, $B = 44°$, $a = 33.3$
7. $C = 80°$, $b = 293$, $c = 329$
8. $B = 85°$, $b = 452$, $c = 115$
9. $A = 87°$, $a = 4.22$, $c = 2.09$
10. $A = 99°10'$, $a = 0.450$, $b = 0.284$
11. $C = 82°16'$, $b = 3.137$, $c = 3.596$
12. $A = 78°38'$, $a = 856.2$, $b = 617.9$
13. $B = 29°41'$, $A = 46°22'$, $a = 8.974$
14. $C = 101°43'$, $a = 0.45$, $c = 0.6036$
15. $C = 67°44'$, $a = 58.91$, $c = 63.42$
16. $A = 62°40'$, $B = 73°20'$, $b = 309$
17. $A = 45°30'$, $a = 723$, $b = 578$
18. $B = 45°14'$, $C = 37°23'$, $b = 43.21$
19. $A = 35°$, $B = 68°$, $a = 25$
20. $A = 125°40'$, $C = 48°50'$, $c = 275$

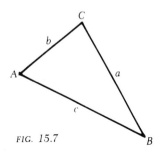

FIG. 15.7

15.2   THE AMBIGUOUS CASE

Remember that in the last example there was one point at which it looked as though we were going to have two values for A—either 27°54' or 152°6'. Then, luckily, the 152°6' turned out to be impossible because it would have made the angles add up to more than 180°.

But what if we weren't so lucky? If we were to interchange the 7 and the 10 and solve the triangle in Figure 15.8, we would come up with

$$\sin A = 0.9559$$

so

$$A = 72°54' \quad \text{or} \quad 107°6'$$

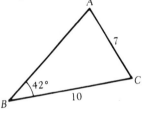

FIG. 15.8

Either of these values of A is possible, and they lead to the two possible triangles shown in Figure 15.9. Both of these triangles are answers to the problem.

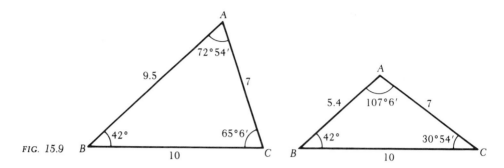

FIG. 15.9

You can see why this occurs if you try to draw the triangle *ABC* accurately from the information given. First draw the line *BC* 10 units long, and then draw a line through *B* at an angle of 42° to *BC*; *A* has to be somewhere on this line. You also know that *A* is 7 units from *C*, so put a compass at *C* and make arcs 7 units from *C*, cutting the line *BA*. There will be two points where the arcs cut *AB*, and so two possible positions for *A*, which are shown in Figure 15.10.

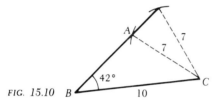

FIG. 15.10

Since this problem has two answers, it is called *the ambiguous case*.

Now let's see what the construction would have given us in the previous example when $B = 42°$, $a = 7$, $b = 10$. See Figure 15.11. In this case, only one of the two arcs gives a triangle which makes $B = 42°$, and therefore there is only one answer to the problem.

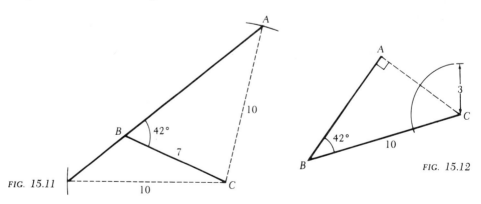

FIG. 15.11

FIG. 15.12

Just for completeness, we should consider a third case when $B = 42°$, $a = 10$, $b = 3$. The construction looks like Figure 15.12. Clearly, there is no point *A* on the 42° line through *B* which is also 3 units away from *C*, and therefore there is no triangle satisfying these conditions.

In summary, knowing two sides of a triangle and an angle which does not lie between them may or may not determine the triangle uniquely. If they do not, there may be two or no triangles which fulfill the specifications. You will be glad to hear that knowing any other three pieces of information, except three angles, does determine a triangle uniquely.

PROBLEM SET   15.2

Find the missing parts of each triangle.

1. $a = 190$, $b = 140$, $A = 55°$
2. $a = 715$, $b = 440$, $B = 23°$
3. $a = 0.287$, $b = 0.342$, $A = 69°$
4. $b = 0.21$, $c = 0.24$, $C = 41°$
5. $b = 420$, $c = 790$, $B = 71°40'$
6. $a = 53$, $b = 76$, $A = 40°30'$
7. $a = 6.5$, $b = 12$, $A = 36°50'$
8. $a = 70$, $b = 30$, $A = 47°30'$
9. $a = 31.5$, $b = 51.8$, $A = 33°40'$
10. $a = 320$, $c = 475$, $A = 35°20'$
11. $a = 62.5$, $b = 51.5$, $B = 40°40'$

15.3   THE LAW OF COSINES

Suppose we want to solve a triangle in which we are not given an angle and the side opposite to it. For example, consider the triangle in Figure 15.13. In this case the law of sines gives us

$$\frac{\sin A}{5} = \frac{\sin B}{2} = \frac{\sin 59°}{c}$$

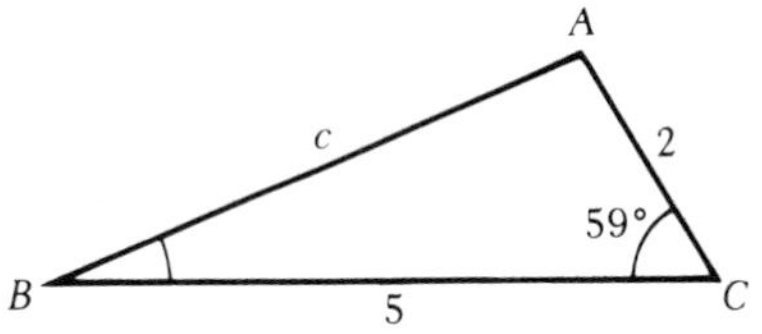

FIG. 15.13

Unfortunately, any equation we get out of this will have two unknowns in it and so can't be solved. But the information given about the triangle is enough to allow us to draw it accurately, and so it must be possible to find the other sides and angles. Indeed it is, but it can't be done using the law of sines. Instead we need the law of cosines:

The Law of Cosines

For any triangle,

$$c^2 = a^2 + b^2 - 2ab \cos C$$

and

$$b^2 = a^2 + c^2 - 2ac \cos B$$

and

$$a^2 = b^2 + c^2 - 2bc \cos A$$

Each one of these equations is called the *law of cosines*.

Before proving this, let me point out why it is not unreasonable. Let us apply the law of cosines to a right triangle in which C is the right angle, as shown in Figure 15.14. Then

$$\cos C = \cos 90° = 0$$

so

$$c^2 = a^2 + b^2 - 2ab \cos C$$

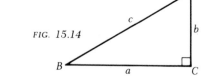

FIG. 15.14

reduces to

$$c^2 = a^2 + b^2 \qquad \text{which is Pythagoras's theorem}$$

Therefore the law of cosines is a generalization of Pythagoras's theorem, and the "$-2ab \cos C$" is a kind of "correction factor" for the case in which the triangle is not right-angled.

Notice also that when $C < 90°$, $\cos C$ is positive, and so the "correction factor" $-2ab \cos C$ is negative, meaning that

$$c^2 = a^2 + b^2 - 2ab \cos C$$

gives a smaller value for c than

$$c^2 = a^2 + b^2$$

This is as you would expect, since c is less than the hypotenuse of a right triangle with the same a and b—see Figure 15.15.

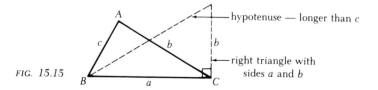

FIG. 15.15

On the other hand, when $C > 90°$, $\cos C$ is negative, and the correction factor is positive. Then the c from the law of cosines is greater than from Pythagoras's theorem, which fits in with the picture in Figure 15.16.

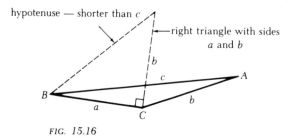

FIG. 15.16

Proof of the Law of Cosines

We will assume that all the angles are acute. The proof for an obtuse-angled triangle is an adaption of the one given here.

Draw a perpendicular as shown in Figure 15.17, and let $AD = h$ and $DC = x$ so that $BD = a - x$. Then, applying Pythagoras's theorem:

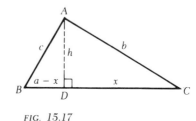

FIG. 15.17

$$\text{In } \triangle ADC: \quad b^2 = h^2 + x^2$$

$$\text{In } \triangle ADB: \quad c^2 = h^2 + (a - x)^2$$

Multiplying out gives:

$$c^2 = h^2 + a^2 - 2ax + x^2$$

or

$$c^2 = h^2 + x^2 + a^2 - 2ax$$

Substituting b^2 for $h^2 + x^2$:

$$c^2 = b^2 + a^2 - 2ax$$

Looking at $\triangle ADC$ again, we see that

$$\cos C = \frac{x}{b}$$

so

$$x = b \cos C$$

Substituting for x in

$$c^2 = b^2 + a^2 - 2ax$$

we get

$$c^2 = a^2 + b^2 - 2ab \cos C \qquad \text{which is what we want.}$$

Use of the Law of Cosines

In order to use the law of cosines, you must know three of the four things appearing in one version of the law. If you know a, b, and c, say, you can use the law to solve for C. If you know a, b, and C, you can solve for c. However, if you know a, c, and C, you can get an equation for b, but it will be quadratic and probably very hard to solve. In this case, you would be much better off using the law of sines, which you can do since you know an angle and the side opposite it (C and c) and one other quantity (a). Therefore, *in order to use the law of cosines, you should know either three sides, or two sides and the angle between them.*

EXAMPLE: *Solve the triangle in which a = 2, b = 3, c = 4.*

Draw the triangle, as in Figure 15.18. Now, the law of cosines tells you that

$$a^2 = b^2 + c^2 - 2bc \cos A$$

$$2^2 = 3^2 + 4^2 - 2(3)(4) \cos A$$

so

$$\cos A = \frac{21}{24}$$

that is,

$$\cos A = 0.8750$$

From the tables,

$$\cos 29° = 0.8750$$

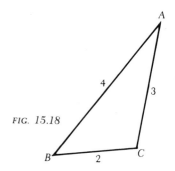

FIG. 15.18

so 29° is the related angle to A.

Since the cosine is positive in the first and fourth quadrants, A is 29° or 331° (see Figure 15.19). But A is in a triangle, and so can't possibly be 331°. Therefore

$$A = 29°$$

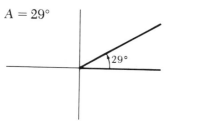
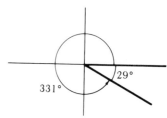

FIG. 15.19

At this stage we could use the law of cosines again to get B or C, or we could use the law of sines since we have an angle and the side opposite it. Usually the law of sines is easier because you don't have to do all that squaring. However, to get more practice in the law of cosines, we will say:

$$c^2 = a^2 + b^2 - 2ab \cos C$$

$$4^2 = 2^2 + 3^2 - 2(2)(3) \cos C$$

So

$$\cos C = -\frac{1}{4}$$

$$= -0.2500$$

Since

$\cos 75°30' = 0.2500$

$75°30'$ must be the related angle to C.

The cosine is negative in quadrants II and III (see Figure 15.20), but since C is in a triangle, it must be in quadrant II. Therefore

$C = 104°30'$

Since the angles of a triangle add up to $180°$,

$B = 180° - 104°30' - 29°$

$\quad = 46°30'$

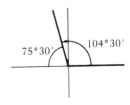

FIG. 15.20

EXAMPLE: *Solve the triangle in which* $a = 5$, $b = 2$, $C = 59°$.

Here we have two sides and the angle between them; see Figure 15.21. Using the law of cosines, we get:

$c^2 = a^2 + b^2 - 2ab \cos C$

$\quad = 5^2 + 2^2 - 2(5)(2) \cos 59°$

giving

$c^2 = 25 + 4 - 20(0.5150)$

or

$c^2 = 18.7$

So

$c = 4.3$

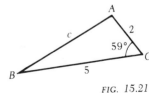

FIG. 15.21

Now let us try the law of sines to get B (the law of cosines would work too):

$\dfrac{\sin B}{b} = \dfrac{\sin C}{c}$

$\dfrac{\sin B}{2} = \dfrac{\sin 59°}{4.3}$

$\sin B = \dfrac{2(0.8572)}{4.3}$

$\sin B = 0.3987$

The related angle is therefore 23°30′, so B is 23°30′ or 156°30′ (see Figure 15.22). Now, $B = 156°30′$ is clearly impossible, because then B and C ($C = 59°$) would add up to more than 180°, so

$B = 23°30′$

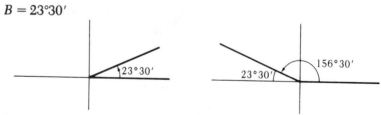

FIG. 15.22

Since the angles all must add to 180°,

$A = 180° - 23°30′ - 59°$

$\quad = 97°30′$

PROBLEM SET   15.3

Solve the triangle ABC

1. $a = 20,\ b = 24,\ C = 30°$
2. $a = 13,\ b = 1.7,\ C = 45°$
3. $a = 6,\ b = 8,\ c = 13$
4. $a = 2,\ b = 3,\ C = 40°$
5. $a = 3.9,\ c = 5.9,\ B = 130°$
6. $b = 1,\ c = 0.8,\ A = 120°$
7. $a = 12,\ b = 10,\ C = 120°$
8. $a = 10,\ b = 15,\ c = 21$
9. $a = 6,\ b = 7,\ c = 10$
10. $b = 10.5,\ c = 20,\ A = 53°8′$

11. $a = 24,\ c = 35,\ B = 143°$
12. $a = 7,\ b = 6,\ c = 2$
13. $a = 16,\ b = 25,\ C = 129°$
14. $a = 100,\ b = 45,\ C = 66°$
15. $a = 4.4,\ b = 9.8,\ c = 10.8$
16. $a = 2,\ b = 3,\ c = 4$
17. $b = 28.62,\ c = 22.34,\ A = 52°$
18. $a = 5,\ b = 7,\ c = 9$
19. $a = 10,\ b = 12,\ C = 36°$
20. $a = 12,\ b = 7,\ C = 21°$

15.4   WORD PROBLEMS

If a word problem involves finding a side or an angle of a triangle, then it can usually be done using trigonometry. To solve a trig word problem you should follow the steps of Section 13.4 with one addition—you need to know:

Which Method to Use to Solve a Triangle

Right Triangle: the triangle definitions

General triangle:
   Given an angle and the side opposite to it: the law of sines
   Given three sides: the law of cosines
   Given two sides and the angle between them: the law of cosines

EXAMPLE: *In a storm over New York, a plane's compass is damaged so that it reads to within 3° of the true reading (i.e., if the true reading is N 60° W, it might read anywhere from N 57° W to N 63° W). The pilot doesn't have time to have it mended and so heads for Boston, 225 miles away. If she sets a straight course as though the compass were working correctly, what is the most by which she could miss Boston?*

The faulty compass means that she may set her plane on a course that is at most 3° to either side of the proper course; the problem wants to know by how much she misses Boston when her course is 3° off the correct course. A picture is essential: Look at Figure 15.23.

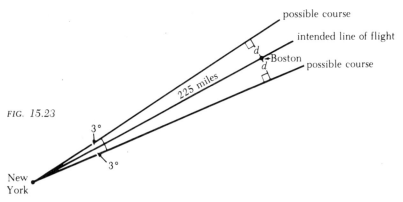

FIG. 15.23

If the plane is on one of the possible courses shown, then it passes within a distance d of Boston, where d is measured along a line perpendicular to the line of flight (see Figure 15.23). Since d is contained in a right triangle, it can be found directly from the triangle definitions:

$$\sin 3° = \frac{d}{225}$$

$$d = 225 \sin 3°$$

$$d = 225(0.0523)$$

$$d \cong 11.8 \text{ miles}$$

Therefore the plane will not miss Boston by more than 11.8 miles.

EXAMPLE: *In a never-to-be published study, it has been shown that the amount of cement dust deposited at a given place in a given time is inversely proportional to the square of the distance from that point to the cement factory. Two towns, Weston and Easton, each with a cement factory, are being sued by the owners of a baseball field near the towns. The owners want the towns to pay*

*for cleaning up the cement dust that is continually being depos-
ited on their field. Each town of course claims that it is all the
other one's fault, but eventually an argreement is reached that
each town shall pay in proportion to the amount of dust its factory deposits. What percentage is paid by Weston and what by
Easton?*

*You are given that Weston is 20 miles due west of Easton; that
the bearing of the baseball field from Weston is N 62° E and from
Easton is N 32° E. Realize also that the rate of deposit of dust at
a point d away from a factory is $k \cdot \dfrac{1}{d^2}$, where k is a constant
(because it is inversely proportional to the square of the distance). You are given that k = 1220 if d is in miles.*

To find out how much dust is deposited by each factory, it is necessary to know how far the factory is from the baseball field.
First, draw a picture, like Figure 15.24.

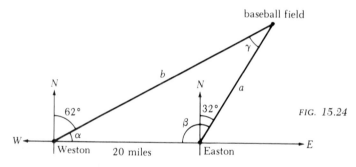

FIG. 15.24

$\alpha = 90° - 62° = 28°$

$\beta = 90° + 32° = 122°$

Since the angles of a triangle add up to 180°,

$\gamma = 180° - 28° - 122° = 30°$

We know all the angles of this triangle, but only one side, and
therefore this is a case for the law of sines:

$$\frac{\sin \alpha}{a} = \frac{\sin \gamma}{20}$$

giving

$$a = \frac{20 \sin \alpha}{\sin \gamma}$$

$$= \frac{20(0.4695)}{(0.5)}$$

$$\cong 18.8 \text{ miles}$$

Similarly,

$$\frac{\sin \beta}{b} = \frac{\sin \gamma}{20}$$

$$b = \frac{20 \sin \beta}{\sin \gamma}$$

Since $\beta = 122°$, the related angle $= 58°$, so

$$\sin \beta = \sin 122° = \sin 58°$$

$$b = \frac{20(0.8480)}{0.5}$$

$$b \cong 33.9 \text{ miles}$$

Therefore the amount of dust deposited by the Easton factory is

$$\frac{k}{a^2} = \frac{1220}{(18.8)^2} \cong 3.45$$

and by the Weston factory is

$$\frac{k}{b^2} = \frac{1220}{(33.9)^2} \cong 1.06$$

The total amount deposited is

$$3.45 + 1.06 = 4.51$$

Therefore the Weston factory deposited $\frac{1.06}{4.51}$ of the dust or

$$\frac{1.06}{4.51} \cdot 100 = 23.5\% \text{ of the dust}$$

So Easton deposited 76.5% of the dust.

EXAMPLE: *A plane takes off at an angle of 8°6' and flies due north at a speed of 420 miles per hour for a minute before turning westward to go out to sea. A man who lives 5 miles due north of the point of take-off wants to set up his telescope to observe the plane as it changes direction. At what angle should he point his telescope, and at what distance should he focus it?*

The telescope owner is interested in where the plane is after 1 minute of flight; the fact that it is then going to turn westward is quite irrelevant. The picture (not to scale) is shown in Figure 15.25. The problem is to find α, the angle at which the telescope should be pointed, and x, the distance at which it should be focused.

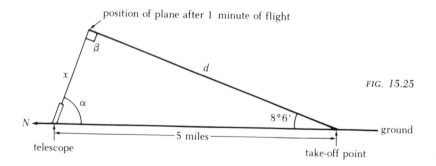

FIG. 15.25

So far the picture contains all the information that we were given except that the plane flies for 1 minute at 420 mph before turning. This tells us that

$$d = 420 \cdot \tfrac{1}{60} = 7 \text{ miles} \qquad (1 \text{ minute} = \tfrac{1}{60} \text{ hour})$$

We now know two sides of the triangle, plus the angle between them—which means that it's time for the law of cosines:

$$
\begin{aligned}
x^2 &= d^2 + 5^2 - 2 \cdot d \cdot 5 \cos 8°6' \\
&= 7^2 + 5^2 - 2 \cdot 7 \cdot 5 \cdot (0.99) \\
&= 49 + 25 - 69.3 \\
&= 4.7
\end{aligned}
$$

So

$$x = 2.17 \text{ miles}$$

The angle α is best found from the law of sines, since now we know an angle and the side opposite it (8°6' and $x = 2.39$). Therefore,

$$\frac{\sin \alpha}{d} = \frac{\sin 8°6'}{2.17}$$

$$\frac{\sin \alpha}{7} = \frac{0.1409}{2.17}$$

$$\sin \alpha = 0.4545$$

so

$$\alpha = 27° \quad \text{or} \quad 153°$$

Since we have two values for α, this is the ambiguous case. Fortunately, it turns out that by sketching the triangles we can eliminate one of the possibilities. First we calculate the third angle β:

If $\alpha = 27°$, $\quad \beta = 180° - \alpha - 8°6' = 144°54'$

If $\alpha = 153°$, $\quad \beta = 180° - \alpha - 8°6' = 18°54'$

The two possible triangles are shown in Figure 15.26. The first of these triangles could not exist because the 7-mile side would have to be shorter than the 5-mile side—clearly impossible. Therefore the second triangle must be the correct solution.

FIG. 15.26

The fact that α is greater than 90° means that the plane has passed over the man's head by the time it makes its turn. You should also notice the cheerful fact that although the original picture was somewhat misleading because it didn't include the fact that the plane was to the north of the man by the time it turned, this picture nonetheless led us to the right solution—and in fact to correcting the original picture.

The solution, therefore, is that the telescope should be trained northward at an angle of elevation of 180° − 153° = 27°, and should be focused at a distance of 2.17 miles.

PROBLEM SET   15.4

1. Suppose the keeper of a lighthouse takes two sightings ½ hour apart of a ship at sea. By radio-electronic means suppose the keeper further determines that the ship's distance from the lighthouse is 15 miles at first sight, and $\sqrt{3} \cdot 15$ miles upon second sighting. Given that the sightings were separated by a 30° (interior) angle, how fast is the ship traveling?

2. Romeo must rescue Juliet from her wicked uncle. Juliet is trapped in a tower, which rises perpendicularly from the ground and has a window 10 feet above the ground. Romeo needs to know what length of ladder he needs to prop up against Juliet's window, to carry her away from the tower. The angle of elevation between the ladder and the ground must be no greater than 25° or Romeo might lose his balance and drop Juliet. Find the length of the shortest ladder that would work. Give your answer to two significant digits.

3. Adele and Barbara are throwing snowballs at each other by heaving them up in the air and letting them fall on their opponent's head. To do this, they must know accurately how far away their target is. Adele is 50 feet from Barbara. All of a sudden, Michael appears. Adele sees Michael 35° to the right of Barbara, while Barbara sees Michael 55° to the left of Adele. How far from Adele and Barbara is Michael? (Give both distances.)

4. Two children were playing on a set of swings which hung from 8-foot chains on a horizontal pole 12 feet off the ground. If one child reached a height of 6 feet off the ground at the top of his swing (see Figure 15.27), what angle was formed, at that instant, between the swing chains and the vertical? (Find α.)

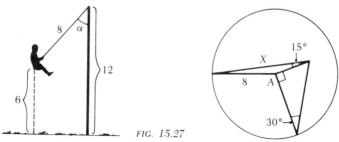

FIG. 15.27   FIG. 15.28

5. The second child (a little more rambunctious), starting at the bottom, swung completely over the top of the pole and back up a little more, swinging a total of 400° around the pole. At what height off the ground did she end up? How far forward from the pole had she swung after 400°?

6. Given the circle diagram in Figure 15.28 with center at A, find X.

7. A Soviet trawler is sailing from port *M* to port *N*. (*N* is 1 mile northeast of *M*). If there is a lighthouse on a hill at *P* such that the top of the lighthouse is ½ mile above sea level, and *P* is in a direction N 15° W from *M*, and N 60° W from *N*, what is the angle of elevation of the lighthouse from the ship anchored at port *M*?

8. Three freshmen, Reed, Brian, and Jim, are playing frisbee in front of the library. Reed and Brian stand 60 feet apart. Reed takes a turn through an angle θ and faces Jim, who is 80 feet away. If the distance from Jim to Brian is 50 feet, how large is the angle θ?

9. A duke was riding in a gondola through the canals of Venice. Just before reaching the landing, he dropped a precious gold medallion out of the boat. The gondolier landed the boat quickly, and then jumped into the water to recover the medal. From where the duke lies on the dock peering anxiously out over the water, he finally sees his medal at a 40° angle of depression. The gondolier, his head at the same level as the duke's, sees it at a 65° angle of depression. If the gondolier is 20 feet from the duke and between him and the medal, find the depth of the medal. (Assume there is no refraction of light.)

10. Two ships leave the Port of Miami at the same time. One sails at 12 mph on a bearing at 60° straight to Key Biscayne, which is 6 miles from the Port of Miami. The second boat sails on a bearing of 15° straight to Miami Beach, which is 5 miles from Key Biscayne. They both reach their destinations at the same time. How fast was the second ship sailing?

11. A ship is swinging at anchor with 40 feet of the anchor line paid out. The captain of the ship observes that the angle of depression from the water's

surface to the taut anchor line is 41°18. Assume that the lake bottom is a flat plane parallel to the water's surface, and that the anchor line leaves the ship at the water's surface. Find the depth of the water to three significant digits.

12. At the Fireman's Fair is a ferris wheel. There are 12 seats, equally spaced on the wheel. If the radius of the ferris wheel is 15 feet, find the distance between any two adjacent seats on the wheel to two significant digits.

13. The height of a seat on the ferris wheel of problem 12 is related to the angle it makes with the vertical as shown in Figure 15.29. This relationship is

$$h = 15(1 - \cos \theta)$$

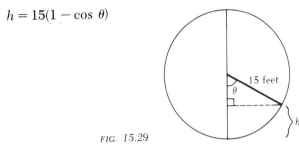

FIG. 15.29

What is the angle θ if a seat is 4 feet from the ground?

14. It is common knowledge that Phileas Fogg and Passepartout started their 80-day journey around the world in a balloon. Relatively unknown, though, is the fact that during the first minute of their journey they had a stowaway aboard whose name was Clyde. After untying the balloon, which then went straight up, Mr. Verne bid the travelers adieu and started back to his library. After walking a hundred feet in 1 minute, he turned around, pulled out his sextant, and measured the ballon's angle of elevation; it was 25°. At that very instant Fogg and Passepartout, unhappy about their slow ascent, threw overboard the trunk containing Clyde. Poor Clyde was most grievously harmed. After another minute, when he had walked another hundred feet, Mr. Verne repeated his measurement; it was now 32°. How much faster did the balloon rise in the second minute than in the first? [The preceding information comes from *Verne's Works* (vol. 2) and *The Complete Clyde* (vol. 68).]

In *The Road Not Taken*, Robert Frost sees two roads diverge in the woods. After some thought, he takes the less traveled one. Suppose the two roads of which the poet speaks are represented by the lines $y = 3x$ and $y = \frac{1}{2}x$.

15. What is the measure of the angle between the two roads?

16. If Mr. Frost starts at the origin and travels 100 miles down the $y = \frac{1}{2}x$ road (in the upward direction), how far is he from the other road at its nearest point to him?

17. Two students are sitting one behind the other in the same row of an examination hall. Both look up at the clock which is on the wall directly in front of them. Person A sits 20 feet from the wall with the clock and must look up at an angle of 36°52' to see the time. How far from her does person B sit if he

must look up at an angle of 30°58' to see the clock? Do either of them finish the exam? (Assume that persons A and B are the same height.)

18. From the top of a 100-foot-tall fire tower near the edge of a river, the angle of depression of the nearest bank is 30°, and of the furthest bank is 25°. How wide is the river?

19. From the top of a mountain a man looks down at the valley 2000 feet below. He sights one house at an angle of 70° (to the vertical) and another at an angle of 45°. How far apart are the two houses?

20. A boat is sailing at 12 mph from P to Q. (See Figure 15.30.) When the ship is at A the captain sights a lobster pot L and finds that the angle from his course to the lobster pot is 30°. One hour later the ship is at B and the angle from the course to the lobster pot is 70°. Find the distance of the lobster pot from B to two significant digits.

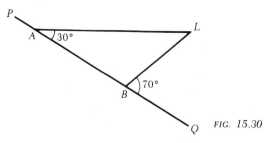

FIG. 15.30

CHAPTER 15 REVIEW

1. Given the triangle in Figure 15.31,

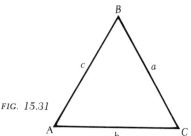

FIG. 15.31

(a) Express b in terms of angle A, angle B, and side a.
(b) Express cos C in terms of a, b, and c.
(c) If a = c, and angle C = 60°, show that b = c.

2. Show that the area of triangle ABC in Figure 15.32 is

$$\frac{1}{2}c^2 \cdot \left(\frac{\sin A \sin B}{\sin C}\right)$$

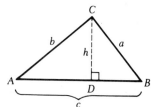

FIG. 15.32

(*Hint:* The area of a triangle is $\frac{1}{2}$ · altitude · base = $\frac{1}{2}ch$.)

3. Show that the area of a △*ABC* is ½*ab* sin *C*.
(*Hint:* area = ½ base · height.)

4. Find the area of a triangle with sides 9 inches, 8 inches, and 15 inches.
5. Find the area of the triangle shown in Figure 15.33.

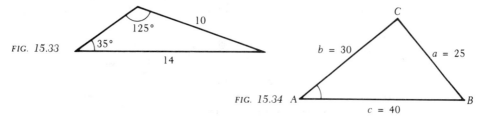

FIG. 15.33

6. Find the value of angle *A* in Figure 15.34
7. Find the measure of *β* in the triangle in
Figure 15.35.

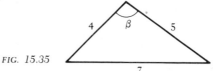

FIG. 15.35

8. A ship is traveling at 19.5 knots due NE. At 8:22 a.m. a mountain peak has a bearing of 149°, and at 9:05 a.m. it has a bearing of 154°. How far is the ship from the mountain peak at the time of the second observation? (A knot = one nautical mile per hour. A nautical mile = 6080 feet.)

9. Eratosthenes (ca. 230 B.C.) made a famous measurement of the earth. He observed that at noon at the summer solstice (the day on which the sun is farthest north) a vertical stick had no shadow at Syene, while at Alexandria (on the same meridian or longitude with Syene), the sun's rays were inclined at 7°12′ (or 7⅕° or 7.2°) to the vertical. See Figure 15.36. He then calculated the circumference of the earth from the known distance of 5000 stades between Alexandria and Syene. (*Note:* Because the earth is so far from the sun, you can always assume that the rays of the sun are parallel to each other whenever they strike the earth.)

(a) Find the angle *α* in the center of the earth.
(b) What fraction of 360° is *α*, the angle at the center of the earth?
(c) Assuming that the distance from Alexandria to Syene is that same fraction of the circumference, what is the circumference of the earth in stades?

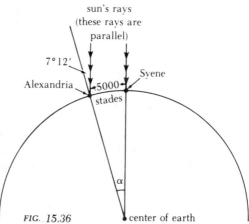

FIG. 15.36

10. Two planes leave the same airport at the same time. One plane flies on a bearing of 10° at 235 mph, the other flies on a bearing of 121° at 305 mph. How far apart are the planes at the end of 1 hour 20 minutes of flying?

A surveyor wants to measure the distance between two points *R* and *S* which are on level ground. For some reason he is unable to do this directly—for example, there is a lake or a mountain in between. Here are two methods which he might use. In each case, describe exactly how he would make the calculation of the distance *RS* from the information he gathers, and what law or formula he would use at each stage.

11. Find a point *H* from which *R* and *S* can be seen; measure distances from *H* to *R* and from *H* to *S*; measure angle *RHS*.

12. Find two points *T* and *W* from which *R* and *S* can be seen. Measure distance *TW*, and the angles *RSW*, *RTW*, *SWR*, and *SWT*.

13. Prove the law of sines for a triangle with an obtuse angle.

14. Prove the law of cosines for a triangle with an obtuse angle. (That is, prove that $c^2 = a^2 + b^2 - 2ab \cos C$ when *C* is acute and *B* is obtuse.)

Solve the following triangles:

15. $a = 9$, $b = 11$, $c = 13$

16. $a = 344$, $B = 17°50'$, $C = 43°30'$

17. $c = 680$, $B = 40°$, $C = 90°20'$

18. A field geologist standing 200 feet from a cliff with three exposed rock formations entered the sketch shown in Figure 15.37 into his notebook. Back in his office, he would like to know the thickness of each layer. Would you?

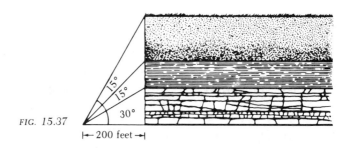

FIG. 15.37

|← 200 feet →|

19. Geophysicists often determine the subsurface geological structure by analyzing the way in which a shock wave travels through the rock. An explosion is set off in one location, the shock wave propagates downward and is partially reflected by the boundaries where layers change, and the time at which the wave reaches the recorder is noted. See Figure 15.38. If the rock transmits the shock at 1000 feet per second and pulses are recorded at 1.155, 1.414, and 2.000 seconds, how deep is each of the three boundaries?

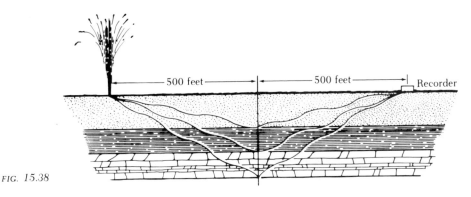

FIG. 15.38

20. A track must be banked toward the outside to make it easier to run on, and the angle at which it should be banked depends on the size of the track and on the average speed of the runners. The university planning office wanted me to determine at what angle the new track they were building should be banked. A little investigation told me that I could assume the track to be circular and that the formula relating the banking angle, θ, to the average speed of the runners, v, and the radius, r, of the track is (see Figure 15.39)

$$\tan \theta = \frac{v^2}{gr}$$

(g is a constant; $g = 32$ if v is in feet per second and r is in feet).

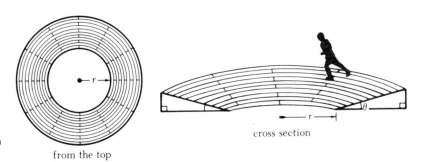

FIG. 15.39

from the top

cross section

(a) If the track has 8 laps (complete circuits) to a mile, what is its radius in feet? (1 mile = 5280 feet.) Don't work out the arithmetic.

(b) If $v = 20$ feet per second, find θ. Use $\pi = 3.2$. [Note for those who are interested: This is the speed for a 4.24-minute mile.]

(c) The reason the planning office needed to know the banking angle was that they were worried about the material with which they had been intending to surface the track. They had been told that the material they wanted to use would slip off if the track were banked with a slope of more than 1 in 10 (see Figure 15.40). Assuming the banking angle found in (b), did they have to find a new surfacing material? (Believe it or not, this is a *true* story!)

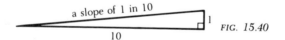

FIG. 15.40

21. Two ships have radio equipment with a range of 175 miles. One is 150 miles N 42°40′ E and the other is 160 miles N 45°10′ W of a shore station. Can the two ships communicate directly?

22. A ship sails 15.0 miles on a course N 40°10′ W and then 21.0 miles on a course N 28°20′ W. Find the distance and direction of the last position from the first.

23. A lighthouse is 20 miles W 20° N of a dock. A ship leaves the dock and steams west at 24 mph. At what time will it be 8 miles from the lighthouse?

24. A tower 150 feet high is situated at the top of a hill. At a point 650 feet down the hill the angle between the surface of the hill and the line of sight to the top of the tower is 12°30′. Find the inclination of the hill to a horizontal plane.

25. Far off in the Tibetan Himalayas is a unique mountain with a steep slope of smooth ice rising toward the summit, and making an angle of exactly 106° with the level ground below. In order to scale this peak, two climbers (wearing all the necessary ice-climbing equipment) have roped themselves together with 500 feet of rope, and one has begun to lead the way while his companion waits below at point C (see Figure 15.41). When the climber reaches point B, the rope is pulled taut, and the angle of elevation of the climber from his companion at C is 46°. Measured along the slope, how far has the climber ascended, from A to B?

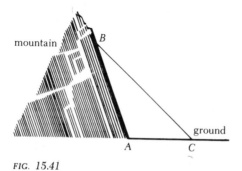

FIG. 15.41

16   RADIANS

16.1   DEFINITION OF RADIANS

Just as length can be measured in feet, centimeters, or inches, and weight in
pounds or kilograms, so angles can be measured in different units. Geometry
always uses degrees, and so far we have done the same. However, there is
one other unit, the radian, which turns out to be of the greatest importance in
any field which depends more on the behavior of the trigonometric functions
than on the trigonometry of triangles. In other words, if trigonometry is being
used as part of geometry, degrees are more useful; if it is contributing to a
study of functions and graphs, then radians are better. Since calculus is concerned exclusively with the study of functions, and since there are aspects of
physics—electronics, orbits, and periodic motion, for example—which rely
heavily on trigonometric functions, radians are clearly extremely important.

Before trying to explain what aspects of radians make them so useful, let me
explain how they are defined. First, remember that degrees were defined by
means of a rotation. A complete rotation about a point was rather arbitrarily
defined (by the Babylonians) as 360°; one degree is therefore one three-
sixtieth of a complete rotation.

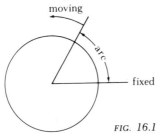

FIG. 16.1

Now imagine carrying out this rotation with the vertex of the angle at the
center of a circle, with one side of the angle fixed, and the other moving, as in
Figure 16.1. Then the length of the arc cut by the sides depends on the angle

between them. In fact, for a given circle, the arc length is proportional to the angle: If the angle doubles, so does the arc length; if the angle is quartered, so is the arc length. This means that, except for one problem, the arc length could be used for measuring the angle. The problem is that the length of the arc depends on the radius of the circle as well as the angle. The way to fix this, however, is to say that the circle must be a unit circle and have a radius of one in whatever units are used for measuring the arc length. Then the arc length will tell you the size of the angle, and we might talk about "an angle which cuts off 2 units," as in Figure 16.2, or "an angle which cuts off −½" (the negative sign meaning that the rotation is clockwise), as in Figure 16.3.

If we do things this way, then the most basic unit of angle will obviously be *that angle which cuts off an arc of length 1 in a unit circle*, and this is called *1 radian*. See Figure 16.4.

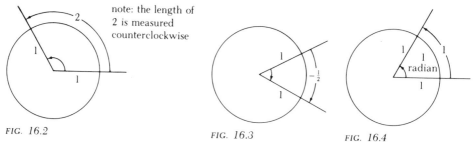

FIG. 16.2 FIG. 16.3 FIG. 16.4

Then the angle which cuts off an arc length of 2 units will be 2 radians, and the one which cuts off −½ units will be −½ radians, provided the radius of the circle is one unit.

Radians are used so often that the word itself is usually dropped. An angle of 2 is therefore understood to mean an angle of 2 radians.

16.2 WHY RADIANS ARE IMPORTANT

When you come to the graphs of the trigonometric functions in the next chapter, you will see that their behavior is quite unlike any we have met before. What makes them special is that they repeat themselves regularly. There are a great many natural phenomena which involve some kind of repeated variation—the motion of a clock pendulum, or the fluctuations in temperature throughout the day—and the trigonometric functions are the obvious candidates for describing such situations.

However, the trigonometric functions have so far all been defined as functions of angles. Now there are a great many physical situations, such as water waves or alternating electric currents, in which some quantity fluctuates regularly as a function of distance or time. In each case the independent variable is a length or a real number, and not an angle. Obviously we would like to describe these oscillations by trigonometric functions also, but how can we if the trigonometric functions are all functions of angles, and the quantity we are interested in depends on a length or a real number? Clearly the only way is somehow to redefine the trigonometric functions so that they are functions of the real numbers, or distances along the number line.

One way to define the sine of a number would be, I suppose, to interpret it as the sine of that many degrees. Then, for example, sin 2 would mean sin 2°. This method has many disadvantages. First, from the point of view of calculus, it ends up giving the sine function some disagreeable properties. Second, and by far the most important, this definition does not make the sine a function of length or distance, and it is very important that we be able to talk about the sine of a length.

The usual way to interpret the sine of a number is to take it to mean the sine of that many radians. In this case, sin 2 means sin(2 radians), and sin(0.5) = sin(0.5 radians). This should make it clear why it is all right to drop the word "radians" when working with radians—since an angle without a degree or a radian sign is understood to be in radians. Fortunately, this way of defining the sine of a number turns out to be very convenient for calculus.

The great thing about this definition is that it enables us to give a reasonable meaning to the sine of a length. When we measure an angle in radians we are measuring it by the length of the arc cut off on the unit circle, and so we could equally well define the sine of the arc length to be the sine of the angle. For example, an angle of 2 radians cuts off an arc of length 2, so we define

$$\sin(2 \text{ units of length}) = \sin(2 \text{ radians}) \qquad \text{(Figure 16.5)}$$

Similarly, the sine of −2 units of length is defined to be the sine of the angle obtained by measuring off an arc length of 2 units clockwise—in other words, the sine of −2 radians—so

$$\sin(-2 \text{ units}) = \sin(-2 \text{ radians}) \qquad \text{(Figure 16.6)}$$

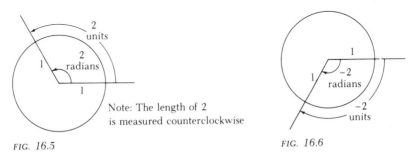

FIG. 16.5   FIG. 16.6

To define the sine (cosine, tangent) of any length: Starting from the positive x axis, measure off the given length along the circumference of the unit circle, counterclockwise for a positive length, clockwise for a negative one. The sine (cosine, tangent) of the angle so formed is the sine (cosine, tangent) of that length.

PROBLEM SET 16.2

Using unit circles, give an approximate representation of the arc length cut off by each of the following angles, measured in radians. (Remember, the circumference of the unit circle is 2π or approximately 6.28.)

1. 3
2. −1
3. 6
4. 11
5. 3.14
6. π
7. 1.57
8. $\dfrac{\pi}{2}$

9. −4
10. 1
11. $\dfrac{\pi}{4}$
12. $\dfrac{3\pi}{4}$
13. $-\dfrac{5\pi}{4}$
14. −1.57

15. −13
16. $\dfrac{\pi}{3}$
17. $-\dfrac{3\pi}{2}$
18. 18
19. 5
20. 62.8

16.3 CONVERSION BETWEEN DEGREES AND RADIANS

If we rotate in a complete circle, through an angle of 360°, the arc length cut off is the whole circumference of the circle. In a unit circle, since $c = 2\pi r$,

$$\text{circumference} = 2\pi \cdot 1 = 2\pi$$

Therefore, using arc lengths, one complete rotation is 2π radians. Since the same angle is also 360°,

$$2\pi \text{ radians} = 360 \text{ degrees}$$

Therefore,

$$1 \text{ radian} = \frac{180}{\pi}\text{degrees}$$

$$1 \text{ degree} = \frac{\pi}{180} \text{ radians}$$

The Size of a Radian

Since

$$1 \text{ radian} = \frac{180}{\pi}\text{degrees}$$

using $\pi \cong 3$, we see that

$$1 \text{ radian is about } 60°$$

The actual value is

$$1 \text{ radian} = 57°18'$$

Also,

$$1 \text{ complete revolution} = 2\pi \text{ radians} \cong 6 \text{ radians}$$

Conversion from degrees to radians can either be done using the conversion factor $\frac{\pi}{180}$ (the usual way) or by thinking of the angle as a fraction of one complete rotation of 2π radians (as in the demonstration that 2π radians $=$ 360°). Let's try this out by converting some of the most common angles to radians.

For example,

$$10° = 10 \cdot \frac{\pi}{180} = \frac{\pi}{18} \text{ radians} \cong 0.17 \text{ radians} \qquad (\text{using } \pi \cong 3.14)$$

In practice we very seldom bother to substitute for π, and so $\frac{\pi}{18}$ radians would be the usual equivalent to 10°. This means that almost any angle we measure in radians will have a π in it. This can make everything somewhat odd-looking, and it is frequently helpful to mentally replace π by 3 to see roughly how many radians are in a given angle. For example:

$$\frac{\pi}{4} \text{ radians } (=45°) \text{ is about } \frac{3}{4} \text{ radian}$$

$$\frac{2\pi}{3} \text{ radians } (=120°) \text{ is about } 2 \text{ radians}$$

EXAMPLE: *Convert 0°, 90°, 180°, 270°, and so on, to radians.*

Thinking of each angle as a fraction of one rotation:

$$0° = 0 \text{ radians}$$

$$90° = \frac{1}{4} \text{ rotation} = \frac{1}{4} \text{ circumference of circle} = \frac{1}{4}(2\pi) = \frac{\pi}{2} \text{ radians}$$

$$180° = \frac{1}{2} \text{ rotation} = \frac{1}{2}(2\pi) = \pi \text{ radians}$$

$$270° = \frac{3}{4} \text{ rotation} = \frac{3}{4}(2\pi) = \frac{3\pi}{2} \text{ radians}$$

$$360° = 2\pi \text{ radians}$$

$$450° = \frac{5\pi}{2} \text{ radians}$$

$$540° = 3\pi \text{ radians}$$

$$630° = \frac{7\pi}{2} \text{ radians}$$

$$720° = 4\pi \text{ radians and so on.}$$

See Figure 16.7.

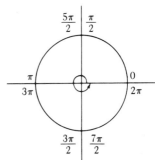

FIG. 16.7

Like angles which differ by 360°, angles which differ by multiples of 2π have their terminal sides at the same point, because 2π represents one complete rotation.

EXAMPLE: *Convert 30°, 45°, 60°, 120°, 135°, 145° to radians.*

Using the conversion factor:

$$30° = 30 \cdot \frac{\pi}{180} \text{ radians} = \frac{\pi}{6} \text{ radians}$$

$$45° = 45 \cdot \frac{\pi}{180} \text{ radians} = \frac{\pi}{4} \text{ radians}$$

$$60° = 60 \cdot \frac{\pi}{180} \text{ radians} = \frac{\pi}{3} \text{ radians}$$

Similarly,

$$120° = \frac{2\pi}{3} \text{ radians}$$

$$135° = \frac{3\pi}{4} \text{ radians}$$

$$150° = \frac{5\pi}{6} \text{ radians}$$

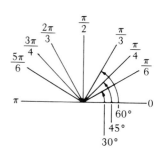

FIG. 16.8

See Figure 16.8.

EXAMPLE: *Convert $-\frac{\pi}{6}, \frac{5\pi}{4}, \frac{\pi}{9}$ radians to degrees.*

This time we are going from radians to degrees, so the conversion factor is $\frac{180}{\pi}$.

$$-\frac{\pi}{6} \text{ radians} = -\frac{\pi}{6} \cdot \frac{180°}{\pi} = -30°$$

$$\frac{5\pi}{4} \text{ radians} = \frac{5\pi}{4} \cdot \frac{180°}{\pi} = 225°$$

$$\frac{\pi}{9} \text{ radians} = \frac{\pi}{9} \cdot \frac{180°}{\pi} = 20°$$

See Figure 16.9.

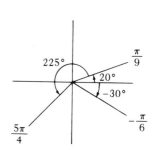

FIG. 16.9

EXAMPLE: *Convert π degrees to radians and 5 radians to degrees.*

$$\pi \text{ degrees} = \pi \cdot \frac{\pi}{180} \text{ radians} = \frac{\pi^2}{180} \text{ radians}$$

This is a good example of how π's can make things look odd. But since $\pi \cong 3$, this just says that $\pi°$ is about equivalent to

$$\frac{3^2}{180} = \frac{1}{20} \text{ radians}$$

Second,

$$5 \text{ radians} = 5 \cdot \frac{180°}{\pi} = \frac{900°}{\pi}$$

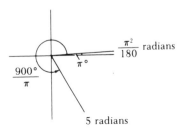

FIG. 16.10

Again this is odd-looking because we are not used to having π's in the degree measure, and certainly not in the denominator. But, using $\pi \cong 3$, this simply says that 5 radians is roughly equivalent to $\frac{900}{3} = 300°$, which corresponds to the fact that a complete revolution is about 6 radians. See Figure 16.10.

PROBLEM SET   16.3

Find the radian measure of each of the following angles given in degrees.

1. $0°$
2. $720°$
3. $-90°$
4. $180°$
5. $-10°$
6. $-360°$
7. $270°$
8. $30°$
9. $-45°$
10. $60°$
11. $135°$
12. $1440°$
13. $573°$
14. $-286.5°$
15. $-100°$
16. $340°$

Express the degree measure of each of the following angles given in radians.

17. 0
18. -2π
19. $\dfrac{3\pi}{2}$
20. $\dfrac{2\pi}{3}$
21. π
22. $-\dfrac{\pi}{12}$
23. $\dfrac{3\pi}{5}$
24. $-\dfrac{\pi}{6}$
25. $\dfrac{\pi}{4}$
26. 3
27. π^2
28. -7
29. $\dfrac{2}{\pi}$

Find all the possible values for the radian measure of an angle whose terminal side:

30. Lies on the positive x axis
31. Lies on the negative x axis
32. Lies on the positive y axis
33. Lies on the negative y axis
34. Is in the first quadrant, making an angle of $30°$ with the initial side

16.4   RADIANS AND
THE TRIGONOMETRIC FUNCTIONS

Everything that can be done with an angle measured in degrees can equally well be done with an angle measured in radians—everything, that is, except to look up its sine or cosine in the tables, which are generally made up only in degrees. But this is as it should be, because radians are important in the study of the trigonometric functions as functions, whereas when trigonometry is being used to solve triangles (which is where the tables usually come in) degrees are more appropriate.

It is important to be able to work as comfortably with radians as with degrees. Initially it is necessary to keep converting radians back to degrees so that you can use what you know about angles in degrees, but in the long run it is most definitely *not* a good idea to solve every problem in degrees and then convert the answer to radians, both because this wastes time and because it tends to reinforce the feeling that radians are not the "natural" way of measuring angles. In fact, from the calculus point of view, radians are more "natural" than degrees.

When we are working in radian measure we will mostly be dealing with the "special angles" (multiples of 30°, 45°, 60°, and 90°). We can find the trigonometric functions of these angles without tables. If we do have an angle which is not one of the special ones, we will first have to convert it into degrees and then look it up in the tables.

EXAMPLE: *Find the sine, cosine, and tangent of* $0, \frac{\pi}{2}, \pi, \frac{3\pi}{2}$.

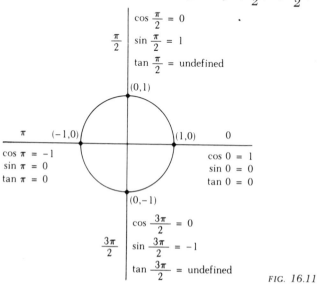

FIG. 16.11

Thinking of the angles in standard position, and using the circle definitions, we get the values in Figure 16.11.

EXAMPLE: *Find the sine, cosine, and tangent of* $\dfrac{\pi}{6}, \dfrac{\pi}{4}, \dfrac{\pi}{3}$.

First, $\dfrac{\pi}{6} = 30°$, and so the angle fits into the right triangle shown in Figure 16.12, with $\dfrac{\pi}{3} = 60°$ as the complementary angle. Therefore,

$$\cos \frac{\pi}{6} = \frac{\sqrt{3}}{2} \qquad \cos \frac{\pi}{3} = \frac{1}{2}$$

$$\sin \frac{\pi}{6} = \frac{1}{2} \qquad \sin \frac{\pi}{3} = \frac{\sqrt{3}}{2}$$

$$\tan \frac{\pi}{6} = \frac{1}{\sqrt{3}} \qquad \tan \frac{\pi}{3} = \frac{\sqrt{3}}{1} = \sqrt{3} \qquad \text{FIG. 16.12}$$

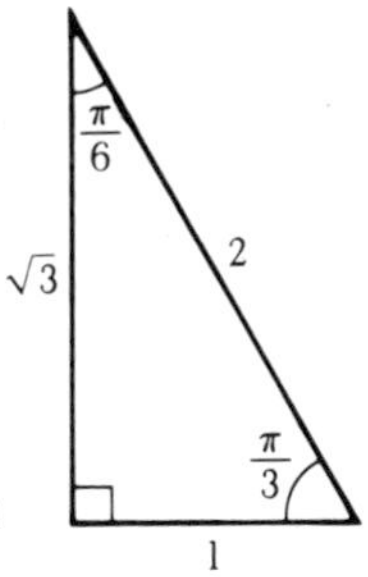

Now, $\dfrac{\pi}{4} = 45°$, which fits into the triangle shown in Figure 16.13. Thus,

$$\cos \frac{\pi}{4} = \sin \frac{\pi}{4} = \frac{1}{\sqrt{2}} \qquad \tan \frac{\pi}{4} = 1$$

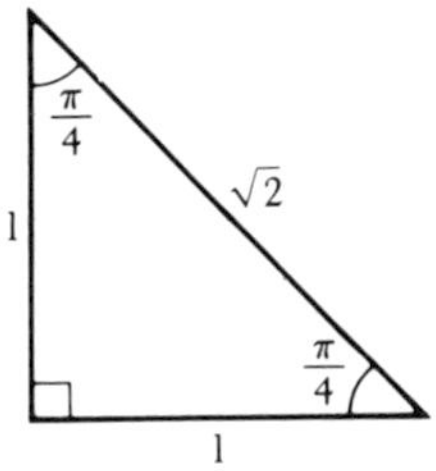

FIG. 16.13

EXAMPLE: *Find* $\cos(7\pi)$, $\cot\left(\dfrac{3\pi}{4}\right)$, $\sin\left(-\dfrac{\pi}{4}\right)$, $\sin\left(\dfrac{7\pi}{6}\right)$.

For $\cos(7\pi)$, notice that even multiples of π lie along the positive x axis and odd multiples along the negative x axis. (see Figure 16.14). Therefore $\cos 7\pi = \cos \pi = -1$.

π, 3π, 5π, 7π, etc.————————————0, 2π, 4π, 6π, etc.

FIG. 16.14

For $\cot\left(\dfrac{3\pi}{4}\right)$, drawing $\dfrac{3\pi}{4}$ in standard position shows that it has a related angle of $\dfrac{\pi}{4}$ (see Figure 16.15).

Therefore,

$$\cot\left(\frac{3\pi}{4}\right) = \pm \cot \frac{\pi}{4} = \pm 1$$

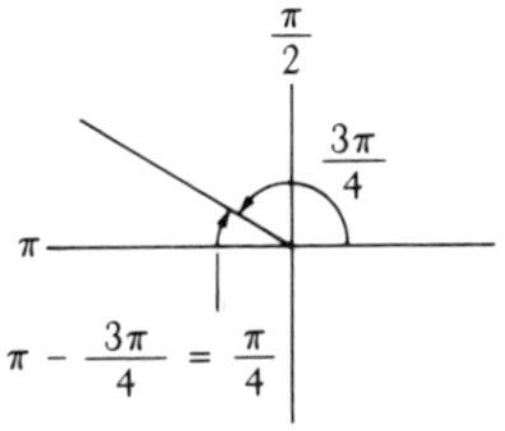

$$\pi - \frac{3\pi}{4} = \frac{\pi}{4}$$

FIG. 16.15

and since $\dfrac{3\pi}{4}$ is in the second quadrant where the cotangent is negative,

$$\cot\left(\frac{3\pi}{4}\right) = -1$$

$$\sin\left(-\frac{\pi}{4}\right) = -\sin\frac{\pi}{4} \quad \text{(because related angle is } \frac{\pi}{4}\text{, and sine is}$$
$$\text{negative in the fourth quadrant)}$$
$$= -\frac{1}{\sqrt{2}} \quad \text{(from the example above)}$$

Finally, for $\sin\left(\dfrac{7\pi}{6}\right)$, look at Figure 16.16 to see that $\dfrac{7\pi}{6}$ has a related angle of $\dfrac{\pi}{6}$. In the third quadrant the sine is negative, so

$$\sin\left(\frac{7\pi}{6}\right) = -\sin\left(\frac{\pi}{6}\right) = -\frac{1}{2}$$

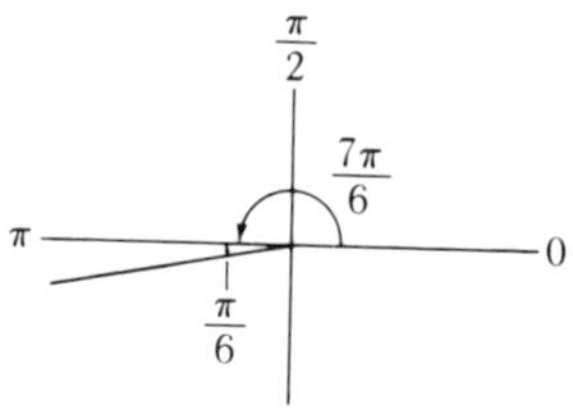

EXAMPLE: *Find* $\cos\left(\dfrac{2\pi}{9}\right)$.

FIG. 16.16

Since $\dfrac{2\pi}{9} = 40°$, which is not a "special angle", we will have to use the tables:

$$\cos\left(\frac{2\pi}{9}\right) = \cos 40° = 0.7660$$

PROBLEM SET 16.4

Draw a diagram showing the following angles in standard position, and find the values of the sine, cosine, and tangent functions of the angle.

1. $\dfrac{\pi}{2}$

2. $\dfrac{3\pi}{2}$

3. 4π

4. $\dfrac{\pi}{6}$

5. $-\dfrac{\pi}{6}$

6. $\dfrac{\pi}{3}$

7. $\dfrac{4\pi}{3}$

8. $-\dfrac{2\pi}{3}$

9. $\dfrac{13\pi}{6}$

10. $-\pi$

11. $-\dfrac{7\pi}{3}$

12. $\dfrac{23\pi}{3}$

13. $-\dfrac{112\pi}{6}$

14. $\dfrac{4\pi}{9}$

15. $-\dfrac{11\pi}{9}$

16. $-\dfrac{19\pi}{9}$

17. $\dfrac{400\pi}{6}$

18. $\dfrac{\pi}{18}$

19. $-\dfrac{2\pi}{15}$

20. $-\dfrac{3\pi}{108}$

16.5   ARC LENGTH

The definition of a radian connects the size of an angle at the center of a unit circle with the arc length it cuts off. Using this definition it is possible to derive an equation connecting the arc length, s, and the angle, θ (measured in radians), in a circle of any radius. Suppose the circle has radius r. Figure 16.17 shows both the circle of radius r, and a concentric circle of radius 1.

First, θ is in radians, so this angle cuts off an arc of length θ on the unit circle, so

$$\text{arc length } AB = \theta$$

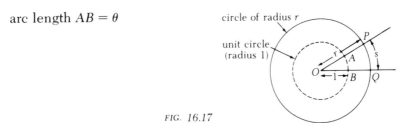

FIG. 16.17

Second, you can see that the two pie-shaped pieces OAB and OPQ are similar—they have exactly the same shape although they are different sizes. This means that ratios of corresponding lengths in the two figures are equal. In other words:

$$\frac{\text{arc length } PQ}{\text{arc length } AB} = \frac{OP}{OA}$$

From Figure 16.17, $OP = r$, $OA = 1$, and the arc length $PQ = s$. Also, arc length $AB = \theta$, so the equation tells us:

$$\frac{s}{\theta} = \frac{r}{1}$$

or

$$s = r\theta$$

arc length = radius · angle subtended

Note: This formula works only when the angle subtended is in radians.

EXAMPLE: *If a pendulum of length 2 feet swings through 5° on either side of the vertical, how long is the arc through which the bob moves from one high point to the other?*

As the bob in Figure 16.18 moves from one side to the other, the string moves through an angle of 10°. We are therefore looking for the arc length of a circle of radius 2 feet cut off by an angle of 10°. First we convert 10° to radians,

$$10° = 10 \cdot \frac{\pi}{180} = \frac{\pi}{18} \text{ radians}$$

which tells us that

arc length = radius · angle subtended

$$= 2 \cdot \frac{\pi}{18} = \frac{\pi}{9} \text{ feet}$$

$$= \frac{\pi}{9} \text{ feet}$$

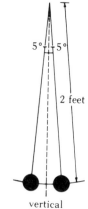

FIG. 16.18

PROBLEM SET   16.5

1. In a circle with a 90-centimeter (cm) radius, an arc 108 cm long subtends an angle of how many radians? How many degrees (to the nearest degree)?

2. In a circle with a 110 cm radius, an arc 99 cm long subtends an angle of how many radians? How many degrees?

3. In a circle with a 10-meter (m) radius, how long is the arc associated with an angle of 0.8 radians?

4. In a circle with a 200 cm radius, how long is the arc associated with an angle of 2.1 radians?

5. In a circle with a 12-millimeter (mm) radius, how long is the arc associated with an angle of 12 radians?

6. In a circle with an 80 cm radius, an arc 64 cm long subtends an angle of how many radians? How many degrees?

7. In a circle with an 11 cm radius, an arc 165 cm long subtends an angle of how many radians? How many degrees?

Through how many radians does the minute hand of a clock rotate in the following time periods?

8. 4 minutes

9. 15 minutes

10. 50 minutes

11. 1 hour 10 minutes

12. 16 minutes

13. 28 minutes

14. 35 minutes

15. 40 minutes

16. 2 hours

17. 30 seconds

18. 7 minutes

In Problems 19–28, take $\pi = \frac{22}{7}$.

19. How far does the tip of the big hand of a clock move in 35 minutes if the hand is 6 inches long?

20. A flywheel of diameter 12.0 feet makes 500 revolutions in an hour.
    (a) How far does a point on the perimeter travel in 1 hour?
    (b) If a bug clinging to the flywheel travels 1 mile in 1 hour, how far is the bug from the center of the wheel?

21. A wheel on a car has a 28-inch radius. Through what angle (in radians) does the wheel turn while the car travels 1 mile?

22. How far does the tip of the little hand of a clock move in 3 hours 10 minutes if the hand is 8 inches long?

23. A curve on a highway subtends an angle of 36° on a circle of radius 1800 feet. How long will it take a car traveling 50 miles per hour to round the curve?

24. A bucket is drawn from a well by pulling a rope over a pulley. Find the radius of the pully if the bucket is raised 126.8 inches while the pulley is turned through 6.94 revolutions.

25. How far does the tip of the big hand of a clock move in 1 hour and 27 minutes if the hand is 2 inches long?

26. Through what angle does a piece of reflective tape on the rim of a bicycle wheel 14 inches in radius turn on a 10-mile bike trip?

27. The end of a 40-inch pendulum describes an arc of 5 inches. Through what angle (in radians) does the pendulum swing?

28. A railroad curve is to be laid out in a circle. What radius should be used if the track is to change direction by 28° in a distance of 140 feet?

CHAPTER 16 REVIEW

(a) Express the following angles in radian measure; (b) find the sine of the angle; and (c) find the tangent of the angle.

1. 90°	5. 225°	9. −240°
2. −270°	6. −300°	10. 540°
3. 135°	7. 30°	
4. 210°	8. 15°	

(a) Express the following angles in degree measures; (b) find the cosine of the angle; (c) find the tangent of the angle.

11. π	15. -3π	19. $-\dfrac{3\pi}{2}$
12. $\dfrac{2\pi}{3}$	16. $\dfrac{5\pi}{2}$	
13. $-\dfrac{\pi}{4}$	17. $-\dfrac{\pi}{3}$	20. $\dfrac{13\pi}{6}$
14. $\dfrac{7\pi}{6}$	18. 21π	

21. In one of the new domed stadia, the distance from home plate to any place along the outfield wall is exactly the same. What is the length of the wall if it subtends an angle of 90° and the distance from home plate is 375 feet?

22. Assuming the earth to be a sphere of radius 3960 miles, find the distance along the surface of the earth of a point at latitude 39°36′N from the equator.

23. Two cities 132 miles apart lie on the same meridian (i.e., longitude line). Find their difference in latitude (in degrees). (Radius = 3960 miles.)

24. The minute hand of a clock is 14 inches long. How far does the tip of the hand move during 40 minutes?

25. The end of a 12-foot pendulum describes an arc of 27 inches. Through what angle does the pendulum swing?

17 THE TRIGONOMETRIC FUNCTIONS AS FUNCTIONS

17.1 GRAPHS OF THE TRIGONOMETRIC FUNCTIONS

So far we have looked at the trigonometric functions mainly as an aid in solving triangles; however, their real importance is as functions. The sine and the cosine are the most useful, but all trig functions have one characteristic—that of periodicity, or continual repetition—that we have seen nowhere else. There are a great many quantities, such as the waves on the sea, an AC electric current, and the output from a heart patient's EKG machine, which are continually repeating and so can only be represented by a periodic function. Thus, besides their usefulness in the geometry and solution of triangles, the trigonometric functions are of the greatest importance for their properties as functions.

The best way to get an idea of how a function behaves is to draw its graph. We will now do this for each of the trigonometric functions in turn.

The Sine Function: $y = \sin x$

First, please note that x and y are here being used in a different way than in the circle definitions of the trig functions. There x and y were the coordinates of the point P and the angle was called θ; here the angle is called x, and y is the value of the trig function in question. The reason for this change is that we want to make it abundantly clear that we are now thinking of the sine as a function, in which the value of the sine, or y (the traditional name for the dependent variable) depends on the angle, or x (the traditional name for the independent variable).

A good way to draw a graph of a function is to make a table of values as in Table 17.1. To do this remember that *x* is an angle, and use the tables, rounding off values to two decimal places. Since we want to use the tables we must have *x* in degrees.

Table 17.1

Negative angles in the fourth quadrant have the same terminal sides, and therefore the same sines, as angles between 270° and 360°.

Beyond 90° use the circle and related angles.

x	−90°	−60°	−30°	0°	30°	60°	90°	120°	150°	180°
Related Angle	—	60°	30°					60°	30°	—
y = sin *x*	−1.00	−0.87	−0.50	0.00	0.50	0.87	1.00	0.87	0.50	0.00

x	210°	240°	270°	300°	330°	360°	390°	420°	450°	⋯
Related Angle	30°	60°	—	60°	30°	—	30°	60°		
y = sin *x*	−0.50	−0.87	−1.00	−0.87	−0.50	0.00	0.50	0.87	1.00	⋯

Same set of related angles as between 90° and 180°; however, we are now in the third or fourth quadrant, so the sine is negative.

Beyond 360° the values of the sine start to repeat. For example, 390° has the same terminal side as 30° and so must have the same sine.

Plotting this massive number of points gives us the graph in Figure 17.1. This graph repeats indefinitely because the terminal side of the angle *x* runs through the same positions over and over again. There it is, with the *x* axis marked in radians as well as degrees.

From Figure 17.1 you can see that if *x* starts at 0, it must increase to 360° before the function starts to repeat itself. In other words, the function takes 360° or 2π to execute one complete cycle. 360°, or 2π, is called the *period* of the sine function.

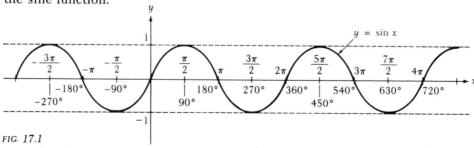

FIG. 17.1

The Cosine Function: y = cos x

We start by constructing Table 17.2, remembering that x is an angle (the independent variable) and y is the value of the cosine (the dependent variable). This time, however, we put both degrees and radians in the table.

Table 17.2

x (radians)	$-\dfrac{\pi}{3}$	$-\dfrac{\pi}{6}$	0	$\dfrac{\pi}{6}$	$\dfrac{\pi}{3}$	$\dfrac{\pi}{2}$	$\dfrac{2\pi}{3}$	$\dfrac{5\pi}{6}$	π
x (degrees)	$-60°$	$-30°$	$0°$	$30°$	$60°$	$90°$	$120°$	$150°$	$180°$
Related Angle	$60°$	$30°$					$60°$	$30°$	$-$
$y = \cos x$	0.50	0.87	1	0.87	0.50	0	-0.50	-0.87	-1

x (radians)	$\dfrac{7\pi}{6}$	$\dfrac{4\pi}{3}$	$\dfrac{3\pi}{2}$	$\dfrac{5\pi}{3}$	$\dfrac{11\pi}{6}$	2π	$\dfrac{13\pi}{6}$	$\dfrac{7\pi}{3}$	$\dfrac{5\pi}{2}$	
x (degrees)	$210°$	$240°$	$270°$	$300°$	$330°$	$360°$	$390°$	$420°$	$450°$	$\cdots$
Related Angle	$30°$	$60°$	$-$	$60°$	$30°$	$-$	$30°$	$60°$	$-$	
$y = \cos x$	-0.87	-0.50	0	0.50	0.87	1	0.87	0.50	0	$\cdots$

The cosine function, just like the sine function, starts to repeat as x increases beyond 360° or decreases below 0° because the terminal side of the angle runs through positions that it has taken before. If we continue the graph for many more values of x it will repeat indefinitely. The graph is shown in Figure 17.2.

So, like the sine function, the cosine function takes 360°, or 2π, to execute a full cycle and therefore has a period of 360° or 2π.

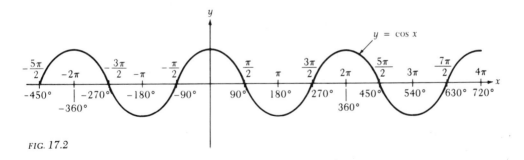

FIG. 17.2

The Tangent Function: $y = \tan x$

The table of values is given in Table 17.3. Being undefined at $\pm 90°$, $\pm 270°$ means that the tangent function has vertical asymptotes at $x = 90°$, $x = -90°$, $x = 270°$, and so on. Looking at the tables shows that $\tan x$ increases very fast as x increases towards $90°$. The graph is shown in Figure 17.3.

Table 17.3

x (radians)	$-\dfrac{\pi}{2}$	$-\dfrac{\pi}{3}$	$-\dfrac{\pi}{6}$	0	$\dfrac{\pi}{6}$	$\dfrac{\pi}{3}$	$\dfrac{\pi}{2}$	$\dfrac{2\pi}{3}$	$\dfrac{5\pi}{6}$	π
x (degrees)	$-90°$	$60°$	$-30°$	$0°$	$30°$	$60°$	$90°$	$120°$	$150°$	$180°$
Related Angle	—	$60°$	$30°$					$60°$	$30°$	—
$y = \tan x$	undef.	-1.73	-0.58	0	0.58	1.73	undef.	-1.73	-0.58	0

x (radians)	$\dfrac{7\pi}{6}$	$\dfrac{4\pi}{3}$	$\dfrac{3\pi}{2}$	$\dfrac{5\pi}{3}$	$\dfrac{11\pi}{6}$	2π	$\dfrac{13\pi}{6}$	$\dfrac{7\pi}{3}$	$\dfrac{5\pi}{2}$	
x (degrees)	$210°$	$240°$	$270°$	$300°$	$330°$	$360°$	$390°$	$420°$	$450°$	$\cdots$
Related Angle	$30°$	$60°$	—	$60°$	$30°$	—	$30°$	$60°$	—	
$y = \tan x$	0.58	1.73	undef.	-1.73	-0.58	0	0.58	1.73	undef.	$\cdots$

The tangent function is unlike the sine and cosine (and any other function we know) because it has infinitely many vertical asymptotes and so is in infinitely many pieces. Like the sine and cosine, however, the tangent function is constantly repeating, but since it takes only $180°$, or π, to carry out one cycle, its period is $180°$ or π.

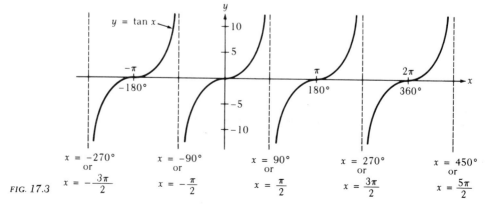

FIG. 17.3

The Cotangent Function: $y = \cot x$

There are two ways of approaching this. We could graph $y = \cot x$ just the same way that we did the previous three functions—by drawing up a table of values and plotting points. Alternatively, we could use the fact (introduced in Section 14.2) that

$$\cot x = \frac{\cos x}{\sin x} \quad \text{and} \quad \tan x = \frac{\sin x}{\cos x}$$

to conclude that

$$\cot x = \frac{1}{\tan x}$$

This means that the y values in $y = \cot x$ are the reciprocals of the corresponding y values for $y = \tan x$. Therefore we should be able to make a rough sketch of the graph of $y = \cot x$ from the graph of $y = \tan x$ using the methods of Section 3.7.

Specifically, we know that:
1. $y = \cot x$ and $y = \tan x$ have the same sign everywhere.
2. When $y = \tan x$ is small, $y = \cot x$ is large, and vice versa.
3. When $y = \tan x$ is zero, $y = \cot x$ is undefined, and vice versa.

Using this method, or a table of values, you should be able to get a graph of $y = \cot x$ that looks like Figure 17.4. Again, the period is 180° or π.

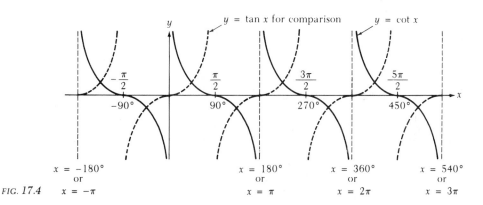

FIG. 17.4

The Secant Function: $y = \sec x$

In Section 14.2 we saw that

$$\sec x = \frac{1}{\cos x}$$

This means that for a given x value, the y value of $y = \sec x$ is the reciprocal of the y value of $y = \cos x$.

The graph of $y = \sec x$ can be derived from that of $y = \cos x$ in the same way that the cotangent comes from the tangent, or it can be done from scratch using a table of values.

The graph has vertical asymptotes at $x = \pm 90°$, $\pm 270°$, and so on, because these are the points at which the cosine is zero. Near these values the graph shoots vertically upwards or downwards—you can find out which by looking at the sign of the cosine. Altogether you should get a graph that looks like Figure 17.5.

Like the cosine function, the secant has a period of 360° or 2π.

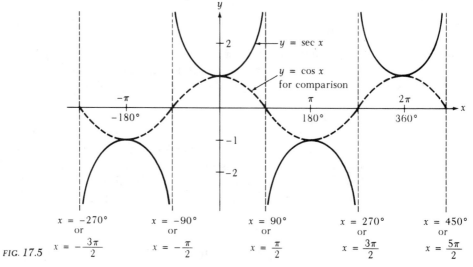

FIG. 17.5

The Cosecant Function: $y = \csc x$

The cosecant is the reciprocal of the sine, so the graph of $y = \csc x$ can be derived from that of $y = \sin x$ in the same way that the secant graph comes from the cosine. The graph is shown in Figure 17.6. The cosecant has a period of 360° or 2π.

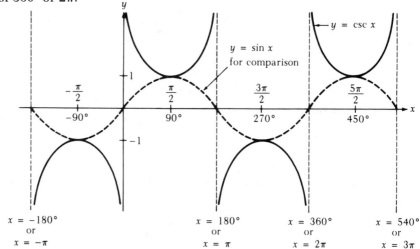

FIG. 17.6

PROBLEM SET   17.1

1. Draw on the same axes graphs of $y = \sin x$ and $y = \cos x$ for x between $-\pi$ and 2π [that is, $-\pi < x < 2\pi$]. Use the graph to find the values of x satisfying $\cos x = \sin x$.

2. On the same axes, draw graphs of $y = \tan x$ and $y = \cot x$ for $-\pi < x < 2\pi$. What values of x satisfy $\tan x = \cot x$?

3. Draw a graph of $y = \sec x$ for $-\pi < x < 2\pi$.
4. Draw a graph of $y = \csc x$ for $-\pi < x < 2\pi$.

Use graphs to answer:

5. If $\sin x = 1$ and $0 \leq x < 2\pi$, find x.
  Is there a value of x for which $\sin x > 1$?

6. If $\sin x = -1$ and $0 \leq x < 2\pi$, find x.
  Is there a value of x for which $\sin x < -1$?

7. If $\cos x = 1$, and $0 \leq x < 2\pi$, find x.
  Is there a value of x for which $\cos x > 1$?

8. If $\cos x = -1$ and $0 \leq x < 2\pi$, find x.
  Is there a value of x for which $\cos x < -1$?

9. If $\cos x = 0$, and $0 \leq x < 2\pi$, find x.
10. If $\sin x = 0$, and $0 \leq x < 2\pi$, find x.
11. Find all the zeros of $F(x) = \cos x$ (that is, find all the values of x for which $\cos x = 0$).

12. Find all the zeros of $g(x) = \sin x$.
13. Find all the zeros of $f(x) = \tan x$.

For what angles between 0 and 2π is

14. $\sin x$ positive? increasing? positive *and* increasing?
15. $\cos x$ positive? decreasing? positive *and* decreasing?
16. $\tan x$ decreasing?

17.2   DOMAINS AND RANGES OF TRIGONOMETRIC FUNCTIONS

The Sine and Cosine Functions

If we consider

$$y = f(x) = \sin x$$

we should be able to determine its domain and range. This means looking at y or $\sin x$ as a number whose value depends on the angle x. The *domain* is all the possible values for the angle, or all the x values that lie on the same vertical line as a point on the graph. The *range* is all the possible values for the sine, or all the y values that lie on the same horizontal line as a point on the graph.

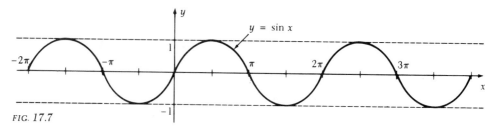

FIG. 17.7

From Figure 17.7 (where *x* is in radians), you can see that the *domain of the sine function* is all real numbers (if *x* is in radians) or all angles (if *x* is in degrees). This is because a vertical line through any *x* value hits a point on the graph, which is the same as saying that you can find the sine of any angle.

The *range of the sine function* is all real numbers between −1 and 1 (inclusive), that is,

range is: $-1 \leqslant y \leqslant 1$

The easiest way to see this is to look at the graph and see that there are no *y* values on the graph above 1 or below −1; a horizontal line above *y* = 1 or below *y* = −1 does not cut the graph. You can also see this directly from the circle definition of the sine (see Section 14.2).

Similarly, the *domain of the cosine function* is all real numbers (or all angles). The *range of the cosine function* is all real numbers between −1 and 1 (inclusive), that is,

range is: $-1 \leqslant y \leqslant 1$

The Tangent and Cotangent Functions

In Figure 17.8, the vertical asymptotes (dashed lines) at $x = \pm\dfrac{\pi}{2}, \pm\dfrac{3\pi}{2}$, and so on, show that there is no *y* value for these *x* values. Therefore the tangent is not defined for $x = \pm\dfrac{\pi}{2}$, $x = \pm\dfrac{3\pi}{2}$, and so on. For every other angle, the tangent is defined, as can be seen from the fact that there is a point on the graph above every other *x* value. Therefore, the *domain of the tangent function* is all real numbers

except $x = \dfrac{\pi}{2} + n\pi$ (*n* an integer)

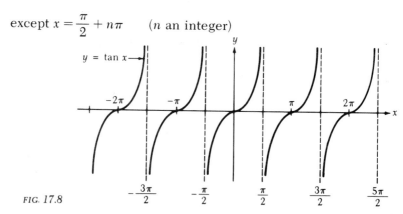

FIG. 17.8

The *range of the tangent function* is all real numbers. Any number can come up as a tangent, as is shown by the fact that any horizontal line cuts the graph.

By exactly the same arguments as those used for the tangent function, the *domain of the cotangent function* is all real numbers

except $x = 0, \pm\pi$, and so on

that is

except $x = n\pi$ (*n* an integer)

The *range of the cotangent function* is all real numbers.

The Secant and Cosecant Functions

Figure 17.9 shows that the secant is not defined for $x = \pm\frac{\pi}{2}, \pm\frac{3\pi}{2}$, and so on, and therefore, the *domain of the secant function* is all real numbers

except $x = \pm\frac{\pi}{2}, \pm\frac{3\pi}{2}$, and so on

that is

except $x = \frac{\pi}{2} + n\pi$ (*n* an integer)

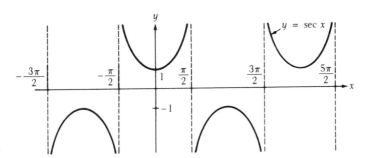

FIG. 17.9

Notice that this is all real numbers except where the cosine is zero, which is as we should expect from the fact that the secant is the reciprocal of the cosine.

Looking at the graph, it is clear that there are no y values between -1 and 1. (There are y values of -1 and 1 on the graph, but none strictly between them). This means that numbers between -1 and 1, such as $\frac{1}{2}$ or 0, are not the secant of any angle. Therefore, the *range of the secant function* is all real numbers greater than 1 (inclusive) or less than -1 (inclusive). In other words, if y is in the range, y is either greater than or equal to 1, or less than or equal to -1, that is,

range is: $y \geq 1$ or $y \leq -1$

These values are exactly the reciprocals of the y values of the cosine function, as we should expect from the fact that

$$\sec x = \frac{1}{\cos x}$$

Similarly, the *domain of the cosecant function* is all real numbers except $x = 0, \pm\pi$, and so on, that is except

$$x = n\pi \quad (n \text{ an integer})$$

Notice that these are all the angles except those for which the sine is zero, which is as you would expect from the fact that the cosecant is the reciprocal of the sine.

The *range of the cosecant function* is all real numbers either greater than or equal to 1 or less than or equal to -1, that is,

$$\text{range is: } y \geqslant 1 \quad \text{or} \quad y \leqslant -1$$

These y values are exactly the reciprocals of all the possible values of the sine function, again as you would expect from the fact that

$$\csc x = \frac{1}{\sin x}$$

17.3   SYMMETRY: ODD AND EVEN TRIGONOMETRIC FUNCTIONS

You may remember that an *even function* is one for which

$$f(-x) = f(x)$$

and whose graph is symmetric about the y axis, as shown in Figure 17.10.

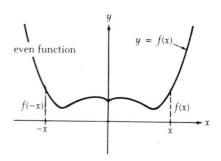

FIG. 17.10

An *odd function* is one for which

$$f(-x) = -f(x)$$

and whose graph is symmetric about the origin, as shown in Figure 17.11.

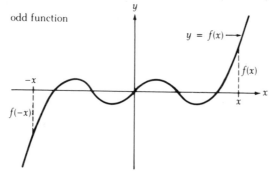

FIG. 17.11

The Sine and Cosine

Looking at the graphs of sine and cosine in Figure 17.12, it should be clear that *the sine function is symmetric about the origin or odd*, $\sin(-x) = -\sin x$; and that *the cosine function is symmetric about the y axis or even*, $\cos(-x) = \cos x$.

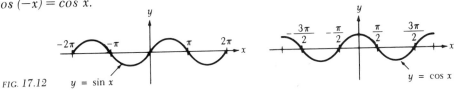

FIG. 17.12 $y = \sin x$

Proof That the Cosine is Even and the Sine Is Odd   It should be possible to derive the fact that the cosine is even directly from the definition, as well as by looking at the graph.

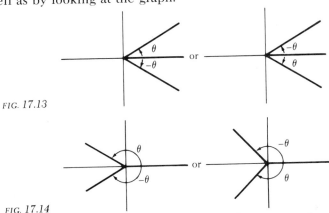

FIG. 17.13

FIG. 17.14

Going back to the circle, think about how an angle and its negative are related to one another. If θ is in the first or fourth quadrant, $-\theta$ is in the other, as shown in Figure 17.13. Similarly, if θ is in the second or third quadrant, $-\theta$ is in the other, as shown in Figure 17.14. By looking at these two figures you can see that θ and $-\theta$ always have the same related angle, and they are always on the *same* side of the y axis but on *opposite* sides of the x axis. This means that if P and Q are the points where the unit circle cuts the terminal sides of θ and $-\theta$, respectively, then P and Q have the same x coordinates, but opposite y coordinates. Now,

$$x \text{ coordinate of } P = \cos \theta$$

$$y \text{ coordinate of } P = \sin \theta$$

and

$$x \text{ coordinate of } Q = \cos (-\theta)$$

$$y \text{ coordinate of } Q = \sin (-\theta)$$

Therefore $\cos(-\theta) = \cos \theta$ and $\sin(-\theta) = -\sin \theta$, so the cosine is even and the sine is odd.

The Tangent, Cotangent, Secant, and Cosecant

Looking at the graphs in Figure 17.15 it should be clear that:

the tangent function is symmetric about the origin, or odd; that is

$$\tan(-x) = -\tan x$$

the cotangent function is symmetric about the origin, or odd; that is

$$\cot(-x) = -\cot x$$

the secant function is symmetric about the y axis, or even; that is

$$\sec(-x) = \sec x$$

the cosecant function is symmetric about the origin, or odd; that is

$$\csc(-x) = -\csc x$$

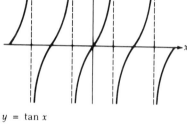

$y = \tan x$

$y = \cot x$

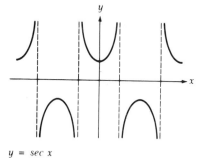

$y = \sec x$

$y = \csc x$

FIG. 17.15

Justification   If we assume that sine is odd and cosine even, we can derive the corresponding results for the other trig functions. Remember that

$$\tan \theta = \frac{\sin \theta}{\cos \theta} \qquad \text{(from Section 14.2)}$$

This relation is true for any angle, in particular $-\theta$, so

$$\tan(-\theta) = \frac{\sin(-\theta)}{\cos(-\theta)}$$

But sin $(-\theta) = -\sin\theta$ and cos $(-\theta) = \cos\theta$, so

$$\tan(-\theta) = \frac{(-\sin\theta)}{\cos\theta} = -\frac{\sin\theta}{\cos\theta} = -\tan\theta$$

So the tangent function is odd.

Similarly, the circle definitions show that

$$\sec\theta = \frac{1}{\cos\theta}$$

so

$$\sec(-\theta) = \frac{1}{\cos(-\theta)} = \frac{1}{\cos\theta} = \sec\theta$$

and the secant function is even.

The cotangent and the cosecant functions can be shown to be odd the same way.

17.4   COMPOSITION INVOLVING THE TRIGONOMETRIC FUNCTIONS

The trigonometric functions can be composed with each other or with any other functions in just the same way that algebraic functions can.

For example, if $f(x) = x + \pi$ and $g(x) = \cos x$, then

$$g(f(x)) = \cos(x + \pi)$$

which means that you add π to x and then take the cosine.

For instance, if $x = \pi$, then

$$g(f(\pi)) = \cos(\pi + \pi) = \cos 2\pi = 1$$

Here $g(f(x))$ represents the function given by applying first f and then g:

$$x \xrightarrow{\quad f \quad} x + \pi \xrightarrow{\quad g \quad} \cos(x + \pi)$$

So f is the inside function and g is the outside function.

In contrast,

$$f(g(x)) = (\cos x) + \pi$$

means that you take the cosine of x and add π to that. In this case if $x = \pi$,

$$f(g(x)) = (\cos\pi) + \pi = -1 + \pi = \pi - 1$$

Here $f(g(x))$ represents the function given by applying first g and then f:

$$x \xrightarrow{\quad g \quad} \cos x \xrightarrow{\quad f \quad} (\cos x) + \pi$$

So g is the inside function and f is the outside function.

It is important to keep clear the distinction between the composition and the product of two functions. The two possible compositions of f and g are those given above; the product of f and g is

$$f(x)g(x) = (x + \pi) \cos x$$

Now if $x = \pi$,

$$f(\pi)g(\pi) = (\pi + \pi) \cos \pi = 2\pi(-1) = -2\pi$$

It is also perfectly possible to compose two trigonometric functions. For example,

$$h(x) = \sin(\cos x)$$

is the result of composing

$$f(x) = \cos x \quad \text{and} \quad g(x) = \sin x$$

with f as the inside function and g as the outside function. The function

$$k(x) = \cos(\sin x)$$

is the same two functions composed in the opposite order.

Then, of course, we could compose the trigonometric functions with the logarithmic or exponential functions. For example, if

$$f(x) = \cos x \qquad g(x) = \log x \qquad h(x) = 10^x$$

then

$$g(f(x)) = \log \cos x \qquad h(f(x)) = 10^{\cos x}$$

EXAMPLE: *Express $k(x) = \cos \tan(x^2 - 7x)$ as a composite function (i.e., as the composition of two or more functions).*

Let

$$f(x) = x^2 - 7x$$

$$g(x) = \tan x$$

$$h(x) = \cos x$$

Then

$$h(g(f(x))) = h(g(x^2 - 7x)) = h(\tan(x^2 - 7x)) = \cos \tan(x^2 - 7x)$$

Special Notation

Composing the cosine function and the square function with the cosine as the inside function leads to

$$f(x) = (\cos x)^2$$

This is usually written as

$$f(x) = \cos^2 x$$

Composing the cosine and the square functions with the square function inside leads to

$$g(x) = \cos(x^2)$$

This is usually written without the parentheses as

$$g(x) = \cos x^2$$

Therefore,

$\cos^2 x$ means find the cosine of x and then square it,

and

$\cos x^2$ means square x and then find its cosine.

Similarly,

$\cos^3 x$ means $(\cos x)^3$, or find the cosine of x and cube it,

and

$\cos x^3$ means $\cos(x^3)$, or cube x and then find its cosine.

This can be extended to other exponents, so that

$$\cos^{1/2} x \quad \text{means} \quad (\cos x)^{1/2} = \sqrt{\cos x}$$

or the square root of the cosine of x, and

$$\cos x^{1/2} \quad \text{means} \quad \cos(x^{1/2}) = \cos(\sqrt{x})$$

or the cosine of the square root of x.

There is one case, however, in which this notation runs into real confusion. Consider $\cos^{-1} x$. You would probably expect that $\cos^{-1} x$ would mean

$$(\cos x)^{-1} \quad \text{or} \quad \frac{1}{\cos x}$$

But remember that $f^{-1}(x)$ was defined to mean the inverse function of $f(x)$, so it is perhaps not surprising that $\cos^{-1} x$ has been reserved to mean the inverse function of the cosine (the inverses of the trigonometric functions will be introduced properly in Chapter 19). The inverse function and the reciprocal are very different. The inverse cosine means that function which "undoes" what the cosine does. Instead of taking angles and giving us a number (their cosine), $\cos^{-1} x$ takes a number that is the cosine of some angle and gives us that angle back. For example, $\cos^{-1} 0$ means an angle that gives you a cosine of 0, say, 90°, so

$$\cos^{-1} 0 = 90°$$

On the other hand,

$$(\cos 0)^{-1} = \frac{1}{\cos 0} = \frac{1}{1} = 1$$

Therefore,

$\cos^{-1} x$ means the inverse function of the cosine

$(\cos x)^{-1}$ means the reciprocal of the cosine, or $\dfrac{1}{\cos x}$

$\cos x^{-1}$, in contrast to both of them, means the cosine of the reciprocal, or $\cos(x^{-1}) = \cos\left(\dfrac{1}{x}\right)$

In summary, $\cos^n x = (\cos x)^n$, except when $n = -1$, and $\cos x^n = \cos(x^n)$.

PROBLEM SET   17.4

Evaluate the following:

1. $\log_2\left(\sin \dfrac{\pi}{4}\right)$ 4. $\log_2\left(\sec \dfrac{\pi}{4}\right)$ 7. $\log_4\left(\sec \dfrac{\pi}{3}\right)$

2. $\log_2\left(\tan \dfrac{\pi}{4}\right)$ 5. $\log_2\left(\sin \dfrac{\pi}{6}\right)$ 8. $\log_4\left(\cot \dfrac{\pi}{2}\right)$

                                                      9. $\log_{10}(\cos 0)$

3. $\log_4(\csc 45°)$ 6. $\log_2(\csc 30°)$ 10. $\log_2(\tan 135°)$

Compute $f\left(g\left(\dfrac{\pi}{2}\right)\right)$ if:

11. $f(x) = \cos x;\ g(x) = \cos x$
12. $f(x) = \sin x;\ g(x) = \cos x$
13. $f(x) = \tan x;\ g(x) = \pi - x$
14. $f(x) = \sin\left(x + \dfrac{\pi}{2}\right);\ g(x) = \cos x$
15. $f(x) = \csc x;\ g(x) = x + \pi$

If $f(x) = \sin x,\ g(x) = \cos x,\ h(x) = x^2$, find:

16. $f(0)$
17. $h\left(f\left(\dfrac{\pi}{2}\right)\right)$
18. $f(h(0))$
19. $h(g(x))$
20. $g(h(x))$
21. $f(g(x)) - f(x)g(x)$

Given the following functions, $h(x)$, find two functions, f and g, such that h can be written as the composition of f and g, that is, $h(x) = f(g(x))$. [Neither $f(x)$ nor $g(x)$ can be x.]

22. $h(x) = \sin^2 x$
23. $h(x) = 1 - \sin^2 x$
24. $h(x) = \sec x$
25. $h(x) = \csc x$
26. $h(x) = \log_a(\sin x)$
27. $h(x) = 10^{\cos x}$

If $f(x) = \sin x$ and $g(x) = \cos x$, solve:

28. $f(x) = g(x)$
29. $f(x) = 0$
30. $[g(x)]^2 = 1$

17.5   AMPLITUDE

Looking at Figure 17.16 you see that the graph of $y = \sin x$ lies entirely between $y = 1$, reached when $x = \dfrac{\pi}{2}, \dfrac{5\pi}{2}$, and so on, and $y = -1$, reached when
$x = -\dfrac{\pi}{2}, \dfrac{3\pi}{2}$, and so on.

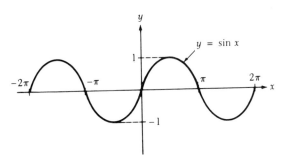

FIG. *17.16*

Now suppose we plot the graph of $y = 3 \sin x$.   The y's here are three
times the size of the corresponding y's in $y = \sin x$, so the graph is as shown
in Figure 17.17.

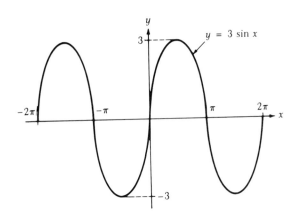

FIG. *17.17*

The maximum value of y is now 3 and the minimum is now -3. The difference between the two functions is expressed by saying that $y = \sin x$ has
amplitude 1 and $y = 3 \sin x$ has *amplitude 3.*

> The *amplitude* of a function is the distance between the maximum
> value of the function and its "average" value (here zero).

It is also possible to talk of the amplitude of a cosine function. The graph of
$y = \cos x$ varies from $y = -1$ to $y = 1$, and that of $y = -\frac{1}{2} \cos x$ varies from
$y = -\frac{1}{2}$ to $y = \frac{1}{2}$.   The latter is just the graph of $y = \cos x$ shrunk by a $\frac{1}{2}$ and
reflected in the x axis. Putting both graphs on the same axes for comparison;

we get Figure 17.18. Therefore the amplitude of $y = \cos x$ is 1, and the amplitude of $y = -\frac{1}{2} \cos x$ is $\frac{1}{2}$.

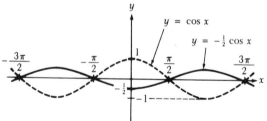

FIG. 17.18

The function $y = 2 \tan x$ looks much the same as $y = \tan x$. It merely climbs a little faster, as shown in Figure 17.19. Since this function does not lie between any particular bounds, we do not define its amplitude. The cotangent, secant, and cosecant functions also vary without bound, and so they do not have amplitudes either.

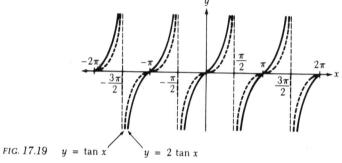

FIG. 17.19 $y = \tan x$ $y = 2 \tan x$

In summary, the amplitude of a function tells you the size of its oscillations. It is defined only for functions that vary within limits ("bounded functions") and therefore only for sines and cosines among the six trigonometric functions.

$y = a \sin x$	has amplitude $|a|$
$y = a \cos x$	has amplitude $|a|$

PROBLEM SET   17.5

Plot the following graphs:

1. $y = 2 \cos x$
2. $y = \frac{1}{2} \cos x$
3. $y = 4 \tan x$
4. $y = -\tan x$
5. $y = -2 \cot x$
6. $y = 3 \csc x$

7. $y = -\frac{1}{3} \sec x$
8. $y = \dfrac{\sin x}{\frac{1}{2}}$
9. $y = 4 \sec x$
10. $y = -\frac{1}{2} \csc x$
11. $2y = -\sin x$

12. $10y = -\sec x$
13. $y = -3 \cot x$
14. $4y = 8 \tan x$
15. $y = -4 \sin x$
16. $3y = -2 \sin x$

17.6   PERIOD

The periods of the six trigonometric functions were introduced in Section 17.1. The period is the amount x must increase before the function starts to repeat. For the sine and cosine, this is 2π, and for the tangent it is π. From Figure 17.20 you can see that adding 2π to any angle does not change the value of its sine or cosine; adding π to any angle does not change its tangent. In other words, for any angle θ

$$\sin(\theta + 2\pi) = \sin\theta \qquad \cos(\theta + 2\pi) = \cos\theta \qquad \tan(\theta + \pi) = \tan\theta$$

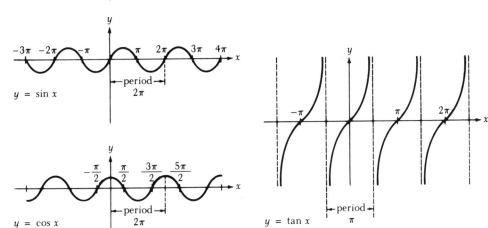

FIG. 17.20

This leads to another way of defining the period of a function:

> The *period* is the smallest positive x value p such that
> $$f(x + p) = f(x) \qquad \text{for all } x$$

Now let us go back to the graphs of the trigonometric functions. In the last section we looked at the graph of things like $y = 2 \sin x$, and saw that the 2 changes the amplitude, or height, of the "waves." Now consider the function

$$y = \sin 2x$$

What effect does the 2 have here?
To draw a graph we will first make a table of values, shown in Table 17.4.

Table 17.4

x (radians)	0	$\dfrac{\pi}{6}$	$\dfrac{\pi}{4}$	$\dfrac{\pi}{3}$	$\dfrac{\pi}{2}$	$\dfrac{2\pi}{3}$	$\dfrac{3\pi}{4}$	$\dfrac{5\pi}{6}$	π
x (degrees)	0°	30°	45°	60°	90°	120°	135°	150°	180°
$y = \sin 2x$	0	0.87	1	0.87	0	−0.87	−1	−0.87	0

We include multiples of $\dfrac{\pi}{4}$ in the table because these are the points at which maximum and minimum values occur. For example, $y = \sin 2x$ reaches its first peak when $2x = \dfrac{\pi}{2}$, that is, when $x = \dfrac{\pi}{4}$.

Plotting $y = \sin 2x$ and $y = \sin x$ on the same axes for comparison, we get Figure 17.21. Clearly $y = \sin 2x$ is going up and down twice as often as $y = \sin x$. The factor of 2 means that when you plug an angle into $y = \sin 2x$ the sine function thinks it is looking at an angle twice the size, and so gives you a point twice as far along the sine curve. As a result, it takes only π instead of 2π for $y = \sin 2x$ to go through a complete cycle. In other words:

The period of $y = \sin 2x$ is π.

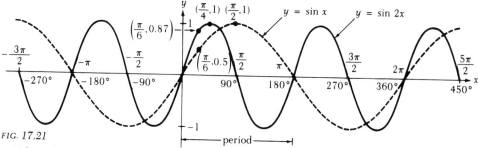

FIG. 17.21

What about $y = \tan 2x$? The table of values is shown in Table 17.5, and the graph in Figure 17.22. Here again, the graph repeats itself twice as quickly as the graph of the tangent function, whose period is π. Therefore:

The period of $y = \tan 2x$ is $\dfrac{\pi}{2}$.

Table 17.5

x (radians)	0	$\dfrac{\pi}{6}$	$\dfrac{\pi}{4}$	$\dfrac{\pi}{3}$	$\dfrac{\pi}{2}$	$\dfrac{2\pi}{3}$	$\dfrac{3\pi}{4}$	$\dfrac{5\pi}{6}$	π
x (degrees)	0°	30°	45°	60°	90°	120°	135°	150°	180°
$y = \tan 2x$	0	1.7	undef.	-1.7	0	1.7	undef.	-1.7	0

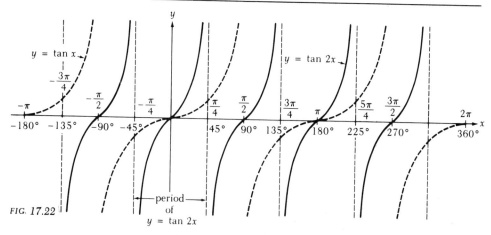

FIG. 17.22

EXAMPLE: *Graph y = sin 3x and find its period.*

The function here always gives you the sine of an angle three times as large as what you put into it. Therefore you get the y value of a point three times as far along the graph as the x value you began with. In other words, we would expect $y = \sin 3x$ to oscillate three times as fast as $y = \sin x$. This would mean that $y = \sin 3x$ takes $\dfrac{2\pi}{3}$ to complete one oscillation, and that

$y = \sin 3x$ has a period of $\dfrac{2\pi}{3}$.

Making a table of values (Table 17.6) and drawing the graph of $y = \sin 3x$ (Figure 17.23) confirms this. You can see that the graph of $y = \sin 3x$ takes $\dfrac{2\pi}{3}$ to get through one full cycle, and this is therefore the period.

Table 17.6

x (radians)	0	$\dfrac{\pi}{6}$	$\dfrac{\pi}{4}$	$\dfrac{\pi}{3}$	$\dfrac{\pi}{2}$	$\dfrac{2\pi}{3}$	$\dfrac{3\pi}{4}$	$\dfrac{5\pi}{6}$	π
x (degrees)	0°	30°	45°	60°	90°	120°	135°	150°	180°
y = sin 3x	0	1	0.7	0	−1	0	0.7	1	0

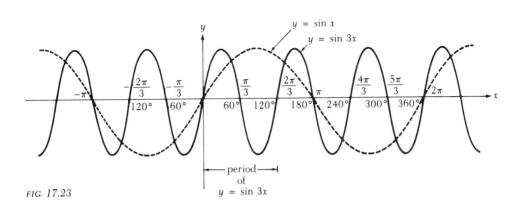

FIG. 17.23

EXAMPLE: *Graph y = cos(½x) and find its period.*

Since the period of $y = \cos 2x$ is $\dfrac{2\pi}{2}$, and that of $y = \cos 3x$ is $\dfrac{2\pi}{3}$, it would seem reasonable for the period of $y = \cos(\tfrac{1}{2}x)$ to be $\dfrac{2\pi}{\frac{1}{2}} = 4\pi$. The table of values (Table 17.7) and the graph (Figure 17.24) confirm this. As expected, $y = \cos(\frac{1}{2}x)$ takes 4π to execute a complete cycle, and therefore:

The period of $y = \cos(\frac{1}{2}x)$ is 4π.

Table 17.7

x (radians)	0	$\frac{\pi}{6}$	$\frac{\pi}{3}$	$\frac{\pi}{2}$	$\frac{2\pi}{3}$	$\frac{5\pi}{6}$	π	$\frac{7\pi}{6}$	$\frac{4\pi}{3}$	$\frac{3\pi}{2}$	$\frac{5\pi}{3}$	$\frac{11\pi}{6}$	2π
x (degrees)	0°	30°	60°	90°	120°	150°	180°	210°	240°	270°	300°	330°	360°
$y = \cos\frac{1}{2}x$	1	0.97	0.87	0.7	0.5	0.26	0	−0.26	−0.5	−0.7	−0.87	−0.97	−1

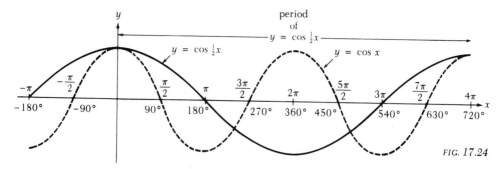

FIG. 17.24

EXAMPLE: *Find the period of* $y = \sin(-2x)$.

Graphing $y = \sin(-2x)$ and $y = \sin 2x$ on the same axes for comparison, we get Figure 17.25. You can see by looking at its graph that $y = \sin(-2x)$ has period π, just like $y = \sin 2x$. Putting a minus in front of the x doesn't change the period. This isn't surprising when you realize that $\sin(-2x) = -\sin 2x$, which means that the minus causes the graph to be reflected in the x axis, but not to oscillate any faster.

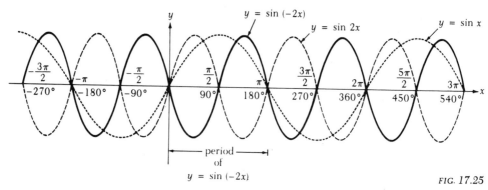

FIG. 17.25

EXAMPLE: *Graph* $y = 3 \sin 2x$.

This can either be done by a table of values or by first drawing $y = \sin 2x$, which has the right period, and then changing the amplitude to 3, giving the graph in Figure 17.26.

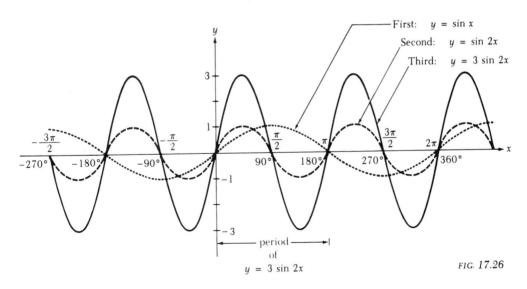

FIG. 17.26

Note: The period of $y = 3 \sin 2x$ is π, the same as the period of $y = \sin 2x$. Changing the amplitude does not change the period.

EXAMPLE: *Graph $y = -\frac{1}{2} \cos (-3x)$.*

Cosine is an even function, so we can ignore the inside negative sign because

$$\cos(-3x) = \cos 3x$$

Therefore

$$y = -\tfrac{1}{2} \cos(-3x) = -\tfrac{1}{2} \cos 3x$$

and the graph of $y = -\frac{1}{2} \cos 3x$ can be drawn by first graphing $y = \cos 3x$, then changing the amplitude to $\frac{1}{2}$ and flipping over. The graph is shown in Figure 17.27.

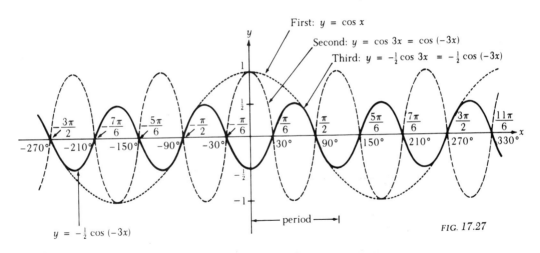

FIG. 17.27

The period of the secant, cosecant, and cotangent functions are defined just as for the other three functions and turn out to be 2π, 2π, and π, respectively. The period can be altered by multiplying x by a constant—for example, $y = \sec 3x$ has period $\frac{2\pi}{3}$ and $y = \cot \frac{x}{2}$ has period 2π.

In summary, the period of a function tells you how long it takes that function to get through one full cycle: The greater the period, the "slower" the function executes a cycle. The period is defined for all six trigonometric functions, and, in the case of the sine and cosine, is unaffected by the amplitude.

$$\left. \begin{array}{l} y = a \sin bx \\ y = a \cos bx \end{array} \right\} \text{ has period } \frac{2\pi}{|b|}, \text{ amplitude } |a|$$

$$\left. \begin{array}{l} y = a \tan bx \\ y = a \cot bx \end{array} \right\} \text{ has period } \frac{\pi}{|b|}, \text{ amplitude is not defined}$$

$$\left. \begin{array}{l} y = a \sec bx \\ y = a \csc bx \end{array} \right\} \text{ has period } \frac{2\pi}{|b|}, \text{ amplitude is not defined}$$

PROBLEM SET   17.6

Plot the graph of $y = \sin x$ and, on the same axes, the graphs of:

1. $y = \sin \frac{x}{2}$

2. $y = \sin 2x$

Plot the following graphs on the same axis with the basic trig function:

3. $y = \cos 3x$

4. $y = 2 \sin \frac{x}{3}$

5. $y = \cos 4x$

6. $y = \cot \left(\frac{x}{2} \right)$

7. $y = -\tan \frac{x}{4}$

8. $y = -2 \sin \frac{x}{2}$

9. $y = \csc 3x$

10. $y = \sec \frac{x}{3}$

11. $y = \cos \left(-\frac{x}{2} \right)$

12. $y = -2 \tan \left(-\frac{x}{2} \right)$

13. $-4y = -2 \cot(-2x)$

14. $2y = 2 \cos(-x)$

15. $y = -\frac{1}{2} \cot \left(-\frac{x}{3} \right)$

16. $3y = 2 \sin 4x$

17. $-2y = \tan 2x$

18. $4 \sin \left(\frac{x}{4} \right) - y = 0$

17.7   PHASE

Having seen how a constant in front of the sine function changes its amplitude (e.g., $y = 2 \sin x$) and a constant in front of the x changes the period (e.g., $y = \sin 3x$), let us look at the effect of a constant in a third position by considering

$$y = \sin\left(x - \frac{\pi}{2}\right)$$

Graphing this equation using the values in Table 17.8, we get the graph in Figure 17.28. We see that the graph of $y = \sin\left(x - \frac{\pi}{2}\right)$ is simply the graph of $y = \sin x$ moved to the right by $\frac{\pi}{2}$. This will not be a surprise if you remember the section on horizontal shifts (Section 3.3), where replacing x by $(x - k)$ shifted the whole graph over to the right by k. Thus replacing x by $\left(x - \frac{\pi}{2}\right)$ will shift $y = \sin x$ to the right by $\frac{\pi}{2}$. The two curves, $y = \sin x$ and $y = \left(x - \frac{\pi}{2}\right)$, are now out of step by $\frac{\pi}{2}$; there is said to be a *phase difference* of $\frac{\pi}{2}$ between them. If the independent variable is time—as it often is—then $y = \sin x$ takes on each y value $\frac{\pi}{2}$ earlier than $y = \sin\left(x - \frac{\pi}{2}\right)$; therefore, $y = \sin x$ is said to lead $y = \sin\left(x - \frac{\pi}{2}\right)$ by $\frac{\pi}{2}$ or a quarter of a cycle; $y = \left(x - \frac{\pi}{2}\right)$ is $\frac{\pi}{2}$ behind $y = \sin x$. [Notice that $y = \sin x$ is "ahead of" or "leading" $y = \sin\left(x - \frac{\pi}{2}\right)$ because $y = \sin x$ is $y = \sin\left(x - \frac{\pi}{2}\right)$ shifted to the left. Also, $y = \sin\left(x - \frac{\pi}{2}\right)$ is "behind" $y = \sin x$ because $y = \sin\left(x - \frac{\pi}{2}\right)$ is $y = \sin x$ shifted to the right.]

Table 17.8

x (radians)	0	$\frac{\pi}{4}$	$\frac{\pi}{2}$	$\frac{3\pi}{4}$	π	$\frac{5\pi}{4}$	$\frac{3\pi}{2}$	2π
$y = \sin\left(x - \frac{\pi}{2}\right)$	-1	-0.7	0	0.7	1	0.7	0	-0.7 -1

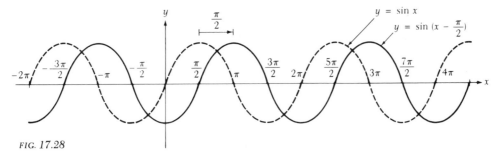

FIG. 17.28

The graph of $y = \cos\left(x + \dfrac{\pi}{2}\right)$ is the graph of $y = \cos x$ shifted $\dfrac{\pi}{2}$ to the left; as shown in Figure 17.29. The phase difference between the two curves is again $\dfrac{\pi}{2}$, but this time $y = \cos\left(x + \dfrac{\pi}{2}\right)$ leads $y = \cos x$ by $\dfrac{\pi}{2}$, or $y = \cos\left(x + \dfrac{\pi}{2}\right)$ is ahead of $y = \cos x$ by $\dfrac{\pi}{2}$.

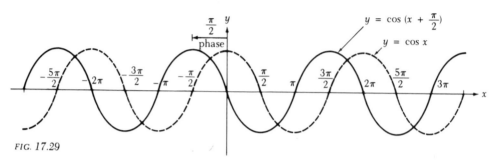

FIG. 17.29

Phase differences could be defined for each of the other functions, though they almost never are. Drawing graphs will show you that $y = \tan\left(x + \dfrac{\pi}{4}\right)$ is $\dfrac{\pi}{4}$ ahead of $y = \tan x$, and that $y = \sec\left(x - \dfrac{\pi}{2}\right)$ is $\dfrac{\pi}{2}$ behind $y = \sec x$.

If we compare $y = \sin x$ and $y = \sin(x - \pi)$, shown in Figure 17.30, we see that they are completely opposite to one another: When one is at its maximum value, or peak, the other is at its minimum, the bottom of a trough. This is expressed by saying that the two functions are *out of phase*, or π *out of phase*.

On the other hand, $y = \sin(x - 2\pi)$ and $y = \sin x$ are the same function (because 2π is the period of the sine function), and they therefore have their maxima and minima at the same values of x. Such functions are said to be *in phase* with one another.

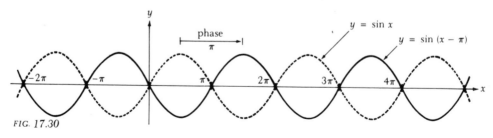

FIG. 17.30

What about $y = \cos x$ and $y = 2\cos(x - \pi)$ or $y = 3\cos(x - \pi)$? The graphs look like Figure 17.31. Whenever $y = \cos x$ has a peak, the other two graphs are at the bottom of a trough, and vice versa. Therefore it makes sense to say that $y = \cos x$ is π out of phase with $y = 2\cos(x - \pi)$ and $y = 3\cos(x - \pi)$ and that the latter are in phase with one another. Clearly, amplitude makes no difference to the phase difference between two functions.

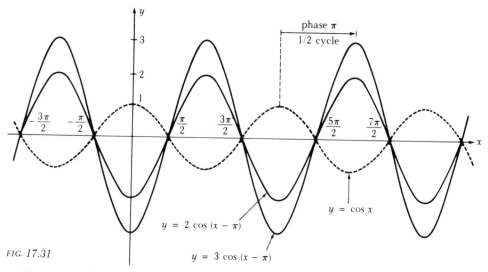

FIG. 17.31

However, the period does make a difference. Compare $y = \sin 2x$ with $y = \sin\left(2x + \dfrac{\pi}{2}\right)$. Since $\sin\left(2x + \dfrac{\pi}{2}\right) = \sin\left[2\left(x + \dfrac{\pi}{4}\right)\right]$, $y = \sin\left(2x + \dfrac{\pi}{2}\right)$ is $y = \sin 2x$ shifted $\dfrac{\pi}{4}$ to the left. This is confirmed by plotting some points (Table 17.9) on the graph in Figure 17.32.

Table 17.9

x	0	$\dfrac{\pi}{4}$	$\dfrac{\pi}{2}$	$\dfrac{3\pi}{4}$	π
$y = \sin\left(2x + \dfrac{\pi}{2}\right)$	1	0	-1	0	1

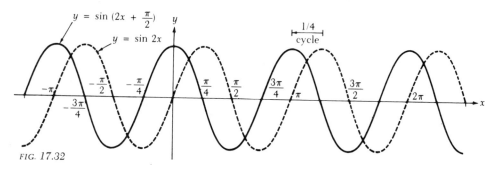

FIG. 17.32

Now, however, we run into trouble. The mathematicians say that since $y = \sin 2x$ is shifted $\dfrac{\pi}{4}$ horizontally to give $y = \sin\left(2x + \dfrac{\pi}{2}\right)$, the phase difference between these two should be defined to be $\dfrac{\pi}{4}$. The $\dfrac{\pi}{4}$ is obtained by dividing $\dfrac{\pi}{2}$ by the coefficient of x, namely, 2.

The physicists, on the other hand, feel that the magnitude of the horizontal shift is less important than the fact that $y = \sin\left(2x + \dfrac{\pi}{2}\right)$ is a quarter of a cycle (a quarter of one complete wave, up and down) ahead of $y = \sin 2x$. A quarter of a cycle corresponds to $\dfrac{\pi}{2}$ on the circle, so the physicists insist that the phase difference between these two functions be defined to be $\dfrac{\pi}{2}$. The $\dfrac{\pi}{2}$ comes immediately from $y = \sin\left(2x + \dfrac{\pi}{2}\right)$. As it turns out, the main users of the idea of phase are the physicists, and so we will adopt their definition.

> The *phase difference* between two functions shows what fraction of a cycle one is out of step with the other. A complete cycle is represented by 2π, half by π, and so on.

This definition of phase for functions with arbitrary periods is used only for sines and cosines; we will just not define the phase difference between things like $y = \tan(2x + \pi)$ and $y = \tan 2x$.

Notice that although we can talk about the phase difference between $y = \sin 2x$ and $y = \sin\left(2x + \dfrac{\pi}{2}\right)$ being $\dfrac{\pi}{2}$, we cannot talk about the phase difference between $y = \sin\ x$ and $y = \sin\ 2x$, or between $y = \sin\ x$ and $y = \sin\left(2x + \dfrac{\pi}{2}\right)$. For example, look at the graphs of $y = \sin x$ and $y = \sin 2x$ in Figure 17.33. You can see that there are some x values at which both functions are at the same point in their cycle (e.g., at 0 and π both functions are zero), and there are other x values at which they are at different points in the cycle (e.g., at $\dfrac{\pi}{2}$, $y = \sin 2x$ is zero while $y = \sin x$ is 1, the top of a wave). Therefore the relationship between the functions changes as x changes and so it is impossible to say that any one number is the phase difference. The same thing will happen with any two functions of different periods, so phase differences are defined only for functions of the same period.

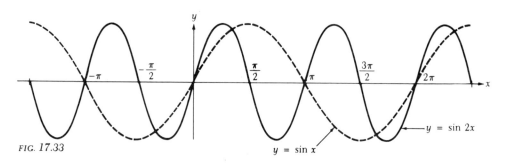

FIG. *17.33* $y = \sin x$

You can, however, make sense of the phase difference between $y = \cos x$ and $y = \sin x$. First of all, drawing them, as in Figure 17.34, makes you think that the phase difference must be $\dfrac{\pi}{2}$, with the cosine leading. In the next chapter we will show that $\cos x = \sin\left(x + \dfrac{\pi}{2}\right)$, so, indeed, it is possible to define the phase difference between $y = \cos x$ and $y = \sin x$; the cosine leads by $\dfrac{\pi}{2}$.

As far as other functions are concerned, it certainly makes sense to say that $y = \tan\left(x + \dfrac{\pi}{2}\right)$ is $\dfrac{\pi}{2}$ ahead of $y = \tan x$. However, we really cannot talk about the phase difference between $y = \tan x$ and $y = \sin x$, say, because the graphs of these functions are so different that it is really impossible to say what would be meant by "being at the same point in the cycle." The only reason that we can talk about the phase difference between $y = \cos x$ and $y = \sin x$ is that their graphs are exactly the same shape and so we can talk about two points being at the same or different positions in the cycle.

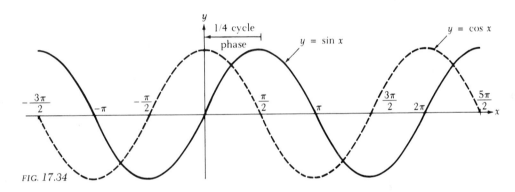

FIG. 17.34

Graphing Functions with Different Amplitudes, Periods, and Phases

EXAMPLE: *Find the period, amplitude, and phase of $y = 2\cos(3x - \pi)$.*

It is perfectly possible to answer this without drawing a thing, but to make it clear what the period, amplitude, and phase mean, we will graph the function first. The easiest way to do this is by stages:
1. Period
2. Phase
3. Amplitude

For $y = 2\cos(3x - \pi)$:

STEP 1. We first draw a function with the right period—that is, we draw $y = \cos 3x$. We know how to do this, because it is just the cosine function, only three times as fast.

STEP 2.   Next we draw a function which is π behind the last one—that is, we draw $y = \cos(3x - \pi)$. This means shifting the graph of $y = \cos 3x$, which we have just drawn, half a cycle to the right (because π is half a cycle).

STEP 3.   Last, we get the amplitude right. This means stretching $y = \cos(3x - \pi)$ so that it goes up twice as high and down twice as low, giving $y = 2 \cos(3x - \pi)$. See Figure 17.35.

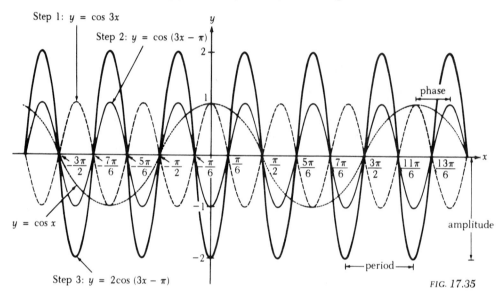

FIG. *17.35*

This question asks for the phase of $y = 2 \cos (3x - \pi)$, but so far we have defined only the phase difference between two functions. The *phase*, or *phase angle* as it is also called, is taken to mean the phase difference between $y = 2 \cos(3x - \pi)$ and $y = \cos 3x$. In general, the phase of a single function is understood to mean the phase difference between the function and some standard function with the same period.

From the graph of $y = 2 \cos(3x - \pi)$, you can see that:

its amplitude is 2

its period is $\dfrac{2\pi}{3}$

its phase is π behind $y = \cos 3x$

You can get this directly from the equation:

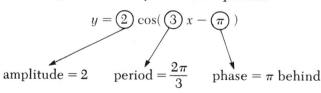

amplitude $= 2$ period $= \dfrac{2\pi}{3}$ phase $= \pi$ behind

EXAMPLE: *Graph* $y = \sin\left(\dfrac{2x}{3} + \dfrac{\pi}{2}\right)$

STEP 1.   Draw $y = \sin\dfrac{2x}{3}$, which has period $\dfrac{2\pi}{\frac{2}{3}} = 3\pi$

STEP 2.   Draw $y = \sin\left(\dfrac{2x}{3} + \dfrac{\pi}{2}\right)$

which means moving $y = \sin\dfrac{2x}{3}$ to the left by a quarter of a cycle,

since the phase difference between these two functions is $\dfrac{\pi}{2}$

(which is a quarter of a cycle), with $y = \sin\left(\dfrac{2x}{3} + \dfrac{\pi}{2}\right)$ leading.

STEP 3.   Since the coefficient of $\sin\left(\dfrac{2x}{3} + \dfrac{\pi}{2}\right)$ is 1, there is no need to correct the amplitude.

From the graph of

$$y = \sin\left(\dfrac{2x}{3} + \dfrac{\pi}{2}\right)$$

in Figure 17.36, we can see that:

the amplitude is 1

the period is 3π

the phase is $\dfrac{\pi}{2}$ ahead of $y = \sin\dfrac{2x}{3}$

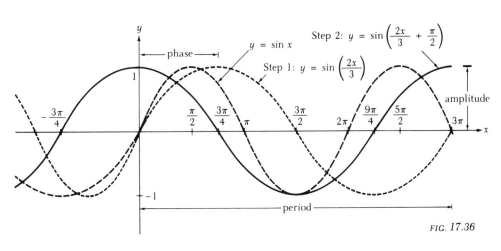

FIG. 17.36

In summary, the phase (phase difference, phase angle) tells you how much out of step one periodic function is with another, and is usually defined only for sine and cosine functions. Graphs can be drawn either using phase or the fact that replacing x by $(x - k)$ shifts a graph k to the right.

$$
\left.
\begin{array}{l}
y = a \sin(bx + c) \\
y = a \cos(bx + c)
\end{array}
\right\}
$$

has amplitude $|a|$

period $\dfrac{2\pi}{|b|}$

phase c $\left\{\begin{array}{l}\text{ahead if } c \text{ is positive} \\ \text{behind if } c \text{ is negative}\end{array}\right.$

PROBLEM SET   17.7

Plot the graph of $y = \sin x$ and, on the same axes, the graphs of:

1. $y = \sin\left(x + \dfrac{\pi}{2}\right)$ 2. $y = \sin(x + \pi)$ 3. $y = \sin\left(x - \dfrac{\pi}{2}\right)$

Plot the graph of $y = \tan x$ and, on the same axes, the graphs of:

4. $y = \tan\left(x + \dfrac{\pi}{2}\right)$ 5. $y = \tan\left(x + \dfrac{\pi}{4}\right)$ 6. $y = \tan(x + \pi)$

Plot the graph of $y = \cos x$ and, on the same axes, the graphs of:

7. $y = 2 \cos x$ 8. $y = \cos\left(x + \dfrac{\pi}{4}\right)$ 9. $y = 2 \cos\left(x + \dfrac{\pi}{4}\right)$

Plot the graph of $y = \sin x$ and, on the same axes, the graphs of:

10. $y = \sin 2x$ 11. $y = \sin(2x + \pi)$ 12. $y = \sin\left(2x + \dfrac{\pi}{2}\right)$

Plot each pair of graphs on the same axes:

13. $y = 3 \sin x$ and $y = 3 \sin\left(x - \dfrac{\pi}{2}\right)$

14. $y = \dfrac{1}{2} \tan x$ and $y = \dfrac{1}{2} \tan\left(x + \dfrac{\pi}{4}\right)$

15. $y = \cos 3x$ and $y = \cos(3x + 90°)$

16. $y = \cos \dfrac{x}{4}$ and $y = \cos\left(\dfrac{x}{4} - 180°\right)$

Give the amplitude, phase, and period of each of the following:

17. $y = 2 \cos 2x$

18. $y = 0.75 \cos\left(\dfrac{\pi x}{2}\right)$

19. $y = -16 \sin\left(x + \dfrac{\pi}{2}\right)$

20 $y = 3 \cos (2x - 1)$

21. $y = 2 \sin[\pi(x + 1)]$

Sketch the graphs of the following equations.

22. $y = 4 \sin 2x$

23. $y = \tan \dfrac{x}{2}$

24. $y = \sin(2x + \pi)$

25. $y = 2 \cos\left(\dfrac{x}{3} - \pi\right)$

26. $y = -\frac{1}{2} \sin(2x + 2\pi)$

27. $2y = 4 \tan\left(\dfrac{x}{2} + 45°\right)$

28. $y = 3 \cos\left(3x - \dfrac{\pi}{2}\right)$

29. $y = 4 \sin\left(4x - \dfrac{\pi}{3}\right)$

30. $y = \frac{1}{2} \sin 2\pi x$

31. $y = \cos\left(\pi x - \dfrac{\pi}{2}\right)$

32. $y = 1 + \cos 2x$

33. $y = 2 + 2 \cos(2x + \pi)$

34. $y = 2 \cot\left(\dfrac{x}{3} - \pi\right)$

35. $y = -3 \csc(3x - 90°)$

17.8   A GRAPHICAL LOOK AT TRIGONOMETRIC EQUATIONS

A *trigonometric equation* is an equation involving the trigonometric functions in which an angle is the unknown. For example,

$$\cos\theta = 0.6428$$

or

$$5 \tan x + \cot x = 2$$

or

$$6 \sin^2 x - 7 \sin x + 2 = 0$$

are all trigonometric equations. On the other hand, things like

$$2x = \cos 63°$$

or

$$(\cos^2 68°)x^2 + (\sin 38°)x = 0$$

are not trigonometric equations, because although they involve trigonometric functions, the unknown is not an angle. These last two equations can be solved by the usual methods for linear and quadratic equations—they merely have rather messy-looking coefficients that happen to involve trigonometric functions.

The first three equations, however, have to be solved by entirely different methods from algebraic equations. You have seen some of these methods before in section 14.3. There we solved things like

$$\cos \theta = 0.6428$$

by the following method:

Look up 0.6428 in the tables to find

$$\cos 50° = 0.6428$$

This tells you that the related angle is 50°. Since cos θ is positive, θ is in the first or fourth quadrants, as shown in Figure 17.37. Therefore,

$$\theta = 50° \quad \text{or} \quad 310°$$

So, the most general possible values for θ are

$$\theta = 50° + 360°n \quad \text{or} \quad \theta = 310° + 360°n \qquad (n \text{ an integer})$$

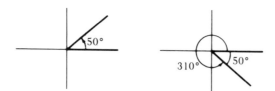

FIG. 17.37

These solutions can be displayed graphically as the points at which the line $y = 0.6428$ cuts the graph of $y = \cos \theta$, as shown in Figure 17.38.

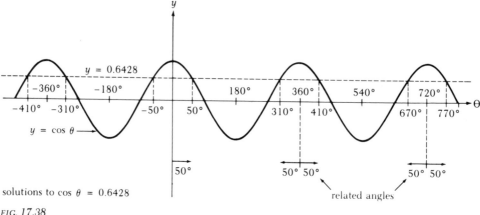

solutions to cos θ = 0.6428

FIG. 17.38

   The shape and symmetry of the graph make it clear exactly why there are infinitely many solutions, and why they are arranged the way they are—all 50° away from multiples of 360°. This is a brilliant example of the power of analytic geometry—the use of graphs to illuminate algebraic problems. Indeed, given the related angle from the tables, it is very convenient, and perhaps quicker, to use the graph to find the solutions to an equation. We'll do the next example that way.

   One important note: Remember that the related angle is always measured from the *x* axis, (i.e., 180°, 360°) and never the *y* axis. Therefore, *the related angles on the graph must be measured from multiples of 180° or 360°, and never from 90° or 270°*, and so on.

EXAMPLE: *Solve cos x = −0.9063.*

From the tables,

cos 25° = 0.9063   so the related angle is 25°.

On the graph we must draw the line $y = -0.9063$ and measure the related angle from multiples of 180°. Here we measure from ±180°, ±540°, and so on, because these are the multiples of 180° with negative cosines. See Figure 17.39.

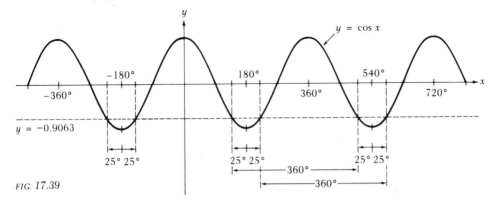

FIG. 17.39

You can see that the angles that are 25° from ±180°, ±540°, and so on, satisfy

cos x = −0.9063

Therefore the solutions to cos x = −0.9063 include

180° − 25° = 155°   180° + 25° = 205°

540° − 25° = 515°   540° + 25° = 565°

as well as

−155°, −205°, −515°, −565°, etc.

Looking at the graph again, notice that all the solutions that arose by subtracting 25° are 360° apart, and therefore of the form

155° + 360°n   (n an integer)

The solutions that arose by adding 25° are also 360° apart and therefore of the form

205° + 360°n   (n an integer)

Therefore the general solution to cosx = −0.9063 is

155° + 360°n   or   205° + 360°n

Graphs can be equally useful for solving equations in radians.

EXAMPLE: *Solve sin x = 0.*

The graph in Figure 17.40 shows that the equation $f(x) = \sin x$ has a zero at every multiple of π. Therefore the solutions to sin $x = 0$ are all multiples of π; that is,

the solutions are $n\pi$ (n an integer).

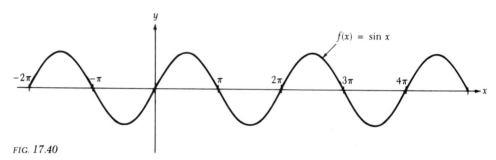

FIG. 17.40

EXAMPLE: *Solve cos x = −1.*

The graph of $f(x) = \cos x$, shown in Figure 17.41, shows that the cosine function is −1 at all odd multiples of π. Therefore, solutions to cos $x = -1$ are all odd multiples of π; that is,

the solutions are $(2n + 1)\pi$
(n an integer, so $(2n + 1)$ is always odd).

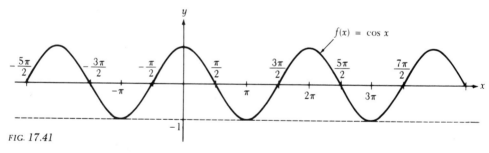

FIG. 17.41

EXAMPLE: *Solve sin x = $\dfrac{1}{\sqrt{2}}$.*

Since

$$\sin \frac{\pi}{4} = \frac{1}{\sqrt{2}}$$

the related angle is $\dfrac{\pi}{4}$. Draw the line $y = \dfrac{1}{\sqrt{2}}$ and measure the related angle from multiples of π (because in degrees you measure from multiples of 180°). From the graph, shown in Figure 17.42, you can see that $\dfrac{\pi}{4}$ and $\dfrac{3\pi}{4}$ are both solutions. Since adding multiples of 2π to either of these gives another solution,

$$\frac{\pi}{4} + (2\pi)n \quad \text{and} \quad \frac{3\pi}{4} + (2\pi)n$$

are the general solutions.

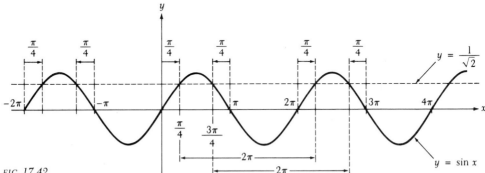

FIG. 17.42

EXAMPLE: *Solve* $\tan x = \dfrac{1}{\sqrt{3}}$.

Since $\tan \dfrac{\pi}{6} = \dfrac{1}{\sqrt{3}}$, the related angle is $\dfrac{\pi}{6}$. These solutions are π apart, as shown in Figure 17.43; therefore, the general solution is

$$\frac{\pi}{6} + n\pi \qquad (n \text{ an integer})$$

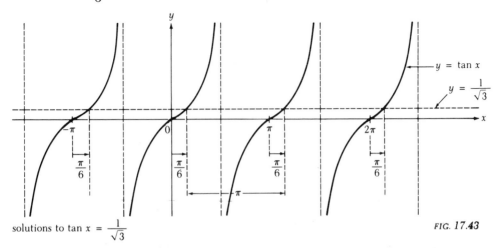

solutions to $\tan x = \dfrac{1}{\sqrt{3}}$

FIG. 17.43

Graphical methods can be used in more complicated problems, such as the following.

EXAMPLE: *Solve sin x = cos x.*

This is asking where the graphs of $y = \sin x$ and $y = \cos x$ intersect. Let us draw them in Figure 17.44 and look. The symmetry between the two graphs in the region between 0 and $\dfrac{\pi}{2}$ tells us that the point of intersection must be in the middle, namely at $\dfrac{\pi}{4}$. Another point of intersection clearly lies midway between π and $\dfrac{3\pi}{2}$, that is, at $\dfrac{5\pi}{4}$, and a third at $\dfrac{9\pi}{4}$, and in the negative direction at $-\dfrac{3\pi}{4}$, $-\dfrac{7\pi}{4}$, and so on. The solutions are π apart, so the general solution is

$$\frac{\pi}{4} + n\pi$$

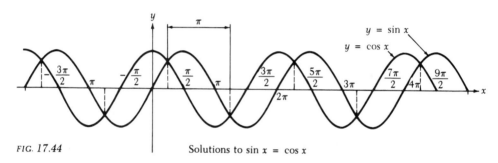

FIG. 17.44   Solutions to sin x = cos x

EXAMPLE: *Solve tan x = cos x.*

We draw graphs of $y = \tan x$ and $y = \cos x$ in Figure 17.45 and look for their points of intersection. There are clearly points of intersection of these two graphs between 0 and $\dfrac{\pi}{2}$, between $\dfrac{\pi}{2}$ and π, between 2π and $\dfrac{5\pi}{2}$, and so on. Unfortunately, however, there is no symmetry or related angle to tell us *exactly* where those points are, and to find them requires further algebraic tricks, which will be covered in the next chapter (Section 18.4).

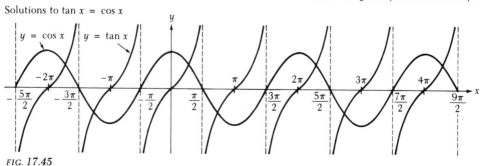

Solutions to tan x = cos x

FIG. 17.45

This problem, therefore, is a good example of what graphical methods cannot do. A graph can give you a brilliantly clear idea of what or where to calculate, but it can't do the calculations for you. It can make you acutely aware of the pattern lying behind a problem, and of any symmetries involved, and it can give you a clear idea what to expect from a problem. Graphs thereby move a great many problems from the do-able-with-much-pain-and-algebra category into the plain do-able category; so if in doubt, draw one!

PROBLEM SET   17.8

Solve the following: (for x between $0°$ and $360°$ or between 0 and 2π)

1. $\sin x = 0.5000$
2. $\cos x = 0.5000$
3. $\tan x = 1.000$
4. $\cot x = 5.05$
5. $\cos x = \dfrac{1}{\sqrt{2}}$
6. $\tan x = \cot x$
7. $\sin 2\theta = \cos 2\theta$
8. $\sin x = 2 \cos x$
9. $\sin x = x$
10. $\tan x = \sec x$
11. $\sin \theta = 0.8192$
12. $\cos \theta = -0.2924$
13. $\tan \theta + 0.3640 = 0$
14. $2 \cos \theta + \sin \theta = 0$
15. $\sec \alpha = 1.5557$
16. $5 \cot \theta = 1$
17. $4 \sec x + 3 \csc x = 0$

18. How many solutions does the equation $\tan x = x$ have? Approximately where are they?

19. How many solutions does the equation $\tan x = |x|$ have? Approximately where are they?

20. How many solutions does the equation $|\tan x| = x$ have? Approximately where are they?

17.9   APPLICATIONS OF THE TRIGONOMETRIC FUNCTIONS

Because of their periodicity, the trigonometric functions are used to represent sound waves, electric currents, the motion of a pendulum, and anything else that repeats regularly. On a day without too much wind the waves on the ocean's surface trace out a sine or cosine wave if viewed horizontally.

EXAMPLE: *Find the equation describing the surface of waves that are 4 feet from crest to trough and 8 feet from one crest to the next.*

First we must decide what the variables should be. Let x be the horizontal distance, measured from a fixed point, and y the vertical distance. Then we have to decide on the axes. We'll put the x axis where the surface of the water would be if it were dead calm, then we'll position the y axis so that is goes through the highest point on a wave. This means that the equation giving y in terms of x involves a cosine. The graph is shown in Figure 17.46.

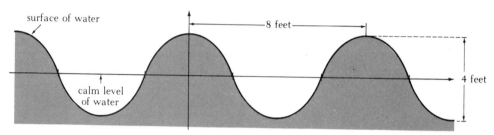

FIG. 17.46

The amplitude of these waves must be 2 if the distance from crest to trough is to be 4 feet. What about the period? Since the period of $y = \cos bx$ is $\frac{2\pi}{b}$ (b positive) and we want a period of 8, we will have to have $\frac{2\pi}{b} = 8$, or $b = \frac{\pi}{4}$. There is no phase shift, so the equation is

$$y = 2 \cos\left(\frac{\pi x}{4}\right)$$

Notice that it is extremely important that we use radians here. Without them we could not make sense of $\cos\left(\frac{\pi x}{4}\right)$, which is the cosine of a length, since x is a length.

Now here is an example that does use degrees.

EXAMPLE: *The range, r, of a cannon with muzzle velocity v, fired at an angle of elevation θ, is given by*

$$r = \frac{v^2}{g} \sin 2\theta \qquad where\ g = constant$$

Sketch a graph of r against θ. Find a formula for the maximum possible range, and the angle at which this is achieved.

The angle of elevation is the angle between the cannon and the horizontal. θ clearly cannot be less than 0° or more than 90°; therefore we will only draw the graph between 0° and 90°. See Figure 17.47.

FIG. 17.47

To graph $r = \dfrac{v^2}{g} \sin 2\theta$ first sketch $r = \sin 2\theta$ and then alter the amplitude. From Figure 17.48 you can see that the maximum possible range is $\dfrac{v^2}{g}$, which occurs at $\theta = 45°$.

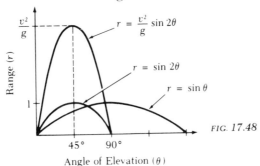

FIG. 17.48

PROBLEM SET 17.9

1. A snapshot of the ocean's surface reveals that the waves on the surface trace out sinusoidal curves if viewed horizontally. Representing a dead-calm surface as the x axis, write the equation of water waves that are 10 feet from crest to trough and 30 feet in wavelength. Draw a graph.

Sound is nothing more than the sinusoidal variation of air pressure over time. This means that the pressure at any point is a sine function of the time. A "middle A" note causes the air pressure to fluctuate through a complete cycle exactly 440 times per second. A note of moderate intensity might have amplitude of 0.1 torr (a unit of pressure).

2. Draw a pressure-versus-time graph of this wave (time on the horizontal axis), marking off the time axis in units of 440ths of a second.

3. Write an equation representing this sound wave.

4. Write an equation representing a louder wave of the same pitch (same frequency).

5. Write an equation representing a lower-pitched sound of the same loudness (same amplitude).

The population of many animals varies sinusoidally about an average value as the seasons change. As a function of time, the population of such animals is given by the equation:

$$P(t) = A + V \sin(2\pi t)$$

where A is the average population, V is the maximum seasonal increase (or decrease) in the population. The time, t, is measured in years and $t = 0$ is taken to be some spring, when the population is at its average value. Then $t = \frac{1}{4}$ is summer, $t = \frac{1}{2}$ is autumn, and $t = \frac{3}{4}$ is summer a year later. $P(t)$ is the total population at time t.

6. A herd of deer has its average size of 4000 in the spring, and has a maximum increase in size of 500. What is the appropriate equation for this herd?

7. What is the herd's size in summertime? Autumn? Winter?

8. Herds become extinct when their total population becomes zero. If a herd disappears sometime during a year, what can be said about A and V?

9. Relative humidity is a measure of how much moisture there is in the air. Obviously the relative humidity over a desert is much lower than over a rain forest, but if you find the average relative humidity, R, at a given latitude, you'll find that R depends on the latitude, θ, in a very regular way:

$$R = 78 + 7 \cos(6\theta)$$

Plot R versus θ. (Note: θ varies from 90° at the North Pole to −90° at the South Pole.)

10. The voltage, V, across a power outlet is given by the formula

$$V = V_0 \cos wt$$

where V_0 and w are constants and t is time. What constant is determined by knowing that the voltage fluctuates through 60 complete cycles in one second?

CHAPTER 17 REVIEW

1. Graph $y = 2 \sin\left(2x + \dfrac{\pi}{2}\right)$.

2. Graph $y = 2 \cos\left(\dfrac{1}{2}x - \dfrac{\pi}{4}\right)$.

3. Graph the function $y = \frac{3}{4} \sin\left(\frac{1}{2}x + \pi\right)$. Include at least one full cycle.
   (a) What is the amplitude?
   (b) What is the period?

   (c) What is the value of the function when $x = \dfrac{\pi}{2}$? When $x = -\dfrac{13\pi}{3}$?

4. Suppose that $f(x) = 2 \cos(3x - 180°)$
   (a) What are the amplitude and period of $f(x)$?
   (b) Graph the function for $0 \leqslant x \leqslant 360°$. Mark the intercepts clearly.
   (c) Compute $f(190°)$.

5. Given $f(x) = 3 \sin(2x - 180°)$
   (a) What is the amplitude of this function?
   (b) What is the period of this function?
   (c) What is the phase shift of this function?
   (d) What are the zeros of this function with $0° \leqslant x \leqslant 360°$?
   (e) Sketch the function.

6. Graph the function $y = \frac{1}{2} \sin(2x + \pi)$. Include at least one full cycle.
   (a) What is the amplitude?
   (b) What is the period?
   (c) What is the value of the function when $x = -\dfrac{2\pi}{3}$? When $x = -\dfrac{\pi}{12}$?

7. (a) What is the amplitude of the graph in Figure 17.49?
   (b) What is its period?
   (c) Write an equation for it.

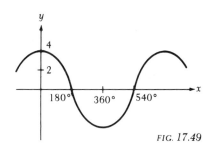

FIG. 17.49

8. Write an equation for the graph in Figure 17.50.

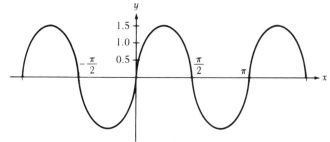

FIG. 17.50

9. What is the equation of the graph in Figure 17.51?

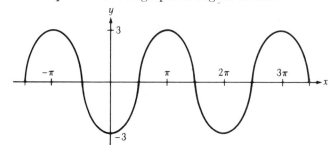

FIG. 17.51

10. Write an equation that describes the function graphed in Figure 17.52.

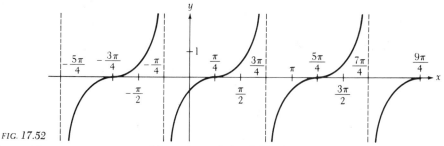

FIG. 17.52

What is the maximum possible value of the following expressions:

11. $-4 \cos(2\theta + \pi)$

12. $4 \sin x$

13. $4 \sin \theta + 5 \sin \phi$

14. $12 \cos(x - \phi)$

For what values of θ is:

15. $\cos \theta > \sin \theta$ 16. $\tan \theta > \cot \theta$ 17. $\sin \theta > \sec \theta$

18. Draw a graph with amplitude and period half that of the graph of $y = \cos x$. (Draw the graph from $-90°$ to $360°$.) What is the equation of the graph you have drawn?

19. Graph $y = 1 + \cos x$.

20. (a) Graph one period of $h(x) = \cot\left(\dfrac{x}{2}\right)$.

    (b) Graph one period of $g(x) = -\tan\left(\dfrac{x}{4}\right)$.

21. Write an equation for the graph in Figure 17.53.

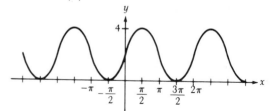

FIG. 17.53

22. Write the equation for the graph in Figure 17.54.

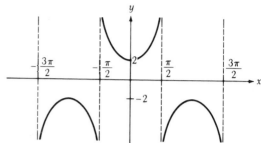

FIG. 17.54

23. Solve *graphically* for the maximum x value such that

$$\begin{cases} x = \log_\pi y \\ y = \cos x \end{cases}$$

24. For what values of θ between 0 and 360° is

$$3 \sin(2\theta + 90°) > -2 \cos(2\theta + 180°)?$$

25. Simplify $\ln(\ln e^{\cos \theta})$, where $\ln = \log_e$. For what values of θ is the value of this expression defined?

26. (a) In a unit circle, prove that the length of the chord cut off by the angle θ is equal to

$$\sqrt{2 - 2 \cos \theta}$$

(*Note:* See Figure 17.55. In the unit circle, you know that $x = \cos \theta$, $y = \sin \theta$. Further, you know a formula for the distance between two points.)

(b) For what angle θ is the length of the chord equal to the radius of the circle?

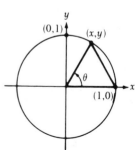

FIG. 17.55

18 TRIGONOMETRIC IDENTITIES

18.1 THE BASIC IDENTITIES: RECIPROCAL, QUOTIENT, PYTHAGOREAN

We have already run into the fact that several of the trigonometric functions are related to one another (Section 14.2). For example, the graph of the secant was derived from the graph of the cosine by using

$$\sec \theta = \frac{1}{\cos \theta}$$

Because this is true for every θ, this is an *identity*, not an equation.

There are six different trigonometric functions, which are all defined in terms of one point P, so it is not surprising that these functions are interrelated. In fact, there are a great many relationships among them. These relationships, or identities, turn out to be of the greatest importance in working with the trigonometric functions, and so this chapter is devoted to their development.

The basic identities were all derived from the definitions which are shown again in Figure 18.1 (page 414). From these we get the *reciprocal identities*,

$$\sec \theta = \frac{1}{\cos \theta} \qquad \csc \theta = \frac{1}{\sin \theta} \qquad \cot \theta = \frac{1}{\tan \theta}$$

and the *quotient identities*,

$$\tan \theta = \frac{\sin \theta}{\cos \theta} \qquad \cot \theta = \frac{\cos \theta}{\sin \theta}$$

The useful thing about the reciprocal and quotient identities is that they enable us to write all the trigonometric functions in terms of just sines and cosines. This turns out to be very important for reducing confusion in trigonometric expressions (see the next section).

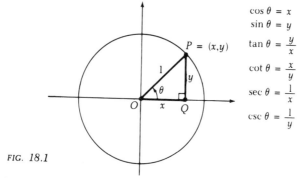

$$\cos \theta = x$$
$$\sin \theta = y$$
$$\tan \theta = \frac{y}{x}$$
$$\cot \theta = \frac{x}{y}$$
$$\sec \theta = \frac{1}{x}$$
$$\csc \theta = \frac{1}{y}$$

FIG. 18.1

We have not yet used the fact that P lies on a unit circle. This tells us that in triangle OPQ, $|OP| = 1$. Therefore by Pythagoras's theorem,

$$x^2 + y^2 = 1$$

(Alternatively, realize that this is the equation of a circle of radius 1, which the coordinates of P must satisfy.)

Now $x = \cos \theta$ and $y = \sin \theta$, so

$$(\cos \theta)^2 + (\sin \theta)^2 = 1$$

Since $(\cos \theta)^2$ is often written $\cos^2 \theta$ (Section 17.4), this usually appears as

$$\boxed{\cos^2 \theta + \sin^2 \theta = 1}$$

This is perhaps the most important identity you will ever meet!

Going back to $x^2 + y^2 = 1$ and dividing through by x^2 we get:

$$\frac{x^2}{x^2} + \frac{y^2}{x^2} = \frac{1}{x^2}$$

or

$$1 + \left(\frac{y}{x}\right)^2 = \left(\frac{1}{x}\right)^2$$

Now

$$\tan \theta = \frac{y}{x} \quad \text{and} \quad \sec \theta = \frac{1}{x}$$

so this tells us that:

$$1 + (\tan \theta)^2 = (\sec \theta)^2$$

or

$$\boxed{1 + \tan^2 \theta = \sec^2 \theta}$$

Another important identity!

Symmetry should make you feel that there was nothing special about dividing $x^2 + y^2 = 1$ through by x^2, and so we had better divide by y^2 also, giving:

$$\frac{x^2}{y^2} + \frac{y^2}{y^2} = \frac{1}{y^2}$$

or

$$\left(\frac{x}{y}\right)^2 + 1 = \left(\frac{1}{y}\right)^2$$

Since

$$\cot \theta = \frac{x}{y} \quad \text{and} \quad \csc \theta = \frac{1}{y}$$

we have

$$(\cot \theta)^2 + 1 = (\csc \theta)^2$$

or

$$\boxed{1 + \cot^2 \theta = \csc^2 \theta}$$

Together these three are called the *Pythagorean identities* (because they come from the Pythagorean theorem):

$$\cos^2 \theta + \sin^2 \theta = 1$$
$$1 + \tan^2 \theta = \sec^2 \theta$$
$$1 + \cot^2 \theta = \csc^2 \theta$$

If you have trouble remembering which function goes on which side in the last two identities (I always do), remember that the secant and cosecant functions are always greater than 1 or less than -1. This means that their squares are always greater than 1, and so can be expressed as: $1 +$ (something positive). Therefore the $\sec^2 \theta$ and $\csc^2 \theta$ must be on the opposite side of the equals sign from the 1.

18.2   SIMPLIFICATION OF TRIGONOMETRIC EXPRESSIONS

Expressions involving the trigonometric functions occur frequently in calculus, so it is obviously important that we be able to write them in the simplest possible form. The reason that the identities are so useful is that they enable us to do this simplification. For example, the function

$$f(x) = \cos x \tan x$$

looks complicated, so it might very well be helpful to know that $f(x)$ is really just the sine function [at least wherever $f(x)$ is defined], because

$$\cos x \tan x = \cancel{\cos x} \cdot \frac{\sin x}{\cancel{\cos x}} \qquad \left(\text{using the quotient identity } \tan x = \frac{\sin x}{\cos x}\right)$$

$$= \sin x$$

So

$$f(x) = \sin x$$

It is obviously easier to work with $\sin x$ than with $\cos x \tan x$.

Along the same lines, if you ended up with

$$\frac{\sin x \cot x \sec x}{\cos^2 x + \sin^2 x}$$

as the answer to some problem, you might be very relieved to know that this horrible fraction is in fact just 1. The reasoning goes like this: The denominator of the fraction is

$$\cos^2 x + \sin^2 x = 1 \qquad \text{(by the Pythagorean identity: } \cos^2 x + \sin^2 x = 1\text{)}$$

The numerator of the fraction is

$$\sin x \cot x \sec x = \cancel{\sin x} \cdot \frac{\cancel{\cos x}}{\cancel{\sin x}} \cdot \frac{1}{\cancel{\cos x}} \qquad \left(\text{using } \cot x = \frac{\cos x}{\sin x};\right.$$

$$\left. \sec x = \frac{1}{\cos x}\right)$$

$$= 1$$

So

$$\frac{\sin x \cot x \sec x}{\cos^2 x + \sin^2 x} = \frac{1}{1} = 1$$

EXAMPLE: *Simplify* $(\cos \theta + \sin \theta)^2 + (\cos \theta - \sin \theta)^2$.

There doesn't seem to be much we can do with this except multiply it out, so let's try that:

$(\cos \theta + \sin \theta)^2 + (\cos \theta - \sin \theta)^2$

$= \cos^2 \theta + \cancel{2 \cos \theta \sin \theta} + \sin^2 \theta + \cos^2 \theta - \cancel{2 \cos \theta \sin \theta} + \sin^2 \theta$

$= \underbrace{\cos^2 \theta + \sin^2 \theta}_{1} + \underbrace{\cos^2 \theta + \sin^2 \theta}_{1}$

$= \quad\quad 1 \quad\quad + \quad\quad 1$

$= 2$

So

$(\cos \theta + \sin \theta)^2 + (\cos \theta - \sin \theta)^2 = 2$

A pleasant surprise.

EXAMPLE: *Simplify*

$$\frac{sec\ x}{csc\ x} \cdot tan\ x + 1$$

In an expression like this, in which you have several functions appearing, it is often helpful to write everything in terms of sine and cosine (which can always be done using the reciprocal and quotient identities). Doing this gives you a chance to see exactly what you do have and whether anything will cancel. So:

$$\frac{\sec x}{\csc x} \cdot \tan x + 1 = \frac{\dfrac{1}{\cos x}}{\dfrac{1}{\sin x}} \cdot \frac{\sin x}{\cos x} + 1$$

$$= \frac{1}{\cos x} \cdot \frac{\sin x}{1} \cdot \frac{\sin x}{\cos x} + 1$$

$$= \frac{\sin x}{\cos x} \cdot \frac{\sin x}{\cos x} + 1$$

$$= \frac{(\sin x)^2}{(\cos x)^2} + 1$$

Unfortunately, nothing has cancelled, but things certainly look more organized. Suppose we combine everything into one term:

$$\frac{(\sin x)^2}{(\cos x)^2} + 1 = \frac{(\sin x)^2 + (\cos x)^2}{(\cos x)^2}$$

$$= \frac{1}{(\cos x)^2} \qquad \text{(by the Pythagorean identity: } \sin^2 x + \cos^2 x = 1)$$

$$= \sec^2 x \qquad \left(\text{since } \frac{1}{\cos x} = \sec x \right)$$

So our original expression turns out to be just $\sec^2 x$—a definite improvement! In other words,

$$\frac{\sec x}{\csc x} \cdot \tan x + 1 = \sec^2 x$$

EXAMPLE: *Simplify* $sin\ x - cos^2\ x\ sin\ x$.

This is already all in terms of sines and cosines and so the reciprocal and quotient identities are of no help. There are no fractions to add nor multiplying out to do, but there is a common factor of $\sin x$, so we will try factoring:

$$\sin x - \cos^2 x\ \sin x = \sin x\ (1 - \cos^2 x)$$

If you now realize that the identity

$$\cos^2 x + \sin^2 x = 1$$

can be rewritten as

$$\sin^2 x = 1 - \cos^2 x$$

then factoring our original expression becomes extremely helpful, because

$$\sin x - \cos^2 x \sin x = \sin x \, (1 - \cos^2 x)$$

$$= \sin x \cdot \sin^2 x$$

$$= \sin^3 x$$

So

$$\sin x - \cos^2 x \sin x = \sin^3 x$$

Note on Simplifying Trigonometric Expressions: Remember that you are working with *expressions* (not equations) and so you can only do things with them that you can do with algebraic expressions—namely, things that preserve the value of the expression. In other words, if you add something, you must also subtract it again; if you multiply the top of a fraction by something, you must multiply the bottom by the same thing.

PROBLEM SET 18.2

Simplify the following expressions.

1. $\sin \theta \cot \theta$

2. $\dfrac{\sin \theta}{\tan \theta}$

3. $\dfrac{\sec \theta}{\sin \theta} - \dfrac{\sin \theta}{\cos \theta}$

4. $\dfrac{\csc \theta}{\sec \theta}$

5. $\sin \theta + \cos \theta \cot \theta$

6. $\dfrac{\csc \theta}{\cos \theta} - \dfrac{\cos \theta}{\sin \theta}$

7. $\dfrac{\cos \theta}{\cot \theta} + \dfrac{\cot \theta}{\sec \theta}$

8. $\tan \theta \, (\sin \theta + \cot \theta \cos \theta)$

9. $\sin^2 \theta \cos^2 \theta + \cos^4 \theta$

10. $\dfrac{4 + \tan^2 \theta - \sec^2 \theta}{\csc^2 \theta}$

11. $(1 + \cos \theta) \, (\csc \theta - \cot \theta)$

12. $\cos^3 \theta + \cos \theta \sin^2 \theta$

13. $\dfrac{\cos^2 \theta}{1 - \sin \theta} - \sin \theta$

14. $\dfrac{\cos \theta + \sin \theta}{\cos \theta} + \dfrac{\cos \theta - \sin \theta}{\sin \theta}$

15. $1 + (\cos \theta + \sin \theta) \, (\cos \theta - \sin \theta)$

16. $\dfrac{\sin^2 \theta}{1 - \cos \theta}$

17. $\tan^2 \theta + \cos^2 \theta + \sin^2 \theta$

18. $(\cos \theta + \sin \theta)^2 - 1$

19. $\dfrac{\csc \theta}{1 + \cot^2 \theta}$

20. $(\cos \theta + \sin \theta)^2 + (\cos \theta - \sin \theta)^2$

21. $\dfrac{\sin \theta}{\cos \theta + 1} + \dfrac{\cos \theta - 1}{\sin \theta}$

22. $\dfrac{(\tan \theta - \sin \theta)(1 + \cos \theta)}{\sin^3 \theta}$

23. $\sin^2\alpha \cos^2 \beta - \cos^2 \alpha \sin^2 \beta - \sin^2 \alpha + \sin^2 \beta$

24. $\dfrac{\tan \alpha + \tan \alpha}{\cot \alpha + \tan \alpha}$

Express the following functions in terms of sin θ.

25. cos θ 27. cot θ

26. tan θ 28. sec θ

Express the following functions in terms of tan θ.

29. sec θ 31. sin θ

30. cos θ 32. csc θ

Express the following functions in terms of cos θ.

33. sin θ 35. tan θ

34. cot θ 36. csc θ

Express the following functions in terms of cot θ.

37. csc θ 39. cos θ

38. sin θ 40. sec θ

18.3   VERIFICATION OF IDENTITIES

An identity tells you that two expressions are equal for all values of the variables involved. Thus

$$\tan x = \frac{\sin x}{\cos x}$$

is an identity because it is true for *every* x for which both sides are defined. On the other hand,

$$\tan x = 1$$

is true only for *certain* values of x, and so this is an equation.

Proving that something is in fact an identity is called verifying the identity. In order to do this we have to work separately with the expressions on either side of the equals sign, and somehow show that they are equal to one another. The problem is that an identity looks very like an equation—both have an equals sign set between two expressions—and so it is very easy to slip into doing things that are O.K. for solving equations but not for verifying identities. For example, in solving an equation we frequently move things from one side of the equation to the other. But when we're verifying an identity, we're trying to prove that the two expressions on either side of the equals sign are identical, and we can't possibly do that if we move things from one side to the other. Again, in solving an equation we often multiply both sides by some quantity, but in verifying an identity we can't do that, as it changes the values of the expressions that we're trying to prove equal.

Therefore:

To Verify an Identity:

Treat each side separately, and manipulate each as an expression until they are both in the same form.

Don't change the value of the expression on either side by
   Moving things from one side to the other
   Adding/subtracting/multiplying/dividing by anything

This means you can
   Factor
   Multiply out
   Add fractions
   Simplify
   Substitute
   Use the basic identities

If in doubt, it is usually helpful to write everything in terms of sines and cosines.

EXAMPLE: *Verify* $\tan \theta + \cot \theta = \sec \theta \csc \theta$.

To emphasize the fact that the two sides are to be treated separately, it is probably a good idea to separate them by a vertical line until they have been shown equal. We will start by turning both sides into sines and cosines:

$$\tan \theta + \cot \theta \quad \Big| \quad \sec \theta \quad \csc \theta$$

This tells you that $\tan \theta + \cos \theta$ equals this line $\longrightarrow$
$$= \frac{\sin \theta}{\cos \theta} + \frac{\cos \theta}{\sin \theta} \quad \Big| \quad = \frac{1}{\cos \theta} \cdot \frac{1}{\sin \theta}$$

Adding fractions $\longrightarrow$
$$= \frac{\sin^2 \theta + \cos^2 \theta}{\cos \theta \sin \theta} \quad \Big| \quad = \frac{1}{\cos \theta \sin \theta} \longleftarrow$$ Multiplying fractions

Pythagorean identity says $\cos^2 \theta + \sin^2 \theta = 1$ $\longrightarrow$
$$= \frac{1}{\cos \theta \sin \theta}$$

So both sides have turned out to be

$$\frac{1}{\cos\theta\,\sin\theta}$$

and therefore they must be equal to one another. So

$$\tan\theta + \cot\theta \equiv \sec\theta\,\csc\theta$$

where $\equiv$ means that this is an identity, not an equation.

EXAMPLE: *Verify* $\dfrac{\sin\theta}{1-\cos\theta} \equiv \dfrac{1+\cos\theta}{\sin\theta}$

Everything here is already in terms of sines and cosines, so it is not easy to see how to begin. Since both sides are equally simple (or equally complicated), it is also not clear whether we should be working on both sides or just one.

For want of a better idea, we will start by trying to turn the left side into the right by inserting a factor of $(1 + \cos\theta)$ top and bottom (since the right side has such a factor):

$$\frac{\sin\theta}{1-\cos\theta} \qquad\qquad\qquad \frac{1+\cos\theta}{\sin\theta}$$

Multiplying top and bottom by $(1+\cos\theta)$ to match right side $\longrightarrow$
$$= \frac{\sin\theta}{(1-\cos\theta)}\cdot\frac{(1+\cos\theta)}{(1+\cos\theta)}$$

Multiplying out the bottom $\longrightarrow$
$$= \frac{\sin\theta\,(1+\cos\theta)}{1-\cos^2\theta}$$

Pythagorean identity $\sin^2\theta = 1 - \cos^2\theta$ $\longrightarrow$
$$= \frac{\sin\theta\,(1+\cos\theta)}{\sin^2\theta}$$

Cancelling $\sin\theta$ $\longrightarrow$
$$= \frac{1+\cos\theta}{\sin\theta}$$

Therefore the left side has worked out to be the right side, so

$$\frac{\sin\theta}{1-\cos\theta} \equiv \frac{1+\cos\theta}{\sin\theta}$$

EXAMPLE: *Verify* $\dfrac{1}{\sec\theta + \tan\theta} = \sec\theta - \tan\theta$

$$\dfrac{1}{\sec\theta + \tan\theta} \qquad\qquad \sec\theta - \tan\theta$$

$$= \dfrac{1}{\dfrac{1}{\cos\theta} + \dfrac{\sin\theta}{\cos\theta}} \qquad = \dfrac{1}{\cos\theta} - \dfrac{\sin\theta}{\cos\theta}$$

Adding fractions $\longrightarrow$ $= \dfrac{1}{\dfrac{1 + \sin\theta}{\cos\theta}}$ $\qquad = \dfrac{1 - \sin\theta}{\cos\theta}$ $\longleftarrow$ Adding fractions

Simplifying complex fraction $\longrightarrow$ $= \dfrac{\cos\theta}{1 + \sin\theta}$

The two sides now look very much like the two sides of the previous identity—only with sines and cosines and signs interchanged—so we will try verifying it by the same method. We multiply the left side top and bottom by $(1 - \sin\theta)$:

$$\dfrac{\cos\theta}{1 + \sin\theta} \qquad\qquad \dfrac{1 - \sin\theta}{\cos\theta}$$

$$= \dfrac{\cos\theta}{(1 + \sin\theta)} \cdot \dfrac{(1 - \sin\theta)}{(1 - \sin\theta)}$$

$$= \dfrac{\cos(1 - \sin\theta)}{1 - \sin^2\theta}$$

$$= \dfrac{\cos\theta(1 - \sin\theta)}{\cos^2\theta}$$

$$= \dfrac{1 - \sin\theta}{\cos\theta}$$

Both sides have now reduced to the same thing and therefore

$$\dfrac{1}{\sec\theta + \tan\theta} \equiv \sec\theta - \tan\theta$$

EXAMPLE: *Verify* $\dfrac{\cos x\,(1 + \sin x)}{\cos(-x) - \dfrac{\tan(-x)}{\sec^2(-x)}} = 1$

Obviously, the thing to do here is to work on the left-hand side and hope it reduces to 1.

We will need what we know about even and odd trigonometric functions:

	Even	*Odd*	*Odd*

$$\cos(-x) = \cos x \qquad \sin(-x) = -\sin x \qquad \tan(-x) = -\tan x$$
$$\sec(-x) = \sec x \qquad \csc(-x) = -\csc x \qquad \cot(-x) = -\cot x$$

$$\frac{\cos x \,(1 + \sin x)}{\cos(-x) - \dfrac{\tan(-x)}{\sec^2(-x)}} \qquad \bigg| \; 1$$

$-\tan(-x) = +\tan x \longrightarrow \; = \dfrac{\cos x \,(1 + \sin x)}{\cos x + \dfrac{\tan x}{\sec^2 x}}$

Rewriting in sines and cosines $\longrightarrow \; = \dfrac{\cos x \,(1 + \sin x)}{\cos x + \dfrac{\sin x}{\dfrac{\cos x}{\dfrac{1}{\cos^2 x}}}}$

Simplifying $\longrightarrow \; = \dfrac{\cos x \,(1 + \sin x)}{\cos x + \dfrac{\sin x}{\cancel{\cos x}} \cdot \cos^{\cancel{2}} x}$

$$= \dfrac{\cos x \,(1 + \sin x)}{\cos x + \cos x \sin x}$$

Factoring $\longrightarrow \; = \dfrac{\cos x \,(1 + \sin x)}{\cos x \,(1 + \sin x)}$

$$= 1$$

Therefore:

$$\frac{\cos x \,(1 + \sin x)}{\cos(-x) - \dfrac{\tan(-x)}{\sec^2(-x)}} \equiv 1$$

PROBLEM SET   18.3

Verify the following identities.

1. $\cos^2 \theta = \dfrac{\cot \theta \, \sin \theta}{\sec \theta}$

2. $\cot \theta + \dfrac{\sin \theta}{1 + \cos \theta} = \csc \theta$

3. $\tan^2 \theta - \sin^2 \theta = \tan^2 \theta \, \sin^2 \theta$

4. $\cot \theta + \tan \theta = \sec \theta \csc \theta$

5. $\dfrac{1 - \tan \theta}{1 - \cot \theta} = -\tan \theta$

6. $\sec^2 \theta + \csc^2 \theta = \sec^2 \theta \csc^2 \theta$

7. $(1 + \tan \theta)^2 + (1 - \tan \theta)^2 = 2 \sec^2 \theta$

8. $\dfrac{\sin \theta}{\csc \theta - \cot \theta} + \dfrac{\sin \theta}{\csc \theta + \cot \theta} = 2$

9. $\dfrac{\cos \theta}{1 + \sin \theta} + \dfrac{1 + \sin \theta}{\cos \theta} = 2 \sec \theta$

10. $\cot \theta \, (1 + \tan^2 \theta) = \tan \theta \, (1 + \cot^2 \theta)$

11. $\dfrac{\sec \theta + \csc \theta}{\tan \theta + \cot \theta} = \sin \theta + \cos \theta$

12. $(\tan^2 \theta) \, (\csc^2 \theta) \, (\cot^2 \theta) \, (\sin^2 \theta) = 1$

13. $1 - \dfrac{\cos^2 \theta}{1 + \sin \theta} = \sin \theta$

14. $\dfrac{1}{1 - \sin \theta} + \dfrac{1}{1 + \sin \theta} = 2 \sec^2 \theta$

15. $\dfrac{\sin \theta}{\sin \theta + \cos \theta} = \dfrac{\sec \theta}{\sec \theta + \csc \theta}$

16. $\cos^4 \theta + 2 \cos^2 \theta \sin^2 \theta + \sin^4 \theta = 1$

17. $\dfrac{\sin \theta \cot \theta + \cos \theta}{2 \cot \theta} = \sin \theta$

18. $(1 + \tan \theta) \left(\dfrac{\sin \theta}{\sin \theta + \cos \theta} \right) = \tan \theta$

19. $\sin \theta (\cot \theta + \sin \theta \sec \theta) = \sec \theta$

20. $\dfrac{1 + \cos \theta}{\sin \theta} + \dfrac{\sin \theta}{\cos \theta} = \dfrac{\cos \theta + 1}{\sin \theta \cos \theta}$

21. $\sec \theta - \sin \theta \tan \theta = \cos \theta$

22. $\dfrac{1}{\sin \theta \cos \theta} - \dfrac{\cos \theta}{\sin \theta} = \dfrac{\sin \theta \cos \theta}{1 - \sin^2 \theta}$

23. $\dfrac{\cos^2 \theta + \cot \theta}{\cos^2 \theta - \cot \theta} = \dfrac{\cos^2 \theta \tan \theta + 1}{\cos^2 \theta \tan \theta - 1}$

24. $\csc^4 \theta - \cot^4 \theta - \csc^2 \theta = \cot^2 \theta$

25. $1 + \cot^2 \theta \csc^2 \theta + \cot^2 \theta = \csc^4 \theta$

26. $\sin^6 \theta + 3 \sin^2 \theta \cos^2 \theta + \cos^6 \theta = 1$

27. $\dfrac{\sin \theta - \cos \theta + 1}{\sin \theta + \cos \theta - 1} = \dfrac{\sin \theta + 1}{\cos \theta}$

28. $\dfrac{\cos \theta \cot \theta - \sin \theta \tan \theta}{\csc \theta - \sec \theta} = 1 + \sin \theta \cos \theta$

29. $\dfrac{\tan \theta - \sin \theta}{\sin^3 \theta} = \dfrac{\sec \theta}{1 + \cos \theta}$

30. $\sec^4 \theta - \sec^2 \theta = \tan^4 \theta + \tan^2 \theta$

31. $\dfrac{1-\cos\theta}{1+\cos\theta}=(\cot\theta-\csc\theta)^2$

32. $\tan\theta-\csc\theta\sec\theta(1-2\cos^2\theta)=\cot\theta$

33. $\dfrac{\sin^3\theta+\cos^3\theta}{\sin\theta+\cos\theta}=1-\sin\theta\cos\theta$

34. $(\tan\alpha+\tan\beta)(1-\cot\alpha\cot\beta)+(\cot\alpha+\cot\beta)(1-\tan\alpha\tan\beta)=0$

35. Show that if the coordinates of a point are given by

$$x=r\cos\theta,\ y=r\sin\theta\qquad (r\text{ is a constant})$$

then as θ varies from 0 to 2π the points sweep out a circle.

36. Show that if

$$x=a\cos\theta,\ y=b\sin\theta\qquad (a\text{ and }b\text{ are constants})$$

then as θ varies from 0 to 2π, you get an ellipse.

37. Show that if

$$x=a\sec\theta,\ y=b\tan\theta\qquad (a\text{ and }b\text{ are constants})$$

then as θ varies from 0 to 2π, you get a hyperbola.

For values of θ for which the logarithms are defined, verify the following identities.

38. $\log(\sec\theta)+\log(\tan\theta)+\log(\csc\theta)=-2\log(\cos\theta)$

39. $\log(1+\sin\theta)+\log(1-\sin\theta)=2\log(\cos\theta)$

40. $\log(\csc\theta-\cot\theta)+\log(1+\cos\theta)=\log(\sin\theta)$

41. $\log(\cot\theta-\tan\theta)-\log(\cot\theta-1)=\log(1+\tan\theta)$

42. $\log(\cot\theta+\tan\theta)+\log(\csc\theta-\sin\theta)+\log(\sec\theta-\cos\theta)=0$

18.4 IDENTITIES IN TRIGONOMETRIC EQUATIONS

Trigonometric equations are defined as equations containing trigonometric functions in which the unknown is the angle (see Section 17.8). We have solved things like

$$\sin x=\frac{1}{\sqrt{2}}$$

and slightly more complicated equations like:

$$\sin x=\cos x$$

In this section identities will turn out to be useful for solving equations like

$$6 \sin^2 x - 7 \sin x + 2 = 0$$

and

$$2 \sin^2 \theta + 3 \cos \theta = 0$$

Trigonometric equations of the type

$$\sin x = \frac{1}{\sqrt{2}}$$

can be solved either by using the circle definitions (Section 14.3) or graphically (Section 17.8).

New Types of Trigonometric Equations

Consider

$$\cos^2 x = \cos x$$

Although it might look like a promising way to start solving this equation, we cannot cancel cos x from both sides any more than we can cancel y from both sides of $y^2 = y$ (because cos x or y might be zero). However to solve $y^2 = y$ we put everything on one side of the equation, factor, and then set each factor equal to zero. Suppose we try that here:
If

$$\cos^2 x = \cos x$$

then

$$\cos^2 x - \cos x = 0$$

which factors into

$$\cos x(\cos x - 1) = 0$$

Therefore either cos x = 0 or cos x − 1 = 0; that is, either

$$\cos x = 0 \quad \text{or} \quad \cos x = 1$$

The solutions can best be visualized by looking at the graph, shown in Figure 18.2.

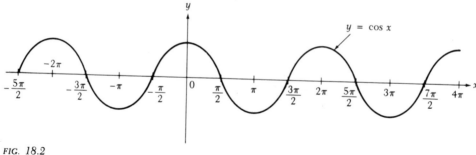

FIG. 18.2

Suppose we are looking for solutions in the range $0 \leq x \leq 2\pi$. Then we see that

$$\cos x = 0 \quad \text{means} \quad x = \frac{\pi}{2} \text{ or } \frac{3\pi}{2}$$

and

$$\cos x = 1 \quad \text{means} \quad x = 0 \text{ or } 2\pi$$

Therefore the solutions are

$$x = 0, \frac{\pi}{2}, \frac{3\pi}{2}, 2\pi$$

EXAMPLE: *Solve $6 \sin^2 x - 7 \sin x + 2 = 0$ for $0° \leq x < 360°$.*

Since we are asked for x between $0°$ and $360°$, we must solve this in degrees.

$$6 \sin^2 x - 7 \sin x + 2 = 0$$

can be factored by analogy with

$$6y^2 - 7y + 2 = (3y - 2)(2y - 1)$$

to give

$$(3 \sin x - 2)(2 \sin x - 1) = 0$$

Therefore, either $3 \sin x - 2 = 0$ or $2 \sin x - 1 = 0$; that is, either

$$\sin x = \frac{2}{3} = 0.6667 \quad \text{or} \quad \sin x = \frac{1}{2} = 0.5$$

If $\sin x = 0.6667$, from the tables, the related angle is $41°48'$.
If $\sin x = 0.5$, the related angle is $30°$. See Figure 18.3.

So solutions between $0°$ and $360°$ are

$x = 30°$ and $180° - 30° = 150°$
$x = 41°48'$ and $180° - 41°48' = 138°12'$

that is,

$x = 30°, 41°48', 138°12',$ and $150°$

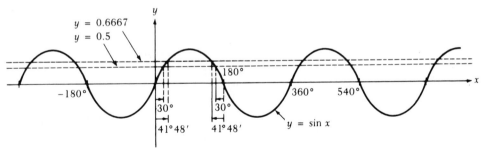

FIG. 18.3

EXAMPLE: *Solve $sec^2 x = 2 \tan x$ for $-\dfrac{\pi}{2} < x < \dfrac{\pi}{2}$.*

This looks hopeless at the moment, as there are two different functions involved, so even if we do put everything on one side to get

$$sec^2 x - 2 \tan x = 0$$

we can't factor it. However there's a way of fixing this. We can write everything in terms of tangents using the Pythagorean identity: $sec^2 x = 1 + \tan^2 x$. This transforms the equation to

$$(1 + \tan^2 x) - 2 \tan x = 0$$

or

$$\tan^2 x - 2 \tan x + 1 = 0$$

By analogy with $y^2 - 2y + 1 = (y - 1)^2$, our equation factors to

$$(\tan x - 1)^2 = 0$$

Therefore

$$\tan x - 1 = 0$$

or

$$\tan x = 1$$

From the tables or the section on special angles, we see that

$$\tan\frac{\pi}{4} = 1$$

And if x is to be between $-\dfrac{\pi}{2}$ and $\dfrac{\pi}{2}$, it must be between 0 and $\dfrac{\pi}{2}$ because the tangent function is negative between $-\dfrac{\pi}{2}$ and 0. Therefore,

$$x = \frac{\pi}{4}$$

EXAMPLE: *Solve $2 \sin^2 x + 3 \cos x = 0$ for $0 \le x < \pi$.*

This again cannot be factored as it stands, because it contains both sines and cosines. But using

$$\cos^2 x + \sin^2 x = 1 \quad \text{or} \quad \sin^2 x = 1 - \cos^2 x$$

the equation can be rewritten in terms of cosines only giving:

$$2(1 - \cos^2 x) + 3 \cos x = 0$$

or

$2 - 2\cos^2 x + 3\cos x = 0$

$2\cos^2 x - 3\cos x - 2 = 0$

Since

$2y^2 - 3y - 2 = (y - 2)(2y + 1)$

the cosine equation factors to

$(\cos x - 2)(2\cos x + 1) = 0$

Therefore either $\cos x - 2 = 0$ or $2\cos x + 1 = 0$; that is, either

$$\cos x = 2 \quad \text{or} \quad \cos x = -\frac{1}{2}$$

But cos x cannot equal 2 no matter what x is, because the cosine function is never greater than 1 (or less than -1). Therefore the equation cos $x = 2$ has no solutions.

If cos $x = -\frac{1}{2}$, the related angle $= \frac{\pi}{3}$; so if x is between 0 and π, then, looking at the graph in Figure 18.4, you can see that

$$x = \pi - \frac{\pi}{3} = \frac{2\pi}{3}$$

So $x = \dfrac{2\pi}{3}$ is the only solution between 0 and π.

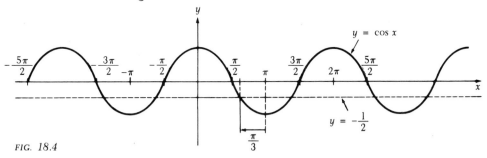

FIG. 18.4

PROBLEM SET 18.4

Solve for θ, $0° \le \theta < 360°$.

1. $\sin \theta = -0.7128$
2. $5\cos \theta = 4.55$
3. $2\cot \theta + 1 = 9$
4. $2\cos^2 \theta = 1$
5. $\tan^2 \theta = 3$
6. $3\cot^2 \theta = 1$
7. $\cos^2 \theta + 1 = 2\cos \theta$
8. $2\cos^2 \theta = \sin \theta + 1$
9. $\sin \theta = \frac{1}{4}\csc \theta$
10. $3\sin^2 \theta = 2 + \sin \theta$

Solve for x, $0 \leq x < 2\pi$.
11. $4 \cos x \sin x + 2 \sin x - 2 \cos x - 1 = 0$
12. $\sin^2 x + 4 \sin x + 4 = 0$
13. $2 \cos^2 x + \cos x - 1 = 0$
14. $\sec^2 x + \sec x - 2 = 0$
15. $\sqrt{3} \sin x = 3 \cos x$
16. $2 \sec^2 x - 3 \sec x = 2$
17. $2 \cos^2 x = 3(1 + \sin x)$
18. $\sin x \sec x = 1$
19. $\tan x = \cot x$
20. $\sin x = \tan x$
21. $\sqrt{6} \sin x = \sqrt{2} \cos x$
22. $1 - \sin^2 x = \frac{1}{4}$
23. $3 \cos^2 x = \sin^2 x$

Solve for θ, $0° \leq \theta < 360°$.
24. $2 \sin \theta \cos \theta - \cos \theta = 0$
25. $\cos \theta = 6 \sin^2 \theta - 5$
26. $2 \sin \theta \cos \theta = \tan \theta$
27. $2 \tan^2 \theta + 5 \tan \theta = -2$
28. $3 \sin^2 \theta + \sin \theta = 0$
29. $1 - 3 \sin \theta - 3 \sin \theta \cos \theta + \cos \theta = 0$
30. $\sin^2 \theta - 3 \cos \theta = 3$
31. $2(2 + \tan^2 \theta) = 5 \sec \theta$
32. $1 - 4 \sin \theta = 4 \cos^2 \theta$
33. $\sec^2 \theta + 2 = 4 \tan \theta$
34. $\sec^2 \theta - 3 = -\tan \theta$
35. $-\sec^2 \theta = \cos^2 \theta - 2$

18.5   IDENTITIES USED TO EVALUATE TRIGONOMETRIC FUNCTIONS

The identities specify the relationships among the trigonometric functions and therefore should enable us to find, without tables, the values of one trigonometric function from another. For example, given $\sin \theta = 0.6$ and $\cos \theta = 0.8$, we can find $\tan \theta$ using

$$\tan \theta = \frac{\sin \theta}{\cos \theta} = \frac{0.6}{0.8} = \frac{3}{4} = 0.75$$

The point is that we have done this without tables—we obviously could also have done it with tables. Since θ must be in the first quadrant, we could have looked up 0.6 in the sine tables to find θ, and then looked up that value of θ in the tangent tables.

EXAMPLE: *Given that sin $\theta = \frac{12}{13}$ and that θ is in the first quadrant, find cos θ.*

Here we have sin θ and want cos θ, so we need an identity connecting them. The obvious candidate is

$$\cos^2 \theta + \sin^2 \theta = 1$$

so we'll use that. Substituting for sin θ:

$$\cos^2 \theta + \left(\frac{12}{13}\right)^2 = 1$$

Solving for cos θ:

$$\cos^2 \theta = 1 - \left(\frac{12}{13}\right)^2 = 1 - \frac{144}{169} = \frac{25}{169}$$

So

$$\cos \theta = \pm\sqrt{\frac{25}{169}} = \pm\frac{5}{13}$$

This is where the information that θ is in the first quadrant comes in—it tells us that cos θ must be positive. Therefore,

$$\cos \theta = \frac{5}{13}$$

Note: If we had been told that θ was in the fourth quadrant, where the cosine is also positive, the answer would have again been $\frac{5}{13}$. If, on the other hand, we had been told that θ was in the second or third quadrant, where the cosine is negative, then cos θ would have been $-\frac{5}{13}$.

EXAMPLE: *Find cot α, given that sec $\alpha = P$ and α is in the second quadrant.*

There is no identity connecting sec α and cot α directly. However, there is an identity connecting sec α and tan α (namely, $1 + \tan^2 \alpha = \sec^2 \alpha$), from which we can find tan α, the reciprocal of cot α.

Substituting into

$$1 + \tan^2 \alpha = \sec^2 \alpha$$

gives

$$1 + \tan^2 \alpha = P^2$$

so

$$\tan^2 \alpha = P^2 - 1$$

$$\tan \alpha = \pm\sqrt{P^2 - 1}$$

Since α is in the second quadrant where the tangent is negative,

$$\tan \alpha = -\sqrt{P^2 - 1}$$

Now

$$\cot \alpha = \frac{1}{\tan \alpha}$$

So

$$\cot \alpha = \frac{-1}{\sqrt{P^2 - 1}}$$

EXAMPLE: *Given that $\tan \theta = \frac{7}{24}$, where θ is in the first quadrant, find $\sin \theta$ and $\cos \theta$.*

This can be done several ways. One of them would be to use

$$1 + \tan^2 \theta = \sec^2 \theta$$

to find $\sec \theta$, and use that to get $\cos \theta$. Then using

$$\cos^2 \theta + \sin^2 \theta = 1$$

we could find $\sin \theta$.

However, there is a much quicker way—namely, drawing a triangle. In any right triangle containing the angle θ.

$$\frac{\text{length of side opposite } \theta}{\text{length of side adjacent to } \theta} = \frac{7}{24}$$

by the triangle definition of tangent. Conversely, any triangle where this ratio does have the value $\frac{7}{24}$ must contain the angle θ (otherwise we'd have two angles between 0° and 90° with the same tangent, which doesn't happen). One such triangle has legs of length 7 and 24, as shown in Figure 18.5. To work out the sine and cosine we need to know the length of the hypotenuse, c. By Pythagoras's theorem:

$$c^2 = 7^2 + 24^2$$
$$= 49 + 576$$
$$= 625$$

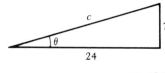

FIG. *18.5*

so

so $c = \sqrt{625} = 25$ (c can't be negative, so we take the positive square root)

Now

$$\sin \theta = \frac{\text{opposite}}{\text{hypotenuse}} \quad \text{and} \quad \cos \theta = \frac{\text{adjacent}}{\text{hypotenuse}}$$

Therefore

$$\sin \theta = \frac{7}{25} \quad \text{and} \quad \cos \theta = \frac{24}{25}$$

Pitfall: Given $\tan \theta = \frac{7}{24}$, it is perfectly reasonable to assume that θ is in a right triangle with sides of 7 and 24 because the trigonometric functions are determined by the shape, not the size, of a triangle. However, it is *not* reasonable to conclude from the fact that

$$\tan \theta = \frac{\sin \theta}{\cos \theta} \quad \text{and} \quad \tan \theta = \frac{7}{24}$$

that $\sin \theta = 7$ and $\cos \theta = 24$. For one thing, this is impossible because $\sin \theta$ and $\cos \theta$ must both be less than 1. For another, knowing that $\sin \theta$ and $\cos \theta$ are in the ratio 7:24 (which is what $\tan \theta = \frac{7}{24}$ tells you) just does not mean that $\sin \theta$ is actually 7 and $\cos \theta$ is actually 24. In general, if

$$\frac{a}{b} = \frac{7}{24}$$

you cannot conclude that $a = 7$ and $b = 24$ because, for example, a might be 14 while b was 48, or you might have $a = 0.7$ and $b = 2.4$.

PROBLEM SET   18.5

Solve the following without using tables.

1. Find the value of cos A, where A is the angle between 0° and 90° for which $\sin A = \frac{4}{5}$.

2. Find the value of sin B, where B is the angle between 0° and 90° for which $\cos B = \frac{3}{5}$.

3. Find the value of sec C, where C is the angle between 180° and 270° for which $\tan C = \frac{5}{12}$.

4. Find the value of cot A, where A is the angle between 90° and 180° for which $\csc A = \frac{13}{5}$.

5. Find the value of tan B, where B is the angle between 270° and 360° for which $\sec B = \frac{17}{15}$.

6. Find the value of csc C, where C is the first-quadrant angle for which $\cot C = \frac{7}{24}$.

7. Find the value of cos A, where A is the second-quadrant angle for which $\sin A = \frac{12}{13}$.

8. Find the value of sin B, where B is the third-quadrant angle for which $\cos B = -\frac{7}{25}$.

9. Find the value of csc θ, where θ is the angle between 0° and 90° for which cos $\theta = \frac{12}{13}$.

10. Find the value of sin θ, where θ is the second-quadrant angle for which tan $\theta = -\frac{15}{8}$.

11. Find the value of sec θ, where θ is the third-quadrant angle for which cot $\theta = \frac{21}{4}$.

12. Find the value of cos θ, where θ is the fourth-quadrant angle for which csc $\theta = -\frac{25}{24}$.

18.6   SUM AND DIFFERENCE FORMULAS

In the earlier part of this chapter we have been concerned with the relationships between different trigonometric functions of the same angle, for example,

$$\cot \theta = \frac{\cos \theta}{\sin \theta} \quad \text{or} \quad 1 + \tan^2 \theta = \sec^2 \theta$$

In this section we will look at the relationship among the values of one trigonometric function of different angles. For example, what is the relationship between sin 30° and sin 60°? Is sin 60° twice sin 30°? What is the connection between cos 10°, cos 40°, and cos 50°? Do the first two add to give the third?

For reasons that are not the least apparent at the moment, such relationships between sines and cosines of different angles are of the greatest importance in calculus. In particular, the connection between the cosine of an angle and the cosine of twice the angle, and between the sine of an angle and the sine of twice the angle, come up over and over again. The next three sections will therefore be devoted to the development of these identities, and the fourth to some rather contrived applications. But as I'm sure you will realize when you read them, such applications are not the real reason for the amount of emphasis on these identities—the real justification and the real applications are all in the calculus books.

Now, back to the actual relationships between the sines and cosines of different angles.

First we will look at the sines and cosines of the sums and differences of angles. It is tempting to guess that sin $(\alpha + \beta) = \sin \alpha + \sin \beta$, but a quick look at the tables should convince you that is not true—sin 70° is just not the sum of sin 30° and sin 40°. Even clearer is the fact that while sin 180° = 0, sin 90° + sin 90° = 2, so sin (180°) $\neq$ sin 90° + sin 90° and therefore, in general,

$$\sin (\alpha + \beta) \neq \sin \alpha + \sin \beta$$

By exactly the same reasoning, you can convince yourself that

$$\cos (\alpha + \beta) \neq \cos \alpha + \cos \beta$$

and that

$$\sin (2\theta) \neq 2 \sin \theta \quad \text{etc.}$$

Having debunked most of the obvious-looking formulas for $\sin(\alpha + \beta)$, $\cos(\alpha + \beta)$, $\sin(\alpha - \beta)$, and $\cos(\alpha - \beta)$ in terms of the sines and cosines of α and β, let me state what the real formulas are, and then prove them. These formulas are anything but obvious, so you should not expect to have guessed them.

The Sum and Difference Formulas for Sine and Cosine

$$\sin(\alpha + \beta) = \sin \alpha \cos \beta + \cos \alpha \sin \beta$$

$$\sin(\alpha - \beta) = \sin \alpha \cos \beta - \cos \alpha \sin \beta$$

$$\cos(\alpha + \beta) = \cos \alpha \cos \beta - \sin \alpha \sin \beta$$

$$\cos(\alpha - \beta) = \cos \alpha \cos \beta + \sin \alpha \sin \beta$$

Now you can see that the actual relationship between the cosines of $10°$, $40°$, and $50°$ is

$$\cos 50° = \cos(40° + 10°) = \cos 40° \cos 10° - \sin 40° \sin 10°$$

(which you can check using tables).

Also,

$$\sin 60° = \sin(30° + 30°) = \sin 30° \cos 30° + \cos 30° \sin 30°$$

so

$$\sin 60° = 2 \sin 30° \cos 30°$$

So indeed $\sin 60°$ is not twice $\sin 30°$, and you cannot get $\cos 50°$ by adding $\cos 40°$ and $\cos 10°$.

Let us check these formulas in a couple of simple cases.

EXAMPLE: *Check the formula for* $\cos(\alpha - \beta)$ *using* $\alpha = 60°$, $\beta = 30°$

$$\sin 60° = \frac{\sqrt{3}}{2} \qquad \cos 60° = \frac{1}{2} \qquad \sin 30° = \frac{1}{2} \qquad \cos 30° = \frac{\sqrt{3}}{2}$$

so

$$\cos(60° - 30°) = \cos 30° = \frac{\sqrt{3}}{2}$$

and

$$\cos 60° \cos 30° + \sin 60° \sin 30° = \frac{1}{2} \cdot \frac{\sqrt{3}}{2} + \frac{\sqrt{3}}{2} \cdot \frac{1}{2} = \frac{2\sqrt{3}}{4}$$

$$= \frac{\sqrt{3}}{2}$$

so

$$\cos(60° - 30°) = \cos 60° \cos 30° + \sin 60° \sin 30°$$

Therefore the formula for $\cos(\alpha - \beta)$ holds when $\alpha = 60°$, $\beta = 30°$.

EXAMPLE: *Check the formula for sin $(\alpha + \beta)$ using $\alpha = \beta = \dfrac{\pi}{4}$*

$$\sin \frac{\pi}{4} = \cos \frac{\pi}{4} = \frac{1}{\sqrt{2}}$$

so

$$\sin \left(\frac{\pi}{4} + \frac{\pi}{4} \right) = \sin \frac{\pi}{2} = 1$$

and

$$\sin \frac{\pi}{4} \cos \frac{\pi}{4} + \cos \frac{\pi}{4} \sin \frac{\pi}{4} = \frac{1}{\sqrt{2}} \cdot \frac{1}{\sqrt{2}} + \frac{1}{\sqrt{2}} \cdot \frac{1}{\sqrt{2}} = \frac{1}{2} + \frac{1}{2} = 1$$

So

$$\sin \left(\frac{\pi}{4} + \frac{\pi}{4} \right) = \sin \frac{\pi}{4} \cos \frac{\pi}{4} + \cos \frac{\pi}{4} \sin \frac{\pi}{4}$$

Therefore the formula for sin $(\alpha + \beta)$ holds when $\alpha = \beta = \dfrac{\pi}{4}$.

How to remember these formulas: Mostly you just have to remember them, because learning the proofs is certainly worse than learning the formulas. However, there are a few rays of hope, one being that it is very easy to get any three of the formulas from the fourth (see next section), so you really have to remember only one of them.

Second, there's a fairly easy way to get the signs, which is to remember that, between 0° and 90°, as the angle increases the sine increases but the cosine decreases. Therefore, adding β to α makes the sine larger, hence the $+$ sign:

$$\sin(\alpha + \beta) = \sin \alpha \cos \beta \; \oplus \; \cos \alpha \sin \beta$$

and the cosine smaller, hence the $-$ sign:

$$\cos(\alpha + \beta) = \cos \alpha \cos \beta \; \ominus \; \sin \alpha \sin \beta$$

The signs for $(\alpha - \beta)$ are the opposite way round, because subtracting β from α makes its sine smaller:

$$\sin(\alpha - \beta) = \sin \alpha \cos \beta \; \ominus \; \cos \alpha \sin \beta$$

and its cosine larger:

$$\cos(\alpha - \beta) = \cos \alpha \cos \beta \; \oplus \; \sin \alpha \sin \beta$$

PROBLEM SET   18.6

Use tables to verify the sum and difference formulas for the following pairs of angles.

1. $27°$ and $13°$
2. $49°$ and $30°$
3. $51°$ and $-10°$
4. $10°$ and $-20°$
5. $81°$ and $312°$
6. $-107°$ and $107°$

Write the following as a single term.

7. $\sin 35° \cos 10° - \cos 35° \sin 10°$
8. $\cos 3x \cos x - \sin 3x \sin x$
9. $\sin(\alpha + \beta) \cos \beta - \cos(\alpha + \beta) \sin \beta$

Applying the sum and difference formulas to $\dfrac{\pi}{3}$ and $\dfrac{\pi}{4}$, find:

10. $\cos \dfrac{7\pi}{12}$

11. $\sin \dfrac{\pi}{12}$

12. $\sin \dfrac{11\pi}{12}$

Simplify:

13. $\dfrac{\sin(\alpha + \beta) + \sin(\alpha - \beta)}{\cos \alpha \cos \beta}$

14. $\dfrac{\cos\left(\dfrac{\pi}{3} + \theta\right) + \cos \theta}{\sin\left(\dfrac{\pi}{3} + \theta\right) - \sin \theta}$

15. $\cos \left(\dfrac{3\pi}{2} + \theta\right) + \cos \left(\dfrac{3\pi}{2} - \theta\right)$

Prove each of the following by using the sum and difference formulas.

16. $\sin(\pi - \theta) = \sin \theta$
17. $\sin(\pi + \theta) = -\sin \theta$
18. $\sin(2\pi - \theta) = -\sin \theta$
19. $\sin\left(\dfrac{\pi}{2} + \theta\right) = \cos \theta$
20. $\sin\left(\dfrac{\pi}{2} - \theta\right) = \cos \theta$
21. $\cos(\pi - \theta) = -\cos \theta$
22. $\cos(\pi + \theta) = -\cos \theta$
23. $\cos(2\pi - \theta) = \cos \theta$
24. $\cos\left(\dfrac{\pi}{2} + \theta\right) = -\sin \theta$
25. $\cos\left(\dfrac{\pi}{2} - \theta\right) = \sin \theta$
26. Prove that for any triangle (not just right triangles) with angles A, B, C,

$$\sin(A + B) = \sin C$$

18.7   PROOF OF THE SUM AND DIFFERENCE FORMULAS

This section will come in two parts: first the derivation of three of the formulas from the fourth, and second, the proof of the fourth from scratch.

Derivation of Three Sum and Difference Formulas from the Fourth

Suppose we start by *assuming*

$$\cos(\alpha - \beta) = \cos \alpha \cos \beta + \sin \alpha \sin \beta$$

I choose this one because it's easiest to prove from scratch.

Formula for cos($\alpha + \beta$)   We can derive the formula for $\cos(\alpha + \beta)$ by saying that $\cos(\alpha + \beta) = \cos[\alpha - (-\beta)]$ and using the above formula with β replaced by $(-\beta)$:

$$\cos(\alpha + \beta) = \cos[\alpha - (-\beta)]$$

$$= \cos \alpha \cos(-\beta) + \sin \alpha \sin(-\beta) \qquad \text{(using the formula we assumed, with } \beta \text{ replaced by } -\beta)$$

Now use the fact that cosine is an even function, that is, $\cos(-\beta) = \cos \beta$ and sine is an odd function, that is, $\sin(-\beta) = -\sin \beta$ so that

$$\cos(\alpha + \beta) = \cos \alpha \cos(-\beta) + \sin \alpha \sin(-\beta)$$

$$= \cos \alpha \cos \beta + \sin \alpha \, (-\sin \beta)$$

$$= \cos \alpha \cos \beta - \sin \alpha \sin \beta$$

Therefore,

$$\cos(\alpha + \beta) = \cos \alpha \cos \beta - \sin \alpha \sin \beta$$

Formula for sin ($\alpha + \beta$)   To derive the formula for $\sin(\alpha + \beta)$, we will need the fact that

$$\sin \theta = \cos\left(\frac{\pi}{2} - \theta\right) \quad \text{and} \quad \cos \theta = \sin\left(\frac{\pi}{2} - \theta\right)$$

These equations have been proved already for θ between 0 and $\frac{\pi}{2}$—they are results about the sines and cosines of complementary angles. However, these equations are true for any value of θ, and can be derived from our formula for $\cos(\alpha - \beta)$, as follows.

Let us start with $\cos\left(\dfrac{\pi}{2} - \theta\right)$, and substitute $\dfrac{\pi}{2}$ for α and θ for β:

$$\cos\left(\frac{\pi}{2} - \theta\right) = \cos\frac{\pi}{2}\cos\theta + \sin\frac{\pi}{2}\sin\theta$$

$$= 0\cdot\cos\theta + 1\cdot\sin\theta \qquad \left[\text{since }\cos\frac{\pi}{2} = 0,\ \sin\frac{\pi}{2} = 1\right]$$

$$= \sin\theta$$

Therefore

$$\sin\theta = \cos\left(\frac{\pi}{2} - \theta\right)$$

Using this, and substituting $\left(\dfrac{\pi}{2} - \theta\right)$ for θ, we can say that:

$$\sin\left(\frac{\pi}{2} - \theta\right) = \cos\left[\frac{\pi}{2} - \left(\frac{\pi}{2} - \theta\right)\right]$$

$$= \cos\left(\frac{\pi}{2} - \frac{\pi}{2} + \theta\right)$$

$$= \cos\theta$$

Therefore

$$\cos\theta = \sin\left(\frac{\pi}{2} - \theta\right)$$

Now we are able to derive the formula for $\sin(\alpha + \beta)$. First

$$\sin(\alpha + \beta) = \cos\left[\frac{\pi}{2} - (\alpha + \beta)\right] \qquad \text{(by the results we just proved)}$$

$$= \cos\left[\left(\frac{\pi}{2} - \alpha\right) - \beta\right]$$

$$= \cos\left(\frac{\pi}{2} - \alpha\right)\cos\beta + \sin\left(\frac{\pi}{2} - \alpha\right)\sin\beta$$

$$\text{[using the formula for }\cos(\alpha - \beta)$$
$$\text{with }\alpha\text{ replaced by }\left(\frac{\pi}{2} - \alpha\right)]$$

$$= \sin\alpha\cos\beta + \cos\alpha\sin\beta$$

$$\left[\text{since }\cos\left(\frac{\pi}{2} - \alpha\right) = \sin\alpha,\ \sin\left(\frac{\pi}{2} - \alpha\right) = \cos\alpha\right]$$

Therefore

$$\sin(\alpha + \beta) = \sin\alpha\cos\beta + \cos\alpha\sin\beta$$

Formula for sin(α − β)   This can be derived from the formula for sin(α + β) by replacing β by (−β):

$$\sin(\alpha - \beta) = \sin\,[\alpha + (-\beta)]$$

$$= \sin\,\alpha\,\cos\,(-\beta) + \cos\,\alpha\,\sin(-\beta)$$

$$= \sin\,\alpha\,\cos\,\beta + \cos\,\alpha\,(-\sin\,\beta) \qquad [\cos(-\beta) = \cos\,\beta$$
$$\sin(-\beta) = -\sin\,\beta]$$

$$= \sin\,\alpha\,\cos\,\beta - \cos\,\alpha\,\sin\,\beta$$

Therefore

$$\sin(\alpha - \beta) = \sin\,\alpha\,\cos\,\beta - \cos\,\alpha\,\sin\,\beta$$

So this is how the other three formulas can be derived from the one for cos(α − β). Now all that is left is to prove this formula from scratch.

Proof of Formula
for cos(α − β) from Scratch

Draw α and β in standard position, assuming for the sake of definiteness that α > β.   (If β is the larger one, switch the order using cos(α − β) = cos[−(β − α)] = cos(β − α), and then relabel the angles, calling the larger one α and the smaller one β.) Let P and Q be the points where the terminal sides of α and β cut the unit circle, as shown in Figure 18.6.

Now, if the coordinates of P and Q are

$$P = (x, y)$$

$$Q = (w, z)$$

then

$$\cos\,\alpha = x \quad \text{and} \quad \sin\,\alpha = y$$

$$\cos\,\beta = w \quad \text{and} \quad \sin\,\beta = z$$

so

$$P = (\cos\,\alpha,\,\sin\,\alpha)$$

$$Q = (\cos\,\beta,\,\sin\,\beta)$$

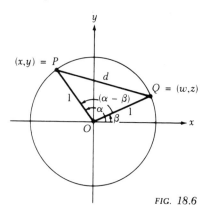

FIG. 18.6

The distance formula tells us that d, the distance from P to Q, is given by

$$d = \sqrt{(x - w)^2 + (y - z)^2}$$

or

$$d^2 = (x - w)^2 + (y - z)^2$$

so

$$d^2 = (\cos \alpha - \cos \beta)^2 + (\sin \alpha - \sin \beta)^2$$

$$= \cos^2 \alpha - 2 \cos \alpha \cos \beta + \cos^2 \beta + \sin^2 \alpha - 2 \sin \alpha \sin \beta + \sin^2 \beta \qquad \text{(multiplying out)}$$

$$= \underbrace{\cos^2 \alpha + \sin^2 \alpha}_{1} + \underbrace{\cos^2 \beta + \sin^2 \beta}_{1} - 2 \cos \alpha \cos \beta - 2 \sin \alpha \sin \beta$$

$$= \qquad\quad 1 \qquad + \qquad 1 \qquad\quad - 2 \cos \alpha \cos \beta - 2 \sin \alpha \sin \beta$$

$$= 2 - 2 \cos \alpha \cos \beta - 2 \sin \alpha \sin \beta$$

But we can also find d^2 by applying the law of cosines to triangle OPQ. For a triangle ABC, the law of cosines says that

$$c^2 = a^2 + b^2 - 2ab \cos C$$

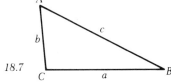

FIG. 18.7

See Figure 18.7. Therefore, in triangle OPQ, with $C = (\alpha - \beta)$, $a = b = 1$, we have

$$d^2 = 1^2 + 1^2 - 2 \cdot 1 \cdot 1 \cos(\alpha - \beta)$$
$$= 2 - 2 \cos(\alpha - \beta)$$

Equating these two formulas for d^2, we see that:

$$2 - 2 \cos(\alpha - \beta) = 2 - 2 \cos \alpha \cos \beta - 2 \sin \alpha \sin \beta$$

So, subtracting 2 from both sides:

$$-2 \cos(\alpha - \beta) = -2 \cos \alpha \cos \beta - 2 \sin \alpha \sin \beta$$

and, dividing through by -2, we find that

$$\cos(\alpha - \beta) = \cos \alpha \cos \beta + \sin \alpha \sin \beta$$

PROBLEM SET   18.7

1. Prove:

$$\tan(\alpha + \beta) = \frac{\tan \alpha + \tan \beta}{1 - \tan \alpha \tan \beta}$$

$$\left[\textit{Hint: } \tan(\alpha + \beta) = \frac{\sin(\alpha + \beta)}{\cos(\alpha + \beta)}; \text{ use the sum formulas and divide} \atop \text{top and bottom by } \cos \alpha \cos \beta. \right]$$

2. Prove:

$$\cot(\alpha + \beta) = \frac{\cot \alpha \cot \beta - 1}{\cot \beta + \cot \alpha}$$

3. Prove:

$$\tan(\alpha - \beta) = \frac{\tan \alpha - \tan \beta}{1 + \tan \alpha \tan \beta}$$

(*Hint:* $\tan(\alpha - \beta) = \tan[\alpha + (-\beta)]$)

4. Prove:

$$\cot(\alpha - \beta) = \frac{\cot \alpha \cot \beta + 1}{\cot \beta - \cot \alpha}$$

5. Prove:

$$\sin x + \sin y = 2 \sin\left(\frac{x + y}{2}\right) \cos\left(\frac{x - y}{2}\right)$$

[*Hint:* Add $\sin(\alpha + \beta)$ to $\sin(\alpha - \beta)$, then let $\alpha + \beta = x$ and $\alpha - \beta = y$, and substitute for α and β.]

6. Prove:

$$\sin x - \sin y = 2 \cos\left(\frac{x + y}{2}\right) \sin\left(\frac{x - y}{2}\right)$$

7. Prove:

$$\cos x + \cos y = 2 \cos\left(\frac{x + y}{2}\right) \cos\left(\frac{x - y}{2}\right)$$

8. Prove:

$$\cos x - \cos y = -2 \sin\left(\frac{x + y}{2}\right) \sin\left(\frac{x - y}{2}\right)$$

18.8 THE DOUBLE ANGLE FORMULAS

The most frequent application of the sum formulas occurs when $\alpha = \beta$, which gives the *double angle formulas*. Putting $\alpha = \beta$ turns

$$\sin(\alpha + \beta) = \sin \alpha \cos \beta + \cos \alpha \sin \beta$$

into

$$\sin(\alpha + \alpha) = \sin \alpha \cos \alpha + \cos \alpha \sin \alpha$$

or

$$\boxed{\sin 2\alpha = 2 \sin \alpha \cos \alpha}$$

Similarly,

$$\cos(\alpha + \beta) = \cos \alpha \cos \beta - \sin \alpha \sin \beta$$

becomes

$$\cos(\alpha + \alpha) = \cos \alpha \cos \alpha - \sin \alpha \sin \alpha$$

or

$$\cos 2\alpha = \cos^2 \alpha - \sin^2 \alpha$$

Using the fact that $\sin^2\alpha = 1 - \cos^2\alpha$, the formula for $\cos 2\alpha$ can be rewritten just in terms of $\cos \alpha$:

$$\cos 2\alpha = \cos^2 \alpha - (1 - \cos^2 \alpha)$$

or

$$\cos 2\alpha = 2 \cos^2 \alpha - 1$$

Similarly, using $\cos^2 \alpha = 1 - \sin^2 \alpha$ enables you to write $\cos 2\alpha$ just in terms of $\sin \alpha$:

$$\cos 2\alpha = (1 - \sin^2 \alpha) - \sin^2 \alpha$$

or

$$\cos 2\alpha = 1 - 2 \sin^2 \alpha$$

Notice that no amount of maneuvering will get the formula for $\sin 2\alpha$ in terms of just $\sin \alpha$ or just $\cos \alpha$. These formulas for $\sin 2\alpha$ and $\cos 2\alpha$ are the *double angle formulas.*

Let us check them in a few cases.

EXAMPLE: *Check the formula for* $\sin 2\alpha$ *when* $\alpha = \dfrac{\pi}{6}$

We know that $\cos \dfrac{\pi}{6} = \dfrac{\sqrt{3}}{2}$, $\sin \dfrac{\pi}{6} = \dfrac{1}{2}$, $\sin \dfrac{\pi}{3} = \dfrac{\sqrt{3}}{2}$. Now,

$$2 \sin \dfrac{\pi}{6} \cos \dfrac{\pi}{6} = 2 \cdot \dfrac{1}{2} \cdot \dfrac{\sqrt{3}}{2} = \dfrac{\sqrt{3}}{2} = \sin \dfrac{\pi}{3}$$

Therefore

$$\sin \frac{\pi}{3} = 2 \sin \frac{\pi}{6} \cos \frac{\pi}{6}$$

and so the double angle formula for the sine holds when $\alpha = \dfrac{\pi}{6}$.

EXAMPLE: *Check the formulas for cos 2α when* $\alpha = \dfrac{\pi}{4}$

$$\sin \frac{\pi}{4} = \cos \frac{\pi}{4} = \frac{1}{\sqrt{2}}$$

Now

$$\cos^2 \frac{\pi}{4} - \sin^2 \frac{\pi}{4} = \left(\frac{1}{\sqrt{2}}\right)^2 - \left(\frac{1}{\sqrt{2}}\right)^2 = 0$$

and

$$2 \cos^2 \frac{\pi}{4} - 1 = 2\left(\frac{1}{\sqrt{2}}\right)^2 - 1 = 2 \cdot \frac{1}{2} - 1 = 0$$

and

$$1 - 2 \sin^2 \frac{\pi}{4} = 1 - 2\left(\frac{1}{\sqrt{2}}\right)^2 = 1 - 2 \cdot \frac{1}{2} = 0$$

Also, $\cos\left(\dfrac{\pi}{2}\right) = 0$, so

$$\cos \frac{\pi}{2} = \cos^2 \frac{\pi}{4} - \sin^2 \frac{\pi}{4}$$

$$= 2 \cos^2 \frac{\pi}{4} - 1$$

$$= 1 - 2 \sin^2 \frac{\pi}{4}$$

So all the double angle formulas for cosine hold when $\alpha = \dfrac{\pi}{4}$.

PROBLEM SET   18.8

1. Show that $\tan 2\alpha = \dfrac{2 \tan \alpha}{1 - \tan^2 \alpha}$.

2. Show that $\cot 2\alpha = \dfrac{\cot^2 \alpha - 1}{2 \cot \alpha}$.

3. Show that $\sin^2 \theta = \dfrac{1 - \cos 2\theta}{2}$.

4. Show that $\cos^2 \theta = \dfrac{1 + \cos 2\theta}{2}$.

5. Prove that $\sin \dfrac{\theta}{2} = \pm \sqrt{\dfrac{1 - \cos \theta}{2}}$.

   [*Hint:* $\sin^2 \beta = \dfrac{1 - \cos 2\beta}{2}$; let $\beta = \dfrac{\theta}{2}$.]

6. Prove that $\cos \dfrac{\theta}{2} = \pm \sqrt{\dfrac{1 + \cos \theta}{2}}$.

7. Prove that $\tan \dfrac{\theta}{2} = \pm \sqrt{\dfrac{1 - \cos \theta}{1 + \cos \theta}}$.

Verify the following identities.

8. $\tan 2\theta = \dfrac{2}{\cot \theta - \tan \theta}$

9. $\dfrac{1 + \cos 2\theta}{\sin 2\theta} = \cot \theta$

10. $\dfrac{2 \tan \theta}{\sec^2 \theta - 1} = \dfrac{\sin 2\theta}{(1 + \cos \theta)(1 - \cos \theta)}$

11. $\cos(\alpha + \beta) + \cos(\alpha - \beta) = 2 \cos \alpha \cos \beta$

12. $\cot^2 \theta = \dfrac{1 + \cos 2\theta}{1 - \cos 2\theta}$

13. $(\sin \theta + \cos \theta)^2 = 1 + \sin 2\theta$

14. $1 + \sin 2\theta = \left(\dfrac{\cos 2\theta}{\cos \theta - \sin \theta} \right)^2$

15. Find a formula for $\sin 3\theta$ in terms of $\sin \theta$ only.
16. Find a formula for $\cos 3\theta$ in terms of $\cos \theta$ only.

Verify the following identities.

17. $\dfrac{\cos 2\theta}{\sin \theta} - \dfrac{\sin 2\theta}{\cos \theta} = \dfrac{2 \cos 3\theta}{\sin 2\theta}$

18. $\dfrac{\sin 3\theta}{\sin \theta} - \dfrac{\cos 3\theta}{\cos \theta} = 2$

19. $\dfrac{\cos 3\theta}{\cos \theta} = 1 - 4 \sin^2 \theta$

20. $1 + \dfrac{\sin 3\theta}{\sin \theta} = 4 \cos^2 \theta$

18.9   **APPLICATIONS OF SUM, DIFFERENCE, AND DOUBLE ANGLE FORMULAS**

The sum, difference, and double angle formulas can be used to simplify expressions, verify identities, solve equations, and evaluate functions just as the basic identities were used at the beginning of this chapter. Examples of their use in each kind of problem follow.

Simplification

EXAMPLE: *Simplify*

$$\frac{\cos x(1 + \cos 2x)}{2 \cos x + \sin 2x}$$

Notice that we have two different angles here, x and $2x$, and we must first write everything in terms of the same angle. This is similar to the idea that, when trying to simplify an expression containing several different trig functions, it helps to write everything in terms of sines and cosines. That way we can at least see what we've got, and there's a chance some of it will cancel. Therefore we will substitute

$$\sin 2x = 2 \sin x \cos x$$

and

$$\cos 2x = 1 - 2 \sin^2 x$$

into the expression. [Starting with one of the other identities for $\cos 2x$ will work out o.k. too, though you will end up having to turn everything into sines at the end.] This gives:

$$\frac{\cos x(1 + \cos 2x)}{2 \cos x + \sin 2x} = \frac{\cos x(1 + 1 - 2 \sin^2 x)}{2 \cos x + 2 \sin x \cos x}$$

$$= \frac{\cancel{\cos x}(2 - 2 \sin^2 x)}{2 \cancel{\cos x}(1 + \sin x)} \qquad \text{(Factoring the denominator)}$$

$$= \frac{\cancel{2}(1 - \sin^2 x)}{\cancel{2}(1 + \sin x)} \qquad \text{(Factoring again)}$$

$$= \frac{(1 - \sin x)\cancel{(1 + \sin x)}}{\cancel{(1 + \sin x)}} \qquad \begin{bmatrix} \text{Factoring} \\ (1 - \sin^2 x) \end{bmatrix}$$

$$= 1 - \sin x$$

Now, $(1 - \sin x)$ is certainly simpler than what we started with and since it cannot be simplified any further,

$$\frac{\cos x(1 + \cos 2x)}{2 \cos x + \sin 2x} = 1 - \sin x \qquad \text{is the answer.}$$

Verification

EXAMPLE: *Verify*

$$\frac{sin(\alpha + \beta) + sin(\alpha - \beta)}{sin(\alpha + \beta) - sin(\alpha - \beta)} \equiv tan\ \alpha\ cot\ \beta$$

As in the previous verification problems, we will work on each side separately until they come down to the same thing. To start with, it's helpful to write everything in terms of sines and cosines (using the quotient identities) of α and β (using the sum and difference formulas).

Left-hand side

$$= \frac{sin(\alpha + \beta) + sin(\alpha - \beta)}{sin(\alpha + \beta) - sin(\alpha - \beta)}$$

Using the sum and difference formulas we get:

$$= \frac{sin\ \alpha\ cos\ \beta + cos\ \alpha\ sin\ \beta + sin\ \alpha\ cos\ \beta - cos\ \alpha\ sin\ \beta}{sin\ \alpha\ cos\ \beta + cos\ \alpha\ sin\ \beta - (sin\ \alpha\ cos\ \beta - cos\ \alpha\ sin\ \beta)}$$

$$= \frac{2\ sin\ \alpha\ cos\ \beta}{2\ cos\ \alpha\ sin\ \beta}$$

$$= \frac{sin\ \alpha}{cos\ \alpha} \cdot \frac{cos\ \beta}{sin\ \beta}$$

Right-hand side

$$= tan\ \alpha\ cot\ \beta$$

$$= \frac{sin\ \alpha}{cos\ \alpha} \cdot \frac{cos\ \beta}{sin\ \beta} \qquad \text{by the Quotient formula}$$

So

$$\frac{sin(\alpha + \beta) + sin(\alpha - \beta)}{sin(\alpha + \beta) - sin(\alpha - \beta)} \equiv tan\ \alpha\ cot\ \beta$$

Solution of Equations

EXAMPLE: *Solve for x between 0° and 360°:*

$$5\ sin\ 2x + 8\ cos\ x = 0$$

We first get everything in terms of x, using

$$sin\ 2x = 2\ sin\ x\ cos\ x$$

which makes the equation read:

$$5(2\ sin\ x\ cos\ x) + 8\ cos\ x = 0$$

So

$$5 \sin x \cos x + 4 \cos x = 0 \qquad \text{(dividing through by 2)}$$

Factoring out $\cos x$ gives

$$\cos x(5 \sin x + 4) = 0$$

Therefore

either $\cos x = 0$ or $5 \sin x + 4 = 0$

that is,

either $\cos x = 0$ or $\sin x = -0.8$

If $\cos x = 0$, and $0 \le x \le 360°$, then

$$x = 90° \quad \text{or} \quad x = 270°$$

If $\sin x = -0.8$ and $0 \le x \le 360°$, then, from the tables,

$$\sin(53°6') = 0.8$$

so the related angle is 53°6'. Looking at the graph of $y = \sin x$ in Figure 18.8, we see that

$$x = 233°6' \quad \text{or} \quad 306°54'$$

Therefore the solutions to the equation are

$$x = 90°, \ 233°6', \ 270°, \ 306°54'$$

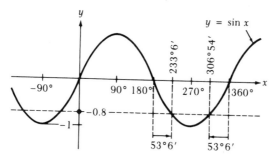

FIG. 18.8

Evaluation of Trigonometric Functions (Without Tables)

EXAMPLE: *Find cos 15° without tables.*

The only angles whose sines and cosines we can find without tables so far are 0°, 30°, 45°, 60°, 90°, and angles which have these as related angles. We can, however, use our knowledge of these angles to find the cosine and sine of 15°, because $15° = 45° - 30°$.

Now

$$\cos 45° = \sin 45° = \frac{1}{\sqrt{2}}$$

and

$$\cos 30° = \frac{\sqrt{3}}{2} \qquad \sin 30° = \frac{1}{2}$$

so we can use

$$\cos(\alpha - \beta) = \cos \alpha \cos \beta + \sin \alpha \sin \beta$$

with $\alpha = 45°$, $\beta = 30°$, to show that

$$\cos 15° = \cos(45° - 30°) = \cos 45° \cos 30° + \sin 45° \sin 30°$$

$$= \frac{1}{\sqrt{2}} \cdot \frac{\sqrt{3}}{2} + \frac{1}{\sqrt{2}} \cdot \frac{1}{2}$$

$$= \frac{\sqrt{3} + 1}{2\sqrt{2}}$$

Therefore,

$$\cos 15° = \frac{\sqrt{3} + 1}{2\sqrt{2}}$$

This problem can also be done using the double angle formulas with $\alpha = 15°$.

$$\cos 2\alpha = 2 \cos^2 \alpha - 1$$

gives

$$\cos 30° = 2 \cos^2 15° - 1$$

or

$$\frac{\sqrt{3}}{2} = 2 \cos^2 15° - 1 \qquad \left(\text{because } \cos 30° = \frac{\sqrt{3}}{2} \right)$$

Solving for $\cos 15°$:

$$2 \cos^2 15° = \frac{\sqrt{3}}{2} + 1 = \frac{\sqrt{3} + 2}{2}$$

$$\cos 15° = \pm \sqrt{\frac{\sqrt{3} + 2}{4}}$$

Since $\cos 15°$ must be positive, we will take the $+$ sign:

$$\cos 15° = \pm \sqrt{\frac{\sqrt{3} + 2}{4}}$$

At this point you are probably struck by the horrible realization that this does not seem to be the answer that we got last time, namely,

$$\frac{\sqrt{3}+1}{2\sqrt{2}}$$

Fortunately,

$$\left(\frac{\sqrt{3}+1}{2\sqrt{2}}\right)^2 = \frac{3+2\sqrt{3}+1}{4\cdot 2} = \frac{\sqrt{3}+2}{4}$$

so

$$\frac{\sqrt{3}+1}{2\sqrt{2}} = \sqrt{\frac{\sqrt{3}+2}{4}}$$

and they are in fact the same.

EXAMPLE: *Express* $\dfrac{1}{\sqrt{2}} \cos x + \dfrac{1}{\sqrt{2}} \sin x$ *as the cosine of some angle.*

Notice that

$$\cos \frac{\pi}{4} = \sin \frac{\pi}{4} = \frac{1}{\sqrt{2}}$$

so we can write

$$\frac{1}{\sqrt{2}} \cos x + \frac{1}{\sqrt{2}} \sin x = \cos \frac{\pi}{4} \cos x + \sin \frac{\pi}{4} \sin x$$

$$= \cos \left(\frac{\pi}{4} - x\right) \qquad \text{[by the formula for } \cos (\alpha - \beta)\text{]}$$

PROBLEM SET   18.9

Simplify the following.

1. $\dfrac{2(\tan \theta + 2 \sec^2 \theta)}{1 + \tan^2 \theta} - \sin 2\theta$

2. $\sin(2\alpha - \beta) \cos \beta + \cos(2\alpha - \beta) \sin \beta$

3. $\dfrac{\sin 2\theta}{1 - \cos 2\theta}$

4. $\cos^2\left(\dfrac{\pi}{4} - \theta\right) - \sin^2\left(\dfrac{\pi}{4} - \theta\right)$

5. $\dfrac{1 + \sin\left(\dfrac{\pi}{2} - 2\theta\right)}{1 - \sin\left(\dfrac{\pi}{2} - 2\theta\right)}$

6. $\dfrac{1 + \sin \theta - \cos 2\theta}{\cos \theta + \sin 2\theta}$

7. $\dfrac{\sin(\alpha - 2\beta) + \sin 2\beta}{\cos(\alpha - 2\beta) + \cos 2\beta}$ (*Hint:* Use Problems 5 and 7 in Section 18.7.)

8. $\dfrac{1 + \cos 2\theta}{\sin 2\theta}$

9. $\dfrac{\tan\left(\dfrac{\pi}{3} - \theta\right) + \tan \theta}{1 - \tan\left(\dfrac{\pi}{3} - \theta\right)\tan \theta}$ (*Hint:* Use Problem 1 in Section 18.7.)

Verify the following identities.

10. $(4 \sin \theta \cos \theta - 1)(\cos^2 \theta - \sin^2 \theta) = \sin 4\theta - \cos 2\theta$

11. $\dfrac{\sin 2\theta}{\cos 2\theta} - \dfrac{\sin \theta}{\cos \theta} = \tan \theta \sec 2\theta$

12. $\dfrac{2 \sin 2\theta}{(1 + \cos 2\theta)^2} = \tan \theta + \tan^3 \theta$

13. $2 \cot 2\theta = \cot \theta - \tan \theta$

14. $\dfrac{\tan \theta + \sin \theta}{2 \tan \theta} = \cos^2 \dfrac{\theta}{2}$

15. $\left(\dfrac{1 + \tan \theta}{1 - \tan \theta}\right)^2 = \dfrac{1 + \sin 2\theta}{1 - \sin 2\theta}$

16. $\dfrac{\tan \theta - \sin \theta}{2 \tan \theta} = \sin^2 \dfrac{\theta}{2}$

17. $2 \sin \theta \cos^3 \theta + 2 \sin^3 \theta \cos \theta = \sin 2\theta$

18. $\log(\cos \theta - \sin\theta) + \log(\cos \theta + \sin \theta) = \log(\cos 2\theta)$

Solve for x between $0°$ and $360°$.

19. $\cos 2x = \cos x$

20. $2 \sin x = \cos(45° - x)$

21. $2 \sin(x + 10°) = 1$

22. $\sin 2x = 3 \sin x$

23. $\sin(60° - x) = 2 \sin x$

24. $\cos(60° + x) + \cos(60° - x) = 3 \sin x$

Evaluate the following without using tables.

25. $\cos 75°$

26. $\sin 165°$

27. $\tan 105°$

28. $\sin(-15°)$

29. $\cos 22°30'$ [*Hint:* Use double-angle formula or problem 6 of Section 18.8.]

30. $\tan 67°30'$

CHAPTER 18 REVIEW

Simplify the following.

1. $\sin\left(\dfrac{3\pi}{2} + \theta\right)$

2. $\dfrac{\sin \theta + \sin \theta \tan^2 \theta}{\tan \theta}$

3. $\dfrac{\dfrac{1 - \cos \theta}{\sin \theta}}{\dfrac{\sin \theta}{1 + \cos \theta}}$

4. $\dfrac{\sin \theta \csc \theta}{\sin^2 \theta + \cos^2 \theta}$

Are the following true? If so, why? If not, why not?
  5. $\sin 90° = \sin (60° + 30°)$
  6. $\sin 90° = \sin 60° + \sin 30°$
  7. $\tan 40° = 2 \tan 20°$
  8. $\tan 40° = \tan 2(20°)$
  9. $3 \sin 90° = \sin 270°$
  10. $\tan (360° + 30°) = \tan 360° + \tan 30°$
  11. $\cos (360° + 30°) = \cos 360° + \cos 30°$
  12. $\cos 45° = \cos 90° \cos 45° + \sin 90° \sin 45°$

Given that $\sin 32° = 0.53$ (to two decimal places), find:

13. $\cos 32°$	17. $\csc 58°$	21. $\sin 212°$
14. $\tan 32°$	18. $\sin 148°$	22. $\sin(-752°)$
15. $\sec 32°$	19. $\cos(-32°)$	23. $\sin(-688°)$
16. $\sin 58°$	20. $\sin 122°$	

Verify the following identities.
  24. $\dfrac{\sin^3 \theta + \cos^3 \theta}{\sin \theta + \cos \theta} = \dfrac{2 - \sin 2\theta}{2}$
  25. $\cos^4 \theta - \sin^4 \theta = \cos 2\theta$

Solve for x, $0 \le x < 2\pi$:
  26. $2 \cot x + 9 = 7$
  27. $4 \sin^2 x - 3 \sin x = 1$
  28. $\sin 2x = 2 \sin x$
  29. $(\sin x \sec x)^2 = 3$
  30. $\tan^2 x + 2 \cos^2 x = 2$

Using the double angle formulas, you find that
$$\sin^2 \theta = \frac{1 - \cos 2\theta}{2}$$

In the same way, rewrite the following expressions to eliminate all powers.
  31. $\cos^2 \theta$ 32. $\sin^4 \theta$ 33. $\cos^4 \theta$

34. Suppose you want to show that $\csc 40° = \dfrac{a}{2b^2 - 1}$, given that $\tan 50° = a$ and that $\cos 20° = b$.
  (a) If $\tan 50° = a$, what is $\cot 40°$? What, then, is $\tan 40°$ equal to?
  (b) If $\cos 20° = b$, what is $\cos 40°$?
  (c) Find a relationship between $\csc \theta$, $\tan \theta$, and $\cos \theta$.
  (d) Show from the above that
$$\csc 40° = \frac{a}{2b^2 - 1}$$

35. If $\sec \theta = t$, write $\cos\theta \sin\theta$ in terms of t $(0 \le \theta < 90°)$.
36. If $x = a \tan \theta$, write $\cos^5 \theta \sin^7 \theta$ in terms of x and a.
37. If $\cos \theta = \dfrac{t}{2}$, write $\tan^2 \theta$ in terms of t.
38. If $\tan \theta = x + 1$, write $\sec^2 \theta \, (1 - \cot^2 \theta)$ in terms of x.

Using trig identities, simplify the following and hence graph:
  39. $y = \cos^2 x \sec x$ 42. $y = \cos^2 x - \sin^2 x$
  40. $y = \cos x \cos 45° - \sin x \sin 45°$ 43. $y = \sin 30° \cos x + \cos 30° \sin x$
  41. $y = 2 \sin x \cos x$

19 THE INVERSE TRIGONOMETRIC FUNCTIONS

19.1 DEFINITION OF THE INVERSE TRIGONOMETRIC FUNCTIONS

Arc Cosine or Inverse Cosine

When we solve an equation like

$$\cos x = 0.5$$

we are looking for those values of x that have a cosine of 0.5. These can be found from the points at which the line $y = 0.5$ cuts the graph of $y = \cos x$, as shown in Figure 19.1. Since there are infinitely many points of intersection, there are infinitely many solutions to $\cos x = 0.5$, namely,

$$\pm\frac{\pi}{3},\ \pm\frac{5\pi}{3},\ \pm\frac{7\pi}{3},\ \text{etc.}$$

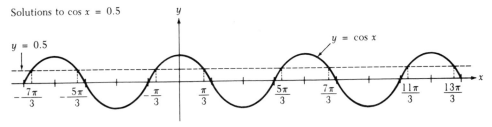

FIG. 19.1

But if we restrict the x values, we get fewer solutions. For example, if we are interested only in x's between 0 and 2π, then the only solutions are

$$\frac{\pi}{3} \quad \text{and} \quad \frac{5\pi}{3}$$

If we are interested only in solutions between $-\pi$ and π, then they are

$$-\frac{\pi}{3} \quad \text{and} \quad \frac{\pi}{3}$$

Suppose that we wanted to ensure that the equation

$$\cos x = 0.5$$

has exactly one solution. This could certainly be achieved by restricting x to between 0 and $\frac{\pi}{2}$, but also to between 0 and π. Restricting x to between 0 and 2π is no good, because it gives the equations two solutions.

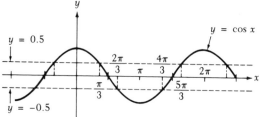

FIG. 19.2

Now suppose we wanted to restrict x in such a way that *every* equation of the form

$$\cos x = \text{some number between } -1 \text{ and } 1$$

has only one solution. (The number must be between -1 and 1 or the equation will have no solutions at all, because no angles have cosines of, say, 2 or -3.) From Figure 19.2, you can see that if we restrict x to between 0 and $\frac{\pi}{2}$, equations like $\cos x = -0.5$ would have no solution. If we allow x's beyond π, equations will start having two solutions. Therefore, if $\cos x = -0.5$ is to have exactly one solution, we should restrict x to between 0 and π. Looking at the graph, it should come as no surprise that:

> Any equation of the form
>
> $\cos x = \text{some number between } -1 \text{ and } 1$
>
> has exactly one solution between 0 and π.

The reason for this is that every horizontal line between -1 and 1 cuts the part of the cosine graph between $x = 0$ and $x = \pi$ exactly once; that is, $y = \cos x$ takes on every value exactly once as x varies between 0 and π.

Therefore each y value is the cosine of a unique angle (x value) between 0 and π, and that angle is called either Arc cos y, read "the arc cosine of y," or $\cos^{-1} y$, read "the inverse cosine of y," where these two names are entirely interchangeable. So

Arc cos y or $\cos^{-1} y$ is the angle (between 0 and π) whose cosine is y.

For example,

$$\left.\begin{array}{l} \text{Arc cos } 1 \\ \cos^{-1} 1 \end{array}\right\} = \text{the angle whose cosine is } 1 = 0$$

$$\left.\begin{array}{l} \text{Arc cos } 0 \\ \cos^{-1} 0 \end{array}\right\} = \text{the angle whose cosine is } 0 = \frac{\pi}{2}$$

$$\left.\begin{array}{l} \text{Arc cos } 0.5 \\ \cos^{-1} 0.5 \end{array}\right\} = \text{the angle whose cosine is } 0.5 = \frac{\pi}{3}$$

$$\left.\begin{array}{l} \text{Arc cos}(-0.5) \\ \cos^{-1}(-0.5) \end{array}\right\} = \text{the angle whose cosine is } -0.5 = \frac{2\pi}{3}$$

Remember:

If $x = \text{Arc cos } y = \cos^{-1} y$, then x is an *angle*, y is a *number*, and y is the cosine of x.

Notation: Arc cos versus arc cos   The notation "arc cos y" is usually used to mean *any* angle (not necessarily between 0 and π) whose cosine is y, and "Arc cos y" to mean just the one angle between 0 and π. Thus arc cos y can have many possible values, and Arc cos y has one specific value. For example,

$$\text{arc cos } 1 = \text{any angle whose cosine is } 1 = 0, \pm 2\pi, \pm 4\pi, \text{ etc.}$$

$$\text{Arc cos } 1 = \text{the angle between } 0 \text{ and } \pi \text{ whose cosine is } 1 = 0$$

Notation: Arc cos y versus $\cos^{-1} y$ The name Arc cos comes from the fact that the angle is often measured in radians using the arc of a circle. The name $\cos^{-1}$ arises from the fact that $\cos^{-1} y$ is the inverse function to the cosine. The only problem, as we saw in Section 17.4, is that $\cos^{-1} y$ runs the risk of being confused with $(\cos y)^{-1}$. They are in fact wildly different, since

$$(\cos y)^{-1} = \frac{1}{\cos y} = \sec y$$

and

$$\cos^{-1} y = \text{inverse cosine of } y$$

Lastly if we want to talk about $\cos\left(\dfrac{1}{y}\right)$ (which is different from both of them), we have to write it as $\cos(y^{-1})$.

I will use the $\cos^{-1}$ rather than the Arc cos notation because it is shorter and because it points out the fact that $\cos^{-1}$ is an inverse function.

Other inverse trigonometric functions are defined in the same way, except that x may have to be restricted differently.

Arc Sine or Inverse Sine

Suppose we want to solve equations like

$$\sin x = 0.5 \quad \text{and} \quad \sin x = -0.5$$

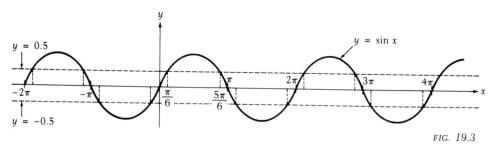

FIG. *19.3*

Looking at the graph in Figure 19.3, it is clear that if we restrict x to between 0 and π, as we did for the cosine, then

$$\sin x = 0.5$$

has two solutions, $\dfrac{\pi}{6}$ and $\dfrac{5\pi}{6}$, whereas

$$\sin x = -0.5$$

has none. If, however, we restrict x to between $-\dfrac{\pi}{2}$ and $\dfrac{\pi}{2}$, then both equations have just one solution. In fact, $y = \sin x$ takes on every y value between -1 and 1 exactly once as x varies between $-\dfrac{\pi}{2}$ and $\dfrac{\pi}{2}$, so:

> Any equation of the form
>
> $\sin x = $ some number between -1 and 1
>
> has exactly one solution between $-\dfrac{\pi}{2}$ and $\dfrac{\pi}{2}$.

As before, we call this solution Arc sin y, the "arc sine of y," or $\sin^{-1} y$, the "inverse sine of y," and again the names are equivalent. So

> Arc sin y or $\sin^{-1} y$ is the angle $\left(\text{between } -\dfrac{\pi}{2} \text{ and } \dfrac{\pi}{2}\right)$ whose sine is y.

For example,

$$\left.\begin{array}{r} \text{Arc sin } 1 \\ \sin^{-1} 1 \end{array}\right\} = \text{the angle whose sine is } 1 = \frac{\pi}{2}$$

$$\left.\begin{array}{r} \text{Arc sin } 0 \\ \sin^{-1} 0 \end{array}\right\} = \text{the angle whose sine is } 0 = 0$$

$$\left.\begin{array}{r} \text{Arc sin } -1 \\ \sin^{-1} -1 \end{array}\right\} = \text{the angle whose sine is } -1 = -\frac{\pi}{2}$$

$$\left.\begin{array}{r} \text{Arc sin } \dfrac{1}{\sqrt{2}} \\[2ex] \sin^{-1} \dfrac{1}{\sqrt{2}} \end{array}\right\} = \text{the angle whose sine is } \frac{1}{\sqrt{2}} = \frac{\pi}{4}$$

Arc Tangent or Inverse Tangent

Suppose we are looking for an angle whose tangent is some number y. From the graph in Figure 19.4, you can see that if y is positive there is one such angle between 0 and $\frac{\pi}{2}$; if y is negative there is such an angle between $-\frac{\pi}{2}$ and 0. As x varies between $-\frac{\pi}{2}$ and $\frac{\pi}{2}$, $y = \tan x$ takes on every value just once and therefore any number turns out to be the tangent of exactly one angle in this interval. We say

> Arc tan y or $\tan^{-1} y$ is the angle $\left(\text{between } -\frac{\pi}{2} \text{ and } \frac{\pi}{2}\right)$ whose tangent is y.

For example,

$$\tan^{-1} 1 = \text{Arc tan } 1 = \frac{\pi}{4} \quad \text{since} \quad \tan \frac{\pi}{4} = 1$$

The other inverse trig functions can be defined similarly.

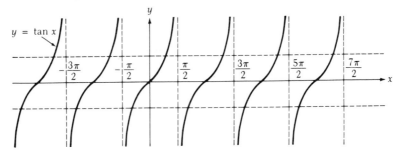

FIG. 19.4

Other Inverse Trigonometric Functions

$$\left.\begin{array}{l} \text{Arc cot } y \\ \cot^{-1} y \end{array}\right\} \text{is the angle (between 0 and } \pi) \text{ whose cotangent is } y.$$

For example,

$$\cot^{-1} \sqrt{3} = \text{Arc cot } \sqrt{3} = \frac{\pi}{6} \quad \text{since} \quad \cot \frac{\pi}{6} = \sqrt{3}$$

$$\left.\begin{array}{l} \text{Arc sec } y \\ \sec^{-1} y \end{array}\right\} \text{is the angle (between 0 and } \pi) \text{ whose secant is } y.$$

For example,

$$\sec^{-1} 2 = \text{Arc sec } 2 = \frac{\pi}{3} \quad \text{since} \quad \sec \frac{\pi}{3} = \frac{1}{\cos \frac{\pi}{3}} = 2$$

$$\left.\begin{array}{l} \text{Arc csc } y \\ \csc^{-1} y \end{array}\right\} \text{is the angle } \left(\text{between } -\frac{\pi}{2} \text{ and } \frac{\pi}{2}\right) \text{ whose cosecant is } y.$$

For example

$$\csc^{-1} (-\sqrt{2}) = \text{Arc csc } (-\sqrt{2}) = -\frac{\pi}{4} \quad \text{since} \quad \csc\left(-\frac{\pi}{4}\right) = -\sqrt{2}$$

Note: The main thing to remember with all these definitions is that if, say,

$$a = \sin^{-1} b$$

then *a is an angle, b is a number,* and

$$b = \sin a$$

Having claimed that we have defined the inverse trig functions, we should show that they do what inverse functions are supposed to.

Proof That $\sin^{-1} x$ and $\sin x$ Are Inverse Functions

This means that we want to show that

$$\sin(\sin^{-1} x) = x \quad \text{and} \quad \sin^{-1}(\sin x) = x$$

Now $\sin^{-1} x$ means the angle whose sine is x, so if you take the sine of that angle, namely, $\sin(\sin^{-1} x)$, you automatically get x. Therefore,

$$\sin(\sin^{-1} x) = x$$

(*Note:* x must be between -1 and 1 for this to make sense.) Now $\sin^{-1}(\sin x)$ means the angle between $-\dfrac{\pi}{2}$ and $\dfrac{\pi}{2}$ whose sine is $\sin x$. If x is between $-\dfrac{\pi}{2}$ and $\dfrac{\pi}{2}$, this is obviously x. So

$$\sin^{-1}(\sin x) = x \quad \text{if } -\frac{\pi}{2} \le x \le \frac{\pi}{2}$$

Similarly, each of the other functions can be shown to be the inverse to its corresponding trig function.

PROBLEM SET   19.1

Evaluate the following:

1. $\cos^{-1} 0$
2. $\tan^{-1} 0$
3. $\sin^{-1}\left(-\dfrac{1}{2}\right)$
4. $\cos^{-1}\left(-\dfrac{1}{\sqrt{2}}\right)$
5. $\cot^{-1}(-1)$
6. $\tan^{-1}\sqrt{3}$
7. $\sin^{-1}(-0.3007)$
8. $\cos^{-1}(0.7642)$
9. $\cot^{-1}(1.220)$
10. $\sec^{-1} 1.524$
11. $\csc^{-1}(-4.9313)$
12. $\tan^{-1} 1$
13. $\sin^{-1}\left(-\dfrac{\sqrt{3}}{2}\right)$
14. $\csc^{-1} 1.9940$
15. $\sec^{-1}(-1.9940)$

Are the following statements true? If so, why? If not, why not?

16. $2 \sin^{-1} u = \sin^{-1} 2u$
17. $\tan^{-1} u = \dfrac{\sin^{-1} u}{\cos^{-1} u}$
18. $\cot^{-1} v = \dfrac{1}{\tan^{-1} v}$
19. $\cot^{-1} v = \tan^{-1}\left(\dfrac{1}{v}\right)$
20. $\cos^{-1}(-u) = \cos^{-1}(u)$
21. $\sin^{-1}(w) = -\sin^{-1}(w)$
22. $\cos^{-1}(-u) = \pi - \cos^{-1}(u)$

19.2   GRAPHS OF THE INVERSE TRIGONOMETRIC FUNCTIONS

If we are given a number, x, between -1 and 1, we can always find y, the angle between $-\frac{\pi}{2}$ and $\frac{\pi}{2}$ whose sine is x. In other words, we can always calculate

$$y = \sin^{-1} x$$

Let us plot a graph to show how y depends on x. Realizing that x can only be between -1 and 1, we use the tables to draw up a list of values. This will have to be in degrees, since that's what the tables give us:

x	-1	-0.8	-0.6	-0.4	-0.2	0	0.2	0.4	0.6	0.8	1
y	$-90°$	$-53°$	$-37°$	$-24°$	$-12°$	$0°$	$12°$	$24°$	$37°$	$53°$	$90°$

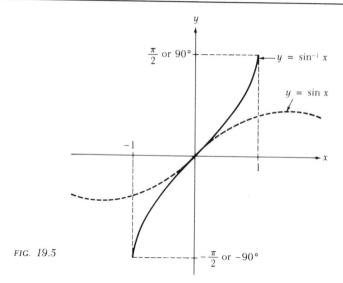

FIG. 19.5

Notice that the graph of $y = \sin^{-1} x$ in Figure 19.5 is what you get by reflecting part of the sine curve in the diagonal line $y = x$. This is as you would expect, because $y = \sin x$ and $y = \sin^{-1} x$ are inverse functions.

The *domain* of $y = \sin^{-1} x$, or all possible x values, is all numbers between -1 and 1; that is,

$$\text{domain of } \sin^{-1}\!: \quad -1 \leq x \leq 1$$

Notice that this is the range of the sine function. This is because the inverse sine function tells you what angle has a given number for its sine, and therefore it can only work on numbers that are the sine of some angle.

The *range* of $y = \sin^{-1} x$, or all y values that are produced, is angles described by real numbers between $-\frac{\pi}{2}$ to $\frac{\pi}{2}$; that is,

$$\text{range of } \sin^{-1}\!: \quad -\frac{\pi}{2} \leq y \leq \frac{\pi}{2}$$

Now let's look at the graph of

$$y = \cos^{-1} x$$

The graph can be obtained by reflecting the part of $y = \cos x$ between 0 and π in the diagonal, as shown in Figure 19.6.

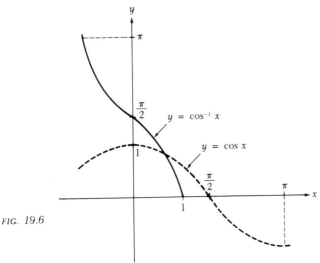

FIG. 19.6

The *domain* of $y = \cos^{-1} x$ is all x's between -1 and 1, that is,

$$\text{domain of } \cos^{-1}: \quad -1 \le x \le 1$$

which is the range of the cosine function.

The *range* of $y = \cos^{-1} x$ is all angles described by real numbers between 0 and π, that is,

$$\text{range of } \cos^{-1}: \quad 0 \le y \le \pi$$

Finally, let's consider

$$y = \tan^{-1} x$$

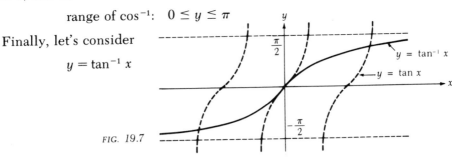

FIG. 19.7

By reflection, we get the graph in Figure 19.7.

The *domain* of $y = \tan^{-1} x$ is all real numbers.

The *range* of $y = \tan^{-1} x$ is all angles described by numbers between $-\frac{\pi}{2}$ and $\frac{\pi}{2}$, but not including $-\frac{\pi}{2}$ and $\frac{\pi}{2}$. The tangent of these two angles is not defined, and so they cannot be the inverse tangent of anything. So

$$\text{range of } \tan^{-1}: \quad -\frac{\pi}{2} < y < \frac{\pi}{2}$$

We can get the graphs of the other inverse functions in the same way.

PROBLEM SET 19.2

Graph the following inverse functions, first in degrees and then in radians.
State an appropriate domain and range for each.

1. $y = \cot^{-1} x$
2. $y = \sec^{-1} x$
3. $y = \csc^{-1} x$
4. $y = \sin^{-1} 2x$
5. $y = 2 \sin^{-1} x$
6. $y = \frac{1}{2} \sin^{-1} x$
7. $y = (\sin^{-1} x) + 1$
8. $y = \sin^{-1} (x + 1)$
9. $y = \frac{1}{2} \cos^{-1} x$

10. $y = 2 \tan^{-1} x$
11. $y = \cot^{-1} \frac{x}{2}$
12. $y = 4 \sec^{-1} x$
13. $y = 3 \sin^{-1} \frac{x}{2}$
14. $y = 2 \cos^{-1} 3x + 2$
15. $y = \frac{1}{2} \cos^{-1} (2x + 2)$

19.3   EXAMPLES USING INVERSE TRIGONOMETRIC FUNCTIONS

The inverse trig functions have surprisingly many applications, primarily in calculus. Here is a mixed bag of problems involving them.

EXAMPLE: *Find* $\tan(\cos^{-1} 1)$, $\cos^{-1}\left(\sin \frac{2\pi}{3}\right)$, $\sin^{-1}(\tan 30°)$.

These are all examples of composite functions that are evaluated by doing the inside function first:

$$\tan(\cos^{-1} 1) = \tan(0) = 0$$

$$\cos^{-1}\left(\sin \frac{2\pi}{3}\right) = \cos^{-1}\left(\frac{\sqrt{3}}{2}\right) = \frac{\pi}{6}$$

and

$$\sin^{-1}(\tan 30°) = \sin^{-1}\left(\frac{1}{\sqrt{3}}\right)$$

but since we don't know offhand what angle has a sine of $\frac{1}{\sqrt{3}}$, we will have to use the fact that $\frac{1}{\sqrt{3}} = 0.5774$ and then check the sine tables to see what angle has a sine of 0.5774. It is 35°18', so

$$\sin^{-1}(\tan 30°) = \sin^{-1}(0.5774) = 35°18'$$

Many examples that look exactly like the ones above are in fact not so well behaved.

EXAMPLE: *Find* $tan(cos^{-1} 0)$, $sin^{-1}\left(tan \dfrac{\pi}{3}\right)$.

$$tan(cos^{-1} 0) = tan\left(\frac{\pi}{2}\right)$$

which is one of the values for which tangent is not defined. So $tan(cos^{-1} 0)$ is not defined.

$$sin^{-1}\left(tan \frac{\pi}{3}\right) = sin^{-1}(\sqrt{3}) = sin^{-1}(1.7321)$$

which is not defined because there are no angles whose sines are greater than 1. Therefore $sin^{-1}\left(tan \dfrac{\pi}{3}\right)$ is not defined.

EXAMPLE: *Find* $sin^{-1}\left(sin \dfrac{3\pi}{4}\right)$.

Remembering the formula $sin^{-1}(sin x) = x$, it is tempting to say $sin^{-1}\left(sin \dfrac{3\pi}{4}\right) = \dfrac{3\pi}{4}$. But doing this out in detail, we see:

$$sin^{-1}\left(sin\frac{3\pi}{4}\right) = sin^{-1}\left(\frac{1}{\sqrt{2}}\right)$$

Now $sin^{-1}\left(\dfrac{1}{\sqrt{2}}\right)$ means the angle *between* $-\dfrac{\pi}{2}$ *and* $\dfrac{\pi}{2}$ whose sine is $\dfrac{1}{\sqrt{2}}$, namely $\dfrac{\pi}{4}$. So:

$$sin^{-1}\left(sin\frac{3\pi}{4}\right) = \frac{\pi}{4}$$

This points out the importance of saying that the formula

$$sin^{-1}(sin x) = x$$

is true *only* when x is between $-\dfrac{\pi}{2}$ and $\dfrac{\pi}{2}$.

EXAMPLE: *Find a formula for* $sin(cos^{-1} u)$, *where* $u > 0$.

$Sin(cos^{-1} u)$ is asking us for $sin \theta$, where $\theta = cos^{-1} u$, that is, θ is an angle whose cosine is u. This is a problem like those in Section 18.5 where we were given the cosine of some angle, and had to find its sine—without finding the angle. We can do it by using

$$cos^2 \theta + sin^2 \theta = 1$$

Now $\theta = cos^{-1} u$ means $u = cos \theta$, so substituting and solving for $sin \theta$:

$$u^2 + sin^2 \theta = 1$$
$$sin \theta = \pm\sqrt{1 - u^2}$$

To find out which sign is appropriate we have to use the fact that $u > 0$. This means that

$$0 \le \theta < \frac{\pi}{2}$$

because, of the angles between 0 and π, only those between 0 and $\frac{\pi}{2}$ have positive cosines. And if θ is in the first quadrant, then $\sin \theta > 0$, so

$$\sin \theta = \sqrt{1 - u^2}$$

that is,

$$\sin(\cos^{-1} u) = \sqrt{1 - u^2}$$

EXAMPLE: *Show that* $\cos(2 \cos^{-1} u) = 1 - 2u^2$.

Let $\theta = \cos^{-1} u$, so $u = \cos \theta$. Then

$$\cos(2 \cos^{-1} u) = \cos 2\theta = 1 - 2 \cos^2 \theta \qquad \text{(the double angle formula)}$$

$$= 1 - 2u^2 \qquad \text{(since } u = \cos \theta)$$

So

$$\cos(2 \cos^{-1} u) = 1 - 2u^2$$

PROBLEM SET 19.3

Evaluate the following:

1. $\sin\left(\sin^{-1} \frac{1}{2}\right)$

2. $\cos\left[\cos^{-1}\left(-\frac{1}{2}\right)\right]$

3. $\cos\left[\sin^{-1}\left(-\frac{\sqrt{2}}{2}\right)\right]$

4. $\cos^{-1}\left(\sin \frac{\pi}{6}\right)$

5. $\sin^{-1}\left(\sin \frac{2\pi}{3}\right)$

6. $\cos^{-1}[\cos(-\pi)]$

7. $\tan^{-1}(\cos 0)$

8. $\cot^{-1}\left(\sin \frac{3\pi}{4}\right)$

9. $\cos\left(\sin^{-1} \frac{3}{5}\right)$

10. $\sin\left[\cos^{-1}\left(-\frac{2}{3}\right)\right]$

11. $\tan[\cot^{-1}(-1)]$

12. $\cot^{-1}\left[\tan\left(-\frac{\pi}{4}\right)\right]$

13. $\cos^{-1}\left(\tan \frac{5\pi}{4}\right)$

14. $\tan^{-1}\left(\sin \frac{\pi}{2}\right)$

15. $\tan^{-1}\left[\sin\left(-\frac{\pi}{2}\right)\right]$

16. $\sin\left(\pi + \sin^{-1} \frac{1}{2}\right)$

17. $\tan(\pi - \tan^{-1} 1)$

18. $\csc^{-1}\left(\tan \frac{5\pi}{4}\right)$

19. $\sec^{-1}\left(\cot \frac{2\pi}{3}\right)$

20. $\csc^{-1}\left(\sec \frac{\pi}{2}\right)$

Verify the following statements:

21. $\sin^{-1} \dfrac{2}{\sqrt{5}} + \sin^{-1} \dfrac{1}{\sqrt{5}} = \dfrac{\pi}{2}$

22. $\cos^{-1} \dfrac{-3}{\sqrt{13}} - \sin^{-1} \dfrac{3}{\sqrt{13}} = \dfrac{\pi}{2}$

23. $\sin^{-1} \dfrac{\sqrt{2}}{2} - \sin^{-1} \dfrac{1}{2} = \dfrac{\pi}{12}$

24. $\cos\left(\tan^{-1} \dfrac{15}{8} - \sin^{-1} \dfrac{7}{25}\right) = \dfrac{297}{425}$

25. $\sin(2 \tan^{-1} 3) = \dfrac{3}{5}$

26. $\tan^{-1} \dfrac{1}{2} + \tan^{-1} \dfrac{1}{3} = \dfrac{\pi}{4}$

27. $\sin^{-1} \dfrac{4}{5} + \tan^{-1} \dfrac{3}{4} = \dfrac{\pi}{2}$

28. $\sin\left(2 \sin^{-1} \dfrac{4}{5} - \cos^{-1} \dfrac{12}{13}\right) = \dfrac{323}{325}$

29. $\tan^{-1} x + \tan^{-1} \dfrac{1}{x} = \dfrac{\pi}{2} \qquad (x > 0)$

30. $\cos^{-1} \dfrac{12}{13} + \tan^{-1} \dfrac{1}{4} = \cot^{-1} \dfrac{43}{32}$

CHAPTER 19 REVIEW

Evaluate the following:

1. $\tan^{-1} \dfrac{\sqrt{3}}{3}$

2. $\cos^{-1} \dfrac{1}{2}$

3. $\cot^{-1} \dfrac{\sqrt{3}}{3}$

4. $\cos^{-1}\left(-\dfrac{1}{2}\right)$

5. $\cos^{-1} 0.0436$
6. $\sin^{-1} 0.9432$
7. $\cot^{-1} 5.0045$
8. $\cot^{-1} (-0.3561)$
9. $\tan^{-1}(-10.99)$
10. $\csc^{-1}(1.0167)$

Graph the following functions. State an appropriate domain and range for each.

11. $y = 2 \csc^{-1} x$
12. $y = \frac{1}{2} \csc^{-1} x$
13. $y = \csc^{-1} 2x$
14. $y = 4 \cot^{-1} x$

15. $2y = \tan^{-1} x$
16. $y = \sin^{-1}(x - 1) + 1$
17. $\dfrac{y}{2} = \cos^{-1}\left(\dfrac{x}{2}\right) + \dfrac{7}{2}$

Evaluate the following:

18. $\cot^{-1}(\csc 60°)$

19. $\cot^{-1}(\csc 6°)$

20. $\csc(\cot^{-1} 6)$

21. $\sin\left(\sin^{-1}\dfrac{1}{2} - \cos^{-1}\dfrac{1}{3}\right)$

22. $\tan\left(\sin^{-1}\dfrac{5}{13} + \sin^{-1}\dfrac{4}{5}\right)$

23. Show that $\sin^{-1}(-x) = -\sin^{-1} x$.

24. Show that $\cos^{-1}(-x) = \pi - \cos^{-1} x$.

25. Show that $\cot^{-1}(-x) = \pi - \cot^{-1} x$.

26. Show that $\sin^{-1} x = \tan^{-1}\dfrac{x}{\sqrt{1 - x^2}}$.

27. Prove that $\tan(\sin^{-1} x) = \dfrac{x}{\sqrt{1 - x^2}}$.

28. Prove that $\sin(\tan^{-1} x) = \dfrac{x}{\sqrt{1 + x^2}}$.

29. Prove that $2 \tan^{-1} x = \tan^{-1}\left[\dfrac{2x}{1 - x^2}\right]$.

Evaluate the following:

30. $\sin\left(2 \cos^{-1}\dfrac{3}{5}\right)$

31. $\tan\left(\dfrac{1}{2} \sin^{-1}\dfrac{1}{2}\right)$

32. $\cos\left(2 \cos^{-1}\dfrac{4}{5}\right)$

33. $\sin(\sin^{-1}\dfrac{1}{2} - \cos^{-1}\dfrac{4}{5})$

34. $\sin\left(\sin^{-1}\dfrac{3}{5} + \cos^{-1}\dfrac{\sqrt{2}}{2}\right)$

35. $\cos\left(\sin^{-1}\dfrac{\sqrt{2}}{2} - \cos^{-1}\dfrac{1}{2}\right)$

20 THE BINOMIAL THEOREM AND INDUCTION

20.1   THE BINOMIAL THEOREM

This chapter will include several topics that are really part of algebra but that are somewhat more theoretical than what we have been doing and that provide an excellent introduction to calculus.

The first topic is the binomial theorem, which tells you how to expand, or multiply out, $(x + y)^n$ for any positive integer n. Let's work out some simple examples:

$$(x + y)^1 = x + y$$

$$(x + y)^2 = x^2 + 2xy + y^2$$

Now,

$$(x + y)^3 = (x + y)^2(x + y) = (x^2 + 2xy + y^2)\,(x + y)$$

and multiplying this out using the distributive law gives

$$(x + y)^3 = x^3 + 3x^2y + 3xy^2 + y^3$$

Similarly, multiplying $x^3 + 3x^2y + 3xy^2 + y^3$ by $x + y$ gives

$$(x + y)^4 = x^4 + 4x^3y + 6x^2y^2 + 4xy^3 + y^4$$

and so on.

The question is, how can we figure out the expansion for $(x + y)^5$, $(x + y)^6$, or worse, $(x + y)^{10}$, *without* doing all that multiplication?

Let's see what general pattern we can extract from the answers above.

First, notice that the expansion of $(x + y)^2$ has three terms; $(x + y)^3$ has four terms, $(x + y)^4$ has five terms, and so on. Therefore you might guess (and you'd be right) that the expansion of $(x + y)^n$ has $n + 1$ terms.

Second, notice that in the expansion of $(x + y)^3$, the exponents of x and y in each term add to 3. For example, in x^3 there are three x's and no y's, so the total is three. In $3x^2y$, there are two x's and one y, giving three altogether. Similarly, in $(x + y)^4$ the exponents of x and y in each term add to 4. So you might guess (and again you'd be right) that the expansion of $(x + y)^n$ is the sum of terms of the form *constant* $\cdot x^ry^s$, where $r + s = n$; or, in other words, terms of the form   *constant* $\cdot x^ry^{n-r}$.

Third, notice that the first term in each expansion has all x's and no y's, the next has one x replaced by a y, the next has one more x replaced by a y, and so on. Therefore the terms are arranged so that as you go from one to the next, the exponent of x drops by 1 and the exponent of y increases by 1.

Fourth, we have to figure out what those constants in front of each term are. The pattern relating the coefficients becomes a bit more obvious if we write them out alone (without the x's and y's):

$$1 \quad 1 \qquad \text{coefficients of } (x + y)^1$$

$$1 \quad 2 \quad 1 \qquad \text{coefficients of } (x + y)^2$$

$$1 \quad 3 \quad 3 \quad 1 \qquad \text{coefficients of } (x + y)^3$$

$$1 \quad 4 \quad 6 \quad 4 \quad 1 \qquad \text{coefficients of } (x + y)^4$$

This display is called *Pascal's triangle* (after its inventor). You can see that the first and last number in each row is 1, and in the row for $(x + y)^n$, the number one away from each end is n. What about the other numbers? The trick is to notice that each number in the triangle is the sum of the two numbers on either side of it in the row above. We can use this to calculate the coefficients in the expansion of $(x + y)^5$:

$$1 \quad 1$$
$$1 \quad 2 \quad 1$$
$$1 \quad 3 \quad 3 \quad 1$$
$$1 \quad 4 \quad 6 \quad 4 \quad 1$$
$$1 \quad 5 \quad 10 \quad 10 \quad 5 \quad 1$$

From the triangle you can see that the coefficients are:

$$1 \quad 5 \quad 10 \quad 10 \quad 5 \quad 1$$

which means that

$$(x + y)^5 = x^5 + 5x^4y + 10x^3y^2 + 10x^2y^3 + 5xy^4 + y^5$$

You can check this by multiplying out $(x + y)^5$.

Using Pascal's triangle, it would obviously be possible to calculate the expansion for $(x + y)^n$ for any particular n. The trouble is that if n is a large number, you'd have to write out a great many rows of the triangle and it might take rather a long time. For example, imagine doing $(x + y)^{20}$ or $(x + y)^{100}$! It would be really nice if we had a way of calculating the coefficients for $(x + y)^{20}$ or $(x + y)^n$ without writing down the earlier rows of the triangle. Fortunately there is a way of doing this, which involves getting each coefficient from the one to the left of it in the same row. Since we know the first coefficient in each row (it's 1), we can get the whole row using this method. Here's how it works:

To get a coefficient from the one to the left of it, multiply the one to the left of it by the exponent of x in the term to the left, and divide by the exponent of y in the term whose coefficient you're looking for.

EXAMPLE: *Find the coefficient of x^3y^2 in the expansion of $(x + y)^5$.*

Before doing any calculations it helps to write out the terms without the coefficients so that you can tell what is to the left of what. For $(x + y)^5$ these terms are

$$x^5 \quad x^4y \quad x^3y^2 \quad x^2y^3 \quad xy^4 \quad y^5$$

The term to the left of x^3y^2 contains x^4y, and being one from the end, its coefficient is 5, so the term itself is $5x^4y$. The exponent of x in $5x^4y$ is 4, and the exponent of y in x^3y^2 is 2, so $5 \cdot \frac{4}{2} = 10$ is the coefficient we want (which agrees with Pascal's triangle).

EXAMPLE: *Expand $(x + y)^6$*

We'll do this two ways: first by getting each coefficient from the one before it, and second, by using Pascal's triangle.

First, ignoring the coefficients, $(x + y)^6$ has terms of the following form:

$$x^6 \quad x^5y \quad x^4y^2 \quad x^3y^3 \quad x^2y^4 \quad xy^5 \quad y^6$$

Now x^6, being the first term, has coefficient 1.

The term in x^5y has coefficient $1 \cdot \frac{6}{1} = 6$ (6 is the exponent of x in x^6; 1 is the exponent of y in x^5y).

The term x^4y^2 has coefficient $6 \cdot \frac{5}{2} = 15$ (6 is the coefficient of x^5y; 5 is the exponent of x in x^5y; 2 is the exponent of y in x^4y^2).

The term in x^3y^3 has coefficient $15 \cdot \frac{4}{3} = 20$.

The term in x^2y^4 has coefficient $20 \cdot \frac{3}{4} = 15$.

The term in xy^5 has coefficient $15 \cdot \frac{2}{5} = 6$.

The term in y^6 has coefficient $6 \cdot \frac{1}{6} = 1$.

So the entire expansion is

$$(x + y)^6 = x^6 + 6x^5y + 15x^4y^2 + 20x^3y^3 + 15x^2y^4 + 6xy^5 + y^6$$

Now we'll check this by writing out the row for $(x + y)^5$ in Pascal's triangle (which we already have) and getting the coefficients of the new row by addition:

$$
\begin{array}{ccccccc}
1 & 5 & 10 & 10 & 5 & 1 & \\
\end{array}
\quad \text{coefficients of } (x+y)^5
$$

$$
\begin{array}{ccccccc}
1 & 6 & 15 & 20 & 15 & 6 & 1
\end{array}
\quad \text{coefficients of } (x+y)^6
$$

This is what we got before. Notice that the expansion is symmetric—meaning that the coefficients to the right of the middle term are the same as those to the left [in the expansion of $(x + y)^6$, the middle term is $20x^3y^3$].

Expansion of $(x + y)^n$

Using the method in which each coefficient is obtained from the one before, it should be possible to get a general expansion for $(x + y)^n$. Here goes:

$(x + y)^n$ starts with x^n, with coefficient 1

The term in $x^{n-1}y$ has coefficient $1 \cdot \dfrac{n}{1} = n$.

The term in $x^{n-2}y^2$ has coefficient $\dfrac{n(n-1)}{2}$.

The term in $x^{n-3}y^3$ has coefficient $\dfrac{n(n-1)(n-2)}{2 \cdot 3}$, etc.

This method will give the coefficients of all the terms in the expansion, ending with nxy^{n-1} and y^n.

Therefore:

$$
(x + y)^n = x^n + nx^{n-1}y + \frac{n(n-1)}{2} x^{n-2}y^2 + \frac{n(n-1)(n-2)}{2 \cdot 3} x^{n-3}y^3 + \cdots
$$
$$
+ nxy^{n-1} + y^n
$$

This is the *binomial theorem*, and the expression to the right of the equals sign is called the binomial expansion of $(x + y)^n$.

Now any term in the expansion is of the form

$$
\frac{n(n-1)(n-2) \cdots [n-(r-1)]}{2 \cdot 3 \quad \cdots \quad r} x^{n-r}y^r
$$

(for the appropriate value of r), and hence this is called the *general term*. At the moment this rather monstrous-looking formula is going to remain unproved, and we will just see what it does for us. However, it will be proved after we have introduced the idea of mathematical induction in Section 20.3.

So now for some examples:

EXAMPLE: *Expand $(x + y)^8$.*

Substituting $n = 8$ into the binomial theorem:

$$(x + y)^8 = x^8 + 8x^{8-1}y + \frac{8(8-1)}{2}x^{8-2}y^2 + \frac{8(8-1)\,(8-2)}{2 \cdot 3}x^{8-3}y^3$$
$$+ \frac{8(8-1)\,(8-2)\,(8-3)}{2 \cdot 3 \cdot 4}x^{8-4}y^4 + \cdots + 8xy^{8-1} + y^8$$

Working this out gives:

$$(x + y)^8 = x^8 + 8x^7y + \frac{8 \cdot 7}{2}x^6y^2 + \frac{8 \cdot 7 \cdot 6}{2 \cdot 3}x^5y^3$$
$$+ \frac{8 \cdot 7 \cdot 6 \cdot 5}{2 \cdot 3 \cdot 4}x^4y^4 + \cdots + 8xy^7 + y^8$$

Since the coefficients beyond the middle term are the same as those before, only in the opposite order, the whole expansion is

$$(x + y)^8 = x^8 + 8x^7y + 28x^6y^2 + 56x^5y^3 + 70x^4y^4 + 56x^3y^5$$
$$+ 28x^2y^6 + 8xy^7 + y^8$$

The skeptics among you can check that these coefficients are what you would get from Pascal's triangle. You should notice that the formula for the general term does give all the terms in the expansion—even those beyond the middle and at the right-hand end. For example, with $n = 8$ and $r = 6$, the general term gives:

$$\frac{8(8-1)\,(8-2)\cdots(8-5)}{2 \cdot 3 \cdots 6}x^{8-6}y^6$$

which reduces to

$$\frac{\overset{4}{\cancel{8}} \cdot 7 \cdot \cancel{6} \cdot \cancel{5} \cdot \cancel{4} \cdot \cancel{3}}{\cancel{2} \cdot \cancel{3} \cdot \cancel{4} \cdot \cancel{5} \cdot \cancel{6}}x^2y^6 = 28x^2y^6 \qquad \text{as we got above}$$

Similarly, $n = 8$ and $r = 8$ gives

$$\frac{8(8-1)\,(8-2)\cdots(8-7)}{2 \cdot 3 \quad \cdots \quad 7 \cdot 8}x^{8-8}y^8 = \frac{8 \cdot 7 \cdot 6 \cdots \cdot 1}{2 \cdot 3 \cdots \cdot 7 \cdot 8}x^0y^8 = y^8$$

as before.

EXAMPLE: *Expand $(A^2 - 2B)^3$.*

The binomial theorem tells us that

$$(x + y)^3 = x^3 + 3x^2y + 3xy^2 + y^3$$

Replacing x by A^2 and y by $-2B$ gives

$$(A^2 - 2B)^3 = (A^2)^3 + 3(A^2)^2(-2B) + 3(A^2)(-2B)^2 + (-2B)^3$$
$$= A^6 - 6A^4B + 12A^2B^2 - 8B^3$$

Notice the alternating signs in this expansion: this always happens when one (but not both) of x or y is negative.

PROBLEM SET   20.1

Expand the following:

1. $(x + y)^6$
2. $(A + 2B)^4$
3. $(1 - t)^3$
4. $(4 - 3x)^5$
5. $(t^2 + 1)^6$
6. $\left(t + \dfrac{1}{t}\right)^6$
7. $(\sqrt{x} + \sqrt{y})^{10}$
8. $\left(\dfrac{1}{2} - \dfrac{1}{x}\right)^8$

Find the first four terms in each of the following expansions:

9. $(\sqrt{x} + \sqrt[3]{y})^{12}$
10. $\left(ax - \dfrac{b^2}{x^3}\right)^{11}$
11. $\left(\dfrac{b^5}{z^2} - \dfrac{z^3 b}{\sqrt{c}}\right)^{15}$

Find the indicated term in each of the following expansions:

12. $\left(\dfrac{1}{2} + \dfrac{2}{3}\right)^{10}$; 8th term
13. $\left(x + \dfrac{1}{y}\right)^8$; middle term
14. $(3x^2 - 2y^3)^6$; 4th term
15. $(x\sqrt{2} + 2\sqrt{x})^{12}$; 3rd term
16. $(\frac{1}{2}x\sqrt{y} - 1)^5$; 3rd term
17. $\left(-A + \dfrac{2}{B^2}\right)^7$; 5th term

Find the coefficient of the indicated term in the following expansions:

18. $(u + v)^{11}$; u^8v^3
19. $(x^2 - 2x^3)^5$; x^{13}
20. $\left(a + \dfrac{4}{a}\right)^4$; a^0

21. Approximate the value of $(1.02)^{12}$ by computing the first three terms of $(1 + 0.02)^{12}$.
22. Compute 99^3 by using the fact that $99 = 100 - 1$.
23. Approximate to five-decimal-place accuracy: $(0.99)^{10}$.
24. Prove that $(1.01)^{10} > 1.1$. (*Hint:* $1.01 = 1 + 0.01$; now expand the first few terms.)
25. Prove that $(1.002)^{10} > 1.02$.
26. Use the binomial theorem to expand $(p + q + r)^3$ (by substituting $q + r$ for y).
27. Find $(A - 2B + C)^3$ using the binomial theorem.
28. Find $(0.97)^5$ to two decimal places.
29. Approximate $(1.98)^4$ to two decimal places.
30. Show that $(2 + \sqrt{3})^4 + (2 - \sqrt{3})^4$ is an integer.

20.2   FACTORIALS AND THE Σ NOTATION

This section will introduce two pieces of notation that often crop up in connection with the binomial theorem and that are used to write it more compactly.

Factorials

In the binomial theorem there are several places where things like

$$2 \cdot 3 \cdot 4 \quad \text{or} \quad n(n-1)(n-2)(n-3)$$

occur. Inspired by these products of consecutive integers, we define

$$2! \text{ (read "two factorial")} = 2 \cdot 1$$

$$3! \text{ (read "three factorial")} = 3 \cdot 2 \cdot 1$$

$$4! = 4 \cdot 3 \cdot 2 \cdot 1, \text{ etc.}$$

For any positive integer n,

$$n! = n(n-1)(n-2) \cdots 3 \cdot 2 \cdot 1$$

To tidy up the ends, we also define

$$1! = 1 \quad \text{and} \quad 0! = 1$$

Rewriting the Binomial Theorem Using Factorials

Coefficients in the binomial theorem can be rewritten using factorials. For example, suppose we look at

$$\frac{n!}{3!(n-3)!}$$

Writing this out in full and cancelling as much as possible:

$$\frac{n!}{3!(n-3)!} = \frac{n(n-1)(n-2)(n-3)(n-4)\cdots 3 \cdot 2 \cdot 1}{3 \cdot 2 \cdot 1(n-3)(n-4)\cdots 3 \cdot 2 \cdot 1}$$

$$= \frac{n(n-1)(n-2)}{3 \cdot 2}$$

This is exactly the coefficient of $x^{n-3}y^3$ in the expansion of $(x+y)^n$. Similarly,

$$\frac{n!}{2!(n-2)!} = \frac{n(n-1)(n-2)(n-3)(n-4)\cdots 3 \cdot 2 \cdot 1}{2 \cdot 1(n-2)(n-3)(n-4)\cdots 3 \cdot 2 \cdot 1}$$

$$= \frac{n(n-1)}{2}$$

This is the coefficient of $x^{n-2}y^2$ in the expansion of $(x + y)^n$. You might guess that the coefficient of $x^{n-r}y^r$ would be

$$\frac{n!}{r!(n - r)!}$$

Writing this out in full shows that

$$\frac{n!}{r!(n - r)!}$$
$$= \frac{n(n - 1)(n - 2) \cdots (n - r + 1)\cancel{(n-r)(n-r-1)}\cdots\cancel{3\cdot 2\cdot 1}}{r(r - 1)\cdots 3 \cdot 2 \cdot 1\cancel{(n-r)(n-r-1)}\cdots\cancel{3\cdot 2\cdot 1}}$$
$$= \frac{n(n - 1)(n - 2) \cdots [n - (r - 1)]}{r(r - 1)\cdots 3 \cdot 2}$$

which is indeed the coefficient of $x^{n-r}y^r$. So it turns out that the binomial theorem can be rewritten as:

$$(x + y)^n = x^n + \frac{n!}{1!(n - 1)!}x^{n-1}y + \frac{n!}{2!(n - 2)!}x^{n-2}y^2 + \frac{n!}{3!(n - 3)!}x^{n-3}y^3$$
$$+ \cdots + \frac{n!}{r!(n - r)!}x^{n-r}y^r + \cdots + y^n$$

The coefficient $\frac{n!}{r!(n - r)!}$ often comes up in probability and statistics. It is called the *binomial coefficient* and is usually abbreviated to $\binom{n}{r}$.

Σ Notation

The Σ notation can be used to write the sum of a string of similar looking terms. Σ, which is read "sigma" (it is the Greek capital "S") means sum and is used as follows:

$$\sum_{i=1}^{5} i^2 \text{ means add up all the numbers } i^2 \text{ from } i = 1 \text{ to } i = 5.$$

That is,

$$\sum_{i=1}^{5} i^2 = 1^2 + 2^2 + 3^2 + 4^2 + 5^2$$

Similarly,

$$\sum_{i=1}^{n} i^2 = 1^2 + 2^2 + 3^2 + \cdots + (n - 1)^2 + n^2$$

Therefore $\sum_{i=1}^{n} i^2$ means that you work out the expression i^2 for each of the integers between the value at the bottom of the Σ sign (here that value is 1) and the value at the top of the Σ sign (here n), and add them up. Similarly:

$$\sum_{s=2}^{7} x^s = x^2 + x^3 + x^4 + x^5 + x^6 + x^7$$

Here we substitute the values 2 through 7 for s in the expression x^s, and add the results. Similarly:

$$\sum_{k=1}^{4} a_k x^k = a_1 x^1 + a_2 x^2 + a_3 x^3 + a_4 x^4$$

Notice that we substitute only for k here, not for anything else, and again add the results.

Now let's look again at the binomial theorem:

$$(x + y)^n = x^n + \frac{n!}{1!(n-1)!} x^{n-1}y + \frac{n!}{2!(n-2)!} x^{n-2}y^2 + \cdots$$

$$+ \frac{n!}{r!(n-r)!} x^{n-r}y^r + \cdots + y^n$$

It turns out that *all* the coefficients in the expansion can be written using factorials. From the pattern, the coefficient of x^n looks as though it ought to be

$$\frac{n!}{0!(n-0)!}$$

and, since $0! = 1$, this works out to be

$$\frac{n!}{1(n)!} = 1,$$

which is right. Similarly, the coefficient of y^n looks as though it ought to be

$$\frac{n!}{n!(n-n)!}$$

which works out to be

$$\frac{n!}{n!0!} = \frac{1}{0!} = 1,$$

as hoped. Therefore, we can write:

$$(x + y)^n = \frac{n!}{0!(n-0)!} x^n y^0 + \frac{n!}{1!(n-1)!} x^{n-1}y^1$$

$$+ \frac{n!}{2!(n-2)!} x^{n-2}y^2 + \cdots$$

$$+ \frac{n!}{r!(n-r)!} x^{n-r}y^r \cdots + \frac{n!}{n!(n-n)!} x^0 y^n$$

Now this looks horribly complicated, but it can be made to look much more reasonable using the Σ notation. Notice that each term in the expansion is of the same form except with a different value of r, and the values for r range from 0 all the way up to n. Therefore,

$$(x + y)^n = \sum_{r=0}^{n} \frac{n!}{r!(n-r)!} \, x^{n-r} y^r$$

because if you expand the right-hand side you will get exactly what we had above.

PROBLEM SET   20.2

Find:
1. $4!$
2. $3!$
3. $(5-2)!$
4. $5! - 3!$
5. $\dfrac{5!}{3!}$
6. $\dfrac{5!}{0!}$

7. $\dfrac{9!}{6!}$
8. $\dfrac{30!}{28!}$
9. $\dfrac{10!}{6! \, 4!}$
10. $\dfrac{50!}{45! \, 5!}$

11. $\dfrac{6! - 5!}{3!}$
12. $\dfrac{10! - 6!}{4!}$
13. $\dfrac{0! \, 1!}{0! + 1!}$

Evaluate the following sums:

14. $\displaystyle\sum_{i=1}^{5} i$

15. $\displaystyle\sum_{k=2}^{4} \frac{(-1)^k}{k}$

16. $\displaystyle\sum_{p=0}^{3} 2^p$

Write out the following sums in full (but do not evaluate):

17. $\displaystyle\sum_{k=3}^{7} k(k-2)$

18. $\displaystyle\sum_{p=0}^{8} (-1)^p p^r$

19. $\displaystyle\sum_{n=0}^{5} n! x^n$

20. $\displaystyle\sum_{p=1}^{n} p x^p$

21. $\displaystyle\sum_{j=1}^{2p} j^3$

Write in summation notation:
22. $3x^2 + 4x^3 + 5x^4 + 6x^5$
23. $a_0 - a_1 x + a_2 x^2 - a_3 x^3 + \cdots + a_{12} x^{12}$
24. $\dfrac{2}{x} + \dfrac{4}{x^2} + \dfrac{8}{x^3} + \cdots + \dfrac{2^p}{x^p}$

Show that:

25. $\displaystyle\sum_{p=0}^{7} p x^{p+1} = \sum_{i=1}^{8} (i-1) x^i$

26. $\displaystyle\sum_{i=2}^{n} \frac{2^i}{i-1} = \sum_{k=1}^{n-1} \frac{2(2^k)}{k}$

27. What should the limits of summation on $\sum_{p=?}^{p=?} (2p-1)^2$ be to make it equal in value to $\sum_{k=0}^{7} (2k+1)^2$?

28. (a) Show that $4! = 4 \cdot 3!$; $5! = 5 \cdot 4!$; $6! = 6 \cdot 5!$
    (b) Write a general formula expressing this relationship.

29. (a) Give a specific example of the formula

$$\frac{n!}{(n-3)!} = n(n-1)(n-2)$$

   (b) Prove this relationship.

30. If $\binom{n}{r}$ is the binomial coefficient, show that

   (a) $\binom{n}{r} = \binom{n}{n-r}$ $(0 \leq r \leq n)$

   (b) $\binom{n+1}{r+1} = \binom{n}{r+1} + \binom{n}{r}$ $(0 \leq r \leq n-1)$

31. Prove, using the binomial theorem, that

$$\sum_{r=0}^{n} \frac{n!}{r!(n-r)!} = 2^n$$

20.3   MATHEMATICAL INDUCTION

This section is concerned with a particularly clever way of proving various mathematical results. We will use it, among other things, to prove the binomial theorem.

Let's look at an example. Suppose we wanted to prove that $2^n > n$ for any positive integer n. How could we do this ? First, we might check this result for $n = 1$ and $n = 2$ (the result is true in both cases because $2^1 = 2$, which is certainly bigger than 1, and $2^2 = 4$, which is bigger than 2). By substituting for n, you can check this result for any number of different values and, indeed, you will find it is true for any values that you try. But this doesn't amount to a *proof*, because there always might be some n somewhere which we hadn't checked and for which the result is not true.

We are clearly going to have to think of another approach, so how about this: Suppose we assume that the result is true for some particular value of n called N, that is,

$$2^N > N$$

Then what about the next value of n, namely, $N + 1$? Is the result true for it, too ? If $2^N > N$, then multiplying by 2 gives $2 \cdot 2^N > 2N$ or

$$2^{N+1} > 2N.$$

Now $2N$ is always larger than $N + 1$ (as long as $N > 1$), so

$$2^{N+1} > N + 1$$

But this says that our result is true for the next integer, namely $N + 1$.

Now let's stop for a minute and see what we've done. In the last paragraph we showed that if we assumed the result was true for N (note: we did not *prove* it true for N, but only *assumed* it true), then the result had to be true for $N + 1$ also.

What good is this? Well, we already *know* the result holds for $n = 2$ (because we checked it). Therefore, from what we've just done we can deduce that the result holds true for $n = 3$, too.

But if the result is true for $n = 3$, then it is true for $n = 4$, because we just proved that if the result is true for some integer, then it's true for the next one, too. Clearly we can go on this way forever, and so can prove that the result is true for any value of n (greater than 2) in which we happen to be interested. Since we already know the result is true for $n = 1$ and 2, we know that

$$2^n > n \qquad \text{for } every \text{ positive integer } n$$

So we have proved exactly what we wanted!

Now let's look back and see how and why this proof worked. There are really three stages to it.

1. We checked a couple of values of n, to show that there were some that made the result true. Actually we only used the fact that it was true for one value of n, namely, 2.

2. We showed that if we assumed the result to be true for some integer N, then we could deduce that it is also true for the next integer, $N + 1$.

3. Having proved that whenever the result is true for some integer, then it's true for the next one, and knowing that it is true for $n = 2$ told us that the result must be true for $n = 3$. Exactly the same reasoning told us that it must also be true for $n = 4$ and for any other positive integer beyond 2. We already know the result is true for $n = 1$ and 2, so we have proved what we wanted.

A proof like this can often be used to prove that some result holds for all positive integers n. It is called a proof by mathematical induction, and such proofs always have these steps:

Proof by Mathematical Induction

1. Prove the statement is true for some value of n (usually $n = 0$ or $n = 1$).
2. Assume that the result holds for some value N, and prove that it holds for $N + 1$.
3. Deduce that the result holds for all integers greater than or equal to the one you checked in step 1.

Let's do another example.

EXAMPLE: *Prove that $n^2 + n$ is even for all positive integers n.*

There are many different ways of doing this, but we will do it by mathematical induction.

STEP 1.   The result is true for $n = 1$ because then

$$n^2 + n = 1^2 + 1 = 2$$

which is even.

STEP 2.   We assume that $N^2 + N$ is even, and we want to deduce that $(N + 1)^2 + (N + 1)$ is even. Now

$$(N + 1)^2 + (N + 1) = N^2 + 2N + 1 + N + 1$$
$$= (N^2 + N) + (2N + 2)$$

But we assumed that $(N^2 + N)$ was even, so it must be divisible by 2, and $(2N + 2)$ is obviously divisible by 2. Hence

$$(N^2 + N) + (2N + 2)$$

is also divisible by 2 and therefore is even, which means that

$$(N^2 + 1)^2 + (N + 1)$$

is even.

STEP 3.   Since the result is true for $n = 1$, step 2 allows us to deduce that it is true for $n = 2$. From there you can get that it is true for $n = 3$, and then for any positive integer; that is,

$$n^2 + n \text{ is even for any positive integers n.}$$

EXAMPLE: *Prove that*

$$\sum_{i=1}^{n} i = \frac{n(n + 1)}{2}$$

Here we're being asked to prove that

$$\sum_{i=1}^{n} i = 1 + 2 + 3 + \cdots + (n - 1) + n = \frac{n(n + 1)}{2}$$

Again induction isn't the only way to prove this, but it's a good way.

STEP 1.   When $n = 1$, the sum only has one term, so

$$\sum_{i=1}^{1} i = 1$$

and

$$\frac{n(n + 1)}{2} = \frac{1(2)}{2} = 1$$

So indeed,

$$\sum_{i=1}^{n} i = \frac{n(n + 1)}{2} \qquad \text{for } n = 1$$

STEP 2.   We assume that the result is true for some integer N, that is,

$$\sum_{i=1}^{N} i = 1 + 2 + \cdots + N = \frac{N(N+1)}{2}$$

We want to prove that the result is true for $(N + 1)$. Substituting $(N + 1)$ for n in the original equation tells us that we're trying to prove:

$$\sum_{i=1}^{N+1} i = \frac{(N+1)\,((N+1)+1)}{2}$$

or

$$\sum_{i=1}^{N+1} i = \frac{(N+1)\,(N+2)}{2}$$

To do this, let's look at $\sum_{i=1}^{N+1} i$ in a bit more detail.

$$\sum_{i=1}^{N+1} i = \underbrace{1 + 2 + 3 + \cdots + N} + (N+1)$$

The quantity marked by the brace is $\sum_{i=1}^{N} i$, and so must equal $\frac{N(N+1)}{2}$ since we assumed the result true for N. Therefore,

$$\sum_{i=1}^{N+1} i = \frac{N(N+1)}{2} + (N+1)$$

Combining terms on the right:

$$\sum_{i=1}^{N+1} i = \frac{N(N+1) + 2(N+1)}{2}$$

Factoring out $(N + 1)$ gives:

$$\sum_{i=1}^{N+1} i = \frac{(N+1)\,(N+2)}{2}$$

which is exactly what we wanted.

STEP 3.   Therefore the result is true for $n = 2$, $n = 3$, and so on, and any other positive integer, and so the result is proved.

Now let's put all this together.

Proof of the Binomial Theorem
Using Mathematical Induction

We are hoping to prove that

$$(x + y)^n = x^n + nx^{n-1}y + \frac{n(n-1)}{2}x^{n-2}y^2 + \cdots + nxy^n + y^n$$

for all positive integers n.

STEP 1.

For $n = 1$,

$$(x + y)^1 = (x + y) = x^1 + y^1$$

So the theorem holds.

STEP 2.

We assume that the theorem holds for N, and want to prove that it holds for $(N + 1)$. This means we had better look at the expansion of $(x + y)^{N+1}$. Now

$$(x + y)^{N+1} = (x + y)^1(x + y)^N \qquad \text{(by the exponent laws)}$$

We are assuming that $(x + y)^N$ can be expanded using the binomial theorem, and therefore

$$(x + y)^{N+1} = (x + y)\left[x^N + Nx^{N-1}y + \frac{N(N-1)}{2}x^{N-2}y^2 + \cdots + Nxy^{N-1} + y^N\right]$$

Multiplying out the right hand side of this equation gives

$$(x + y)^{N+1} = x^{N+1} + x^Ny + Nx^Ny + Nx^{N-1}y^2 + \frac{N(N-1)}{2}x^{N-1}y^2 + \cdots + Nxy^N + xy^N + y^{N+1}$$

Collecting terms gives:

$$(x + y)^{N+1} = x^{N+1} + (1 + N)x^Ny + \left[N + \frac{N(N-1)}{2}\right]x^{N-1}y^2 + \cdots + (N + 1)xy^N + y^{N+1}$$

which comes down to:

$$(x + y)^{N+1} = x^{N+1} + (N + 1)x^Ny + \frac{(N + 1)N}{2}x^{N-1}y^2 + \cdots + (N + 1)xy^N + y^{N+1}$$

Now if we were being really proper about all this, we would have to write in the general term in the expansion for $(x + y)^N$ and show that the general term for $(x + y)^{N+1}$ comes out right. However, we won't because it takes a lot of space, but I suggest you work it out for yourself.

STEP 3.

Since the binomial theorem is true for $n = 1$, and whenever it holds for one integer it holds for the next, we can deduce that it is true for all positive integers.

PROBLEM SET 20.3

Prove by mathematical induction:

1. $3^n > n$

2. $\left(\dfrac{1}{2}\right)^n < \dfrac{1}{n}$

3. $2^n < n!$ for all $n \geq 4$

4. $n^2 + n + 1$ is odd

5. The sum of n even numbers is even

6. $1 + 3 + 5 + \cdots + (2n + 1) = n^2$

7. $1 + 2 + 4 + \cdots + 2^{n-1} = 2^n - 1$

8. $1 + 8 + 27 + \cdots + n^3 = \left[\dfrac{n(n + 1)}{2}\right]^2$

9. $1 + 4 + 9 + \cdots + n^2 = \dfrac{n(n + 1)(2n + 1)}{6}$

10. $\displaystyle\sum_{i=0}^{n-1} 3i = \dfrac{1}{2}(3^n - 1)$

11. $\displaystyle\sum_{i=1}^{n-1} r^i = \dfrac{r^n - r}{r - 1}$

12. $\displaystyle\sum_{p=0}^{n-1} \dfrac{1}{2^p} = \dfrac{2^n - 1}{2^{n-1}}$

13. $\displaystyle\sum_{j=1}^{n} j(j + 1) = \dfrac{1}{3}n(n + 1)(n + 2)$

14. $\log(M_1 M_2 \cdots M_n) = \log M_1 + \log M_2 + \cdots + \log M_n$

15. $4^n - 1$ is divisible by 3 [*Hint:* $4^{n+1} - 1 = (4^{n+1} - 4) + (4 - 1)$.]

CHAPTER 20 REVIEW

Expand:

1. $(b - a)^4$

2. $(2p^2 + 3q^2)^3$

3. $\left(\dfrac{x}{2} - 1\right)^6$

4. $(A^2 - B^2)^{14}$

5. $\left(\dfrac{A}{B} + \dfrac{B}{A}\right)^5$

6. $(t^2 - xs^3)^4$

Find the indicated term:

7. $(p + q)^{17}$; term in q^3

8. $\left(\dfrac{a}{2} + \dfrac{b}{3}\right)^6$; fourth term

9. $(r - 2s)^8$; middle term

10. $(r - 2s)^{11}$; middle terms

11. $\left(p^3 - \dfrac{1}{2p}\right)^{10}$; term in p^6

12. $\left(\dfrac{4}{y} - \dfrac{y^2}{2}\right)^8$; term in y^7

Use the binomial theorem to calculate to three places:

13. 999^2

14. $(1.01)^9$

15. Prove that $(1.03)^{10} > 1.3$.

Evaluate:

16. $\dfrac{8!}{7!}$
17. $\dfrac{7!}{(7-2)!}$
18. $\dbinom{12}{8}$

Express in Σ notation:

19. The sum of the first ten even integers.
20. $a_1 x^3 + a_2 x^4 + a_3 x^5 + a_4 x^6 + a_5 x^7$

Express as the ratio of factorials:

21. $12 \cdot 11 \cdot 10 \cdot 9 \cdot 8$
22. $\dfrac{7 \cdot 8 \cdot 9 \cdot 10}{3 \cdot 2}$

23. Show that

$$\sum_{n=0}^{20} x^n = \sum_{n=0}^{15} x^n + \sum_{n=16}^{20} x^n$$

24. Find k such that

$$\sum_{i=1}^{p} a_i x^i = \sum_{i=7}^{p} a_i x^i + \sum_{i=1}^{k} a_i x^i \qquad i \leq p$$

25. Find the term in x^5 in the expansion $[1 + x(1+x)]^3$.
26. Find the constant term in the expansion $(x^2 - 2x^{-2})^{10}$.
27. (a) Confirm by expanding $(1+1)^5$ that $2^5 = 32$.
    (b) Prove that the sum of the entries in the nth row of Pascal's triangle is 2^n for all n.

28. For what value of n will the coefficients of the 4th and 13th terms in the expansion of $(x + y)^n$ be equal?

29. The formula for the rth term's coefficient in the binomial expansion works even if n is not a positive integer; but the expansion contains an endless number of terms whose sum does not always make sense. Assuming that the result does make sense, expand $(1 + 0.1)^{-1}$ out six terms to approximate $\dfrac{1}{1.1}$.

30. Expand $(100 - 1)^{1/2}$ out four places to approximate $\sqrt{99}$. Does this make sense?

Prove by induction:

31. $\displaystyle\sum_{p=0}^{n} ar^p = \dfrac{a(r^{n+1} - 1)}{r - 1}$
32. $5^n - 1$ is a multiple of 4 [*Hint:* $5^{n+1} - 1 = (5^{n+1} - 5) + (5 - 1)$.]
33. $n^3 - n$ is divisible by 6
34. $x^{a_1} \cdot x^{a_2} \cdot \ldots \cdot x^{a_n} = x^A$, where $A = \displaystyle\sum_{i=1}^{n} a_i$
35. $\left(\dfrac{3}{2}\right)^n \geq n$

APPENDIX

Geometric Formulas

Two-Dimensional Shapes

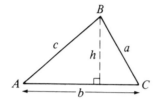

TRIANGLE

Perimeter $= a + b + c$

Area $= \dfrac{b \cdot h}{2}$

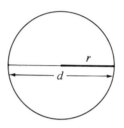

CIRCLE

$d = 2r$

Circumference $= \pi d = 2\pi r$

Area $= \pi r^2$

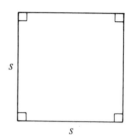

SQUARE

Perimeter $= 4s$

Area $= s^2$

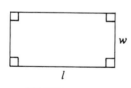

RECTANGLE

Perimeter $= 2(l + w)$

Area $= lw$

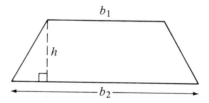

TRAPEZOID

Area $= \dfrac{(b_1 + b_2)h}{2}$

Three-Dimensional Shapes

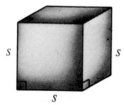

CUBE

Surface Area $= 6s^2$

Volume $= s^3$

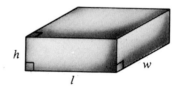

RECTANGULAR BOX

Surface Area $= 2(lw + lh + wh)$

SPHERE

Surface Area $= 4\pi r^2$

Volume $= \dfrac{4}{3}\pi r^3$

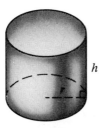

CYLINDER

Surface Area $= 2\pi r^2 + 2\pi rh$

Volume $= \pi r^2 h$

CONE

Volume $= \frac{1}{3}\pi r^2 h$

Triangles: Types and Properties

RIGHT	**ANY**	**EQUILATERAL**	**ISOSCELES**
Pythagorean Theorem: $a^2 + b^2 = c^2$	$\alpha + \beta + \gamma = 180°$ (The sum of the angles of any triangle is 180°.)	All sides equal All angles equal	Two sides equal Two angles equal

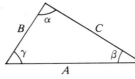

Similar triangles have the same shape and the same angles but are not necessarily the same size.

SIMILAR TRIANGLES

Ratios of corresponding sides are equal:

$$\frac{A}{a} = \frac{B}{b} = \frac{C}{c} = \text{“Magnification factor”}$$

Changing Units

To change (convert) units, multiply by a suitably chosen ratio (which must be equal to 1) so that all units cancel except the ones you want. For example:

$$1 \text{ ft} = 12 \text{ in, so } \frac{12 \text{ in}}{1 \text{ ft}} = 1 \quad \text{Therefore } 5 \text{ ft} = 5 \text{ ft} \cdot \frac{12 \text{ in}}{1 \text{ ft}} = 60 \text{ in}$$

Note that:

1 ft = 12 in   but   1 *square* ft (ft²) = 144 in²   and   1 *cubic* ft (ft³) = 1728 in³

1 cm = 10 mm   but   1 cm² = 100 mm²   and   1 cm³ = 1000 mm³

Table of Common Logarithms

n	0	1	2	3	4	5	6	7	8	9
1.0	.0000	.0043	.0086	.0128	.0170	.0212	.0253	.0294	.0334	.0374
1.1	.0414	.0453	.0492	.0531	.0569	.0607	.0645	.0682	.0719	.0755
1.2	.0792	.0828	.0864	.0899	.0934	.0969	.1004	.1038	.1072	.1106
1.3	.1139	.1173	.1206	.1239	.1271	.1303	.1335	.1367	.1399	.1430
1.4	.1461	.1492	.1523	.1553	.1584	.1614	.1644	.1673	.1703	.1732
1.5	.1761	.1790	.1818	.1847	.1875	.1903	.1931	.1959	.1987	.2014
1.6	.2041	.2068	.2095	.2122	.2148	.2175	.2201	.2227	.2253	.2279
1.7	.2304	.2330	.2355	.2380	.2405	.2430	.2455	.2480	.2504	.2529
1.8	.2553	.2577	.2601	.2625	.2648	.2672	.2695	.2718	.2742	.2765
1.9	.2788	.2810	.2833	.2856	.2878	.2900	.2923	.2945	.2967	.2989
2.0	.3010	.3032	.3054	.3075	.3096	.3118	.3139	.3160	.3181	.3201
2.1	.3222	.3243	.3263	.3284	.3304	.3324	.3345	.3365	.3385	.3404
2.2	.3424	.3444	.3464	.3483	.3502	.3522	.3541	.3560	.3579	.3598
2.3	.3617	.3636	.3655	.3674	.3692	.3711	.3729	.3747	.3766	.3784
2.4	.3802	.3820	.3838	.3856	.3874	.3892	.3909	.3927	.3945	.3962
2.5	.3979	.3997	.4014	.4031	.4048	.4065	.4082	.4099	.4116	.4133
2.6	.4150	.4166	.4183	.4200	.4216	.4232	.4249	.4265	.4281	.4298
2.7	.4314	.4330	.4346	.4362	.4378	.4393	.4409	.4425	.4440	.4456
2.8	.4472	.4487	.4502	.4518	.4533	.4548	.4564	.4579	.4594	.4609
2.9	.4624	.4639	.4654	.4669	.4683	.4698	.4713	.4728	.4742	.4757
3.0	.4771	.4786	.4800	.4814	.4829	.4843	.4857	.4871	.4886	.4900
3.1	.4914	.4928	.4942	.4955	.4969	.4983	.4997	.5011	.5024	.5038
3.2	.5051	.5065	.5079	.5092	.5105	.5119	.5132	.5145	.5159	.5172
3.3	.5185	.5198	.5211	.5224	.5237	.5250	.5263	.5276	.5289	.5302
3.4	.5315	.5328	.5340	.5353	.5366	.5378	.5391	.5403	.5416	.5428
3.5	.5441	.5453	.5465	.5478	.5490	.5502	.5514	.5527	.5539	.5551
3.6	.5563	.5575	.5587	.5599	.5611	.5623	.5635	.5647	.5658	.5670
3.7	.5682	.5694	.5705	.5717	.5729	.5740	.5752	.5763	.5775	.5786
3.8	.5798	.5809	.5821	.5832	.5843	.5855	.5866	.5877	.5888	.5899
3.9	.5911	.5922	.5933	.5944	.5955	.5966	.5977	.5988	.5999	.6010
4.0	.6021	.6031	.6042	.6053	.6064	.6075	.6085	.6096	.6107	.6117
4.1	.6128	.6138	.6149	.6160	.6170	.6180	.6191	.6201	.6212	.6222
4.2	.6232	.6243	.6253	.6263	.6274	.6284	.6294	.6304	.6314	.6325
4.3	.6335	.6345	.6355	.6365	.6375	.6385	.6395	.6405	.6415	.6425
4.4	.6435	.6444	.6454	.6464	.6474	.6484	.6493	.6503	.6513	.6522
4.5	.6532	.6542	.6551	.6561	.6571	.6580	.6590	.6599	.6609	.6618
4.6	.6628	.6637	.6646	.6656	.6665	.6675	.6684	.6693	.6702	.6712
4.7	.6721	.6730	.6739	.6749	.6758	.6767	.6776	.6785	.6794	.6803
4.8	.6812	.6821	.6830	.6839	.6848	.6857	.6866	.6875	.6884	.6893
4.9	.6902	.6911	.6920	.6928	.6937	.6946	.6955	.6964	.6972	.6981
5.0	.6990	.6998	.7007	.7016	.7024	.7033	.7042	.7050	.7059	.7067
5.1	.7076	.7084	.7093	.7101	.7110	.7118	.7126	.7135	.7143	.7152
5.2	.7160	.7168	.7177	.7185	.7193	.7202	.7210	.7218	.7226	.7235
5.3	.7243	.7251	.7259	.7267	.7275	.7284	.7292	.7300	.7308	.7316
5.4	.7324	.7332	.7340	.7348	.7356	.7364	.7372	.7380	.7388	.7396
n	0	1	2	3	4	5	6	7	8	9

n	0	1	2	3	4	5	6	7	8	9
5.5	.7404	.7412	.7419	.7427	.7435	.7443	.7451	.7459	.7466	.7474
5.6	.7482	.7490	.7497	.7505	.7513	.7520	.7528	.7536	.7543	.7551
5.7	.7559	.7566	.7574	.7582	.7589	.7597	.7604	.7612	.7619	.7627
5.8	.7634	.7642	.7649	.7657	.7664	.7672	.7679	.7686	.7694	.7701
5.9	.7709	.7716	.7723	.7731	.7738	.7745	.7752	.7760	.7767	.7774
6.0	.7782	.7789	.7796	.7803	.7810	.7818	.7825	.7832	.7839	.7846
6.1	.7853	.7860	.7868	.7875	.7882	.7889	.7896	.7903	.7910	.7917
6.2	.7924	.7931	.7938	.7945	.7952	.7959	.7966	.7973	.7980	.7987
6.3	.7993	.8000	.8007	.8014	.8021	.8028	.8035	.8041	.8048	.8055
6.4	.8062	.8069	.8075	.8082	.8089	.8096	.8102	.8109	.8116	.8122
6.5	.8129	.8136	.8142	.8149	.8156	.8162	.8169	.8176	.8182	.8189
6.6	.8195	.8202	.8209	.8215	.8222	.8228	.8235	.8241	.8248	.8254
6.7	.8261	.8267	.8274	.8280	.8287	.8293	.8299	.8306	.8312	.8319
6.8	.8325	.8331	.8338	.8344	.8351	.8357	.8363	.8370	.8376	.8382
6.9	.8388	.8395	.8401	.8407	.8414	.8420	.8426	.8432	.8439	.8445
7.0	.8451	.8457	.8463	.8470	.8476	.8482	.8488	.8494	.8500	.8506
7.1	.8513	.8519	.8525	.8531	.8537	.8543	.8549	.8555	.8561	.8567
7.2	.8573	.8579	.8585	.8591	.8597	.8603	.8609	.8615	.8621	.8627
7.3	.8633	.8639	.8645	.8651	.8657	.8663	.8669	.8675	.8681	.8686
7.4	.8692	.8698	.8704	.8710	.8716	.8722	.8727	.8733	.8739	.8745
7.5	.8751	.8756	.8762	.8768	.8774	.8779	.8785	.8791	.8797	.8802
7.6	.8808	.8814	.8820	.8825	.8831	.8837	.8842	.8848	.8854	.8859
7.7	.8865	.8871	.8876	.8882	.8887	.8893	.8899	.8904	.8910	.8915
7.8	.8921	.8927	.8932	.8938	.8943	.8949	.8954	.8960	.8965	.8971
7.9	.8976	.8982	.8987	.8993	.8998	.9004	.9009	.9015	.9020	.9025
8.0	.9031	.9036	.9042	.9047	.9053	.9058	.9063	.9069	.9074	.9079
8.1	.9085	.9090	.9096	.9101	.9106	.9112	.9117	.9122	.9128	.9133
8.2	.9138	.9143	.9149	.9154	.9159	.9165	.9170	.9175	.9180	.9186
8.3	.9191	.9196	.9201	.9206	.9212	.9217	.9222	.9227	.9232	.9238
8.4	.9243	.9248	.9253	.9258	.9263	.9269	.9274	.9279	.9284	.9289
8.5	.9294	.9299	.9304	.9309	.9315	.9320	.9325	.9330	.9335	.9340
8.6	.9345	.9350	.9355	.9360	.9365	.9370	.9375	.9380	.9385	.9390
8.7	.9395	.9400	.9405	.9410	.9415	.9420	.9425	.9430	.9435	.9440
8.8	.9445	.9450	.9455	.9460	.9465	.9469	.9474	.9479	.9484	.9489
8.9	.9494	.9499	.9504	.9509	.9513	.9518	.9523	.9528	.9533	.9538
9.0	.9542	.9547	.9552	.9557	.9562	.9566	.9571	.9576	.9581	.9586
9.1	.9590	.9595	.9600	.9605	.9609	.9614	.9619	.9624	.9628	.9633
9.2	.9638	.9643	.9647	.9652	.9657	.9661	.9666	.9671	.9675	.9680
9.3	.9685	.9689	.9694	.9699	.9703	.9708	.9713	.9717	.9722	.9727
9.4	.9731	.9736	.9741	.9745	.9750	.9754	.9759	.9763	.9768	.9773
9.5	.9777	.9782	.9786	.9791	.9795	.9800	.9805	.9809	.9814	.9818
9.6	.9823	.9827	.9832	.9836	.9841	.9845	.9850	.9854	.9859	.9863
9.7	.9868	.9872	.9877	.9881	.9886	.9890	.9894	.9899	.9903	.9908
9.8	.9912	.9917	.9921	.9926	.9930	.9934	.9939	.9943	.9948	.9952
9.9	.9956	.9961	.9965	.9969	.9974	.9978	.9983	.9987	.9991	.9996
n	0	1	2	3	4	5	6	7	8	9

Table of Trigonometric Functions

Angle	Sin	Cos	Tan	Cot	Sec	Csc	
0°00'	0.00000	1.0000	0.00000	—	1.0000	—	90°00'
06	.00175	1.0000	.00175	573.0	1.0000	572.96	54
12	.00349	1.0000	.00349	286.5	1.0000	286.48	48
18	.00524	1.0000	.00524	191.0	1.0000	190.99	42
24	.00698	1.0000	.00698	143.24	1.0000	143.24	36
30	.00873	1.0000	.00873	114.59	1.0000	114.59	30
36	.01047	0.9999	.01047	95.49	1.0001	95.495	24
42	.01222	.9999	.01222	81.85	1.0001	81.853	18
48	.01396	.9999	.01396	71.62	1.0001	71.622	12
54	.01571	.9999	.01571	63.66	1.0001	63.665	06
1°00'	0.01745	0.9998	0.01746	57.29	1.0002	57.299	89°00'
06	.01920	.9998	.01920	52.08	1.0002	52.090	54
12	.02094	.9998	.02095	47.74	1.0002	47.750	48
18	.02269	.9997	.02269	44.07	1.0003	44.077	42
24	.02443	.9997	.02444	40.92	1.0003	40.930	36
30	.02618	.9997	.02619	38.19	1.0003	38.202	30
36	.02792	.9996	.02793	35.80	1.0004	35.815	24
42	.02967	.9996	.02968	33.69	1.0004	33.708	18
48	.03141	.9995	.03143	31.82	1.0005	31.836	12
54	.03316	.9995	.03317	30.14	1.0006	30.161	06
2°00'	0.03490	0.9994	0.03492	28.64	1.0006	28.654	88°00'
06	.03664	.9993	.03667	27.27	1.0007	27.290	54
12	.03839	.9993	.03842	26.03	1.0007	26.050	48
18	.04013	.9992	.04016	24.90	1.0008	24.918	42
24	.04188	.9991	.04191	23.86	1.0009	23.880	36
30	.04362	.9990	.04366	22.90	1.0010	22.926	30
36	.04536	.9990	.04541	22.02	1.0010	22.044	24
42	.04711	.9989	.04716	21.20	1.0011	21.229	18
48	.04885	.9988	.04891	20.45	1.0012	20.471	12
54	.05059	.9987	.05066	19.74	1.0013	19.766	06
3°00'	0.05234	0.9986	0.05241	19.081	1.0014	19.107	87°00'
06	.05408	.9985	.05416	18.464	1.0015	18.492	54
12	.05582	.9984	.05591	17.886	1.0016	17.914	48
18	.05756	.9983	.05766	17.343	1.0017	17.372	42
24	.05931	.9982	.05941	16.832	1.0018	16.862	36
30	.06105	.9981	.06116	16.350	1.0019	16.380	30
36	.06279	.9980	.06291	15.895	1.0020	15.926	24
42	.06453	.9979	.06467	15.464	1.0021	15.496	18
48	.06627	.9978	.06642	15.056	1.0022	15.089	12
54	.06802	.9977	.06817	14.669	1.0023	14.703	06
4°00'	0.06976	0.9976	0.06993	14.301	1.0024	14.336	86°00'
06	.07150	.9974	.07168	13.951	1.0026	13.987	54
12	.07324	.9973	.07344	13.617	1.0027	13.654	48
18	.07498	.9972	.07519	13.300	1.0028	13.337	42
24	.07672	.9971	.07695	12.996	1.0030	13.035	36
30	.07846	.9969	.07870	12.706	1.0031	12.745	30
36	.08020	.9968	.08046	12.429	1.0032	12.469	24
42	.08194	.9966	.08221	12.163	1.0034	12.204	18
48	.08368	.9965	.08397	11.909	1.0035	11.951	12
54	.08542	.9963	.08573	11.664	1.0037	11.707	06
5°00'	0.08716	0.9962	0.08749	11.430	1.0038	11.474	85°00'
06	.08889	.9960	.08925	11.205	1.0040	11.249	54
12	.09063	.9959	.09101	10.988	1.0041	11.034	48
18	.09237	.9957	.09277	10.780	1.0043	10.826	42
24	.09411	.9956	.09453	10.579	1.0045	10.626	36
30	.09585	.9954	.09629	10.385	1.0046	10.433	30
36	.09758	.9952	.09805	10.199	1.0048	10.248	24
42	.09932	.9951	.09981	10.019	1.0050	10.068	18
48	.10106	.9949	.10158	9.845	1.0051	9.8955	12
54	.10279	.9947	.10334	9.677	1.0053	9.7283	06
6°00'	0.10453	0.9945	0.10510	9.514	1.0055	9.5668	84°00'
	Cos	Sin	Cot	Tan	Csc	Sec	Angle

Angle	Sin	Cos	Tan	Cot	Sec	Csc	
6°00′	0.10453	0.9945	0.10510	9.514	1.0055	9.5668	**84°00′**
06	.10626	.9943	.10687	9.357	1.0057	9.4105	54
12	.10800	.9942	.10863	9.205	1.0059	9.2593	48
18	.10973	.9940	.11040	9.058	1.0061	9.1129	42
24	.11147	.9938	.11217	8.915	1.0063	8.9711	36
30	.11320	.9936	.11394	8.777	1.0065	8.8337	30
36	.11494	.9934	.11570	8.643	1.0067	8.7004	24
42	.11667	.9932	.11747	8.513	1.0069	8.5711	18
48	.11840	.9930	.11924	8.386	1.0071	8.4457	12
54	.12014	.9928	.12101	8.264	1.0073	8.3238	06
7°00′	0.12187	0.9925	0.12278	8.144	1.0075	8.2055	**83°00′**
06	.12360	.9923	.12456	8.028	1.0077	8.0905	54
12	.12533	.9921	.12633	7.916	1.0079	7.9787	48
18	.12706	.9919	.12810	7.806	1.0082	7.8700	42
24	.12880	.9917	.12988	7.700	1.0084	7.7642	36
30	.13053	.9914	.13165	7.596	1.0086	7.6613	30
36	.13226	.9912	.13343	7.495	1.0089	7.5611	24
42	.13399	.9910	.13521	7.396	1.0091	7.4635	18
48	.13572	.9907	.13698	7.300	1.0093	7.3684	12
54	.13744	.9905	.13876	7.207	1.0096	7.2757	06
8°00′	0.13917	0.9903	0.14054	7.115	1.0098	7.1853	**82°00′**
06	.14090	.9900	.14232	7.026	1.0101	7.0972	54
12	.14263	.9898	.14410	6.940	1.0103	7.0112	48
18	.14436	.9895	.14588	6.855	1.0106	6.9273	42
24	.14608	.9893	.14767	6.772	1.0108	6.8454	36
30	.14781	.9890	.14945	6.691	1.0111	6.7655	30
36	.14954	.9888	.15124	6.612	1.0114	6.6874	24
42	.15126	.9885	.15302	6.535	1.0116	6.6111	18
48	.15299	.9882	.15481	6.460	1.0119	6.5366	12
54	.15471	.9880	.15660	6.386	1.0122	6.4637	06
9°00′	0.15643	0.9877	0.15838	6.314	1.0125	6.3925	**81°00′**
06	.15816	.9874	.16017	6.243	1.0127	6.3228	54
12	.15988	.9871	.16196	6.174	1.0130	6.2546	48
18	.16160	.9869	.16376	6.107	1.0133	6.1880	42
24	.16333	.9866	.16555	6.041	1.0136	6.1227	36
30	.16505	.9863	.16734	5.976	1.0139	6.0589	30
36	.16677	.9860	.16914	5.912	1.0142	5.9963	24
42	.16849	.9857	.17093	5.850	1.0145	5.9351	18
48	.17021	.9854	.17273	5.789	1.0148	5.8751	12
54	.17193	.9851	.17453	5.730	1.0151	5.8164	06
10°00′	0.1736	0.9848	0.1763	5.671	1.0154	5.7588	**80°00′**
06	.1754	.9845	.1781	5.614	1.0157	5.7023	54
12	.1771	.9842	.1799	5.558	1.0161	5.6470	48
18	.1788	.9839	.1817	5.503	1.0164	5.5928	42
24	.1805	.9836	.1835	5.449	1.0167	5.5396	36
30	.1822	.9833	.1853	5.396	1.0170	5.4874	30
36	.1840	.9829	.1871	5.343	1.0174	5.4362	24
42	.1857	.9826	.1890	5.292	1.0177	5.3860	18
48	.1874	.9823	.1908	5.242	1.0180	5.3367	12
54	.1891	.9820	.1926	5.193	1.0184	5.2883	06
11°00′	0.1908	0.9816	0.1944	5.145	1.0187	5.2408	**79°00′**
06	.1925	.9813	.1962	5.097	1.0191	5.1942	54
12	.1942	.9810	.1980	5.050	1.0194	5.1484	48
18	.1959	.9806	.1998	5.005	1.0198	5.1034	42
24	.1977	.9803	.2016	4.959	1.0201	5.0593	36
30	.1994	.9799	.2035	4.915	1.0205	5.0159	30
36	.2011	.9796	.2053	4.872	1.0209	4.9732	24
42	.2028	.9792	.2071	4.829	1.0212	4.9313	18
48	.2045	.9789	.2089	4.787	1.0216	4.8901	12
54	.2062	.9785	.2107	4.745	1.0220	4.8496	06
12°00′	0.2079	0.9781	0.2126	4.705	1.0223	4.8097	**78°00′**
	Cos	**Sin**	**Cot**	**Tan**	**Csc**	**Sec**	**Angle**

Angle	Sin	Cos	Tan	Cot	Sec	Csc	
12°00'	0.2079	0.9871	0.2126	4.705	1.0223	4.8097	**78°00'**
06	.2096	.9778	.2144	4.665	1.0227	4.7706	54
12	.2113	.9774	.2162	4.625	1.0231	4.7321	48
18	.2130	.9770	.2180	4.586	1.0235	4.6942	42
24	.2147	.9767	.2199	4.548	1.0239	4.6569	36
30	.2164	.9763	.2217	4.511	1.0243	4.6202	30
36	.2181	.9759	.2235	4.474	1.0247	4.5841	24
42	.2198	.9755	.2254	4.437	1.0251	4.5456	18
48	.2215	.9751	.2272	4.402	1.0255	4.5137	12
54	.2233	.9748	.2290	4.366	1.0259	4.4793	06
13°00'	0.2250	0.9744	0.2309	4.331	1.0263	4.4454	**77°00'**
06	.2267	.9740	.2327	4.297	1.0267	4.4121	54
12	.2284	.9736	.2345	4.264	1.0271	4.3792	48
18	.2300	.9732	.2364	4.230	1.0276	4.3469	42
24	.2317	.9728	.2382	4.198	1.0280	4.3150	36
30	.2334	.9724	.2401	4.165	1.0284	4.2837	30
36	.2351	.9720	.2419	4.134	1.0288	4.2579	24
42	.2368	.9715	.2438	4.102	1.0293	4.2223	18
48	.2385	.9711	.2456	4.071	1.0297	4.1923	12
54	.2402	.9707	.2475	4.041	1.0302	4.1627	06
14°00'	0.2419	0.9703	0.2493	4.011	1.0306	4.1336	**76°00'**
06	.2436	.9699	.2512	3.981	1.0311	4.1048	54
12	.2453	.9694	.2530	3.952	1.0315	4.0765	48
18	.2470	.9690	.2549	3.923	1.0320	4.0486	42
24	.2487	.9686	.2568	3.895	1.0324	4.0211	36
30	.2504	.9681	.2586	3.867	1.0329	3.9939	30
36	.2521	.9677	.2605	3.839	1.0334	3.9672	24
42	.2538	.9673	.2623	3.812	1.0338	3.9408	18
48	.2554	.9668	.2642	3.785	1.0343	3.9147	12
54	.2571	.9664	.2661	3.758	1.0348	3.8890	06
15°00'	0.2588	0.9659	0.2679	3.732	1.0353	3.8637	**75°00'**
06	.2605	.9655	.2698	3.706	1.0358	3.8387	54
12	.2622	.9650	.2717	3.681	1.0363	3.8140	48
18	.2639	.9646	.2736	3.655	1.0367	3.7897	42
24	.2656	.9641	.2754	3.630	1.0372	3.7657	36
30	.2672	.9636	.2773	3.606	1.0377	3.7420	30
36	.2689	.9632	.2792	3.582	1.0382	3.7186	24
42	.2706	.9627	.2811	3.558	1.0388	3.6955	18
48	.2723	.9622	.2830	3.534	1.0393	3.6727	12
54	.2740	.9617	.2849	3.511	1.0398	3.6502	06
16°00'	0.2756	0.9613	0.2867	3.487	1.0403	3.6280	**74°00'**
06	.2773	.9608	.2886	3.465	1.0408	3.6060	54
12	.2790	.9603	.2905	3.442	1.0413	3.5843	48
18	.2807	.9598	.2924	3.420	1.0419	3.5629	42
24	.2823	.9593	.2943	3.398	1.0424	3.5418	36
30	.2840	.9588	.2962	3.376	1.0429	3.5209	30
36	.2857	.9583	.2981	3.354	1.0435	3.5003	24
42	.2874	.9578	.3000	3.333	1.0440	3.4799	18
48	.2890	.9573	.3019	3.312	1.0446	3.4598	12
54	.2907	.9568	.3038	3.291	1.0451	3.4399	06
17°00'	0.2924	0.9563	0.3057	3.271	1.0457	3.4203	**73°00'**
06	.2940	.9558	.3076	3.251	1.0463	3.4009	54
12	.2957	.9553	.3096	3.230	1.0468	3.3817	48
18	.2974	.9548	.3115	3.211	1.0474	3.3628	42
24	.2990	.9542	.3134	3.191	1.0480	3.3440	36
30	.3007	.9537	.3153	3.172	1.0485	3.3255	30
36	.3024	.9532	.3172	3.152	1.0491	3.3072	24
42	.3040	.9527	.3191	3.133	1.0497	3.2891	18
48	.3057	.9521	.3211	3.115	1.0503	3.2712	12
54	.3074	.9516	.3230	3.096	1.0509	3.2535	06
18°00'	0.3090	0.9511	0.3249	3.078	1.0515	3.2361	**72°00'**
	Cos	Sin	Cot	Tan	Csc	Sec	Angle

Angle	Sin	Cos	Tan	Cot	Sec	Csc	
18°00′	0.3090	0.9511	0.3249	3.078	1.0515	3.2361	**72°00′**
06	.3107	.9505	.3269	3.060	1.0521	3.2188	54
12	.3123	.9500	.3288	3.042	1.0527	3.2017	48
18	.3140	.9494	.3307	3.024	1.0533	3.1848	42
24	.3156	.9489	.3327	3.006	1.0539	3.1681	36
30	.3173	.9483	.3346	2.989	1.0545	3.1515	30
36	.3190	.9478	.3365	2.971	1.0551	3.1352	24
42	.3206	.9472	.3385	2.954	1.0557	3.1190	18
48	.3223	.9466	.3404	2.937	1.0564	3.1030	12
54	.3239	.9461	.3424	2.921	1.0570	3.0872	06
19°00′	0.3256	0.9455	0.3443	2.904	1.0576	3.0716	**71°00′**
06	.3272	.9449	.3463	2.888	1.0583	3.0561	54
12	.3289	.9444	.3482	2.872	1.0589	3.0407	48
18	.3305	.9438	.3502	2.856	1.0595	3.0256	42
24	.3322	.9432	.3522	2.840	1.0602	3.0106	36
30	.3338	.9426	.3541	2.824	1.0608	2.9957	30
36	.3355	.9421	.3561	2.808	1.0615	2.9811	24
42	.3371	.9415	.3581	2.793	1.0622	2.9665	18
48	.3387	.9409	.3600	2.778	1.0628	2.9521	12
54	.3404	.9403	.3620	2.762	1.0635	2.9379	06
20°00′	0.3420	0.9397	0.3640	2.747	1.0642	2.9238	**70°00′**
06	.3437	.9391	.3659	2.733	1.0649	2.9099	54
12	.3453	.9385	.3679	2.718	1.0655	2.8960	48
18	.3469	.9379	.3699	2.703	1.0662	2.8824	42
24	.3486	.9373	.3719	2.689	1.0669	2.8688	36
30	.3502	.9367	.3739	2.675	1.0676	2.8555	30
36	.3518	.9361	.3759	2.660	1.0683	2.8422	24
42	.3535	.9354	.3779	2.646	1.0690	2.8291	18
48	.3551	.9348	.3799	2.633	1.0697	2.8161	12
54	.3567	.9342	.3819	2.619	1.0704	2.8032	06
21°00′	0.3584	0.9336	0.3839	2.605	1.0711	2.7904	**69°00′**
06	.3600	.9330	.3859	2.592	1.0719	2.7778	54
12	.3616	.9323	.3879	2.578	1.0725	2.7674	48
18	.3633	.9317	.3899	2.565	1.0733	2.7529	42
24	.3649	.9311	.3919	2.552	1.0740	2.7407	36
30	.3665	.9304	.3939	2.539	1.0748	2.7285	30
36	.3681	.9298	.3959	2.526	1.0755	2.7165	24
42	.3697	.9291	.3979	2.513	1.0763	2.7046	18
48	.3714	.9285	.4000	2.500	1.0770	2.6927	12
54	.3730	.9278	.4020	2.488	1.0778	2.6811	06
22°00′	0.3746	0.9272	0.4040	2.475	1.0785	2.6695	**68°00′**
06	.3762	.9265	.4061	2.463	1.0793	2.6580	54
12	.3778	.9259	.4081	2.450	1.0801	2.6466	48
18	.3795	.9252	.4101	2.438	1.0808	2.6354	42
24	.3811	.9245	.4122	2.426	1.0816	2.6242	36
30	.3827	.9239	.4142	2.414	1.0824	2.6131	30
36	.3843	.9232	.4163	2.402	1.0832	2.6022	24
42	.3859	.9225	.4183	2.391	1.0840	2.5913	18
48	.3875	.9219	.4204	2.379	1.0848	2.5805	12
54	.3891	.9212	.4224	2.367	1.0856	2.5699	06
23°00′	0.3907	0.9205	0.4245	2.356	1.0864	2.5593	**67°00′**
06	.3923	.9198	.4265	2.344	1.0872	2.5488	54
12	.3939	.9191	.4286	2.333	1.0880	2.5384	48
18	.3955	.9184	.4307	2.322	1.0888	2.5282	42
24	.3971	.9178	.4327	2.311	1.0896	2.5180	36
30	.3987	.9171	.4348	2.300	1.0904	2.5078	30
36	.4003	.9164	.4369	2.289	1.0913	2.4978	24
42	.4019	.9157	.4390	2.278	1.0921	2.4879	18
48	.4035	.9150	.4411	2.267	1.0929	2.4780	12
54	.4051	.9143	.4431	2.257	1.0938	2.4683	06
24°00′	0.4067	0.9135	0.4452	2.246	1.0946	2.4586	**66°00′**
	Cos	Sin	Cot	Tan	Csc	Sec	Angle

Angle	Sin	Cos	Tan	Cot	Sec	Csc	
24°00'	0.4067	0.9135	0.4452	2.246	1.0946	2.4586	66°00'
06	.4083	.9128	.4473	2.236	1.0955	2.4490	54
12	.4099	.9121	.4494	2.225	1.0963	2.4395	48
18	.4115	.9114	.4515	2.215	1.0972	2.4300	42
24	.4131	.9107	.4536	2.204	1.0981	2.4207	36
30	.4147	.9100	.4557	2.194	1.0989	2.4114	30
36	.4163	.9092	.4578	2.184	1.0998	2.4022	24
42	.4179	.9085	.4599	2.174	1.1007	2.3931	18
48	.4195	.9078	.4621	2.164	1.1016	2.3841	12
54	.4210	.9070	.4642	2.154	1.1025	2.3751	06
25°00'	0.4226	0.9063	0.4663	2.145	1.1034	2.3662	65°00'
06	.4242	.9056	.4684	2.135	1.1043	2.3574	54
12	.4258	.9048	.4706	2.125	1.1052	2.3486	48
18	.4274	.9041	.4727	2.116	1.1061	2.3400	42
24	.4289	.9033	.4748	2.106	1.1070	2.3314	36
30	.4305	.9026	.4770	2.097	1.1079	2.3228	30
36	.4321	.9018	.4791	2.087	1.1089	2.3144	24
42	.4337	.9011	.4813	2.078	1.1098	2.3060	18
48	.4352	.9003	.4834	2.069	1.1107	2.2976	12
54	.4368	.8996	.4856	2.059	1.1117	2.2894	06
26°00'	0.4384	0.8988	0.4877	2.050	1.1126	2.2812	64°00'
06	.4399	.8980	.4899	2.041	1.1136	2.2730	54
12	.4415	.8973	.4921	2.032	1.1145	2.2650	48
18	.4431	.8965	.4942	2.023	1.1155	2.2570	42
24	.4446	.8957	.4964	2.014	1.1164	2.2490	36
30	.4462	.8949	.4986	2.006	1.1174	2.2412	30
36	.4478	.8942	.5008	1.997	1.1184	2.2333	24
42	.4493	.8934	.5029	1.988	1.1194	2.2256	18
48	.4509	.8926	.5051	1.980	1.1203	2.2179	12
54	.4524	.8918	.5073	1.971	1.1213	2.2103	06
27°00'	0.4540	0.8910	0.5090	1.963	1.1223	2.2027	63°00'
06	.4555	.8902	.5117	1.954	1.1233	2.1952	54
12	.4571	.8894	.5139	1.946	1.1243	2.1877	48
18	.4586	.8886	.5161	1.937	1.1253	2.1803	42
24	.4602	.8878	.5184	1.929	1.1264	2.1730	36
30	.4617	.8870	.5206	1.921	1.1274	2.1657	30
36	.4633	.8862	.5228	1.913	1.1284	2.1584	24
42	.4648	.8854	.5250	1.905	1.1294	2.1513	18
48	.4664	.8846	.5272	1.897	1.1305	2.1441	12
54	.4679	.8838	.5295	1.889	1.1315	2.1371	06
28°00'	0.4695	0.8829	0.5317	1.881	1.1326	2.1301	62°00'
06	.4710	.8821	.5340	1.873	1.1336	2.1231	54
12	.4726	.8813	.5362	1.865	1.1347	2.1162	48
18	.4741	.8805	.5384	1.857	1.1357	2.1093	42
24	.4756	.8796	.5407	1.849	1.1368	2.1025	36
30	.4772	.8788	.5430	1.842	1.1379	2.0957	30
36	.4787	.8780	.5452	1.834	1.1390	2.0890	24
42	.4802	.8771	.5475	1.827	1.1401	2.0824	18
48	.4818	.8763	.5498	1.819	1.1412	2.0757	12
54	.4833	.8755	.5520	1.811	1.1423	2.0692	06
29°00'	0.4848	0.8746	0.5543	1.804	1.1434	2.0627	61°00'
06	.4863	.8738	.5566	1.797	1.1445	2.0561	54
12	.4879	.8729	.5589	1.789	1.1456	2.0498	48
18	.4894	.8721	.5612	1.782	1.1467	2.0434	42
24	.4909	.8712	.5635	1.775	1.1478	2.0371	36
30	.4924	.8704	.5658	1.767	1.1490	2.0308	30
36	.4939	.8695	.5681	1.760	1.1501	2.0245	24
42	.4955	.8686	.5704	1.753	1.1512	2.0183	18
48	.4970	.8678	.5727	1.746	1.1524	2.0122	12
54	.4985	.8669	.5750	1.739	1.1535	2.0061	06
30°00'	0.5000	0.8660	0.5774	1.732	1.1547	2.0000	60°00'
	Cos	Sin	Cot	Tan	Csc	Sec	Angle

Angle	Sin	Cos	Tan	Cot	Sec	Csc	
30°00'	0.5000	0.8660	0.5774	1.7321	1.1547	2.0000	**60°00'**
06	.5015	.8652	.5797	1.7251	1.1559	1.9940	54
12	.5030	.8643	.5820	1.7182	1.1570	1.9880	48
18	.5045	.8634	.5844	1.7113	1.1582	1.9821	42
24	.5060	.8625	.5867	1.7045	1.1594	1.9762	36
30	.5075	.8616	.5890	1.6977	1.1606	1.9703	30
36	.5090	.8607	.5914	1.6909	1.1618	1.9645	24
42	.5105	.8599	.5938	1.6842	1.1630	1.9587	18
48	.5120	.8590	.5961	1.6775	1.1642	1.9530	12
54	.5135	.8581	.5985	1.6709	1.1654	1.9473	06
31°00'	0.5150	0.8572	0.6009	1.6643	1.1666	1.9416	**59°00'**
06	.5165	.8563	.6032	1.6577	1.1679	1.9360	54
12	.5180	.8554	.6056	1.6512	1.1691	1.9304	48
18	.5195	.8545	.6080	1.6447	1.1703	1.9249	42
24	.5210	.8536	.6104	1.6383	1.1716	1.9194	36
30	.5225	.8526	.6128	1.6319	1.1728	1.9139	30
36	.5240	.8517	.6152	1.6255	1.1741	1.9084	24
42	.5255	.8508	.6176	1.6191	1.1753	1.9031	18
48	.5270	.8499	.6200	1.6128	1.1766	1.8977	12
54	.5284	.8490	.6224	1.6066	1.1779	1.8924	06
32°00'	0.5299	0.8480	0.6249	1.6003	1.1792	1.8871	**58°00'**
06	.5314	.8471	.6273	1.5941	1.1805	1.8818	54
12	.5329	.8462	.6297	1.5880	1.1818	1.8766	48
18	.5344	.8453	.6322	1.5818	1.1831	1.8714	42
24	.5358	.8443	.6346	1.5757	1.1844	1.8063	36
30	.5373	.8434	.6371	1.5697	1.1857	1.8612	30
36	.5388	.8425	.6395	1.5637	1.1870	1.8561	24
42	.5402	.8415	.6420	1.5577	1.1883	1.8510	18
48	.5417	.8406	.6445	1.5517	1.1897	1.8460	12
54	.5432	.8396	.6469	1.5458	1.1910	1.8410	06
33°00'	0.5446	0.8387	0.6494	1.5399	1.1924	1.8361	**57°00'**
06	.5461	.8377	.6519	1.5340	1.1937	1.8312	54
12	.5476	.8368	.6544	1.5282	1.1951	1.8263	48
18	.5490	.8358	.6569	1.5224	1.1964	1.8214	42
24	.5505	.8348	.6594	1.5166	1.1978	1.8166	36
30	.5519	.8339	.6619	1.5108	1.1992	1.8118	30
36	.5534	.8329	.6644	1.5051	1.2006	1.8070	24
42	.5548	.8320	.6669	1.4994	1.2020	1.8023	18
48	.5563	.8310	.6694	1.4938	1.2034	1.7976	12
54	.5577	.8300	.6720	1.4882	1.2048	1.7929	06
34°00'	0.5592	0.8290	0.6745	1.4826	1.2062	1.7883	**56°00'**
06	.5606	.8281	.6771	1.4770	1.2076	1.7837	54
12	.5621	.8271	.6796	1.4715	1.2091	1.7791	48
18	.5635	.8261	.6822	1.4659	1.2105	1.7745	42
24	.5650	.8251	.6847	1.4605	1.2120	1.7700	36
30	.5664	.8241	.6873	1.4550	1.2134	1.7655	30
36	.5678	.8231	.6899	1.4496	1.2149	1.7610	24
42	.5693	.8221	.6924	1.4442	1.2163	1.7566	18
48	.5707	.8211	.6950	1.4388	1.2178	1.7522	12
54	.5721	.8202	.6976	1.4335	1.2193	1.7478	06
35°00'	0.5736	0.8192	0.7002	1.4281	1.2208	1.7434	**55°00'**
06	.5750	.8181	.7028	1.4229	1.2223	1.7391	54
12	.5764	.8171	.7054	1.4176	1.2238	1.7348	48
18	.5779	.8161	.7080	1.4124	1.2253	1.7305	42
24	.5793	.8151	.7107	1.4071	1.2268	1.7263	36
30	.5807	.8141	.7133	1.4019	1.2283	1.7221	30
36	.5821	.8131	.7159	1.3968	1.2299	1.7179	24
42	.5835	.8121	.7186	1.3916	1.2314	1.7137	18
48	.5850	.8111	.7212	1.3865	1.2329	1.7095	12
54	.5864	.8100	.7239	1.3814	1.2345	1.7054	06
36°00'	0.5878	0.8090	0.7265	1.3764	1.2361	1.7013	**54°00'**
	Cos	**Sin**	**Cot**	**Tan**	**Csc**	**Sec**	**Angle**

Angle	Sin	Cos	Tan	Cot	Sec	Csc	
36°00'	0.5878	0.8090	0.7265	1.3764	1.2361	1.7013	**54°00'**
06	.5892	.8080	.7292	1.3713	1.2376	1.6972	54
12	.5906	.8070	.7319	1.3663	1.2392	1.6932	48
18	.5920	.8059	.7346	1.3613	1.2408	1.6892	42
24	.5934	.8049	.7373	1.3564	1.2424	1.6852	36
30	.5948	.8039	.7400	1.3514	1.2440	1.6812	30
36	.5962	.8028	.7427	1.3465	1.2456	1.6772	24
42	.5976	.8018	.7454	1.3416	1.2472	1.6733	18
48	.5990	.8007	.7481	1.3367	1.2489	1.6694	12
54	.6004	.7997	.7508	1.3319	1.2505	1.6655	06
37°00'	0.6018	0.7986	0.7536	1.3270	1.2521	1.6616	**53°00'**
06	.6032	.7976	.7563	1.3222	1.2538	1.6578	54
12	.6046	.7965	.7590	1.3175	1.2554	1.6540	48
18	.6060	.7955	.7618	1.3127	1.2571	1.6502	42
24	.6074	.7944	.7646	1.3079	1.2588	1.6464	36
30	.6088	.7934	.7673	1.3032	1.2605	1.6427	30
36	.6101	.7923	.7701	1.2985	1.2622	1.6390	24
42	.6115	.7912	.7729	1.2938	1.2639	1.6353	18
48	.6129	.7902	.7757	1.2892	1.2656	1.6316	12
54	.6143	.7891	.7785	1.2846	1.2673	1.6279	06
38°00'	0.6157	.7880	0.7813	1.2799	1.2690	1.6243	**52°00'**
06	.6170	.7869	.7841	1.2753	1.2708	1.6207	54
12	.6184	.7859	.7869	1.2708	1.2725	1.6171	48
18	.6198	.7848	.7898	1.2662	1.2742	1.6135	42
24	.6211	.7837	.7926	1.2617	1.2760	1.6099	36
30	.6225	.7826	.7954	1.2572	1.2778	1.6064	30
36	.6239	.7815	.7983	1.2527	1.2796	1.6029	24
42	.6252	.7804	.8012	1.2482	1.2813	1.5994	18
48	.6266	.7793	.8040	1.2437	1.2831	1.5959	12
54	.6280	.7782	.8069	1.2393	1.2849	1.5925	06
39°00'	0.6293	0.7771	0.8098	1.2349	1.2868	1.5890	**51°00'**
06	.6307	.7760	.8127	1.2305	1.2886	1.5856	54
12	.6320	.7749	.8156	1.2261	1.2904	1.5822	48
18	.6334	.7738	.8185	1.2218	1.2923	1.5788	42
24	.6347	.7727	.8214	1.2174	1.2941	1.5755	36
30	.6361	.7716	.8243	1.2131	1.2960	1.5721	30
36	.6374	.7705	.8273	1.2088	1.2978	1.5688	24
42	.6388	.7694	.8302	1.2045	1.2997	1.5655	18
48	.6401	.7683	.8332	1.2002	1.3016	1.5622	12
54	.6414	.7672	.8361	1.1960	1.3035	1.5590	06
40°00'	0.6428	0.7660	0.8391	1.1918	1.3054	1.5557	**50°00'**
06	.6441	.7649	.8421	1.1875	1.3073	1.5525	54
12	.6455	.7638	.8451	1.1833	1.3093	1.5493	48
18	.6468	.7627	.8481	1.1792	1.3112	1.5461	42
24	.6481	.7615	.8511	1.1750	1.3131	1.5429	36
40°30'	0.6494	0.7604	0.8541	1.1708	1.3151	1.5398	**49°30'**
	Cos	Sin	Cot	Tan	Csc	Sec	Angle

Angle	Sin	Cos	Tan	Cot	Sec	Csc	
40°30′	0.6494	0.7604	0.8541	1.1708	1.3151	1.5398	**49°30′**
36	.6508	.7593	.8571	1.1667	1.3171	1.5366	24
42	.6521	.7581	.8601	1.1626	1.3190	1.5335	18
48	.6534	.7570	.8632	1.1585	1.3210	1.5304	12
54	.6547	.7559	.8662	1.1544	1.3230	1.5273	06
41°00′	0.6561	0.7547	0.8692	1.1504	1.3250	1.5243	**49°00′**
06	.6574	.7536	.8724	1.1463	1.3270	1.5212	54
12	.6587	.7524	.8754	1.1423	1.3291	1.5182	48
18	.6600	.7513	.8785	1.1383	1.3311	1.5151	42
24	.6613	.7501	.8816	1.1343	1.3331	1.5121	36
30	.6626	.7490	.8847	1.1303	1.3352	1.5092	30
36	.6639	.7478	.8878	1.1263	1.3373	1.5062	24
42	.6652	.7466	.8910	1.1224	1.3393	1.5032	18
48	.6665	.7455	.8941	1.1184	1.3414	1.5003	12
54	.6678	.7443	.8972	1.1145	1.3435	1.4974	06
42°00′	0.6691	0.7431	0.9004	1.1106	1.3456	1.4945	**48°00′**
06	.6704	.7420	.9036	1.1067	1.3478	1.4916	54
12	.6717	.7408	.9067	1.1028	1.3499	1.4887	48
18	.6730	.7396	.9099	1.0990	1.3520	1.4859	42
24	.6743	.7385	.9131	1.0951	1.3542	1.4830	36
30	.6756	.7373	.9163	1.0913	1.3563	1.4802	30
36	.6769	.7361	.9195	1.0875	1.3585	1.4774	24
42	.6782	.7349	.9228	1.0837	1.3607	1.4746	18
48	.6794	.7337	.9260	1.0799	1.3629	1.4718	12
54	.6807	.7325	.9293	1.0761	1.3651	1.4690	06
43°00′	0.6820	0.7314	0.9325	1.0724	1.3673	1.4663	**47°00′**
06	.6833	.7302	.9358	1.0686	1.3696	1.4635	54
12	.6845	.7290	.9391	1.0649	1.3718	1.4608	48
18	.6858	.7278	.9424	1.0612	1.3741	1.4581	42
24	.6871	.7266	.9457	1.0575	1.3763	1.4554	36
30	.6884	.7254	.9490	1.0538	1.3786	1.4527	30
36	.6896	.7242	.9523	1.0501	1.3809	1.4501	24
42	.6909	.7230	.9556	1.0464	1.3832	1.4474	18
48	.6921	.7218	.9590	1.0428	1.3855	1.4448	12
54	.6934	.7206	.9623	1.0392	1.3878	1.4422	06
44°00′	0.6947	0.7193	0.9657	1.0355	1.3902	1.4396	**46°00′**
06	.6959	.7181	.9691	1.0319	1.3925	1.4370	54
12	.6972	.7169	.9725	1.0283	1.3949	1.4344	48
18	.6984	.7157	.9759	1.0247	1.3972	1.4318	42
24	.6997	.7145	.9793	1.0212	1.3996	1.4293	36
30	.7009	.7133	.9827	1.0176	1.4020	1.4267	30
36	.7022	.7120	.9861	1.0141	1.4044	1.4242	24
42	.7034	.7108	.9896	1.0105	1.4069	1.4217	18
48	.7046	.7096	.9930	1.0070	1.4093	1.4192	12
54	.7059	.7083	.9965	1.0035	1.4118	1.4167	06
45°00′	0.7071	0.7071	1.0000	1.0000	1.4142	1.4142	**45°00′**
	Cos	Sin	Cot	Tan	Csc	Sec	Angle

ANSWERS TO ODD-NUMBERED PROBLEMS

Chapter 1 Review

Note: Numbers in parentheses after the answers refer to the chapter in *The Math Workshop: Algebra* in which this type of problem is discussed.

1. $1\frac{19}{20}$ (4)
2. -1.5 (5)
3. 0.2 (6)

4. 0.4 (6)
5. $\frac{1}{10}$ (4)
6. $\frac{13}{30}$ (4)

7. 0.02 (7)
8. $-6.85 \cdot 10^{-22}$ (7)
9. 0.05 (6)

10. $9\frac{1}{11}\%$ (6)
11. 0.152 (6)
12. $-0.52, -\frac{1}{2}, -5.1 \cdot 10^{-3}$ (7)
13. $\frac{1}{8}$ inch   (Appendix)
14. $2\pi R$ (Appendix)
15. $3a^2$ (Appendix)
16. $4\pi 10^{-10}$ (Appendix)
17. 20π square inches   (Appendix)
18. $6x^3$ (Appendix)
19. 1000π cubic inches   (Appendix)
20. $\frac{32}{9}\pi^4$ (Appendix)
21. $(4 + 3a^2)(4 - 3a^2)$ (9)
22. $x^2(2 - x)(3 + x)$ (9)
23. $(5L - 7)(7L - 5)$ (9)
24. $(A^2 + B^2)(A + B)(A - B)$ (9)
25. $(x - 1)(x - 2)(x - 3)$ (9)
26. $(10AT + \pi)(AT - \pi)$ (9)
27. $[3(c + d(-1][(c + d) - 3)]$ (9)
28. $(p + q + r)(p + q - r)$ (9)
29. $x^2 + x + \frac{1}{4} = (x + \frac{1}{2})^2$ (9)
30. $-\frac{9}{8} + 3x - 2x^2 = -2(x - \frac{3}{4})^2$ (9)
31. $\dfrac{1 - 2a}{2 + a}$ (10)

32. $\dfrac{x + y}{x - y}$ (10)

33. $z + 2w$ (10)

34. $\dfrac{1}{(p - 1)^2}$ (10)

35. $\dfrac{d}{d + 1}$ (10)

36. $\dfrac{4(2k - 1)}{(k - 1)}$ (10)

37. $\dfrac{y}{2d - 2c}$(10)

38. $\dfrac{z + 1}{(3z - 1)^2(3z + 1)}$ (10)

39. $-2x$ (10)

40. $\dfrac{5(t - 1)(2t - 1)}{(2t + 1)(t + 2)}$ (10)

41. $a + 2\sqrt{ab} + b$ (9)

42. x^2 (10)

43. 0 (6)

44. $2(2v - 1)^7$ (9)

45. $|p + 1|$ (17)

46. $x = \frac{9}{10}$ (12)

47. $x = 10$ (12)

48. $t = \frac{10}{3}$ (12)

49. $p = -1.22$ (12)

50. $p = -\frac{3}{2}, -\frac{2}{3}$ (14)

51. $x = 2, -3$ (14)

52. $a = \dfrac{3 \pm \sqrt{105}}{4}$ (14)

53. $t = \frac{2}{3}$ (15)

54. $p = -2, 1$ (15)

55. $x = -1, 1, 3$ (15)

56. $x = 0, y = 7$ (16)

57. $x = 1, y = 2$ (16)

58. $A = 0.1, B = 0.9$ or $A = 0.9, B = 0.1$ (16)

59. $p = \pm 2, q = \pm 1$ (16)

60. $x = 3, y = 2$ (16)

61. $p = \dfrac{AT - A^2T^2}{1 + AT}$ (12)

62. $w = \dfrac{y(y + 5z)}{(2y + z)(y + z)}$ (12)

63. $n = \pm \sqrt{\dfrac{r}{q + re}}$ (15)

64. $x = \dfrac{2b - 1}{a}, \dfrac{1}{a}$ (15)

65. $p = d, 2d$ (15)

66. $x > 2$ (17)

67. $-4 \le x \le 4$ (17)

68. $x > 1$ or $x < -1$ (17)

69. $-8 < x < 6$ (18)

70. $x > 6$ or $x < -5$ (18)

71. $z = \frac{1}{3}, \frac{1}{4}$ (18)

72. $x = 0, -4$ (15)

73. $3 < x < 4$ (17)

74. $x < \dfrac{d - c}{a}$ if $a > 0$ (17)

$x > \dfrac{d - c}{a}$ if $a < 0$

75. $x = 1 \pm \sqrt{2}, 1 \pm \dfrac{1}{\sqrt{2}}$ (15)

76. (a) \$18 (13)
(b) \$4
(c) yes

77. 7 (13)

78. Current $= 75$ mi/day (16)
Boat $= 325$ mi/day

79. 84 years old (13)

80. citron $= 8$ (16)
wood apples $= 5$

81. 10 cubits and 8 cubits (13)

82. $\sqrt{0.73}$ (20)

83. $\left(\dfrac{\sqrt{29}}{2}\right)\pi$ (20)

84. $(0.6, -4)$ (20)

85. $\left(1, \dfrac{1 + a^2}{2a}\right)$ (20)

Note: The subject matter of answers 86–95
is discussed in Chapter 19 of *The Math Workshop: Algebra.*

86.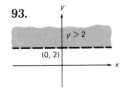

87.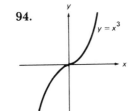

88.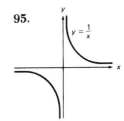

89.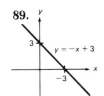

90.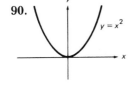

91.

92.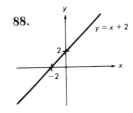

93.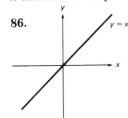

94.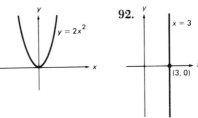

95.

Problem Set 2.1

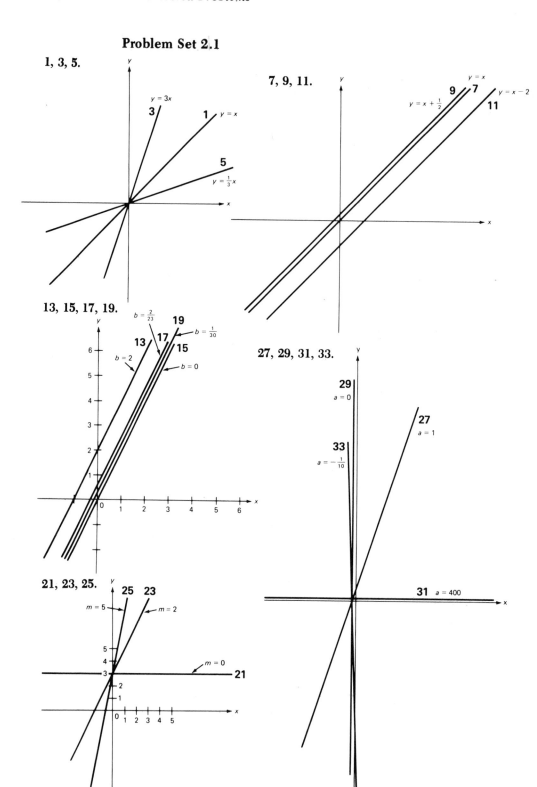

1, 3, 5.

7, 9, 11.

13, 15, 17, 19.

27, 29, 31, 33.

21, 23, 25.

Problem Set 2.2

1. -2 3. $\frac{6}{13}$ 5. $\frac{5}{2}$ 7. $-\frac{5}{4}$ 9. $-\frac{81}{70}$ 11. $-\frac{89}{58}$

13.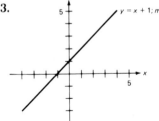

15.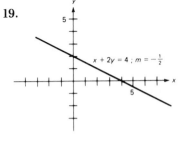

17.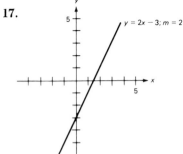

19.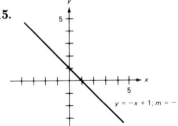

Problem Set 2.3

1. $y = 2x - 1$ 7. $y = 3x - 12$ 13. $y = 2x + b - 2a$
3. $y = 2x - 31$ 9. $y = -x + 5$ 15. $y = cx + b - ca$
5. $y = -\frac{8}{5}x - \frac{9}{5}$ 11. $y = mx - (m + 16)$

16.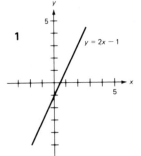

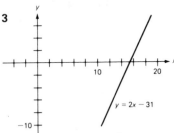

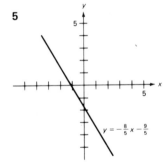

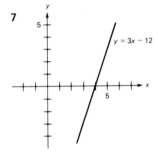

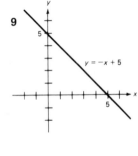

17. Slope $= -2$; x intercept $= \frac{1}{2}$; y intercept $= 1$

19. Slope is undefined since it is a vertical line;
x intercept $= -9$; y intercept does not exist

21. Slope $= \frac{3}{2}$; x intercept $= -\frac{2}{3}$; y intercept $= 1$

23. Slope $= -\frac{1}{3}$; x intercept $= 2$; y intercept $= \frac{2}{3}$

25. Slope $= \frac{5}{3}$; x intercept $= 0$; y intercept $= 0$

27. Slope $= -4$; x intercept $= 0$; y intercept $= 0$

29. Slope $= \sqrt{2}$; x intercept $= 3$; y intercept $= -3\sqrt{2}$

31.

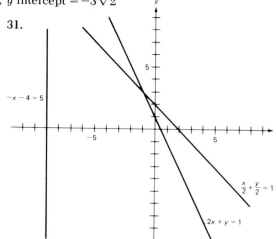

33. $y = 2x - 4$

35. $y = \dfrac{x}{2} - \dfrac{5}{2}$

37. (a) $(3, 0)$, $(0, -\frac{3}{2})$
(b) $(4, 0)$, $(0, 5)$
(c) $(7, 0)$, $(0, \frac{7}{2})$

39. (a) $n > m$ (because n is a positive slope
and m is a negative slope)
(b) $c > d$ (Q crosses the y axis below the origin;
P crosses above it)

41. (a) $m = \frac{9}{5}$
(b) $y = (\frac{9}{5}) x + 32$
(c) $68°F$
(d) $-40°$

Problem Set 2.5

1. $x = 0$

3. $y = -4$

5. $y = -(\frac{2}{3})x$

7. $y = (\frac{1}{3})x$

9. $y = -3x + 7$

11. $y = (\frac{1}{3})x + (\frac{2}{3})$

13. $y = (\frac{8}{3})x - (\frac{41}{3})$

15. $y = x + 3.5$

17. $y = k$

19. $x = a$

21.

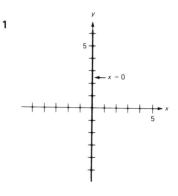

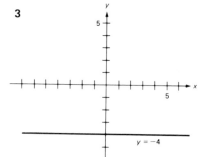

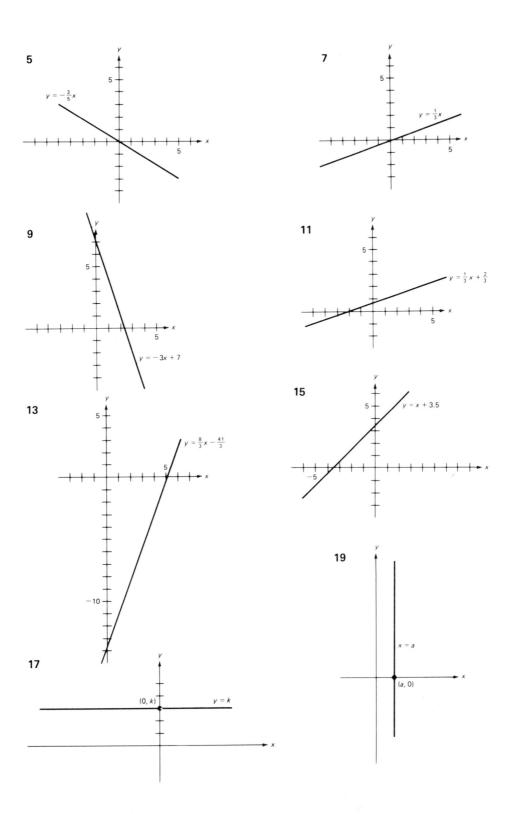

5 $y = -\frac{3}{5}x$

7 $y = \frac{1}{3}x$

9 $y = -3x + 7$

11 $y = \frac{1}{3}x + \frac{2}{3}$

13 $y = \frac{8}{3}x - \frac{41}{3}$

15 $y = x + 3.5$

17 $(0, k)$ $y = k$

19 $x = a$ $(a, 0)$

25. $(-3, 1)$; $(0, 5)$

27. $y = -(\frac{1}{2})x + (\frac{11}{2})$

29. L: $y = -(\frac{1}{3})x + 2$; M: $y = 3x + 2$

31. L: $y = 3x - 3$; M: $y = 3x + 1$

33. $y = -x + \sqrt{2}$

35. $y = (\frac{1}{2})x$

37. (a) $y = -(\frac{1}{2})x + 4$ (b) $(2, 3)$ (c) $2\sqrt{5}$

39. 10

41. $\dfrac{7\sqrt{10}}{10}$

43. $\dfrac{|mx_0 - y_0 + b|}{\sqrt{m^2 + 1}}$

Chapter 2 Review

1. $y = \frac{1}{3}x - \frac{1}{3}$

3. $y = 0.75x - 0.025$

5. $y = 5$

7. $y = \frac{98}{89}x + \frac{5}{8}$

9. $y = 2x + b - 2a$

11.

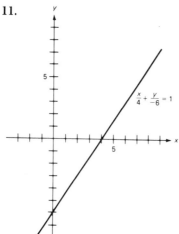

13.

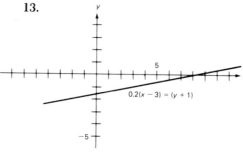

17.

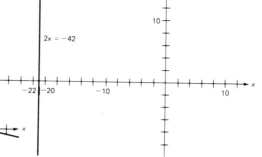

15.

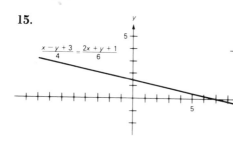

19. $y = -\dfrac{4}{3}x + \dfrac{1}{60}$

21. $\frac{6}{5}$

23. (b) $y = \frac{96}{31}x$ (c) $-\frac{18}{49}$

25. $(-\frac{4}{3}, 0)$

27. (a) yes; $y = 4$
(b) yes; $y = -x$ is perpendicular and negative to $y = x$
(c) no

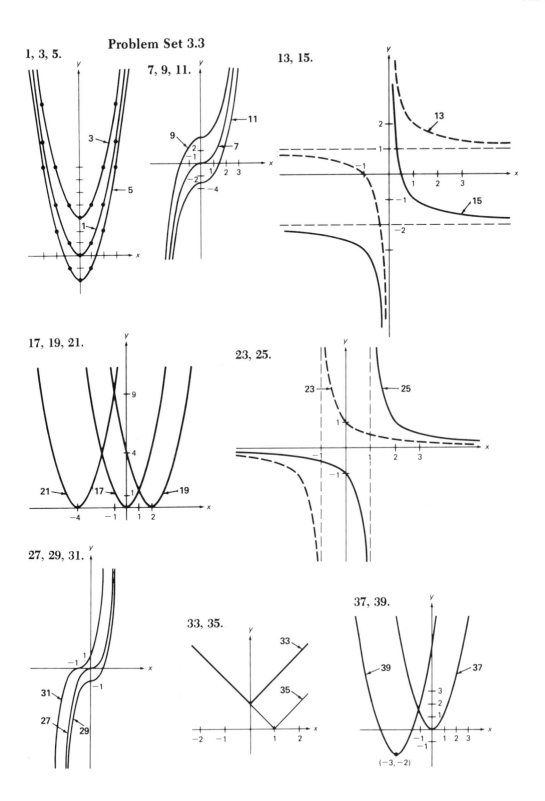

Problem Set 3.3

1, 3, 5.

7, 9, 11.

13, 15.

17, 19, 21.

23, 25.

27, 29, 31.

33, 35.

37, 39.

41, 43.

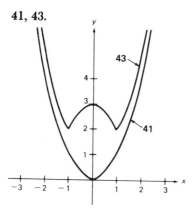

45.

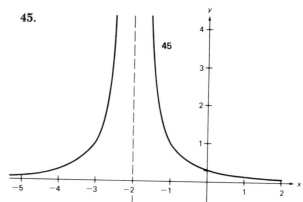

Problem Set 3.4

1, 3, 5.

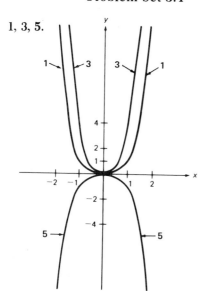

7, 9, 11.

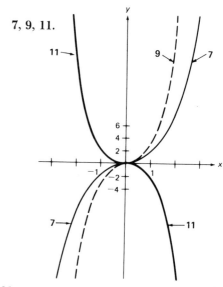

13, 15.

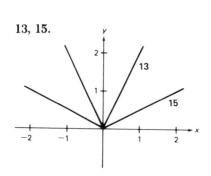

17, 19, 21.

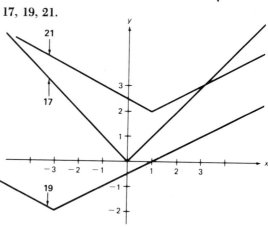

23.

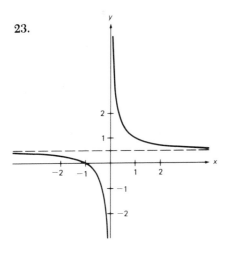

25.

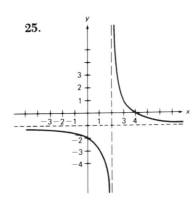

27.

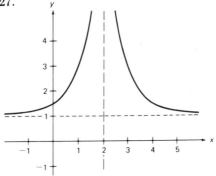

Problem Set 3.5

1, 3, 5.

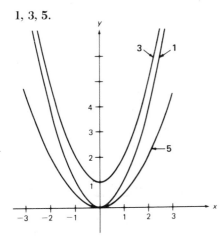

7, 9, 11.

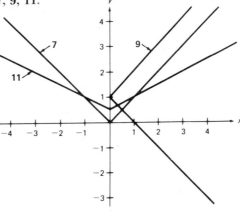

13.

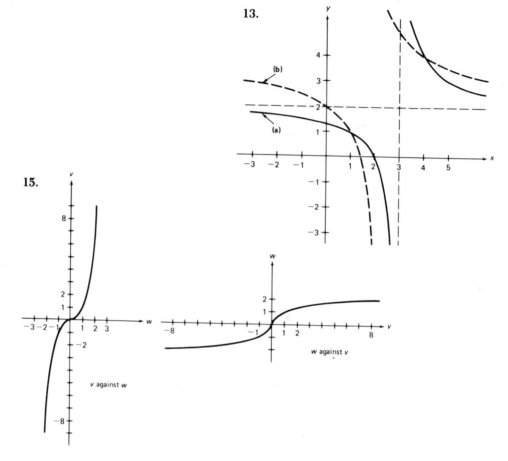

15.

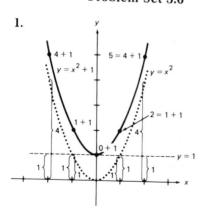

v against w

w against v

Problem Set 3.6

1.

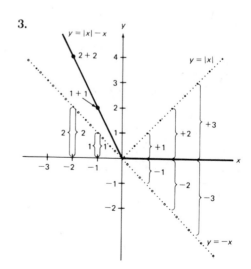

5.

7.

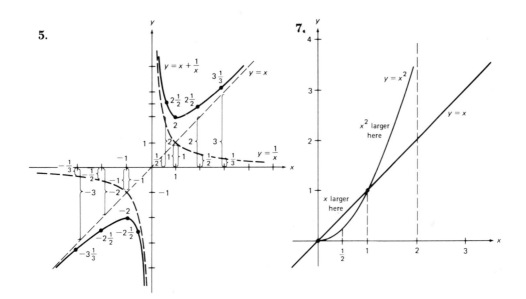

Problem Set 3.7

1.

3.

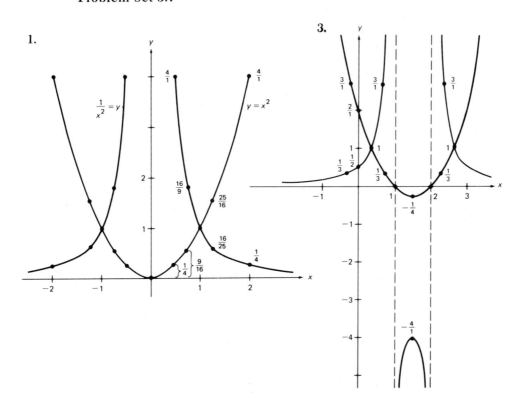

,

5.

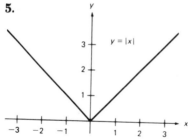

7.

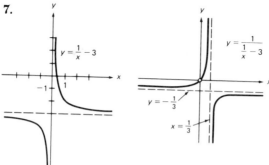

9. (a)

Problem Set 3.8

1.

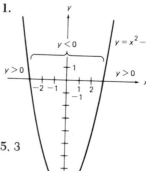

3. (a) $x > 3$; $1.5 < x < 2$ (b) $x < 1.5$; $2 < x < 3$ (c) 1.5, 3
5. (a) $|x| > \sqrt{3}$ (b) $-\sqrt{3} < x < \sqrt{3}$ (c) $x = \pm\sqrt{3}$
7. y intercept $= -6$; x intercepts $= 1, 2, 3$
9. y intercept $= 8$; x intercepts $= 2, 4$

11.

13. (a) All $k \neq 0$ (b) No k (c) $k = 0$

Chapter 3 Review

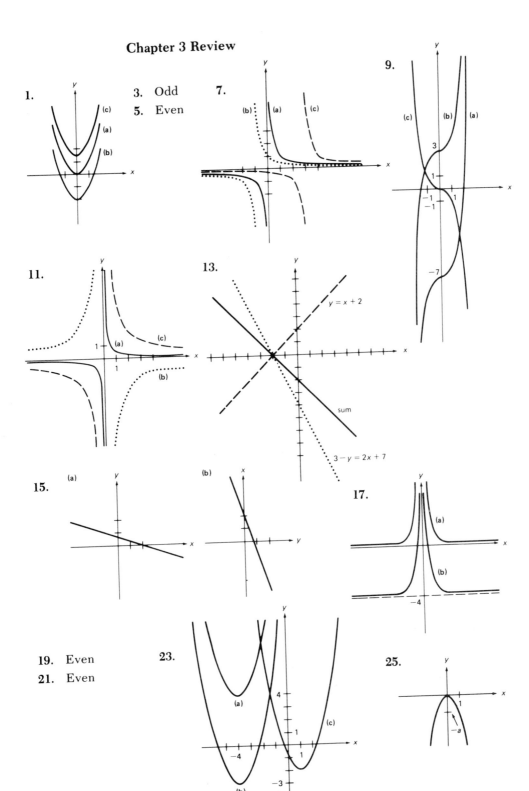

1.

3. Odd

5. Even

7.

9.

11.

13.

$y = x + 2$

sum

$3 - y = 2x + 7$

15.

(a)

(b)

17.

19. Even

21. Even

23.

25.

27.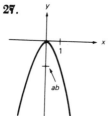

29.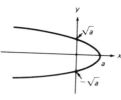

31.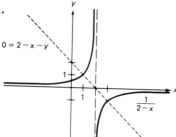

$0 = 2 - x - y$

$\dfrac{1}{2-x}$

33. (a)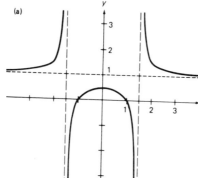

(b)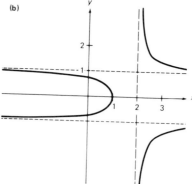

35.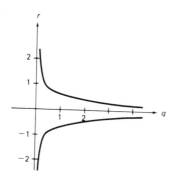

37.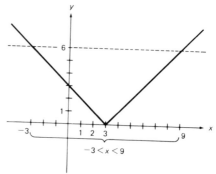

$-3 < x < 9$

Problem Set 4.1

1.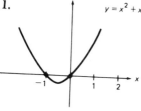

$y = x^2 + x$

3.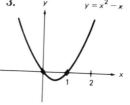

$y = x^2 - x$

5.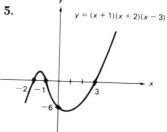

$y = (x + 1)(x + 2)(x - 3)$

7.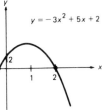

$y = x^2 - 6x + 5$

minimum at
$(3, -4)$

9.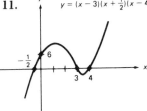

$y = -3x^2 + 5x + 2$

11.

$y = (x - 3)(x + \frac{1}{2})(x - 4)$

13.

$y = 2x^2 - 8x - 42$

15.

$y = -x^4 + 41x^2 - 400$

$(-5, 0)$ $(-4, 0)$ $(4, 0)$ $(5, 0)$

$(0, -400)$

17. (d)

19. (a)

21. (f)

23. $y = x(x + 2)(x - 2)$

25. $y = \frac{1}{3}(x + 3)(x - 2)$

27. $y = \frac{3}{4}(x - 1)(x + 1)(x - 4)(x + 4)$

29. $y = -(x + 6)(x + 4)^2(x + 3)$

31. $y = (x - 4)^2(x - 5)^2(x - 6)$

Problem Set 4.2.

1. (a) $-1, 3$

(b) $(-2, \frac{1}{5})$; $(0, -\frac{1}{3})$; $(1, -\frac{1}{4})$; $(2, -\frac{1}{3})$; $(4, \frac{1}{5})$

(c) $(100, 1.02 \cdot 10^{-4})$; $(1000, 1 \cdot 10^{-6})$; $(-100, 9.8 \cdot 10^{-5})$ $(-1000, 9.98 \cdot 10^{-7})$

(d) $(3.1, 2.44)$; $(3.01, 25)$; $(3.001, 250)$; $(2.9, -2.56)$ $(2.99, -25)$; $(2.999, -250)$

(e) $(-0.99, -25)$; $(-0.999, -250)$; $(-1.01, 25)$; $(-1.001, 250)$

1(f)

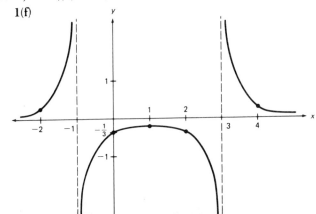

3.

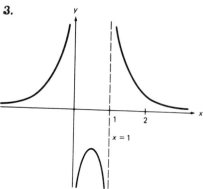

5.

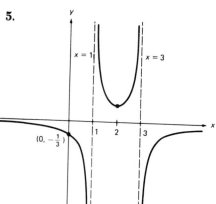

7.

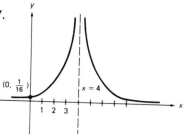

9.

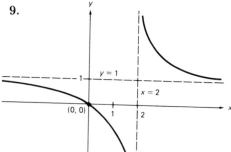

11.

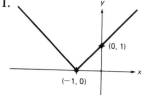

13.

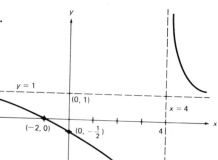

15.

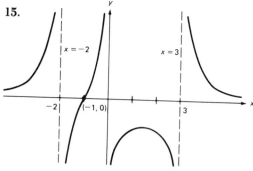

17.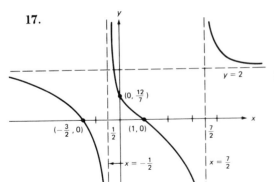

19.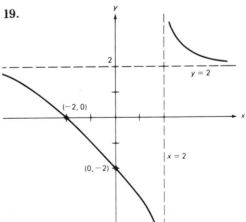

21.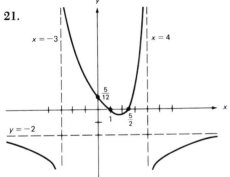

23. $y = \dfrac{(x-1)}{(x-2)}$

25.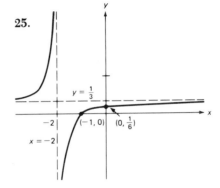

27.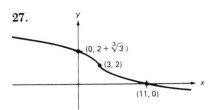

29.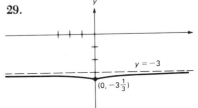

31.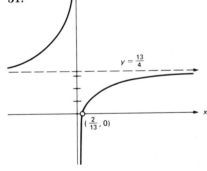

33.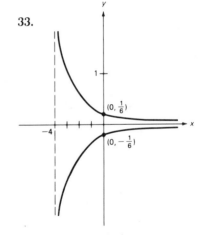

Chapter 4 Review

1.

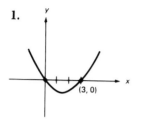

(3, 0)

3.

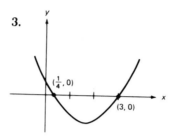

$(\frac{1}{4}, 0)$

(3, 0)

5. $y = \frac{9}{4}(x^2 - 4)$

7.   (a) 1   (b) $\frac{1}{3}$
   (c) approximately $(1000, 3), (-1000, 3)$
   (d) approximately $(1.001, 2000), (0.999, -2000)$

7(e)

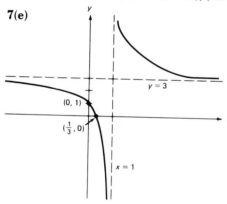

$y = 3$

(0, 1)

$(\frac{1}{3}, 0)$

$x = 1$

9.

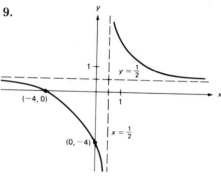

1

$y = \frac{1}{2}$

(−4, 0)

1

(0, −4)

$x = \frac{1}{2}$

11.

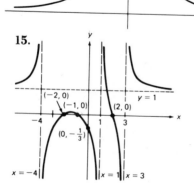

13.

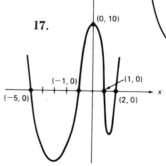

(−5, 0)

$(\frac{1}{3}, 0)$

(0, −5)

15.

(−2, 0)

(−1, 0)

$y = 1$

(2, 0)

−4

1

3

$(0, -\frac{1}{3})$

$x = -4$

$x = 1$ $x = 3$

17.

(0, 10)

(−1, 0)

(1, 0)

(−5, 0)

(2, 0)

19.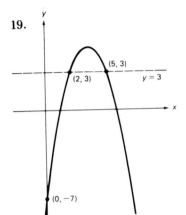

21. $y = x(x + 3)(x - 3)^2$

23.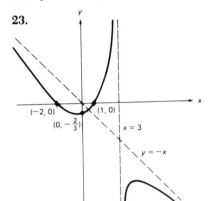

25.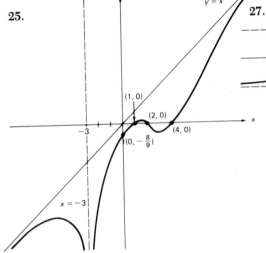

27.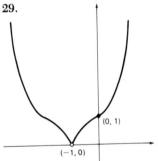

29.

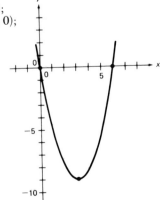

Problem Set 5.1

Graphs are provided for selected problems only.

1. Vertex at $(3, -9)$;
   intercepts at $(0, 0)$;
   $(6, 0)$

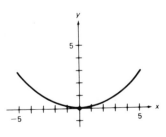

3. Vertex, intercept at $(0, 0)$

5. Vertex at $(1, -9)$; intercepts at $(0, -8)$, $(4, 0)$, $(-2, 0)$

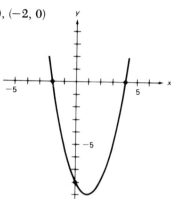

7. Vertex at $(-\frac{3}{8}, -\frac{49}{12})$; intercepts at $(0, -2)$, $(\frac{1}{3}, 0)$, $(-2, 0)$

9. Vertex at $(4, 1)$; intercept at $(0, 17)$

11. Vertex, intercept at $(0, 7)$

13. Vertex at $(-\frac{81}{16}, \frac{1}{8})$; intercepts at $(0, \frac{5}{4})$, $(0, -1)$, $(-5, 0)$

15. Vertex at $(3, 1)$; intercept at $(5, 0)$

17. Vertex at $(2, 4)$; intercepts at $(0, 0)$, $(4, 0)$

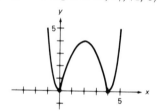

19. Vertex at $(9, 2)$; intercepts at $(-3, 0)$, $(0, 2 + \sqrt{3})$, $(0, 2 - \sqrt{3})$

Problem Set 5.2

Graphs are provided for selected problems only.

1. $(x - 3)^2 + (y - 4)^2 = 25$

3. $(x + 1)^2 + (y + 3)^2 = 36$

5. $(x - 9)^2 + (y - 6)^2 = 64$

7. Center at $(0, 3)$; radius $= 3$

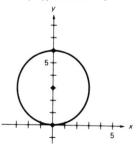

9. Center at $(0, -2)$; radius $= 2$

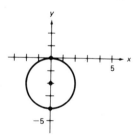

11. Center at $(2, -3)$; radius $= 5$

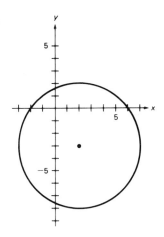

13. Center at $(1, 2)$; radius $= 0$. Thus, the graph is only the point $(1, 2)$.

15. Center at $(1, 1)$; radius $= 1$

17. Center at $(\frac{3}{2}, \frac{1}{2})$; radius $= \frac{3}{2}$

19. Center at $(-2, -10)$; radius $= 2\sqrt{26}$

Problem Set 5.3

Graphs are provided for selected problems only.

1. Center at $(0, 0)$; vertices at
$(0, 4), (0, -4), (3, 0), (-3, 0)$

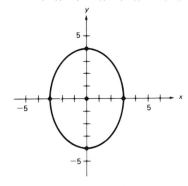

3. Center at $(0, 0)$; vertices at
$(0, 12), (0, -12), (11, 0), (-11, 0)$

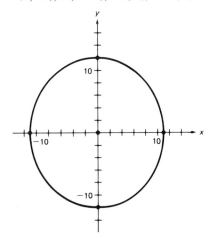

5. Center at $(2, 4)$; vertices at $(2, 13)$, $(2, -5)$, $(8, 4)$, $(-4, 4)$

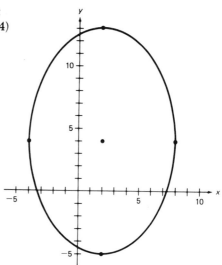

7. Center at $(2, -4)$; vertices at $(2, -4 + \sqrt{8})$, $(2, -4 - \sqrt{8})$, $(2 + \sqrt{13}, -4)$, $(2 - \sqrt{13}, -4)$

9. Center at $(-\frac{1}{2}, \frac{5}{2})$; vertices at $(-\frac{1}{2}, \frac{9}{2})$, $(-\frac{1}{2}, \frac{1}{2})$, $(\frac{9}{2}, \frac{5}{2})$, $(-\frac{11}{2}, \frac{5}{2})$

11. Center at $(-2, 1)$; vertices at $(-2, 3)$, $(-2, -1)$, $(4, 1)$, $(-8, 1)$

13. Center at $(3, 2)$; vertices at $(3, 3)$, $(3, 1)$, $(3 + \sqrt{\frac{1}{2}}, 2)$ $(3 - \sqrt{\frac{1}{2}}, 2)$

15. Center at $(1, 1)$; vertices at $(1, 3)$, $(1, -1)$, $(2, 1)$, $(0, 1)$

17. Center at $(5, 1)$; vertices at $(5, 1 + 3\sqrt{3})$, $(5, 1 - 3\sqrt{3})$, $(5 + 3\sqrt{2}, 1)$, $(5 - 3\sqrt{2}, 1)$

19. Center at $(3, -5)$; vertices at $(3, 1)$, $(3, -11)$, $(3 + \sqrt{11}, -5)$, $(3 + \sqrt{11}, -5)$

Problem Set 5.4

Graphs are provided for selected problems only.

3. Center at $(-3, 1)$; $y - 1 = \pm \frac{5}{2}(x + 3)$

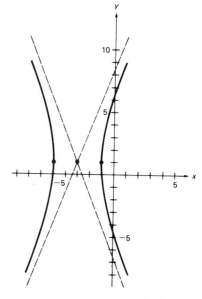

1. Center at $(0, 0)$; $y = \pm \frac{1}{2}x$

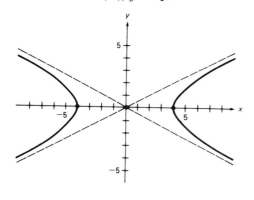

5. Center at $(2, 4)$; $y - 4 = \pm\frac{3}{2}(x - 2)$

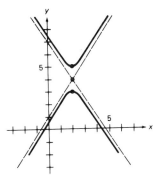

7. Center at $(3, 2)$; $y - 2 = \pm\frac{5}{4}(x - 3)$ **15.** Center at $(3, \frac{1}{3})$; $y - \frac{1}{3} = \pm\frac{1}{6}(x - 3)$

9. Center at $(-3, 1)$; $y - 1 = \pm\frac{5}{3}(x + 3)$ **17.** Center at $(2, -5)$; $y + 5 = \pm\frac{1}{3}(x - 2)$

11. Center at $(2, 4)$; $y - 4 = \pm\frac{3}{2}(x - 2)$ **19.** Center at $(-1, -3)$; $y + 3 = \pm\frac{5}{4}(x + 1)$

13. Center at $(1, -2)$; $y + 2 = \pm(x - 1)$

Chapter 5 Review

1. Ellipse; center at $(0, 0)$;
vertices at $(0, 4), (0, -4), (2, 0), (-2, 0)$

3. Hyperbola; center at $(0, 0)$;
asymptotes $y = \pm 2x$

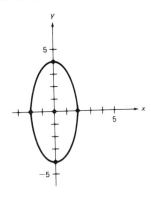

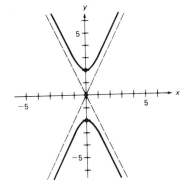

5. Ellipse; center at $(0,0)$; vertices at $(0, \sqrt{\frac{45}{2}}), (0, -\sqrt{\frac{45}{2}}), (\sqrt{15}, 0), (-\sqrt{15}, 0)$

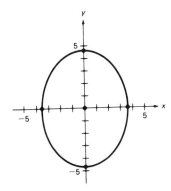

7. Vertex at $(\frac{1}{4}, -\frac{25}{8})$;
intercepts at $(0, -3)$, $(\frac{3}{2}, 0)$ $(-1, 0)$

9. Hyperbola; center at $(0, 0)$;
asymptotes $y = \pm x$

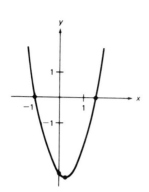

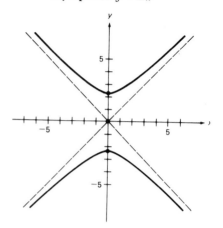

11. Parabola; vertex at $(\frac{1}{2}, \frac{23}{4})$;
intercept at $(0, 6)$

13. Ellipse; center at $(-3, 7)$;
vertices at $(2, 7)$, $(-8, 7)$, $(-3, 3)$, $(-3, 11)$

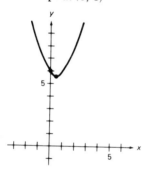

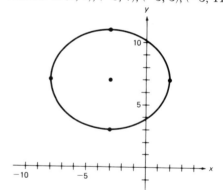

15. Parabola; vertex at $(-2, -3)$;
intercept at $(0, \frac{16}{5})$

17. Hyperbola; center at $(2, -3)$;
asymptotes $y + 3 = \pm \frac{3}{5} (x - 2)$

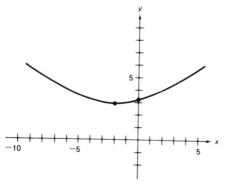

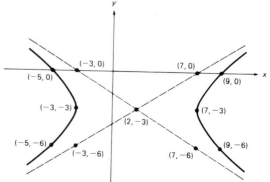

19. $\dfrac{x^2}{16} + \dfrac{y^2}{4} = 1$

21. $\dfrac{x^2}{16} - \dfrac{y^2}{4} = 1$

23. $(x - 2)^2 + (y - 4)^2 = 1$

25. Hyperbola; center at $(2, 4)$;
asymptotes: $y - 4 = \pm\frac{1}{2}(x - 2)$

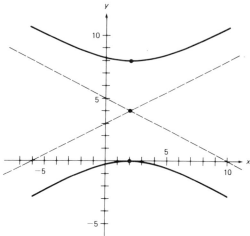

27. Ellipse; center at $(-1, 2)$;
vertices at $(2, 2), (-4, 2), (-1, 6), (-1, -2)$

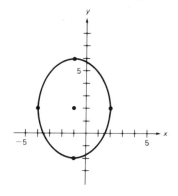

29. Ellipse; center at $(3, -4)$;
vertices at $(5, -4), (1, -4), (3, -3), (3, -5)$

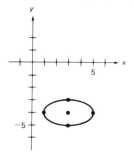

31. Circle; center at $(-4, 3)$; radius $= 2\sqrt{10}$

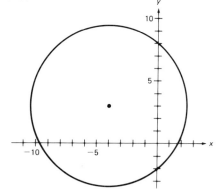

33. Parabola; vertex at $(3, -\frac{16}{3})$;
intercepts at $(0, -\frac{7}{3})$, $(7, 0)$, $(-1, 0)$

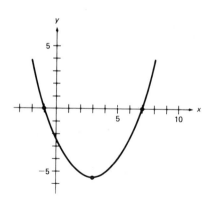

35. Circle; center at $(-\frac{1}{2}, -\frac{1}{2})$;
radius $= \sqrt{\frac{1}{2}}$

37. Circle; center at $(-4, 1)$;
radius $= \sqrt{2}$

39. A

41. C

43. A

45. D

47. $x^2 + y^2 = 4$
$$\frac{x^2}{25} + \frac{y^2}{4} = 1$$

49. (a) Yes
(b) No

Problem Set 6.1

1. $x = 4, y = 7$
3. $x = 1, y = 2$
5. All pairs (x, y) satisfying the first equation will also satisfy the second.
7. $x = 2, y = 3$
9. $x = \frac{1}{2}, y = -\frac{3}{2}$
11. $x = 10, y = 5$
13. $(3, -1)$
15. $(3, -3)$
17. $(3, 3)$

Problem Set 6.2

1. $(0, 0), (1, 1)$
3. $(1, 1)$
5. $(0, 0), (1, 1)$
7. $(0, 0), (\frac{1}{4}, 1)$
9. $(0, 0)$
11. $(\sqrt{\frac{1}{3}}, 3\sqrt{\frac{1}{3}}), (-\sqrt{\frac{1}{3}}, -3\sqrt{\frac{1}{3}})$
13. $(1, 0), (2 + \sqrt{2}, 1 + \sqrt{2}), (2 - \sqrt{2}, 1 - \sqrt{2})$

15. No intersection
17. $(\frac{1}{3}, 2)$
19. $(0, 0), (1, 1), (-1, 1)$
21. $(1, 1)$
23. $(1, 0)$
25. $(-1, 0), (-2, 1)$

Problem Set 6.3

1. No points of intersection
3. $(4, 3), (-4, -3)$
5. $(6, 9), (6, -9), (-6, 9), (-6, -9)$
7. $(2\sqrt{2}, 2\sqrt{2}), (2\sqrt{2}, -2\sqrt{2}), (-2\sqrt{2}, 2\sqrt{2}), (-2\sqrt{2}, -2\sqrt{2})$
9. $(4, 0), (-4, 0)$
11. $(2, 0), (-2, 0)$
13. $(3, 2), (-3, 2), (3, -2), (-3, -2)$
15. $(-2, 0), (\frac{17}{5}, \frac{9}{5})$
17. No points of intersection
19. $\left.\begin{array}{l} x^2 + y^2 = 4 \\ (y - 2) = x^2 \end{array}\right\}$ There are many other possible answers.
21. $\left.\begin{array}{l} x^2 + y^2 = 36 \\ (y + 6) = 10x^2 \end{array}\right\}$ There are many other possible answers.
23. $\left.\begin{array}{l} x^2 + y^2 = 1 \\ (x + 2)^2 + y^2 = 1 \end{array}\right\}$ There are many other possible answers.
25. $\left.\begin{array}{l} 4y^2 - x^2 = 16 \\ (y - 2) = 100x^2 \end{array}\right\}$ There are many other possible answers.

Problem Set 6.4

1.

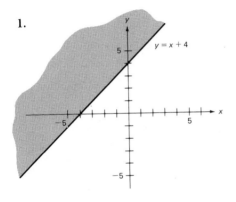

3.

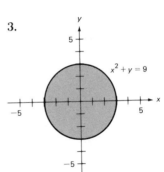

5.

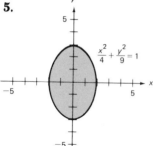

$$\frac{x^2}{4} + \frac{y^2}{9} = 1$$

7.

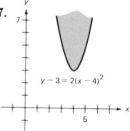

$y - 3 = 2(x - 4)^2$

9.

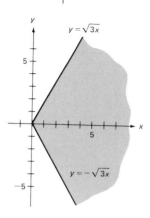

$y = \sqrt{3x}$

$y = -\sqrt{3x}$

11.

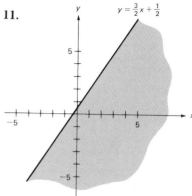

$y = \frac{3}{2}x + \frac{1}{2}$

13.

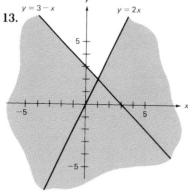

$y = 3 - x$

$y = 2x$

15.

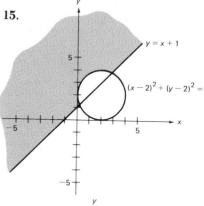

$y = x + 1$

$(x - 2)^2 + (y - 2)^2 = 4$

17.

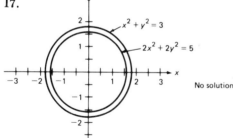

$x^2 + y^2 = 3$

$2x^2 + 2y^2 = 5$

No solution

19.

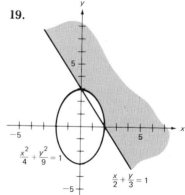

$\frac{x^2}{4} + \frac{y^2}{9} = 1$

$\frac{x}{2} + \frac{y}{3} = 1$

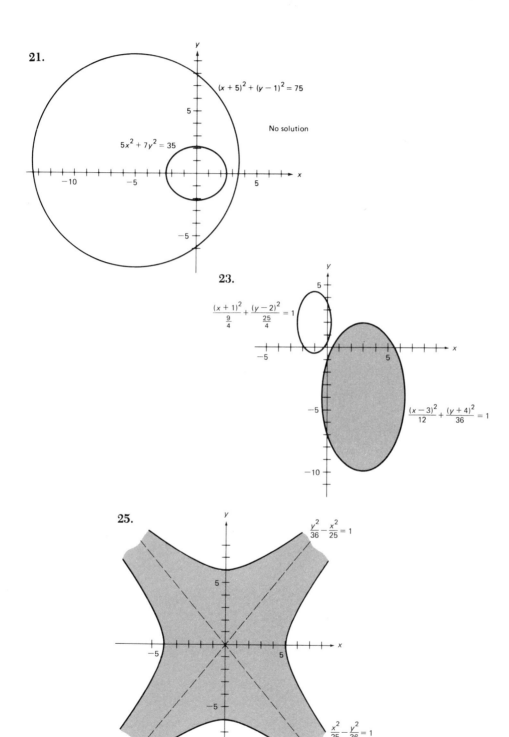

21.

$(x + 5)^2 + (y - 1)^2 = 75$

No solution

$5x^2 + 7y^2 = 35$

23.

$\dfrac{(x + 1)^2}{\frac{9}{4}} + \dfrac{(y - 2)^2}{\frac{25}{4}} = 1$

$\dfrac{(x - 3)^2}{12} + \dfrac{(y + 4)^2}{36} = 1$

25.

$\dfrac{y^2}{36} - \dfrac{x^2}{25} = 1$

$\dfrac{x^2}{25} - \dfrac{y^2}{36} = 1$

Chapter 6 Review

1. $x = 4, y = 5$

3. $x = 6, y = 2$

5. $x = \frac{1}{2}, y = \frac{1}{2}$

7. $x = 2, y = 5$

9. $x = 3, y = -1$

11. $x = 1, y = 1$

13. $x = 3, y = 0;\ x = \frac{25}{8}, y = \frac{1}{4}$

15. No solutions

17. No solutions

19. $x = \dfrac{5 + \sqrt{37}}{2}$ $y = \sqrt{\dfrac{-1 + \sqrt{37}}{2}} + 2;\ x = \dfrac{5 + \sqrt{37}}{2},\ y = \sqrt{\dfrac{-1 + \sqrt{37}}{2}} + 2$

21. $x = 0, y = \frac{1}{2}$

23.

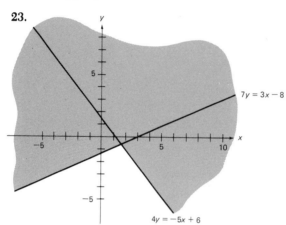

25.

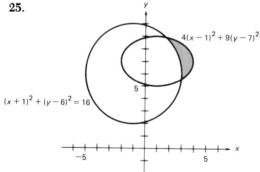

27.

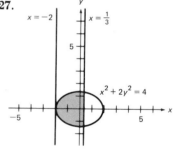

29. A hyperbola never crosses its asymptotes.

31.

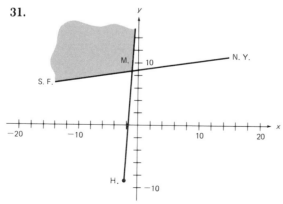

Problem Set 7.1

1. (a) Yes: (b) Yes
3. Yes; no
5. Yes; no
7. Yes; no
9. Yes; yes
11. Yes; yes
13. Domain is all real numbers; range is all real numbers
15. Domain is reals ≥ 7; range is reals ≥ 0
17. Domain is all reals; range is reals ≤ 1
19. Domain is all reals except $\frac{3}{2}$; range is reals except 0
21. Domain is reals > 3; range is reals > 0
23. Domain is reals ≥ -30; range is reals ≤ 0
25. Domain is all reals; range is reals ≥ 0
27. Domain is reals except -2; range is reals except 1
29. Domain is reals ≥ 4; range is reals ≤ 0
31. (a) Yes; (b) Yes
33. Yes; yes
35. No; no
37. No; yes
39. Yes; yes
41. Yes; yes
43. Domain is reals $\geq -\frac{1}{3}$ and $\leq \frac{1}{3}$; range is reals ≥ 0 and $\leq \frac{1}{3}$
45. Domain is reals except 0; range is reals except 0
47. Domain is reals except 0; range is reals except 0
49. Domain is all reals; range is all reals

Problem Set 7.2

1. (a) $\frac{2}{3}$ (b) $\frac{3}{4}$ (c) -2 (d) $-\frac{3}{2}$
3. (a) 4 (b) 0 (c) 4, -4 (d) 0
5. 1
7. 201
9. $2a^2 + 1$
11. $\dfrac{2}{a} + 1 = \dfrac{2 + a}{a}$
13. 2
15. 4, -4
17. 3
19. 0
21. 6, -2
23. $p^4 - 4p^2 + 3$
25. $\frac{1}{13}$
27. $\dfrac{1}{a + 9}$
29. $\left(\dfrac{1}{a^2 + 9}\right)^2$
31. $a^2 + 9$
33. 4, -4
35. $p^2 + 2p + 1$
37. $a^2 + 2ah + h^2$
39. $\dfrac{1}{x + 1}$
41. $\left(\dfrac{1}{x^2 + 1}\right)^2$
43. $\dfrac{1}{x^4 + 2x^2 + 2}$
45. 2
47. $\dfrac{-1}{t^2 + t}$
49. $v = 2x^3$
51. 0.02
53. 0.0011
55. 0.06
57. $a^2(1 + b)$

Chapter 7 Review

1. 1, 7
3. 6, 2
5. 10, −2
7. $2x + a$
9. $3x^2 + 3xa + a^2$
11. $f(0) = 3; f(-\tfrac{3}{2}) = 0$
13. 0
15. 0
17. 0
19. $s = 6d^2 + 28d + 20$

21. 2
23. 56
25. $\dfrac{t + 2}{t - 1}$
27. $\dfrac{t + 3}{t - 3}$
29. $\dfrac{t^2 + t + 1}{t^2 + 2t + 4}$
31. $\dfrac{x^4 - 6x^3 + 9x^2 + 1}{2x^2 - 6x}$

33. $\dfrac{(2v - 1)^2}{2(v + 2)\,(v - 3)}$
35. $\dfrac{[(a + 1)^2 + 1]^2}{2(a + 1)^2}$
37. $(x - 1)$
39. 16
43. No; $(a + c)^2 \neq a^2 + c^2$

Problem Set 8.1

1. $\dfrac{f(c) - f(a)}{c - a}$
3. 4
5. 0, 6
7. $x < 0$ or $x > 6$

9. Reciprocal
11. $\tfrac{1}{3}$
13. $\sqrt{[h(c)]^2 + c^2}$

Problem Set 8.2

1. $f(0) = 3; f(-3) = 0$
3. $f(0) = 0; f(0) = 0$
5. $f(0) = 12; f(-4), f(-3) = 0$
7. $f(0) = -6; f(-\tfrac{1}{2}), f(6) = 0$
9. $f(0) = 0; f(0), f(1), f(-1) = 0$

11. $f(0) = -1$; no solutions to $f(x) = 0$
13. $f(0) = 2; f(2) = 0$
15. $f(0)$ is undefined; $f(\tfrac{1}{4}), f(-\sqrt{\tfrac{3}{4}}) = 0$
17. $f(0) = 9.8 \cdot 10^{-2}; f(3.7 \cdot 10^{-1}) = 0$
19. $f(0) = \tfrac{2}{3}; f(-2) = 0$

Problem Set 8.3

1. Even
3. Neither
5. Odd
7. Even

9. Even
11. Odd
13. Even

Problem Set 8.4

1. Function
3. Function
5. Function
7. Function
9. Not a function

11. Function
13. Not a function
15. Function
17. Function

Problem Set 8.5

1. Domain is all real numbers; range is all real numbers ≥ 0
3. Domain is reals ≥ -3 and ≤ 3; range is reals ≥ -2 and ≤ 2
5. Domain is all reals; range is all integers
7. Domain is reals except $\frac{\pi}{2}x$ where x is an odd integer;

    range is reals ≥ 1 or reals ≤ -1
9. Domain is reals ≥ -2 and ≤ 2; range is reals ≥ -2 and ≤ 0
11. Absolute value function: $y = |x|$
13. $y = -\dfrac{1}{x-2} + 1$
15. $y = \begin{cases} x+3, & x \leq 0 \\ x-3, & x > 0 \end{cases}$

Chapter 8 Review

1. 2
3. All values <3 or >5
5. b
7. e
9. p
11. $(e, 0)$
13. c, i
15. $C = 36t + 16$
17. Integers >0
19. 4
21. (a) 5 (b) No solution (c) Even (d) All reals (e) Reals ≥ 5
23. (a) $\frac{7}{2}$ (b) $\frac{7}{2}, -\frac{7}{2}$ (c) Even (d) All reals (e) Reals $\geq -\frac{7}{2}$
25. (a) $-\frac{1}{3}$ (b) No solution (c) Even (d) Reals except $\pm\sqrt{3}$ (e) Reals except 0
27. (a) -3 (b) $+\sqrt{3}$ (c) Neither (d) Reals except $-\sqrt{3}$ (e) All reals
29. (a) -1 (b) 1 (c) Neither (d) All reals (e) Reals ≤ 0
31. $A(t) = \begin{cases} \$2.05, & t \leq 3 \\ \$2.05 + 0.13(t-3), & t \geq 4 \end{cases}$ (t an integer ≥ 1)
33. 7
35. Numbers that can be expressed as $0.53 + 0.34t$, where t is an integer ≥ 0
37. $C(p) = \dfrac{p^2}{2\pi}$
39. Reals ≤ 400
41. $p \geq 209.8$
43. Circular bullring is larger
45. Domain is positive reals <6; range is reals between 0 and 9
47. $A(x) = x(500 - x)$
49. (a) 0 (b) $0, \dfrac{\sqrt{2}}{2}, -\dfrac{\sqrt{2}}{2}$ (c) Even (d) All reals (e) Reals ≥ 0
51. (a) $\dfrac{-\sqrt{5}}{500}$ (b) $-\sqrt{3}, \sqrt{15}$ (c) Neither (d) Reals except 15, 20
    (e) Reals except $\frac{1}{50}$
53. (a) $\frac{1}{4}$ (b) No solutions (c) Even (d) All reals (e) Reals >0 and $\leq\frac{1}{4}$

Problem Set 9.1

1. 5

3. $v^2 + 1$

5. $\dfrac{1}{x^2} + 1$

7. 1

9. $\dfrac{1}{v}$

11. $\dfrac{1}{x^2 + 1}$

13. 5

15. $w + 1$

17. $\dfrac{x^2 + 2}{x^2 + 1}$

19. $\frac{1}{5}$

21. $\dfrac{1}{x^2 + 2x + 2}$

23. $\left(\dfrac{1}{x - 1}\right)^2$

25. $\dfrac{x^2}{x - 1}$

27. $\dfrac{4x^4}{(x + 1)^2}$

29. $\dfrac{2x^4}{x^2 + 1}$

31. $\dfrac{2}{x^2 - x}$

33. $\left(\dfrac{1}{x - 1}\right)^2$

35. (a) $v(r) = \frac{4}{3}\pi r^3$
(b) $v(t) = \frac{4}{3}\pi t^6$

37. (a) $I(G) = 0.15G$
(b) $S(I) = 0.10I$
(c) $S(G) = 0.015G$

39. $\frac{1}{4}$

41. 9

43. $\frac{10}{9}$

45. $\dfrac{1}{a^2}$

47. $\dfrac{x^2 + 1}{x^2}$

49. $\dfrac{1}{(x + 1)^2}$

51. $\dfrac{(x + 1)^2}{x^2}$

53. $x^2 + x$

55. $\dfrac{1}{x + 2\sqrt{x - 1}}$

57. $\dfrac{\sqrt{-x^2 - 2x}}{x + 1}$

59. $\dfrac{x + 2\sqrt{x - 1}}{x - 1}$

61. $g(x) = \dfrac{1}{x}$

63. $m(x) = \dfrac{1}{\sqrt{x}}$

Problem Set 9.2

1. Domain is reals ≥ 0; range is reals ≥ 1

3. Domain is reals except -1; range is reals > 1

5. Domain is reals except $2, -2$; range is reals > 1

7. Domain is reals > -1; range is reals > 0

9. Domain is reals < -2 or > 2; range is reals > 0

11. Domain is reals except $2, -2, \sqrt{3}, -\sqrt{3}$; range is reals except 0

13. Domain is reals > -1; range is reals > 1

15. Domain is reals except $2, -2, \sqrt{3}, -\sqrt{3}$; range is reasl > 1

17. Domain is reals except $2, -2$; range is reals > 1

19. Domain is reals except $\sqrt{3} \leq x \leq 2$ and $-2 \leq x \leq -\sqrt{3}$; range is reals > 1

21. Domain is reals ≥ 0 except $4, 3$; range is reals > 1

23. Domain is all reals; range is reals $\geq \sqrt{2}$

25. Domain is reals except -1; range is reals > 1

Problem Set 9.4

1. $\dfrac{x}{3}$

3. $\sqrt{\dfrac{t}{2}}$

5. $\sqrt{\dfrac{x}{3} - 12}$

7. $\sqrt{\dfrac{1}{x}}$

9. $\dfrac{2}{x - 1}$

11. $\dfrac{1}{u - 2}$

13. $x^2 - 2$

15. $\sqrt{t - 1}$

17. $(t + 2)^2 + 1$

19. $\dfrac{1}{u^2}$

21. $\dfrac{1}{(s - 2)^2}$

23. $\dfrac{1}{\sqrt{x}}$

25. $\dfrac{4\sqrt{x}}{1 - \sqrt{x}}$

27. $\dfrac{x^2}{x^2 - 1}$

29. $\dfrac{x^2 + 1}{1 - x^2}$

31. $(x^2 + 2)^2 - 2$

33. $\dfrac{3 + \sqrt{x}}{1 - \sqrt{x}}$

35. $\dfrac{4(\sqrt{x} + 2)^2}{1 - (\sqrt{x} + 2)^2}$

37. $\sqrt[4]{\dfrac{1}{x} + 2}$ $x > 0$ or $x < -\frac{1}{2}$

39. $\sqrt{\sqrt{x} - 3}$ $x \geq 9$

41. $x \geq 4$

43. $x \geq -1$

45. $x < 0$ or $x \geq 2$

Problem Set 9.5

1. Domain is reals except 0; range is reals except -2
3. Domain is reals except -3; range is reals except -3
5. Domain is reals ≥ -3; range is reals ≥ -2
7. Domain is reals ≥ 0; range is reals ≥ -1
9. Domain is all reals; range is all reals
11. Restrict domain of original to reals ≥ 1; domain of inverse is reals ≥ -4; range of inverse is reals ≥ 1
13. Restrict domain of original to reals ≥ 0; domain of inverse is reals ≥ -8; range of inverse is reals ≥ 0
15. Restrict domain of original to $x \geq 1$; domain of inverse is reals ≥ 0; range is reals ≥ 1
17. $f^{-1}(x) = \dfrac{x}{2}$; domain is all reals; range is all reals

19. $f^{-1}(x) = \dfrac{-x-2}{2}$; domain is all reals; range is all reals

21. $f^{-1}(x) = -x$; domain is all reals; range is all reals
23. $f^{-1}(x) = \sqrt[3]{x}$; domain is all reals; range is all reals
25. $f^{-1}(x) = \sqrt[3]{x} + 1$; domain is all reals; range is all reals
27. $f^{-1}(x) = \dfrac{\sqrt[3]{x+2}-1}{2}$; domain is all reals; range is all reals

29. $f^{-1}(x) = \sqrt[5]{x+1}$; domain is all reals; range is all reals
31. $f^{-1}(x) = \dfrac{1}{x-3}$; domain is reals except 3; range is reals except 0

33. $f^{-1}(x) = \dfrac{4x+6}{1-x}$; domain is reals except 1; range is reals except -4

35. $f^{-1}(x) = \dfrac{2-x}{x}$; domain is reals except 0; range is reals except -1

37. $f^{-1}(x) = \dfrac{5+\sqrt{9+8x}}{4}$; domain is reals $\geq -\frac{9}{8}$; range is reals $\geq \frac{5}{4}$

39. $f^{-1}(x) = \sqrt{x^2 + \frac{1}{4}} + \frac{3}{2}$; domain is reals ≥ 0; range is reals ≥ 2

Chapter 9 Review

1. (b) $f^{-1}(x) = \dfrac{x}{1-x}$; $f(g(x)) = \dfrac{1}{x+2}$

   (c) $P(x) = \dfrac{1}{x}$, $Q(x) = x+1$

3. $\dfrac{x+3}{x}$

5. $D^{-1}(x) = \dfrac{x}{x-2}$

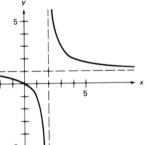

7. $j(x) = x + 2$, $h(x) = x(x - 2)$, $g(x) = \dfrac{x}{5}$

9. $h^{-1}(x) = \dfrac{x + 2}{x - 1}$

11. Reals except $2, -2$

13. $k(x) = \dfrac{x + 4}{x - 6}$, $h(x) = 3x^4$

15. $f(x) = \sqrt{x}$, $g(x) = x^2 + 1$

17. $2x^2 + 23$

19. $\dfrac{x - 2 + \sqrt{x^2 + 4}}{2}$ or $\dfrac{x - 2 - \sqrt{x^2 + 4}}{2}$

21. $(x - 2)^2$

23. $W = 35N$

25. $P(W) = \dfrac{W}{560}$

27. 105 pounds

29. $C(n) = \dfrac{5(n - 88)}{36}$

31. $\dfrac{x + 2}{3}$

33. $F(1) = 1$
$F(2) = 2$
$F(3) = 6$
$F(4) = 24$
$F(n) = n! = n(n - 1)(n - 2) \cdots (2)(1)$

35. $\dfrac{3x + 1}{x - 2}$

37. $\dfrac{x^2 - 9}{x^2 - 4}$

39. $\dfrac{9 - 4x}{x - 1}$

43. $\frac{29}{5}$ mph

45. $\dfrac{5x + 13}{6}$ mph

Problem Set 10.1

1. 3
3. 144
5. $\frac{1}{2}$
7. 5
9. 8
11. $\frac{1}{9}$
13. $\frac{9}{4}$
15. $-\frac{1}{2}$
17. 9
19. $\dfrac{x^3}{2}$
21. $\frac{1}{6}$
23. $3x^3$
25. a

27. $\dfrac{2 - x}{2}$
29. $\dfrac{8a^2}{b^2c^2}$
31. $3 \cdot 10^{-8}$
33. 10^4
35. $5.9 \cdot 10^{199}$
37. $9.0 \cdot 10^{-4}$
39. $3.33 \cdot 10^6$
41. 9
43. 4
45. $\frac{16}{27}$
47. 1.2

49. $\frac{26}{5}$
51. $\dfrac{5x}{y}$
53. $a^{7/6}b$
55. 6^{x+7}
57. $\dfrac{m^x}{m^x - b^x}$
59. $\dfrac{abc}{bc + ac + ab}$
61. $7.8 \cdot 10^{-4}$
63. $4.298 \cdot 10^{-5}$
65. $-4.25 \cdot 10^{-7}$
67. $3.33 \cdot 10^7$

Problem Set 10.2

1. 2^4
3. $3 \cdot 2^4$
5. 2^6
7. 2^{-4}
9. $\frac{1}{2}$
11. -1
13. $-\frac{1}{3}$
15. $\frac{1}{3}$

17. 4
19. $\frac{1}{3}$
21. $-\frac{1}{3}$
23. 2
25. 1
27. $\frac{3}{2}$
29. $\frac{4}{3}$
31. False

33. True
35. False
37. -12
39. 3
41. $\frac{13}{2}$
45. 0
47. $x = \frac{3}{2}$, $y = \frac{1}{2}$
49. 0, 2

Problem Set 10.3

1. $2^{9/2}$
3. $2^{1/2}$
5. $2^{-2/3} + 2^{-5/3}$
7. $3^{1/n}x^{1/n}y^{2/n}$
9. $x^{1/5}y^{2/5}z^{3/5}$
11. $2^{4/3}b^{2/3}c/a$
13. $a(b^2 + c^2)^{1/2}$
15. $2x^{1/2}y^3$
17. 1
19. $\dfrac{x + y - 1}{x^{2/3}y^{2/3}}$

21. $-\frac{3}{4}$
23. $\frac{1}{3}$
25. True
27. True
29. True
31. $\dfrac{14 + 5\sqrt{3}}{11}$
33. $4 - \sqrt{15}$
35. $\dfrac{a(x - \sqrt{y})}{x^2 - y}$
37. $b = c - 2\sqrt{a^2 + ac} + 2a$

39. 1
41. $4\left(\dfrac{3^{1/2} - 2^{1/3}}{3^{2/3}}\right)$
43. $-10a(2x)^{1/3}$
45. $\dfrac{18x^4}{y^{3/2}}$
47. x^9y^{10}
49. $5bx^{1/2}$
51. $\dfrac{x + 1}{x - 1}$

Problem Set 10.4

1.

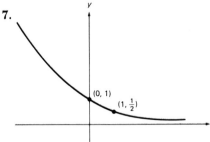

(1, 2)
(0, 1)

3.

(0, 3)

5.

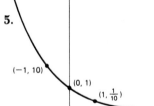

(−1, 10)
(0, 1)
(1, $\frac{1}{10}$)

7.

(0, 1)
(1, $\frac{1}{2}$)

9.

(1, 2A)
(0, A)

11.

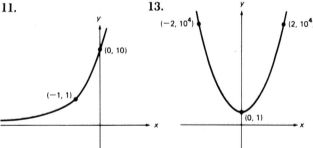

(0, 10)
(−1, 1)

13.

(−2, 10⁴)
(2, 10⁴)
(0, 1)

15.

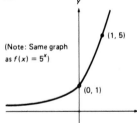

(1, 5)
(Note: Same graph as $f(x) = 5^x$)
(0, 1)

Chapter 10 Review

1. $\frac{9}{4}$
3. 1
5. 1000
7. $5 \cdot 10^2$
9. $\frac{175}{16}$
11. $3 \cdot 2^8$
13. 2^{13}
15. 2^{-7}
17. $\frac{1}{8}$

19. a^2
21. $a + b + 2\sqrt{ab}$
23. >
25. =
27. -0.3010
29. -1
35. 0
37. 1
39. $\dfrac{1}{(a^2 - x^2)^{1/2}}$

41. $\dfrac{1}{p}$
43. $\dfrac{[x^4 - x^2 + a^2(2x^2 + a^2)]}{(x^2 + a^2)^{3/2}}$
45. $\left(\dfrac{u^2 - x^2}{u^2 + x^2}\right)^{1/2}$
47. 2
49. $-\frac{17}{3}$, (undefined for $p = -1$)
51. (a) IV (b) III (c) I (d) II

Problem Set 11.1

1. 0
3. 0
5. 5
7. $-\frac{1}{2}$
9. 1
11. 25.2
13. 0.692
15. 1.4
17. $10^{1.3010}$
19. $10^{0.6990}$

Problem Set 11.2

1. 0.0170
3. 3.6628
5. 3.4871
7. $0.2553 - 4$
9. 2.7050
11. 0.8500
13. $0.4200 - 3$
15. 1.1367
17. 82
19. 8400
21. 0.893
23. $5.82 \cdot 10^{12}$
25. 0.0291
27. $1.65 \cdot 10^{-4}$
29. $3.95 \cdot 10^{-5}$
31. 317
33. 3.16
35. 0.0316
37. 8.44
39. $4.38 \cdot 10^3$
41. 100
43. Undefined
45. $\frac{4}{7}$
47. 12.1
49. 670.32

Problem Set 11.3

1. 0.7782
3. 1.0792
5. -0.0511
7. -0.8293
9. 0.6990
11. False
13. False
15. False
17. False
19. False
21. -2.7050
23. 0.1075
25. 2.3979
27. $\log 64$
29. $\log \frac{1}{81}$
31. $\log 500$
33. $\log 2$
35. $\log \frac{1}{10}$
37. x^2
39. $10x$
45. $\dfrac{A^2 B}{3}$
47. 4
49. 1.6
51. No, yes. $c = \dfrac{a + b}{ab - 1}$
53. $\log d = \log \frac{1}{2} + \log a + 2 \log t$
55. $\log v = \frac{1}{2}(\log 3 + \log k + \log t - \log m)$

Problem Set 11.4

1. 2
3. 25
5. a
7. $\dfrac{pq}{p + q}$
9. $\dfrac{1}{510}$
11. $\frac{1}{5}$
13. $\dfrac{1}{(510)^2}$
15. N
17. 0
19. -1
21. $\frac{1}{8}$
23. 1000
25. $\frac{1}{117}$
27. 10^{ax+b}
29. $\log 5$ or 0.6990
33. $x = 10$
35. $y = \frac{1}{5}$
37. $x = 10$
39. $x = a$
41. 0.7
43. 7
45. x^a
47. $a + x$
49. x

Problem Set 11.5

1. $\log_3 27 = 3$
3. $\log_4 (\tfrac{1}{2}) = -\tfrac{1}{2}$
5. $\log_9 (\tfrac{1}{243}) = -\tfrac{5}{2}$
7. $16^{3/4} = 8$
9. $x^b = a$
11. $x^a = yz$
13. $x^a = \dfrac{y}{p}$

15. $a^2 = 5^{-2}$
17. 3
19. $a + b$
21. $y = 5$
23. $z = 2$
25. $v = 8$
27. $x = 27$
29. $t = \tfrac{1}{8}$

31. -1
33. $\dfrac{1}{x^2}$
35. $\tfrac{1}{5}$
37. 4
39. 1.6
41. 7
43. 2.10

45. 3.92
47. $\tfrac{2}{3}$
49. $\log_{10} 22$
51. 7
53. (a) 4 (b) $2 \log_3 x$
55. 1

Problem Set 11.6

1. $4.13 \cdot 10^3$
3. 7.10
5. 3.16
7. 3.34
9. 23.2
11. 60.3
13. 2.58
15. 1.11
17. 20
19. 1.57
21. $\log_{10} 6$
23. 5
25. $\log_c \tfrac{3}{4}$
27. -1

29. $\log_b \dfrac{2x^2}{(LZ)^{1/2}}$
31. (a) $\log (\sqrt{110} + \sqrt[3]{53}) \neq \log \sqrt{110} + \log \sqrt[3]{53}$
    (b) 14.25
33. $\log_{10} A = n \log_{10} y - \dfrac{p}{q} \log_{10} x$
35. $3.25 \cdot 10^3$ cubic feet
37. $3.38 \cdot 10^4$ grams
39. 0.133 seconds
41. $5.84 \cdot 10^8$ miles; $6.67 \cdot 10^4$ miles
43. $[\log (xy) + 1] [\log (xy) + 2]$
45. 32.5 pounds
47. $5.6 \cdot 10^{776}$
49. $\$538$

Chapter 11 Review

1. 1.9868
3. 12.3
5. 3
7. $0.4314 - 3$
9. 0.2
11. 2

13. 3
15. -0.8
17. 2.5
19. 0.1685
21. 3960
23. 0.0127

25. 0.1059
27. 10
29. 0.6
31. $\log x^2$
33. $A = (CD)^B$
35. x^k

37. $1.77 \cdot 10^6$
39. 64
41. 28 seconds
43. $\$9821$
45. 75

Problem Set 12.1

1. 1.14
3. -1.53
5. 0.404
7. 0.575
9. -5.19
11. 1.32
13. 4
15. 2.26

17. 0.125
19. 1.23
21. 1.16
23. $a = 2.47, b = -0.523$
25. (a) $x = 1.4$ (b) $x = \dfrac{\log P_2 - \log P_1}{\log V_1 - \log V_2}$
27. $k = \dfrac{-\log A}{\log B}$

29. $S = \dfrac{\log(AB - M)}{t \log p} - \dfrac{r}{t}$

31. $x = \dfrac{\log ab}{2}$

33. $a = \tfrac{5}{2}$

35. $v^2 = \dfrac{\log A}{D \log B} - \dfrac{C}{D}$

37. 26.3

39. 2001

41. $a = 4.19, b = 3.00$

43. (a) 10^{-12} grams   (b) $2n \cdot 10^{-12}$ grams   (c) 133   (d) $66\frac{1}{2}$ hours

45. 33.5 years

47. 1.66 hours

49. 8 bounces

Problem Set 12.2

1. 10,000,002

3. 0.4976

5. 1

7. $-5, \tfrac{5}{3}$

9. 199.5

11. 4

13. 13.16

15. $z = 20, 0$

17. $b = \dfrac{1}{\sqrt[3]{ax}}$

19. $x = (np)^m$

23. $x = a$

25. $y = 101$

27. $A = 10^{3/2}$

29. $x = \tfrac{3}{5}$

31. $f^{-1}(x) = 2(2^x) + 1$

33. $0 < p < 100$

35. $-1000 < x < 1000$

37. $10^{1.9} < x < 10^{2.1}$

39. (a) 1.58   (b) 80 windows   (c) To avoid log 0

Problem Set 12.3

1.

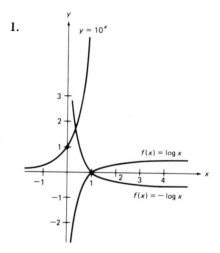

3.

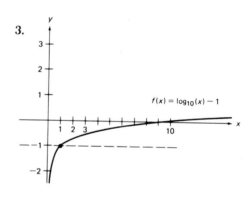

5.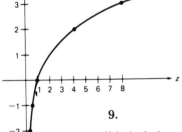

7.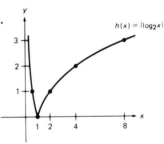

9.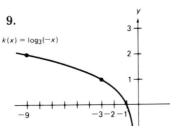

11. 1
13. 100
15. 1
17. 0
19. 0
21. (a) Yes (b) 1/25
(c) M (d) 1 (e) no
(f) $y \to$ infinity; $y \to 0$

21(g)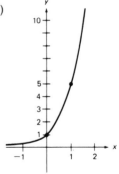

23. $f(x) = 2x + 7$; $g(x) = \log x$
25. $f(x) = x^2$; $g(x) = 10^x$
27. $f(x) = 3x$; $g(x) = 2^x$
29. $f(x) = \log x$; $g(x) = 2^x$

31.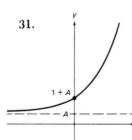

33. $k = 1$

35.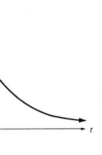

37. (a) $f(x) = 3 \log\left(\dfrac{1+x}{1-x}\right)$
(b) $kf(x)$; $k = 3$

39(a)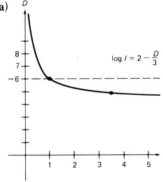

39. (b) 3.9 miles
(c) 0.9 miles

Problem Set 12.4

1.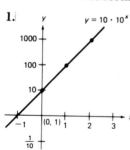

$y = 10 \cdot 10^x$

3.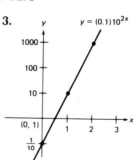

$y = (0.1)10^{2x}$

5.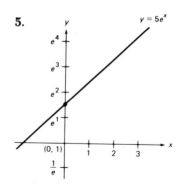

$y = 5e^x$

7.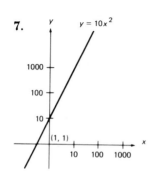

$y = 10x^2$

9.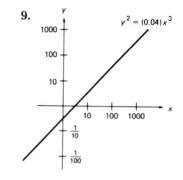

$y^2 = (0.04)x^3$

11. $A = 10,\ c = \dfrac{1}{\log_{10} e}$

13. $A = 10,\ m = 2$

19. (a) $K = \frac{1}{2}$
(c) 316

19(b)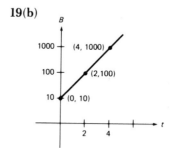

19(d)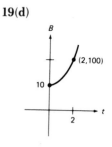

21. (b) $V = 5{,}000;\ L = 2.5 \cdot 10^9$
$V = 10{,}000;\ L = 10^{10}$

Chapter 12 Review

1. 1.71

3. 4.34

5. 1

7. $p = \dfrac{q}{r}$

9. $x = A^2 \sqrt[3]{B}$

11. 14

13. -0.444

15. 5.36

17. $\alpha = \frac{1}{3};\ P_0 = 200$

19. (a) 4.05 (b) 4.05

21. False

23. True

25. False

27. $g[f(x)] = [\log(x - 3)]^2$

29. Domain = all real numbers except $-3 < x < 3$

31.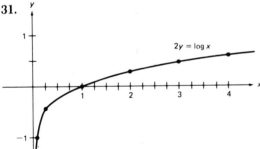

33.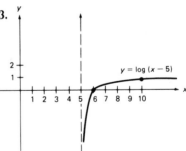

35. (a) y intercept $= 1$ (b) $x = -9$ (d) $f[h(30)] = 2$

35(c)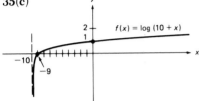

37. $f^{-1}(t) = 5^t$

39. $\sqrt{10}$

41.

43.

45. (a) $2 \log \left(\dfrac{x+1}{x-1} \right)$ (b) $kf(x); k = 2$

51. $t = \dfrac{2s - s^2}{s + 2}$

47. 123.7 hours

53. 9.21 days

49. 5.68 years

55. $R(Y) = 0.1(2^{-x})$

57(a)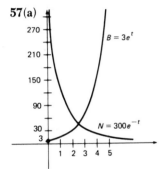

57. (b) $B = 3$, $N = 300$ (c) $t \approx 2.3$ (d) $B \approx 40$

59(a)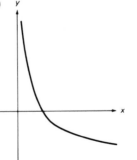

59(b)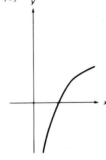

Problem Set 13.2

1. $\frac{12}{13}$
3. $\frac{12}{5}$
5. $\frac{13}{5}$
7. $\frac{7}{25}$
9. $\frac{7}{24}$
11. $\frac{25}{24}$
13. $\frac{1}{\sqrt{2}}$

15. $\frac{1}{\sqrt{2}}$
17. $\sqrt{2}$
19. $\frac{4}{5}$
21. $\frac{4}{3}$
23. $\frac{3}{4}$
25. $\frac{5}{\sqrt{61}}$

27. $\frac{6}{5}$
29. $\frac{\sqrt{61}}{5}$
31. $\frac{2}{\sqrt{13}}$
33. $\frac{2}{\sqrt{13}}$
35. $\frac{3}{2}$

37. $\frac{\sqrt{97}}{9}$
39. $\frac{4}{\sqrt{97}}$
41. $\frac{9}{4}$
43. $\frac{6}{\sqrt{85}}$

45. $\frac{7}{6}$
47. $\frac{\sqrt{85}}{6}$

Problem Set 13.3

1. 0.1736
3. 0.8693
5. 0.9976
7. 0.1736
9. 0.9325

11. 57.29
13. 9.514
15. 0.8028
17. 1.1253
19. 0.7955

21. 1.7929
23. 3.0415
25. 0.0576
27. 5.935
29. 0.9694

31. cot 40°, cos 60°, sin 2°

Problem Set 13.4

1. $A = 73°42'$, $B = 16°18'$, $b = 7$
3. $A = 45°$, $B = 45°$, $c = \sqrt{2}$
5. $A = 14°30'$, $B = 75°30'$, $b = \sqrt{15}$
7. $B = 40°30'$, $b = 2.79$, $c = 4.30$
9. $a = 2.1$, $B = 48°50'$, $c = 3.2$
11. $A = 28°06'$, $B = 61°54'$, $c = 17$
13. $A = 62°48'$, $a = 22.2$, $b = 11.4$
15. $B = 63°18'$, $b = 23.8$, $c = 26.7$

17. $A = 37°18'$, $B = 52°42'$, $b = 26.2$
19. $B = 77°$, $b = 10.72$, $a = 2.47$
21. 532 yards of fence
23. 78.3 feet of wire
25. The speed of the ship is 1145 yards per minute
27. 63°24'
29. 96.5 feet

Problem Set 13.5

1. $B = 60°$, $a = 2\sqrt{3}$, $c = 4\sqrt{3}$
3. $A = 60°$, $a = 4\sqrt{3}$, $c = 8$
5. $A = 30°$, $B = 60°$, $c = 8$
7. $A = 60°$, $B = 30°$, $a = 2\sqrt{3}$,
9. $A = 45°$, $B = 45°$, $c = 5\sqrt{2}$
11. $\sin 30° = \frac{1}{2}$

13. $\tan 30° = \frac{\sqrt{3}}{3}$
15. $\sec 30° = \frac{2\sqrt{3}}{3}$
17. $\sin 45° = \frac{\sqrt{2}}{2}$

19. $\tan 45° = 1$
21. $\sec 45° = \sqrt{2}$
23. $\sin 60° = \frac{\sqrt{3}}{2}$
25. $\tan 60° = \sqrt{3}$
27. $\sec 60° = 2$

Chapter 13 Review

1. $\frac{2\sqrt{13}}{13}$
3. $\frac{2}{3}$
5. $\frac{\sqrt{13}}{3}$

7. $\frac{3\sqrt{13}}{13}$
9. $\frac{2}{3}$
11. $\frac{\sqrt{13}}{3}$

13. 0.3616
15. 0.2736
17. 0.3134
19. 0.2867

21. 0.3739
23. 1.0198

25. $A = 35°$, $B = 55°$, $a = 12$, $b = 17.1$, $c = 20.9$
27. $A = 29°42'$, $B = 60°18'$, $a = 4$, $b = 7$, $c = 8.1$
29. $A = 41°$, $B = 49°$, $a = 2.8$, $b = 3.2$, $c = 4.2$
31. $A = 68°$, $B = 22°$, $a = 9.9$, $b = 4$, $c = 10.7$
33. $A = 48°12'$, $B = 41°48'$, $a = 6.71$, $b = 6$, $c = 9$

35. $y = \left(\dfrac{\sqrt{3}}{3}\right)x$
37. $y = 1.5517x$
39. $26°36'$
41. $63°24'$
43. $78°42'$

45. $x = 1.1668$; $y = 1.6663$
47. 8.41 feet
49. At 4 p.m.
51. $x = 60$ feet; $y = 40$ feet

Problem Set 14.2

1. 0
3. 0
5. 1
7. $\frac{1}{2}$
9. $-\sqrt{3}$
11. 0

13. -0.57
15. -0.7
17. $\sin \theta$ is $+$, $\cos \theta$ is $-$
19. $\sin \theta$ is $-$, $\cos \theta$ is $+$
21. 1, 3
23. 3, 4

25. 1
27. 3
29. 1, 3
31. $0°$, $180°$, $360°$
33. $0°$, $180°$, $360°$
35. $0°$, $360°$

37. $30°$, $150°$
39. $45°$, $225°$
41. $45°$, $315°$
43. $10°$, $170°$
45. $260°$, $280°$

Problem Set 14.3

1. 0.4067
3. 11.43
5. -0.5736
7. -1.064
9. -0.5519
11. 0.0262
13. 0.1219
15. 5.885

17. -0.6184
19. 11.13
21. $75°$, $105°$
23. $115°$, $245°$
25. $34°48'$, $145°12'$
27. $160°$, $340°$
29. $60°$, $300°$
31. $50°$, $310°$

33. $250°54'$, $289°6'$
35. $160°12'$, $340°12'$
37. $\theta = 90° \pm 180°n$
39. $\theta = 30° \pm 360°n$, $150° \pm 360°n$
41. $\theta = 36°54' \pm 360°n$, $143°6' \pm 360°n$
43. $36°54' \pm 180n$
45. Does not exist

Chapter 14 Review

1.

3.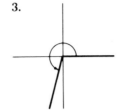

5.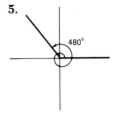

7. $\dfrac{-\sqrt{2}}{2}$
9. 5.6713
11. 1.0002
13. 0.5272
15. 0.9993

17. -57.30
19. -0.7563
21. $-\frac{4}{3}$
23. $\dfrac{\sqrt{5}}{5}$

25. $\frac{4}{5}$
27. $140°$, $220°$
29. $80°$, $260°$

Problem Set 15.1

These answers may differ slightly from yours, depending on the method by which you found them.

1. $A = 133°$, $a = 1.251$, $b = 0.855$
3. $C = 70°$, $b = 32.6$, $c = 33.0$
5. $A = 15°54'$, $B = 145°06'$, $b = 35.85$
7. $A = 38°42'$, $B = 61°18'$, $a = 209$

9. $B = 63°24'$, $C = 29°36'$, $b = 3.78$
11. $A = 37°56'$, $B = 59°48'$, $a = 2.231$
13. $C = 103°57'$, $b = 6.140$, $c = 12.03$
15. $A = 59°18'$, $B = 52°58'$ $b = 54.74$
17. $B = 34°48'$, $C = 99°42'$, $c = 999$
19. $C = 77°$, $b = 40.4$, $c = 42.5$

Problem Set 15.2

1. $B = 37°6'$, $C = 87°54'$, $c = 232$
3. No solution
5. No solution
7. No solution

9. $B = 65°48'$ $C = 80°32'$ $c = 56.0$ *or*
   $B = 114°12'$ $C = 32°08'$ $c = 30.2$
11. $A = 52°16'$ $C = 87°04'$ $c = 78.99$ *or*
   $A = 127°44'$ $C = 11°36°$ $c = 15.89$

Problem Set 15.3

1. $A = 56°16'$, $B = 93°42'$, $c = 12$
3. $A = 18°42'$, $B = 25°18'$, $c = 136°$
5. $A = 19°36'$, $C = 30°24'$, $b = 8.9$
7. $A = 33°$, $B = 27°$, $c = 19$
9. $A = 36°12'$, $B = 43°30'$, $C = 100°18'$

11. $A = 14°54'$, $C = 22°6'$, $b = 56$
13. $A = 19°30'$, $B = 31°30'$, $c = 37.2$
15. $A = 24°00'$, $B = 65°6'$, $C = 90°54'$
17. $B = 78°12'$, $C = 49°48'$, $a = 23.04$
19. $A = 56°24'$, $B = 87°36'$, $c = 7$

Problem Set 15.4

1. 30 mph
3. Michael is 28.7 feet from Barbara, 41.0 feet from Adele.
5. 5.9 feet off the ground, 5.1 feet from the pole.
7. 21°30'
9. 27.6 feet
11. 26.5 feet
13. 42°48'
15. 45°
17. B is 5 feet behind A.
19. 3495 feet

Chapter 15 Review

1. (a) $b = \dfrac{a \sin B}{\sin A}$ (b) $\cos C = \dfrac{a^2 + b^2 - c^2}{2ab}$
3. Adele's distance = 41.0 feet
   Barbara's distance = 28.7 feet
5. 23.9
7. 101°30'
9. (a) 7°12' (b) $\frac{1}{50}$ (c) 250,000 stades
11. Use the law of cosines:
    $|RS|^2 = |RH|^2 + |SH|^2 - 2|RH||SH|\cos (<RHS)$

15. $A = 43°6'$, $B = 56°30'$, $C = 80°24'$
17. $A = 49°40'$, $a = 518$, $b = 437$
19. 289 feet, 500 feet, 866 feet
21. No, distance is 215 miles
23. 0.61 hours and 0.96 hours
25. 374 feet

Problem Set 16.2

Selected solutions only

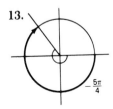

1.

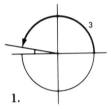

7.

13.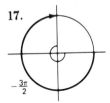

$-\frac{5\pi}{4}$

15.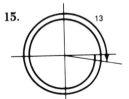

(just over two complete revolutions, clockwise)

17.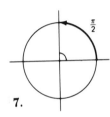

$-\frac{3\pi}{2}$

Problem Set 16.3

1. 0

3. $-\frac{\pi}{2}$

5. $-\frac{\pi}{18}$

7. $\frac{3\pi}{2}$

9. $-\frac{\pi}{4}$

11. $\frac{3\pi}{4}$

13. $\frac{191\pi}{60}$

15. $\frac{5\pi}{9}$

17. $0°$

19. $270°$

21. $180°$

23. $108°$

25. $45°$

27. $180\pi°$

29. $\left(\frac{360}{\pi^2}\right)°$

31. $\dots, -5\pi, -3\pi, -\pi, \pi, 3\pi, 5\pi, \dots, (2n+1)\pi$

33. $\dots, -\frac{5\pi}{2}, -\frac{\pi}{2}, \frac{3\pi}{2}, \frac{7\pi}{2}, \dots, \left(\frac{3\pi}{2} + 2\pi n\right)$

Problem Set 16.4

1. 1, 0; $\tan\frac{\pi}{2}$ is undefined

3. 0, 1, 0

5. $\frac{-1}{2}, \frac{\sqrt{3}}{2}, \frac{-\sqrt{3}}{3}$

7. $\frac{-\sqrt{3}}{2}, \frac{-1}{2}, \sqrt{3}$

9. $\frac{1}{2}, \frac{\sqrt{3}}{2}, \frac{\sqrt{3}}{3}$

11. $\frac{-\sqrt{3}}{2}, \frac{1}{2}, -\sqrt{3}$

13. $-\frac{\sqrt{3}}{2}, \frac{-1}{2}, \sqrt{3}$

15. $0.6428, -0.7660, -0.8391$

17. $\frac{\sqrt{3}}{2}, \frac{-1}{2}, -\sqrt{3}$

19. $-0.4067, 0.9135, -0.4452$

Problem Set 16.5

1. $\frac{6}{5}$ radians, $69°$

3. 8 meters

5. 144 mm

7. 15 radians, $\frac{2700}{\pi} \approx 859°$

9. $\frac{\pi}{2}$

11. $\frac{7\pi}{3}$

13. $\frac{14\pi}{15}$

15. $\frac{4\pi}{3}$

17. $\frac{\pi}{60}$

19. 22 inches

21. $\frac{15840}{7}$ radians

23. $\frac{9}{35}$ minutes

25. $\frac{638}{35}$ inches (≈ 18.2 inches)

27. $\frac{1}{8}$ radian

Chapter 16 Review

1. $\dfrac{\pi}{2}$; 1; undefined

3. $\dfrac{3\pi}{4}$; $\dfrac{1}{\sqrt{2}}$; -1

5. $\dfrac{5\pi}{4}$; $\dfrac{-1}{\sqrt{2}}$; 1

7. $\dfrac{\pi}{6}$; $\dfrac{1}{2}$; $\dfrac{1}{\sqrt{3}}$

9. $\dfrac{4\pi}{3}$; $\dfrac{\sqrt{3}}{2}$; $\sqrt{3}$

11. $180°$; -1; 0

13. $-45°$; $\dfrac{1}{\sqrt{2}}$; -1

15. $-540°$; -1, 0

17. $-60°$; $\dfrac{1}{2}$; $-\sqrt{3}$

19. $-270°$; 0; undefined

21. approximately 589 feet

23. $1.9°$ or $1°54'$

25. 0.188 radian or $10.7°$ or $10°42'$

Problem Set 17.1

1.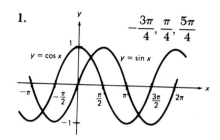

$-\dfrac{3\pi}{4}$, $\dfrac{\pi}{4}$, $\dfrac{5\pi}{4}$

3.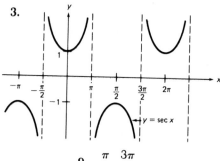

5. $\dfrac{\pi}{2}$; no

7. 0; no

9. $\dfrac{\pi}{2}$, $\dfrac{3\pi}{2}$

11. $\ldots, -\dfrac{\pi}{2}, \dfrac{\pi}{2}, \dfrac{3\pi}{2}, \ldots$ (in general, $\dfrac{\pi}{2} + \pi n$)

13. $\ldots, -\pi, 0, \pi, 2\pi, \ldots$ (in general, πn)

15. Positive for $0 < x < \dfrac{\pi}{2}$ and $\dfrac{3\pi}{2} < x < 2\pi$; decreasing for $0 < x < \pi$; positive *and* decreasing for $0 < x < \dfrac{\pi}{2}$

Problem Set 17.4

1. $-\frac{1}{2}$

3. $\frac{1}{4}$

5. -1

7. $\frac{1}{2}$

9. 0

11. 1

13. undefined

15. -1

17. 1

19. $\cos^2 x$

21. $\sin(\cos x) - (\sin x)(\cos x)$

23. $f(x) = 1 - x^2$; $g(x) = \sin x$

25. $f(x) = \dfrac{1}{x}$; $g(x) = \sin x$

27. $f(x) = 10^x$; $g(x) = \cos x$

29. $\ldots, -\pi, 0, \pi, 2\pi, \ldots$ (in general, $n\pi$)

Problem Set 17.5

1.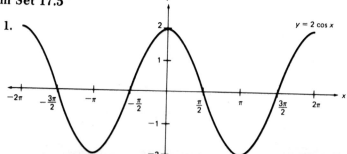

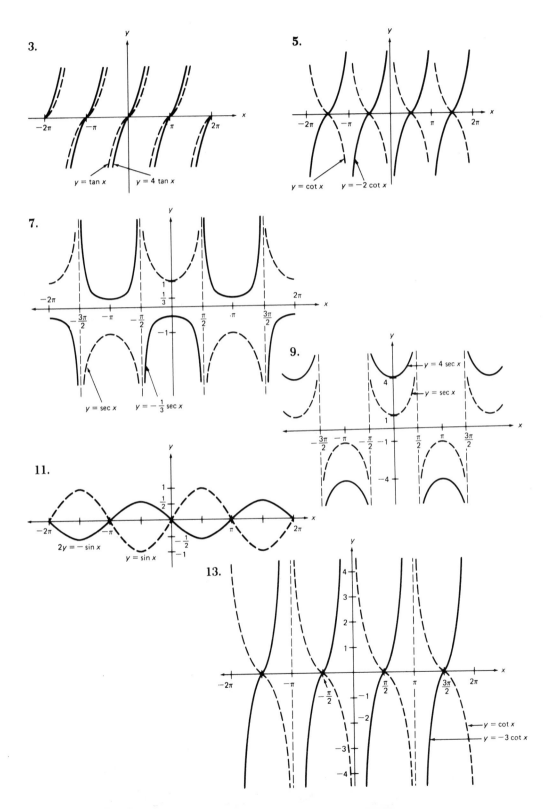

3.

$y = \tan x$ $y = 4 \tan x$

5.

$y = \cot x$ $y = -2 \cot x$

7.

$y = \sec x$ $y = -\frac{1}{3} \sec x$

9.

$y = 4 \sec x$
$y = \sec x$

11.

$2y = -\sin x$
$y = \sin x$

13.

$y = \cot x$
$y = -3 \cot x$

15.

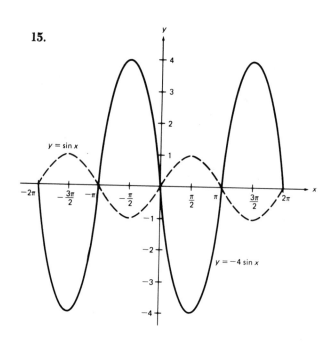

Problem Set 17.6

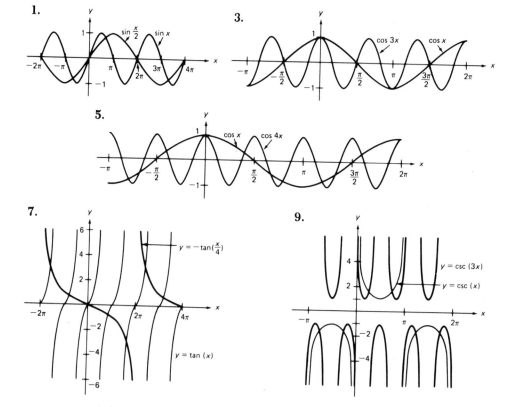

11.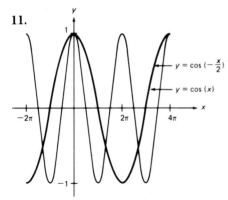

$y = \cos\left(-\frac{x}{2}\right)$

$y = \cos(x)$

13.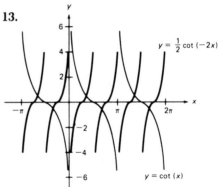

$y = \frac{1}{2}\cot(-2x)$

$y = \cot(x)$

15.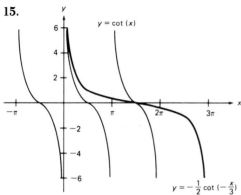

$y = \cot(x)$

$y = -\frac{1}{2}\cot\left(-\frac{x}{3}\right)$

17.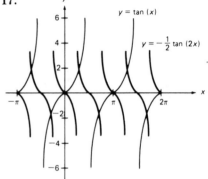

$y = \tan(x)$

$y = -\frac{1}{2}\tan(2x)$

Problem Set 17.7

1.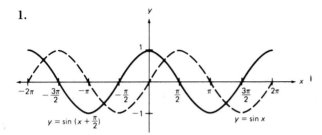

$y = \sin\left(x + \frac{\pi}{2}\right)$

$y = \sin x$

3.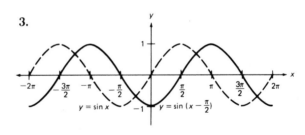

$y = \sin x$

$y = \sin\left(x - \frac{\pi}{2}\right)$

5.

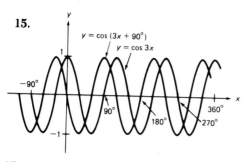

7, 9.

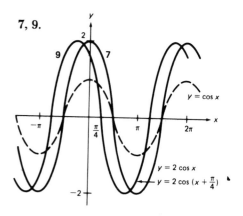

11.

13.

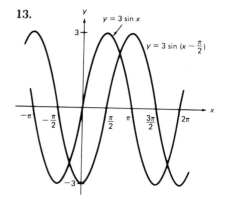

15.

17. Amplitude = 2; phase = π ahead; period = 2
19. Amplitude = 2; phase = no shift; period = π
21. Amplitude = 16; phase = $\dfrac{\pi}{2}$ ahead; period = 2π

23.

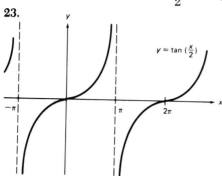

25.

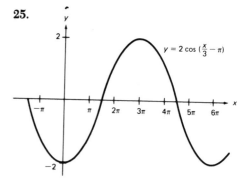

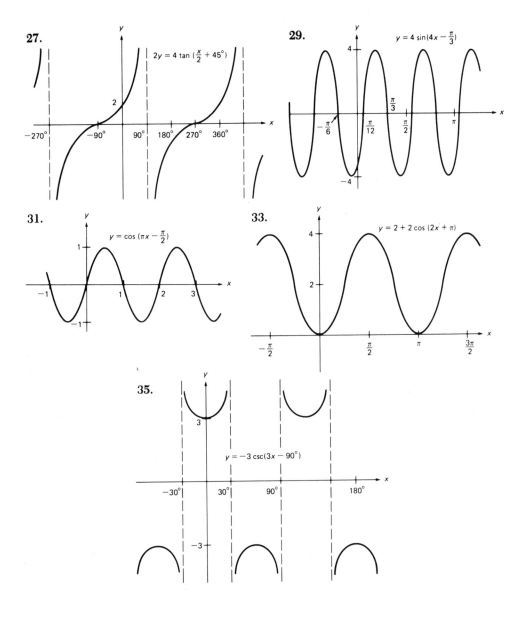

27. $2y = 4 \tan \left(\frac{x}{2} + 45° \right)$

29. $y = 4 \sin(4x - \frac{\pi}{3})$

31. $y = \cos \left(\pi x - \frac{\pi}{2} \right)$

33. $y = 2 + 2 \cos (2x + \pi)$

35. $y = -3 \csc(3x - 90°)$

Problem Set 17.8

1. $30° + 360°n$ or $150° + 360°n$

3. $45° + 180°n$

5. $45° + 360°n$ or $315° + 360°n$

7. $\frac{45°}{2} + 180°n$ or $\frac{225°}{2} + 180°n$

9. $0°$

11. $55° + 360°n$ or $125° + 360°n$

13. $160° + 180°n$

15. $50° + 360°n$ or $310° + 360°n$

17. $143°6' + 180°n$

19. An infinite number of solutions, one in each interval from $180°n$ to $90° + 180°n$

Problem Set 17.9

1. $y = 5 \sin \left(\dfrac{\pi}{15}\right)x$

3. $P = 0.1 \sin (880\pi t)$

5. For example, $P = 0.1 \sin (440\pi t)$

7. Summertime: 4500; autumn: 4000; winter: 3500

9.

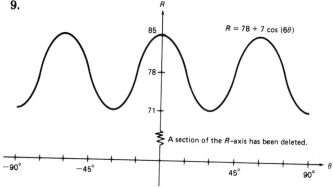

Chapter 17 Review

1.

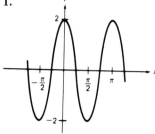

3 (a)

3. (b) $\dfrac{3}{4}$

(c) 4π

(d) $-\dfrac{3\sqrt{2}}{8}; \dfrac{3}{8}$

5. (a) 3   (b) 180°   (c) 180° behind
    (d) 0°, 90°, 180° 270°, 360°

5(e)

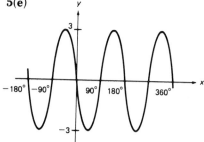

7. (a) 4   (b) 720°   (c) $y = 4 \cos \left(\dfrac{x}{2}\right)$

9. $y = -3 \cos x$

11. 4

13. 9

15. $\dfrac{-3\pi}{4} + 2\pi n < \theta < \dfrac{\pi}{4} + 2\pi n$

17. $\dfrac{\pi}{2} + 2\pi n < \theta < \dfrac{3\pi}{2} + 2\pi n$

19.

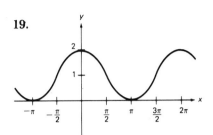

21. $y = 2 \cos \left(x - \dfrac{\pi}{2} \right) + 2$

23. 0

25. ln cos θ; defined from $-\dfrac{\pi}{2} + 2\pi n$ to $\dfrac{\pi}{2} + 2\pi n$

Problem Set 18.2

1. cos θ

3. cot θ

5. csc θ

7. csc θ

9. $\cos^2 \theta$

11. sin θ

13. 1

15. $2 \cos^2 \theta$

17. $\sec^2 \theta$

19. sin θ

21. 0

23. 0

25. $\pm \sqrt{1 - \sin^2 \theta}$ *Note:* In this and following problems, the sign taken for the square root depends on which quadrant θ falls in.

27. $\pm \dfrac{\sqrt{1 - \sin^2 \theta}}{\sin \theta}$

29. $\pm \sqrt{1 + \tan^2 \theta}$

31. $\pm \dfrac{\tan \theta}{\sqrt{1 + \tan^2 \theta}}$

33. $\pm \sqrt{1 - \cos^2 \theta}$

35. $\pm \dfrac{\sqrt{1 - \cos^2 \theta}}{\cos \theta}$

37. $\pm \sqrt{1 + \cot^2 \theta}$

39. $\pm \dfrac{\cot \theta}{\sqrt{1 + \cot^2 \theta}}$

Problem Set 18.4

1. 314°30′, 225°30′

3. 14°, 194°

5. 60°, 120°, 240°, 300°

7. 0°

9. 30°, 150°, 210°, 330°

11. $\dfrac{\pi}{6}, \dfrac{5\pi}{6}, \dfrac{2\pi}{3}, \dfrac{4\pi}{3}$

13. $\dfrac{\pi}{3}, \dfrac{5\pi}{3}, \pi$

15. $\dfrac{\pi}{3}, \dfrac{4\pi}{3}$

17. $\dfrac{7\pi}{6}, \dfrac{11\pi}{6}, \dfrac{3\pi}{2}$

19. $\dfrac{\pi}{4}, \dfrac{3\pi}{4}, \dfrac{5\pi}{4}, \dfrac{7\pi}{4}$

21. $\dfrac{\pi}{6}, \dfrac{7\pi}{6}$

23. $\dfrac{\pi}{3}, \dfrac{2\pi}{3}, \dfrac{4\pi}{3}, \dfrac{5\pi}{3}$

25. 120°, 240°, 70°30′, 289°30′

27. 153°24′, 116°30′, 296°30′, 333°24′

29. 180°, 19°36′, 160°36′

31. 60°, 300°

33. 45°, 71°36′, 225°, 251°36′

35. 0°, 180°

Problem Set 18.5

1. $\frac{3}{5}$

3. $-\frac{13}{12}$

5. $-\frac{8}{15}$

7. $-\frac{5}{13}$

9. $\frac{13}{5}$

11. $-\frac{25}{24}$

Problem Set 18.6

7. $\sin 25°$

9. $\sin \alpha$

11. $\dfrac{\sqrt{3}-1}{2\sqrt{2}}$

13. $2 \tan \alpha$

15. 0

Problem Set 18.8

15. $3 \sin \theta - 4 \sin^3 \theta$

Problem Set 18.9

1. 4

3. $\cot \theta$

5. $\cot^2 \theta$

7. $\tan\left(\dfrac{\alpha}{2}\right)$

9. $\sqrt{3}$

19. 0°, 240°, 120°

21. 20°, 140°

23. 19°6′, 199°6′

25. $\dfrac{\sqrt{3}-1}{2\sqrt{2}}$

27. $-(\sqrt{3}+2)$

29. $\sqrt{\dfrac{2+\sqrt{2}}{4}}$

Chapter 18 Review

1. $-\cos \theta$

3. 1

5. Yes, because $90° = 60° + 30°$

7. No, because $\tan 40° = \dfrac{2 \tan 20°}{1 - \tan^2 20°}$ and $\tan 20° = 0$

9. No, because $3 \sin 90° = 3$, $\sin 270° = -1$

11. No, because $\cos(360° + 30°) = \dfrac{\sqrt{3}}{2}$, $\cos 360° + \cos 30° = 1 + \dfrac{\sqrt{3}}{2}$

13. 0.85

15. 1.18

17. 1.18

19. 0.85

21. -0.53

23. 0.53

27. $\dfrac{\pi}{2}$, 194°30′, 345°30′

29. $\dfrac{\pi}{3}, \dfrac{2\pi}{3}, \dfrac{4\pi}{3}, \dfrac{5\pi}{3}$

31. $\dfrac{1 + \cos 2\theta}{2}$

33. $\frac{1}{8}(3 + 4\cos 2\theta + \cos 4\theta)$

35. $\dfrac{\sqrt{t^2-1}}{t^2}$

37. $\dfrac{4}{t^2} - 1$

39. $y = \cos x$

41. $y = \sin 2x$

43. $y = \sin(x + 30°)$

Problem Set 19.1

1. $\dfrac{\pi}{2}$

3. $-\dfrac{\pi}{6}$

5. $\dfrac{3\pi}{4}$

7. $-17°30'$

9. $39°18'$

11. $-11°42'$

13. $-\dfrac{\pi}{3}$

15. $120°06'$

17. False, example: $u = 1$

$\tan^{-1} u = \dfrac{\pi}{4}$, $\cos^{-1} u = 0$, therefore $\dfrac{\sin^{-1} u}{\cos^{-1} u}$ is undefined

19. True, because $\cot \theta = \dfrac{1}{\tan \theta}$

21. False; $\sin^{-1}(w) = -\sin^{-1}(-w)$

Problem Set 19.2

1. $y = \cot^{-1}x$
Domain: all real numbers
Range: reals between 0 and π

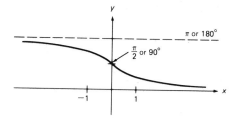

3. $y = \csc^{-1}x$
Domain: reals ≤ -1 or ≥ 1

Range: reals between $-\dfrac{\pi}{2}$ and $\dfrac{\pi}{2}$

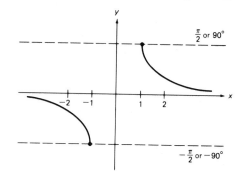

5. $y = 2\sin^{-1}x$
Domain: reals between -1 and 1
Range: reals between $-\pi$ and π

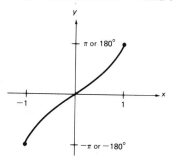

7. $y = \sin^{-1} x + 1$
Domain: reals between -1 and 1

Range: reals between $1 - \dfrac{\pi}{2}$ and $1 + \dfrac{\pi}{2}$

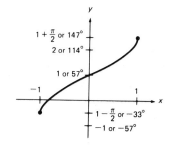

9. $y = \frac{1}{2} \cos^{-1} x$
Domain: reals between -1 and 1
Range: reals between 0 and $\frac{\pi}{2}$

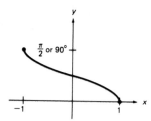

11. $y = \cot^{-1} \frac{x}{2}$
Domain: all real numbers
Range: reals between 0 and π

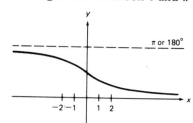

13. $y = 3 \sin^{-1} \frac{x}{2}$
Domain: reals between -2 and 2
Range: reals between $-\frac{3\pi}{2}$ and $\frac{3\pi}{2}$

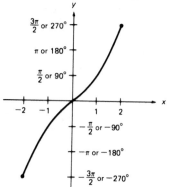

15. $y = \frac{1}{2} \cos^{-1}(2x + 2)$
Domain: reals between $-\frac{3}{2}$ and $-\frac{1}{2}$
Range: reals between 0 and $\frac{\pi}{2}$

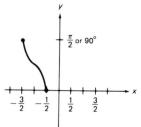

Problem Set 19.3

1. $\dfrac{1}{2}$

3. $\dfrac{\sqrt{2}}{2}$

5. $\dfrac{\pi}{3}$

7. $\dfrac{\pi}{4}$

9. $\dfrac{4}{5}$

11. -1

13. 0

15. $-\dfrac{\pi}{4}$

17. -1

19. Not defined

Chapter 19 Review

1. $30°$

3. $60°$

5. $87°30'$

7. $11°18'$

9. $-84°48'$

11. Domain: all reals: < -1 or > 1
Range: all reals between $-\pi$ and 0
or between 0 and π

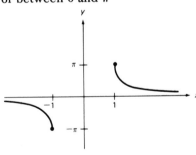

13. Domain: all reals $< -\frac{1}{2}$ or $> \frac{1}{2}$
Range: all reals between $-\dfrac{\pi}{2}$ and 0

or between 0 and $\dfrac{\pi}{2}$

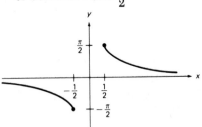

15. Domain: all reals
Range : all reals between $-\pi$ and π

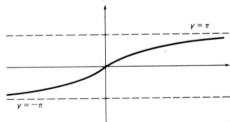

17. Domain: $-2 \le x \le 2$
Range: $7 \le y \le 7 + 2\pi$

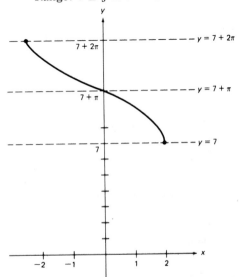

19. $6°$
21. -0.6498
31. 0.2679
33. -0.1201
35. 0.9659

Problem Set 20.1

1. $(x + y)^6 = x^6 + 6x^5y + 15x^4y^2 + 20x^3y^3 + 15x^2y^4 + 6xy^5 + y^6$
3. $(1 - t)^3 = 1 - 3t + 3t^2 - t^3$
5. $(t^2 + 1)^6 = t^{12} + 6t^{10} + 15t^8 + 20t^6 + 15t^4 + 6t^2 + 1$
7. $(\sqrt{x} + \sqrt{y})^{10} = x^5 + 10x^{9/2}y^{1/2} + 45x^4y + 120x^{7/2}y^{3/2} + 210x^3y^2 + 252x^{5/2}y^{5/2}$
    $+ 210x^2y^3 + 120x^{3/2}y^{7/2} + 45xy^4 + 10x^{1/2}y^{9/2} + y^5$
9. $(\sqrt{x} + \sqrt[3]{y})^{12} = x^6 + 12x^{11/2}y^{1/3} + 66x^5y^{2/3} + 220x^{9/2}y + \cdots$
11. $\left(\dfrac{b^5}{z^2} - \dfrac{z^3b}{\sqrt{c}}\right)^{15} = \dfrac{b^{75}}{z^{30}} - \dfrac{15b^{71}}{z^{25}c^{1/2}} + \dfrac{105b^{67}}{z^{20}c} - \dfrac{455b^{63}}{z^{15}c^{3/2}} + \cdots$
13. $70\dfrac{x^4}{y^4}$
15. $8488x^{11}$
17. $-560\dfrac{A^3}{B^8}$
19. -80
21. 1.2664
23. 0.90438
24. $3x^5$
25. $(1.002)^{10} = (1 + 0.002)^{10} = 1 + 0.02 + \text{positive terms} > 1.02$
27. $A^3 - 6A^2B + 3A^2C + 12AB^2 + 3AC^2 - 12ABC + 12B^2C - 6BC^2 + C^3 - 8B^3$
29. 15.37

Problem Set 20.2

1. 24
3. 6
5. 20
7. 504
9. 210
11. 100
13. $\frac{1}{2}$
15. $\frac{5}{12}$
17. $3 \cdot 1 + 4 \cdot 2 + 5 \cdot 3 + 6 \cdot 4 + 7 \cdot 5$
19. $1 + x + 2x^2 + 6x^3 + 24x^4 + 120x^5$
21. $1 + 2^3 + 3^3 + \cdots + (2p - 1)^3 + (2p)^3$
23. $\displaystyle\sum_{j=0}^{12} (-1)^j a_j x^j$
27. $p = 1, p = 8$
29. (a) $\dfrac{5!}{(5 - 3)!} = \dfrac{5!}{2!} = \dfrac{(5 \cdot 4 \cdot 3 \cdot 2 \cdot 1)}{2} = 60$
    (b) $\dfrac{n!}{(n - 3)!} = \dfrac{n(n - 1)(n - 2)(n - 3)!}{(n - 3)!} = n(n - 1)(n - 2)$

Chapter 20 Review

1. $b^4 - 4b^3a + 6b^2a^2 - 4ba^3 + a^4$
3. $\dfrac{x^6}{64} - \dfrac{3x^5}{16} + \dfrac{15x^4}{16} - \dfrac{5x^3}{2} + \dfrac{15x^2}{4} - 3x + 1$
5. $\dfrac{a^5}{b^5} + \dfrac{5a^3}{b^3} + \dfrac{10a}{b} + \dfrac{5b^3}{a^3} + \dfrac{5b^3}{a} + \dfrac{b^5}{a^5}$
7. $680p^{14}q^3$
9. $1120r^4s^4$
11. $\dfrac{105}{32}p^6$
13. $998,000$
17. 42
19. $\displaystyle\sum_{k=1}^{10} 2k$
21. $\dfrac{12!}{7!}$
25. $3x^5$
29. 0.90909

INDEX